INTERNATIONAL UNION OF THEORETICAL AND APPLIED MECHANICS

INSTABILITY OF CONTINUOUS SYSTEMS

SYMPOSIUM HERRENALB (GERMANY)
SEPTEMBER 8–12, 1969

EDITOR

H. LEIPHOLZ

WITH 147 FIGURES

SPRINGER-VERLAG BERLIN HEIDELBERG NEW YORK 1971

Professor Dr. HORST LEIPHOLZ

Solid Mechanics Division, Faculty of Engineering
University of Waterloo, Waterloo, Ontario, Canada

ISBN-13: 978-3-642-65075-8 e-ISBN-13: 978-3-642-65073-4
DOI:10.1007/978-3-642-65073-4

Softcover reprint of the hardcover 1st edition 1971
Library of Congress Catalog Card Number: 72-146 050

Preface

Until recently there was no uniform stability theory. Different approaches to stability problems had been developed in the different branches of mechanics. In the field of elasticity, it was mainly the so-called static method and energy method which were used, while in the field of dynamics it was the kinetic method, which found its perfect expression in the theory of Liapunov. During the last few decades there has been a rapid development in the general theory of stability, stimulated, for example, by the investigations of H. ZIEGLER on elastic systems subject to non-conservative loads, and by the problems arising in aeroelasticity which are closely related to those introduced by ZIEGLER. The need was felt for kinetic methods which could also be used in investigating the stability of deformable systems. Efforts were made to adapt such methods, already known and developed in the stability theory of rigid systems, for application in the stability theory of *continuous* systems. During the last ten years interest was focused mainly on the application of a generalized *Liapunov method* to stability problems of continuous systems. All this was done in attempts to unify the various approaches to stability theory. It was with the idea of encouraging such a tendency, establishing to what extent a uniform physical and mathematical foundation already existed for stability theory in all branches of mechanics, and stimulating the further development of a common stability theory, that a IUTAM-Symposium was devoted to this topic.

In the summer of 1966, the German Association of Applied Mathematics and Mechanics (GAMM) proposed to organize such a Symposium on "Problems of a Non-Conservative Stability and Aeroelasticity". At the same time, the National Committee of the Netherlands suggested a Symposium on "General Methods in the Theory of Stability". In pursuance of the aim to strive for a uniform stability theory, the Bureau of IUTAM decided at its meeting in Vienna in June 1966 that the two symposia should be combined and held at Karlsruhe, Germany. A Study Group was instituted under the chairmanship of W. T. KOITER, with L. J. F. BROER, H. LEIPHOLZ and R. MAZET as members, to settle matters of detail. This Study Group finally suggested that the

Symposium be held in Karlsruhe at the beginning of September, 1969, under the title "Instability of Continuous Systems", a suggestion accepted by the Bureau of IUTAM. H. Leipholz, then in Karlsruhe, was appointed chairman of the Local Organizing Committee. The Symposium took place in Herrenalb, near Karlsruhe, on September 8—12, 1969.

Thanks are due to the German *Bundesministerium für Wissenschaft und Forschung*, to *Kultusministerium des Landes Baden-Württemberg*, and to *Universität Karlsruhe (TH)* for their co-sponsorship of the Symposium. Finally, appreciation has to be expressed to the lecturers and to all those who helped to make the Symposium a success.

Waterloo, Ontario, Canada,
November 1970

H. Leipholz

Scientific Committee

V. V. Bolotin, Moscow, U.S.S.R.
L. J. F. Broer, Eindhoven, Netherlands (chairman)
L. Contri, Padova, Italy
G. Herrmann, Evanston, Ill., U.S.A.
M. D. Kruskal, Princeton, N.J., U.S.A.
H. Leipholz, Karlsruhe, Germany (chairman)
R. Mazet, Châtillon-sous-Bagneux, France
J. T. Stuart, London, U.K.
T. Tatsumi, Kyoto, Japan

Participants

Adams, E., Karlsruhe, Germany
Anton, J., Providence, R.I., U.S.A.
Araldsen, P. O., Oslo, Norway
Ariaratnam, S. T., London, U.K.
Augusti, G., Providence, R.I., U.S.A.

Barta, J., Budapest, Hungary
Barta, T. A., London, U.K.
Beatty, M. F., Lexington, Ky., U.S.A.
Berger, M. S., Minneapolis, Minn., U.S.A.
Bergeron, R. F., Jr., Cambridge, Mass., U.S.A.
Bland, D. R., Manchester, U.K.
Boley, B. A., Ithaca, N.Y., U.S.A.
Bolotin, V. V., Moscow, U.S.S.R.
Broer, L. J. F., Eindhoven, Netherlands
Browning, S. C., Magna, Utah, U.S.A.
Bufler, H., Stuttgart, Germany
Busse, F. H., München, Germany

Christoffersen, J., København, Denmark
Coleman, B. D., Pittsburgh, Pa., U.S.A.
Collins, W. D., Toronto, Canada
Como, M., Napoli, Italy
Cotsaftis, M., Fontenay-aux-Roses, France

Davis, S. H., Baltimore, Md., U.S.A.
Dikmen, M., Ankara, Turkey
DiPrima, R. C., Troy, N.Y., U.S.A.
Dowell, E. H., Princeton, N.J., U.S.A.
Drazin, P. G., Bristol, U.K.

Eckhaus, W., Delft, Netherlands
Eckhoff, K. S., Bergen, Norway
Ellingsen, T., Oslo, Norway

Fischer, W., Karlsruhe, Germany

Gjevik, B., Oslo, Norway
Goedbloed, J. P., Jutphaas, Netherlands
Görtler, H., Freiburg i. Br., Germany
Gradowczyk, M. H., Buenos Aires, Argentinia

GRANNICK, R. N., Chicago, Ill., U.S.A.
GÜNTHER, W., Karlsruhe, Germany

HARVÁNCIK, J., Bratislava, Č.S.S.R.
HERRMANN, G., Evanston, Ill., U.S.A.
HOFF, N. J., Stanford, Cal., U.S.A.
HSU, C. S., Berkeley, Cal., U.S.A.
HUNT, G. W., London, U.K.

JASHMAN, W. E., London, U.K.
JOHNS, K. C., London, U.K.
JOSEPH, D. D., Minneapolis, Minn., U.S.A.
JUHASZ, ST., San Antonio, Tex., U.S.A.

KNETS, I., Riga, U.S.S.R.
KNOPS, R. J., Newcastle upon Tyne, U.K.
KODNÁR, R., Bratislava, Č.S.S.R.
KRUSKAL, M. D., Princeton, N.J., U.S.A.

LANGE, H., Freiburg i. Br., Germany
LEBOVITZ, N. R., Chicago, Ill., U.S.A.
LEIPHOLZ, H., Karlsruhe, Germany
LESLIE, F. M., Glasgow, U.K.
LESSEN, M., Rochester, N.Y., U.S.A.
LIGHTHILL, M. J., London, U.K.
LIND, N. C., Waterloo, Ontario, Canada

MAIER, G., Milano, Italy
MAZET, R., Châtillon-sous-Bagneux, France
METTLER, E., Karlsruhe, Germany
MIKHAILOV, G. K., Moscow, U.S.S.R.
MORENO, F. G., Torrejón de Ardoz, Spain

NASH, W. A., Amherst, Mass., U.S.A.
NEALE, K., Waterloo, Ontario, Canada
NEMAT-NASSER, S., La Jolla, Cal., U.S.A.
NEWELL, A. C., Los Angeles, Cal., U.S.A.
NIORDSON, F. I., København, Denmark

OLSZAK, W., Warszawa, Poland
OOMS, G., Amsterdam, Netherlands

PALM, E., Oslo, Norway
PARKS, P. C., Coventry, U.K.
PAYNE, L. E., Zürich, Switzerland
PETRE, A., Bucuresti, Roumania
PINDERA, J. T., Waterloo, Ontario, Canada
POPESCU, N. D., Petrosani, Roumania

REISSNER, E., Cambridge, Mass., U.S.A.
ROY, M., Paris, France

SCARLAT, A., Bucuresti, Roumania
SCHROEDER, J., Waterloo, Ontario, Canada
SEDOV, L. I., Moscow, U.S.S.R.
SEGEL, L. A., New York, N.Y., U.S.A.
SEWELL, M. I., Reading, U.K.
SIEKMANN, E., Baden-Baden, Germany
SLEMROD, M., Providence, R.I., U.S.A.
STARK, V., Fribourg, Switzerland
STORÅKERS, B., Stockholm, Sweden
STUART, J. T., London, U.K.
STUMPF, H., Aachen, Germany

TATSUMI, T., Kyoto, Japan
THOMAS, G., Stuttgart, Germany
TROCH, I., Wien, Austria

VALENTINE, R. S.
VELTE, W., Würzburg, Germany
VELTHUIZEN, H. G. M., Enschede, Netherlands
VENKAYYA, V. B., Wright-Patterson Air Force Base, Ohio, U.S.A.
VILLAGGIO, P., Pisa, Italy

WANG, P. K. C., Los Angeles, Cal., U.S.A.
WEENINK, M. P. H., Jutphaas, Netherlands
WEIDENHAMMER, F., Karlsruhe, Germany
WILKEN, H., Karlsruhe, Germany
WILKES, E. W., Newcastle upon Tyne, U.K.
WILLE, F., Freiburg i. Br., Germany

ZABUSKY, N. J., Whippany, N.J., U.S.A.
ZIEGLER, H., Zürich, Switzerland
ZIEREP, J., Karlsruhe, Germany
ZIMMERMANN, G., Göttingen, Germany
ŻYCZKOWSKI, M., Kraków, Poland

Authors

Dr.-Ing. E. ADAMS, Institut für Angewandte Mathematik, Universität Karlsruhe, D-75 Karlsruhe, Kaiserstr. 12, Germany

Dipl.-Math. J. ANTON, Division of Applied Mathematics, Brown University, Providence, RI 02912, U.S.A.

Prof. S. T. ARIARATNAM, Department of Civil Engineering, University of Waterloo, Waterloo, Ontario, Canada

Prof. G. AUGUSTI, Istituto di Scienza delle Costruzioni, Università di Napoli, I-80125 Napoli, Italy

Prof. Dr. J. BARTA, Jozsef-körut 35, Budapest, Hungary

Dr. T. A. BARTA, Department of Civil and Municipal Engineering, University College London, Gower Street, London W.C. 1, U.K.

Prof. M. F. BEATTY, Department of Engineering Mechanics, University of Kentucky, Lexington, KY 40506, U.S.A.

Prof. M. S. BERGER, University of Minnesota, Minneapolis, MN 55455, U.S.A.

Dr. R. F. BERGERON, Jr., Engineering Mechanics and Physics Department, Bell Telephone Laboratories, Whippany Road, Whippany, NJ 07981, U.S.A.

Dr. L. BERKE, Air Force Flight Dynamics Laboratory, Wright-Patterson Air Force Base, OH 45433, U.S.A.

Prof. D. R. BLAND, Department of Mathematics, Cranfield Institute of Technology, Cranfield, Bedford, U.K.

Prof. B. A. BOLEY, Department of Theoretical and Applied Mechanics, College of Engineering, Thurston Hall, Cornell University, Ithaca, NY 14850, U.S.A.

Prof. V. V. BOLOTIN, Institute of Mechanics, U.S.S.R. Academy of Sciences, Leningrad Avenue, Moscow A-40, U.S.S.R.

Prof. L. J. F. BROER, Technische Hogeschool Eindhoven, Insulindelaan 2, Eindhoven, Netherlands

Dr. F. H. BUSSE, Institut für Astrophysik, Max-Planck-Institut für Physik und Astrophysik, D-8 München 23, Föhringer Ring 6, Germany

Mr. B. H.-P. CHEN, Department of Mechanical and Aerospace Sciences, College of Engineering and Applied Science, University of Rochester, Rochester, NY 14627, U.S.A.

Prof. A. H. CHILVER, Department of Civil and Municipal Engineering, University College London, Gower Street, London W.C. 1, U.K.

Prof. J. CHRISTOFFERSEN, Department of Solid Mechanics, Technical University of Denmark, Building 404, Lundtoftevej 100, DK-2800 Lyngby, Denmark

Prof. B. D. COLEMAN, Mellon Institute, Carnegie-Mellon University, 4400 Fifth Avenue, Pittsburgh, PA 15213, U.S.A.

Prof. Ing. M. COMO, Istituto di Tecnica delle Costruzioni, Università di Napoli, I-80125 Napoli, Italy

Prof. M. COTSAFTIS, Département de Physique du Plasma et de la Fusion Contrôlée, Centre d'Etudes Nucléaires, Association EURATOM—CEA, B.P. n° 6, F-92 Fontenay-aux-Roses, France

Prof. S. H. DAVIS, Department of Mechanics, Johns Hopkins University, Baltimore, MD 21218, U.S.A.
Mr. G. S. DEEM, Plasma and Computational Physics Research Department, Bell Telephone Laboratories, Whippany Road, Whippany, NJ 07981, U.S.A.
Prof. M. DIKMEN, Middle East Technical University, Ankara, Turkey
Prof. R. C. DIPRIMA, Department of Mathematics, Rensselaer Polytechnic Institute, Troy, NY 12181, U.S.A.
Prof. E. H. DOWELL, Department of Aerospace and Applied Science, Princeton University, Princeton, NJ 08540, U.S.A.
Prof. Dr. ir. W. ECKHAUS, Afdeling der Algemene Wetenschappen, Technische Hogeschool Delft, Julianalaan 132, Delft, Netherlands
Dr. T. ELLINGSEN, Institutt for Matematiske Fag, Universitetet i Oslo, Blindern, Oslo 3, Norway
Mr. P. FALB, Division of Applied Mathematics, Brown University, Providence, RI 02912, U.S.A.
Mr. M. I. FREEDMAN, NASA Electronics Research Center, Cambridge, MA 02138, U.S.A.
Mr. A. GAJEWSKI, Politechnika Krakowska, Kraków, Poland
Dr. B. GJEVIK, Institutt for Matematiske Fag, Universitetet i Oslo, Blindern, Oslo 3, Norway
Mr. K. GOTOH, Department of Physics, Faculty of Science, University of Kyoto, Kyoto, Japan
Dr. J. DE GRAAF, Technische Hogeschool Eindhoven, Insulindelaan 2, Eindhoven, Netherlands
Prof. M. H. GRADOWCZYK, Departamento de Metalurgia, Comisión Nacional de Energía Atómica, Buenos Aires, Argentinia
Prof. R. N. GRANNICK, Department of Mathematics, University of Chicago, Chicago, IL 60637, U.S.A.
Prof. Dr. G. HERRMANN, Department of Applied Mechanics, School of Engineering, Stanford University, Stanford, CA 94305, U.S.A.
Prof. C. S. HSU, Division of Applied Mechanics, Department of Mechanical Engineering, University of California, Berkeley, CA 94720, U.S.A.
Dr. A. HUAUX, Ecole Technique Supérieure de l'Etat d'Anderlecht, Bruxelles 7, Belgium
Prof. G. W. HUNT, Department of Civil and Municipal Engineering, University College London, Gower Street, London W.C. 1, U.K.
Prof. E. F. INFANTE, Center for Dynamical Systems, Division of Applied Mathematics, Brown University, Providence, RI 02912, U.S.A.
Prof. K. C. JOHNS, Ecole d'Ingénieurs, Faculté des Sciences Appliquées, Université de Sherbrooke, Cité Universitaire, Sherbrooke, Québec, Canada
Prof. D. D. JOSEPH, Department of Mathematics, Imperial College of Science and Technology, Exhibition Road, London S.W. 7, U.K.
Mr. H. KANEMATSU, Department of Mechanical and Aerospace Engineering, University of Massachusetts, Amherst, MA 01003, U.S.A.
Dr. N. S. KHOT, Anythesis Group, Solid Mechanics Branch, Structures Division, Wright-Patterson Air Force Base, OH 45433, U.S.A.
Prof. N. A. KILCHEVSKY, Academy of Sciences of the Ukrainian S.S.R., Kiev, U.S.S.R.
Dr. K. KIRCHGÄSSNER, Mathematisches Institut, Ruhr-Universität, D-463 Bochum, Germany
Dr. R. J. KNOPS, University of Newcastle upon Tyne, Newcastle upon Tyne NE1 7RU, U.K.

Dr. R. Kodnár, Institut für Bauwesen und Architektur, Slowakische Akademie der Wissenschaften, Dubravská cesta 4, P.O.B. 1191, Bratislava, Č.S.S.R.

Dr. H. Lange, Institut für Angewandte Mathematik, Albert-Ludwigs-Universität, D-78 Freiburg i.Br., Hebelstr. 40, Germany

Mr. K. H. Lau, Department of Mechanical and Aerospace Sciences, College of Engineering and Applied Science, University of Rochester, Rochester, NY 14627, U.S.A.

Mr. T. H. Lee, Division of Applied Mechanics, Department of Mechanical Engineering, University of California, Berkeley, CA 94720, U.S.A.

Prof. Dr. H. Leipholz, Solid Mechanics Division, Faculty of Engineering, University of Waterloo, Waterloo, Ontario, Canada

Prof. M. Lessen, Department of Mechanical and Aerospace Sciences, College of Engineering and Applied Science, University of Rochester, Rochester, NY 14627, U.S.A.

Prof. Dr. G. Maier, Istituto di Scienza e Tecnica delle Costruzioni, Politecnico di Milano, Piazza Leonardo da Vinci, 32, I-20133 Milano, Italy

Prof. W. A. Nash, Department of Mechanical and Aerospace Engineering, University of Massachusetts, Amherst, MA 01003, U.S.A.

Mr. K. Neale, Department of Civil Engineering, University of Waterloo, Waterloo, Ontario, Canada

Prof. Dr. S. Nemat-Nasser, Department of the Aerospace and Mechanical Engineering Sciences, University of California, San Diego, P.O.B. 109, La Jolla, CA 92037, U.S.A.

Prof. A. C. Newell, Department of Mathematics, Clarkson College of Technology, Potsdam, NY 13676, U.S.A.

Mrs. S. N. Nikulinskaya, Academy of Sciences of the Ukrainian S.S.R., Kiev, U.S.S.R.

Prof. E. Palm, Institutt for Matematiske Fag, Universitetet i Oslo, Blindern, Oslo 3, Norway

Dr. P. C. Parks, School of Engineering Science, University of Warwick, Coventry CV4 7AL, U.K.

Prof. L. E. Payne, Cornell University, Ithaca, NY 14850, U.S.A.

Dr. L. A. Peletier, Mathematics Division, University of Sussex, Falmer Brighton, Sussex, U.K.

Prof. A. Petre, Dtr. Puskin nr. 3, Bucuresti 30, Roumania

Prof. Dr. E. Reissner, Department of the Aerospace and Mechanical Engineering Sciences, University of California, San Diego, P.O.B. 109, La Jolla, CA 92037, U.S.A.

Dr.-Ing. A. Scarlat, Merkaz Klita, Afula — Givat Hamore 25/184, Israel

Prof. J. Schroeder, Department of Civil Engineering, University of Waterloo, Waterloo, Ontario, Canada

Prof. L. A. Segel, Department of Mathematics, Rensselaer Polytechnic Institute, Troy, NY 12181, U.S.A.

Dr. M. Slemrod, Center for Dynamical Systems, Division of Applied Mathematics, Brown University, Providence, RI 02912, U.S.A.

Dr. B. Storåkers, Division of Solid Mechanics, Royal Institute of Technology, S-10044 Stockholm 70, Sweden

Prof. T. Tatsumi, Department of Physics, Faculty of Science, University of Kyoto, Kyoto, Japan

Prof. J. M. T. Thompson, Department of Civil and Municipal Engineering, University College London, Gower Street, London W.C. 1, U.K.

Dr. ir. H. G. M. Velthuizen, Afdeling der Werktuigbouwkunde, Technische Hogeschool Twente, Enschede, Netherlands
Dr. V. B. Venkayya, Air Force Flight Dynamics Laboratory, Wright-Patterson Air Force Base, OH 45433, U.S.A.
Prof. P. Villaggio, Istituto di Scienza delle Costruzioni, Università di Pisa, Pisa, Italy
Prof. P. K. C. Wang, Department of System Science, School of Engineering and Applied Science, University of California, Los Angeles, CA 90024, U.S.A.
Prof. J. A. Whitehead, Institute of Geophysics and Planetary Physics, University of California, Los Angeles, CA 90024, U.S.A.
Prof. Dr. ir. L. van Wijngaarden, Laboratorium voor Warmte- en Stromingsleer, Technische Hogeschool Twente, Enschede, Netherlands
Dr. F. Wille, Institut für Angewandte Mathematik, Albert-Ludwigs-Universität, D-78 Freiburg i. Br., Hebelstr. 40, Germany
Mr. N. J. Zabusky, Plasma and Computational Physics Research Department, Bell Telephone Laboratories, Whippany Road, Whippany, NJ 07981, U.S.A.
Prof. Dr. H. Ziegler, Lehrstuhl für Mechanik, Eidgenössische Technische Hochschule, CH-8006 Zürich, Leonhardstr. 33, Switzerland
Prof. Dr. M. Życzkowski, Politechnika Krakowska, Kraków, Poland

Contents

Application of Liapunov's Direct Method to the Stability Problem of Rods Subject to Follower Forces

By

H. Leipholz

University of Waterloo, Waterloo, Ont., Canada

Introduction

We consider an autonomous, dynamic, continuous system in a given state $(\boldsymbol{q}, t)$. The motion of the system can be represented by a path in the $\boldsymbol{Q}$, T-space, where origin $(\boldsymbol{0}, t)$ may be chosen as the undisturbed state. Our aim is to investigate the conditions under which the undisturbed state is stable in comparison with the disturbed state.

We introduce a metric ϱ_0 to measure the distance of the initial point $(\boldsymbol{q}, t_0)$ of a disturbed path from the origin and another metric ϱ to measure the distance of any other point of $(\boldsymbol{q}, t)$ from the origin. These two metrics are supposed to satisfy the following conditions:

(i) ϱ_0, ϱ positive definite,

(ii) $\varrho_0(\boldsymbol{0}, t) = 0$, $\varrho(\boldsymbol{0}, t) = 0$,

(iii) ϱ continuous function of t,

(iv) ϱ continuous with respect to ϱ_0.

Stability is defined with respect to both metrics:

The undisturbed state $(\boldsymbol{0}, t)$ is stable in comparison with the disturbed state $(\boldsymbol{q}, t)$, if for any $\varepsilon > 0$ a $\delta(\varepsilon) > 0$ can be found such that $\varrho(\boldsymbol{q}, t) < \varepsilon$ if $\varrho_0(\boldsymbol{q}(t_0), t_0) < \delta$.

In order to predict stability, a Liapunov functional $V(\boldsymbol{q}, t)$ has to be introduced. With the help of V the following stability criteria can be formulated:

The undisturbed state $(\mathbf{0}, t)$ *is stable with respect to* ϱ_0 *and* ϱ *if there exists a functional* $V(\mathbf{q}, t)$ *which satisfies the following conditions:*

(i) V *is positive definite,*

(ii) $\dot{V} \leqq 0$,

(iii) V *admits an infinitely small upper bound with respect to* ϱ_0, *i.e. for any* $\varepsilon_1 > 0$ *a* $\delta_1(\varepsilon_1) > 0$ *can be found such that* $V(\mathbf{q}(t_0), t_0) < \varepsilon_1$ *if* $\varrho_0(\mathbf{q}(t_0), t_0) < \delta_1$.

Choice of V and Calculation of $\dot{V}$

For an application of Liapunov's direct method, the main task is to find an appropriate functional V. Fortunately there is the following starting point:

Using the "functional derivative"

$$\begin{aligned}\frac{\delta L}{\delta \mathbf{q}} = \frac{\partial \mathscr{L}}{\partial \mathbf{q}} &- \frac{\partial}{\partial x_1}\left(\frac{\partial \mathscr{L}}{\partial \mathbf{q}_{x_1}}\right) + \frac{\partial^2}{\partial x_1^2}\left(\frac{\partial \mathscr{L}}{\partial \mathbf{q}_{x_1 x_1}}\right) - \cdots + \cdots \\ &- \frac{\partial}{\partial x_2}\left(\frac{\partial \mathscr{L}}{\partial \mathbf{q}_{x_2}}\right) + \frac{\partial^2}{\partial x_2^2}\left(\frac{\partial \mathscr{L}}{\partial \mathbf{q}_{x_2 x_2}}\right) - \cdots + \cdots \\ &\vdots \\ &+ \frac{\partial^2}{\partial x_1\, \partial x_2}\left(\frac{\partial \mathscr{L}}{\partial \mathbf{q}_{x_1 x_2}}\right) + \cdots\end{aligned}$$

of the Lagrangian L of the system ($\mathscr{L}$ is the Lagrangian density, i.e. $L = \int_{V_0} \mathscr{L}\, dV_0$), the equation of motion of the system can be expressed by

$$\frac{d}{dt}\frac{\delta L}{\delta \dot{\mathbf{q}}} - \frac{\delta L}{\delta \mathbf{q}} = \mathbf{Q},$$

where $\mathbf{Q}$ are generalized forces which cannot be derived from a potential, i.e. $\mathbf{Q} \neq 0$ means that the system is non-conservative.

Because of $L(\mathbf{q}, p) = \int_{V_0} p\dot{\mathbf{q}}\, dV_0 - H(\mathbf{q}, p)$, where H is the Hamiltonian and p the generalized momentum, the Lagrangian equation can be transformed such that one gets the canonical system of Hamilton's equations

$$\left.\begin{aligned}\frac{d}{dt}\frac{\delta L}{\delta \dot{\mathbf{q}}} - \frac{\delta L}{\delta \mathbf{q}} &= \dot{p} + \frac{\delta H}{\delta \mathbf{q}} = \mathbf{Q}, \\ \frac{d}{dt}\frac{\delta L}{\delta p} - \frac{\delta L}{\delta p} &= -\dot{\mathbf{q}} + \frac{\delta H}{\delta p} = 0, \\ \frac{\delta H}{\delta \mathbf{q}} &= -\dot{p} + \mathbf{Q}, \\ \frac{\delta H}{\delta p} &= \dot{\mathbf{q}},\end{aligned}\right\} \tag{1}$$

respectively.

It is worthwhile to mention that in (1) Q has to be taken as a Dirac function of appropriate dimension, if it has to represent a force which is not distributed over the whole volume V_0 of the system but acts at a point or along a line or on a bounded surface.

Now let us assume that a certain Liapunov functional V has been chosen. For its derivative with respect to the time t

$$\frac{dV}{dt} = \dot{V} = \int_{V_0} \left(\frac{\delta V}{\delta p}\dot{p} + \frac{\delta V}{\delta q}\dot{q}\right) dV_0 \tag{2}$$

holds. Using (1) we get from (2)

$$\dot{V} = \int_{V_0} \left[-\frac{\delta V}{\delta p}\frac{\delta H}{\delta q} + \frac{dV}{\delta q}\frac{\delta H}{\delta p}\right] dV_0 + \int_{V_0} Q\frac{\delta V}{\delta p} dV_0. \tag{3}$$

Since the expression

$$\int_{V_0} \left[-\frac{\delta V}{\delta p}\frac{\delta H}{\delta q} + \frac{dV}{\delta q}\frac{\delta H}{\delta p}\right] dV_0$$

is the so-called Poisson bracket $[V, H]$, we have finally

$$\dot{V} = [V, H] + \int_{V_0} Q\frac{\delta V}{\delta p} dV_0.$$

For conservative systems one should obviously take $V \equiv H$, because in this case $Q \equiv 0$ and also $[V, H] \equiv [H, H] \equiv 0$. Therefore $\dot{V} \equiv 0$ and stability of the undisturbed state of the system is established if $H = V$ is positive definite. This shows the close connection between Liapunov's stability criterion and the classical energy criterion in the case of conservative systems.

If the system is non-conservative, the stability problem is more complicated. But even then it might be advisable to use H as V. Because of $Q \neq 0$, $V = H$ and (1):

$$\dot{H} = \dot{V} = \int_{V_0} Q\dot{q}\, dV_0. \tag{4}$$

For many practical cases Q and q are not vectors but scalars. For example

$$q = w(x_1, x_2, x_3, t), Q = -\sum_i P_i \frac{\partial w}{\partial x_i}, \tag{5}$$

where w is a deflection, x_i are spatial coordinates, P_i are constants and t is the time. Now

$$\dot{H} = \dot{V} = -\int_{V_0} \dot{w} \sum_i P_i \frac{\partial w}{\partial x_i} dV_0.$$

Furthermore, the system may be linear and such that w can be written as a product:

$$w = f(t)\,\varphi(\boldsymbol{x}).$$

Consequently

$$\dot{H} = \dot{V} = -\iiint\limits_{V_0} \varphi \sum_i P_i \varphi_{x_i} dV_0 .$$

This integral can be rewritten as

$$\dot{H} = \dot{V} = -\frac{1}{2} \iint\limits_{S} \varphi^2 \sum_i P_i (\cos \nu, x_i)\, dS , \tag{6}$$

where S denotes the surface of the system and ν the outer normal of the surface.

Let us assume that

$$S_1 \subset S,\ \ S_2 \subset S,\ \ S_1 + S_2 = S \tag{7}$$

holds and that

$$\varphi \equiv 0 \text{ on } S_1,\ \ \sum_i P_i (\cos \nu, x_i) \equiv 0 \text{ on } S_2 . \tag{8}$$

Then

$$\dot{H} = \dot{V} \equiv 0 ,$$

and the system under non-conservative forces is stable, if $H = V$ is positive definite.

Conditions (8) are indeed not exceptions but occur often enough in practice. Take for example a rod, supported at both ends:

$$S_1 = S,\ \ \varphi \equiv 0 \text{ on } S, \text{ i.e. at both ends},\ \ \dot{H} = \dot{V} \equiv 0 .$$

Or take a plate, supported along its whole edges. Again

$$S_1 = S,\ \ \varphi \equiv 0 \text{ on } S,\ \ \dot{H} = \dot{V} \equiv 0 .$$

Finally, take the plate of Fig. 1. It is supported along the edges

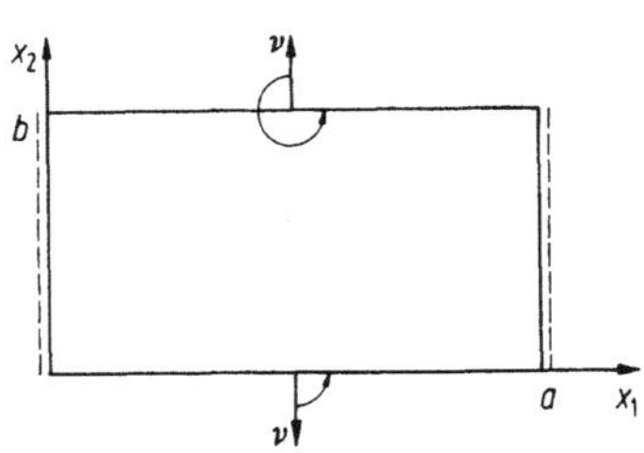

Fig. 1

$$x_1 = 0,\ \ x_1 = a$$

and is free along the edges

$$x_2 = 0,\ \ x_2 = b .$$

Now, on the boundary,

$$S_1 = \begin{cases} x_1 = 0, \\ x_1 = a, \end{cases} \qquad \varphi \equiv 0 .$$

Furthermore, on

$$S_2 = \begin{cases} x_2 = 0, \\ x_2 = b, \end{cases} \qquad \sum_{i=1}^{i=2} P_i \cos (\nu, x_i) \equiv 0 ,$$

if $P_2 = 0$.

Thus, again $\dot{H} = \dot{V} \equiv 0$.

Examples

To apply Liapunov's direct method to the problem of rods subject to non-conservative loads, we introduce two sets of boundary value problems, for which we derive some basic inequalities. These inequalities are of fundamental importance for our further investigations.

There are four cases of buckling of rods subject to conservative loads, each corresponding to different boundary conditions, see Fig. 2.

For all cases the rods have the length l, constant linear mass density μ, constant flexural rigidity α, and they are subject to the constant, distributed, unidirectional load q. All cases are described by the same differential equation

$$\alpha v_{xxxx} + [q(l - x)\, v_x]_x + \mu \ddot{v} = 0\,.$$

The boundary conditions are

$$v(0, t) = v_{xx}(0, t) = v(l, t) = v_{xx}(l, t) = 0 \text{ in case C 1},$$

$$v(0, t) = v_x(0, t) = v(l, t) = v_{xx}(l, t) = 0 \text{ in case C 2},$$

$$v(0, t) = v_x(0, t) = v(l, t) = v_x(l, t) = 0 \text{ in case C 3},$$

$$v(0, t) = v_x(0, t) = v_{xx}(l, t) = v_{xxx}(l, t) = 0 \text{ in case C 4}.$$

Likewise, there are four cases of buckling of rods, under non-conservative load, see Fig. 3. The difference with the preceding cases lies

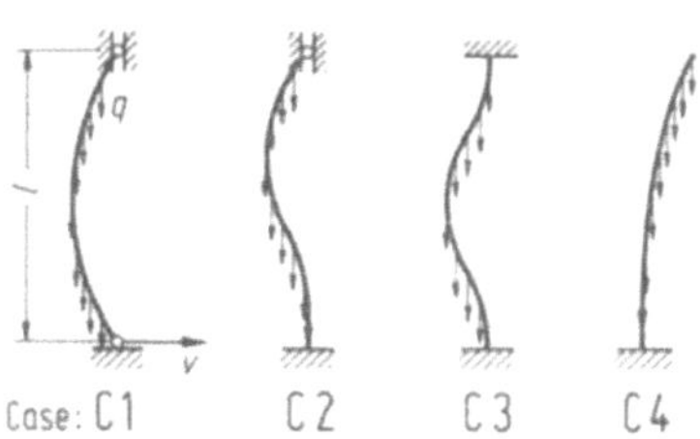

Fig. 2

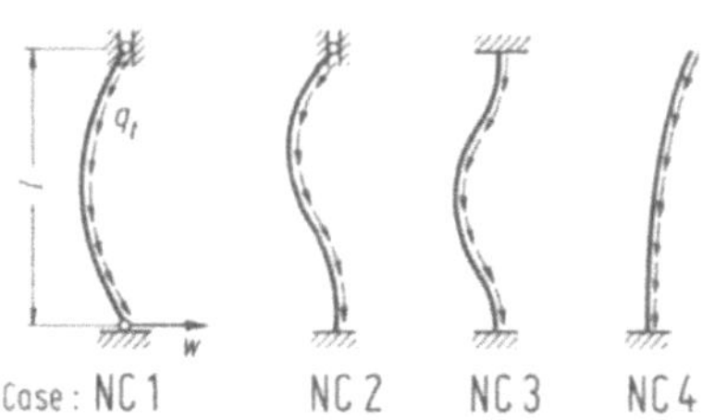

Fig. 3

in the fact, that the load q_t now is a tangential one. For this case, the differential equation of motion is:

$$\alpha w_{xxxx} + q_t(l - x)\, w_{xx} + \mu \ddot{w} = 0\,.$$

The boundary conditions are the same as above:

$$w(0, t) = w_{xx}(0, t) = w(l, t) = w_{xx}(l, t) = 0 \text{ in case NC 1},$$

$$w(0, t) = w_x(0, t) = w(l, t) = w_{xx}(l, t) = 0 \text{ in case NC 2},$$

$$w(0, t) = w_x(0, t) = w(l, t) = w_x(l, t) = 0 \text{ in case NC 3},$$

$$w(0, t) = w_x(0, t) = w_{xx}(l, t) = w_{xxx}(l, t) = 0 \text{ in case NC 4}.$$

For all cases, the solutions of NC 1—4, namely w, satisfy the boundary conditions of C 1—4. Therefore, they are applicable for the calculation of Rayleigh's quotient in the cases C 1—4:

$$\lambda_c = \frac{q_{\text{crit}}}{\alpha} \leqq \frac{\int_0^l w w_{xxxx}\, dx}{\int_0^l - w[(l-x)\, w_x]_x\, dx}.$$

As a result of integration by parts and taking the boundary conditions into account, in all cases

$$\int_0^l w w_{xxxx}\, dx = \int_0^l w_{xx}^2\, dx_1, \quad \int_0^l -w\,[(l-x)\, w_x]_x\, dx = \int_0^l (l-x)\, w_x^2\, dx.$$

Hence,

$$\lambda_c \leqq \frac{\int_0^l w_{xx}^2\, dx}{\int_0^l (l-x)\, w_x^2\, dx}, \quad \int_0^l w_{xx}^2\, dx \geqq \lambda_c \int_0^l (l-x)\, w_x^2\, dx. \tag{9}$$

Besides, in all cases NC 1—4:

$$w = \int_0^x w_x\, dx$$

because of $w(0, t) = 0$. The Schwarz inequality leads to

$$w^2 \leqq x \int_0^x w_x^2\, dx \leqq x \int_0^l w_x^2\, dx.$$

By integration:

$$\int_0^l w^2\, dx \leqq \frac{l^2}{2} \int_0^l w_x^2\, dx, \quad \int_0^l w_x^2\, dx \geqq \frac{2}{l^2} \int_0^l w_x^2\, dx. \tag{10}$$

Finally, in accordance with the mean value theorem,

$$\int_0^l (l-x)\, w_x^2\, dx = M \int_0^l w_x^2\, dx, \quad 0 < M < l. \tag{11}$$

For the cases NC 1—3 we use the Hamiltonian H as a Liapunov functional:

$$H = V = \int_0^l \left[\frac{\mu}{2}\, \dot{w}^2 + \frac{\alpha}{2}\, w_{xx}^2 - \frac{1}{2}\, q_t (l-x)\, w_x^2\right] dx. \tag{12}$$

It is easy to find

$$Q = -q_t w_x \quad \text{and} \quad w = f(t)\, \varphi(x).$$

Furthermore for the cases NC 1—3

$$\varphi(0) = \varphi(l) = 0.$$

Therefore all conditions of the preceding section, namely (5), (8) are satisfied to get $\dot{H} = \dot{V} = 0$. Indeed, from (12) follows

$$\dot{H} = \dot{V} = -\int_0^l q_t w_x \dot{w}\, dx = -\dot{f}f q_t \int_0^l \varphi_x \varphi\, dx = -\dot{f}f q_t \frac{1}{2} [\varphi^2]_0^l = 0. \tag{13}$$

Furthermore

$$V \geqq \int_0^l \left[\frac{\alpha}{2} w_{xx}^2 - \frac{1}{2} q_t (l - x)\, w_x^2\right] dx,$$

and because of (9), (11)

$$V \geqq \frac{\alpha M}{2}\left(\lambda_c - \frac{q_t}{\alpha}\right) \int_0^l w_x^2\, dx. \tag{14}$$

For stability, V has to be positive definite. That is the case for

$$q_t \leqq \alpha\lambda_c = q_{\text{crit}}. \tag{15}$$

We can conclude that in the cases of NC 1—3 the non-conservative load does not destabilize, since the non-conservative loaded rod is at least stable as long as this is the case for the conservative loaded rod.

This is true indeed. The buckling loads, calculated with the help of Galerkin's method prove this fact:

Case	C 1	C 2	C 3
q_{crit}	$18.46\ \alpha/l^3$	$56\ \alpha/l^3$	$71.46\ \alpha/l^3$
Case	NC 1	NC 2	NC 3
$q_{t,\text{crit}}$	$18.96\ \alpha/l^3$	$57.95\ \alpha/l^3$	$81.37\ \alpha/l^3$

Always, $q_{t,\text{crit}} > q_{\text{crit}}$. Thus, (15) holds for stability in the cases NC 1—3.

Now we have to consider the norms with respect to which stability exists. In the preceding considerations we found that

$$w^2 \leqq x \int_0^l w_x^2\, dx.$$

Therefore

$$|w|^2 \leqq l \int_0^l w_x^2\, dx. \tag{16}$$

Hence from (14):

$$V \geqq \frac{\alpha M}{2l}\left(\lambda_c - \frac{q_t}{\alpha}\right) |w|^2.$$

We choose $|w|$ as norm ϱ and have

$$V \geqq \varkappa^* \varrho^2. \tag{17}$$

Besides from (12) follows

$$V \leqq \int_0^l \left[\frac{\mu}{2}\dot{w}^2 + \frac{\alpha}{2} w_{xx}^2\right] dx \leqq \varkappa^{**} \int_0^l (\dot{w}^2 + w_{xx}^2)\, dx.$$

Therefore we choose the norm ϱ_0 such that

$$\int_0^l (\dot{w}^2 + w_{xx}^2)\, dx = \varrho_0^2,$$

and

$$V \leqq \varkappa^{**} \varrho_0^2. \tag{18}$$

Consequently, V admits an infinitely small upper bound with respect to ϱ_0 and $\varrho^2 < \varkappa\varrho_0^2$, i.e. ϱ depends continuously on ϱ_0. It is trivial to mention that all the other conditions imposed on the norms ϱ and ϱ_0 are also satisfied by our choice.

The conclusion is that in the non-conservative cases NC 1—3, the rod will be stable for $q_t \leqq q_{\text{crit}}$ (q_{crit} = buckling load of the corresponding conservative case), and $|w|$ remains small for $t \geqq t_0$ as far as $|\dot{w}|$ and $|w_{xx}|$ were small enough for $t = t_0$, i.e.

$$\varrho \equiv |w| < \varepsilon \quad \text{for} \quad t \geqq t_0,$$

if

$$\varrho_0 \equiv \left[\int_0^l (\dot{w}^2 + w_{xx}^2)\, dx\right]^{1/2} < \delta(\varepsilon) \quad \text{for} \quad t = t_0.$$

The case NC 4 (see Fig. 3) has to be investigated separately, because it is more difficult compared to the cases NC 1—3. This time we cannot set $V = H$, but we can at least use H as the basic term of V. We can take

$$V = H + \int_0^l q_t w w_x\, dx - \int_0^t \left[\int_0^l q_t \dot{w}_x w\, dx\right] dt. \tag{19}$$

Thus

$$\dot{V} = \dot{H} + \frac{d}{dt} \int_0^l q_t w w_x\, dx - \int_0^l q_t \dot{w}_x w\, dx.$$

By using the relation

$$\dot{H} = - \int_0^l q_t w_x \dot{w}\, dx,$$

and performing the differentiation in the equation above,

$$\dot{V} = - \int_0^l q_t w_x \dot{w}\, dx + \int_0^l q_t \dot{w} w_x\, dx + \int_0^l q_t w \dot{w}_x\, dx - \int_0^l q_t \dot{w}_x w\, dx.$$

Therefore

$$\dot{V} \equiv 0.$$

Now we have to investigate V. For this purpose we use

$$w = f(t)\,\varphi(x).$$

With the help of this expression we find, because of $\varphi(0) = 0$,

$$\int_0^l q_t w w_x\,dx - \int_0^t \left[\int_0^l q_t \dot{w}_x w\,dx\right] dt = f^2 q_t \int_0^l \varphi_x \varphi\,dx - \int_0^t f\dot{f}\,dt\; q_t \int_0^l \varphi_x\varphi\,dx$$

$$= \left[f^2 - \frac{1}{2}\left(f^2(t) - f^2(0)\right)\right] q_t \int_0^l \varphi_x\varphi\,dx = \frac{1}{4}\,[f^2(t) + f^2(0)]\,q_t\varphi^2(l) > 0.$$

Hence from (19), and because of $\dot{V} \equiv 0$, i.e. $V(t) \equiv V(0)$,

$$V = H + \frac{1}{4}\,q_t\varphi^2(l)\,[f^2(t) + f^2(0)] = H + \frac{1}{2}\,q_t\varphi^2(l)\,f^2(0). \qquad (20)$$

Therefore, V is positive definite and stability exists, if

$$H = \int_0^l \left[\frac{\mu}{2}\,\dot{w}^2 + \frac{\alpha}{2}\,w_{xx}^2 - \frac{1}{2}\,q_t(l - x)\,w_x^2\right] dx \qquad (21)$$

is positive definite. Since

$$H \geqq \frac{\alpha M}{2}\left(\lambda_c - \frac{q_t}{\alpha}\right)\int_0^l w_x^2\,dx \geqq \frac{\alpha M}{2l}\left(\lambda_c - \frac{q_t}{\alpha}\right)|w|^2, \qquad (22)$$

this is the case for

$$q_t \leqq \alpha\lambda_c q_{\text{crit}},$$

i.e. again (15) is valid as in the cases NC 1—3. Thus, we can draw the same conclusion as in the cases C 1—3. *The non-conservative loaded rod of case* NC 4 *cannot become unstable for a load smaller than the buckling load of the corresponding conservative case* C 4, *i.e. the non-conservative load does not destablize in this case too.* Again, this theorem can be justified by the following results of a numerical calculation with the help of Galerkin's method:

Case	C 4	NC 4
q_{crit}	$7.84\,\alpha/l^3$	$40.7\,\alpha/l^3$

To complete the investigation of case NC 4, the norms have to be considered, which decide in this case about stability. From (20), (21), it follows that

$$V > H > \frac{\alpha M}{2l}\left(\lambda_c - \frac{q_t}{\alpha}\right)|w|^2.$$

Setting

$$\frac{\alpha M}{2l}\left(\lambda_c - \frac{q_t}{\alpha}\right) = \varkappa^* \quad\text{and}\quad |w| = \varrho,$$

$$V \geqq \varkappa^*\varrho^2 \qquad (23)$$

results.

On the other hand it follows from (20) and (21) that

$$V < \int_0^l \left[\frac{\mu}{2}\dot{w}^2 + \frac{\alpha}{2} w_{xx}^2\right] dx + \frac{1}{4} q_t \varphi^2(l)\,[f^2(t) + f^2(0)].$$

As a result of $\dot{V} \equiv 0$,

$$V(t) \equiv V(0),$$

and therefore

$$V < \int_0^l \left[\frac{\mu}{2}\dot{w}^2 + \frac{\alpha}{2} w_{xx}^2\right] dx + \frac{1}{2} q_t [f(0)\,\varphi(l)]^2. \tag{24}$$

Since $w = f(t)\,\varphi(x)$,

$$[f(0)\,\varphi(l)]^2 = w^2(0, l). \tag{25}$$

Furthermore

$$\int_0^l \left[\frac{\mu}{2}\dot{w}^2 + \frac{\alpha}{2} w_{xx}^2\right] dx \leqq \frac{\mu l}{2} \sup |\dot{w}|^2 + \frac{\alpha l}{2} \sup |w_{xx}|^2. \tag{26}$$

Hence, from (24) follows

$$V < \varkappa^{**} \varrho_0 \tag{27}$$

with the help of (25), (26) and by setting

$$\varkappa^{**} = \operatorname{Max}\left[\frac{\mu l}{2}, \frac{\alpha l}{2}, \frac{q_t}{2}\right]$$

and

$$\varrho_0 = \sup |\dot{w}|^2 + \sup |w_{xx}|^2 + |w(0, l)|^2.$$

Because of (27) V admits an infinitely small upper bound with respect to ϱ_0.

Combining (23) with (27)

$$\varkappa^* \varrho^2 < V < \varkappa^{**} \varrho_0$$

results. Consequently, the following statement holds:

If

$$\varrho_0 = \sup |\dot{w}|^2 + \sup |w_{xx}|^2 + |w(0, l)|^2 < \delta \quad \text{for} \quad t = t_0,$$

then

$$\varrho \equiv |w| < \varepsilon \quad \text{for} \quad t \geqq t_0,$$

i.e. stability in the sense of Liapunov.

Frequency Domain Criteria for Stability of Systems Modeled by Certain Partial Differential Equations

By

J. Anton, P. Falb

Brown University, Providence, R.I.
and NASA Electronics Research Center, Cambridge, Mass., U.S.A.

and

M. I. Freedman

NASA Electronics Research Center, Cambridge, Mass., U.S.A.

In recent years many of the methods used to investigate stability of lumped parameter systems have been extended to study the stability of distributed systems. In this work we employ an extension of the *circle-criterion*, cf. [1], to Hilbert space-valued systems, cf. [2], as well as some rather sophisticated results on elliptic boundary value problems, cf. [3]. The stimulus for the approach is the desire to have calculable criteria for stability. In particular, we study the "L_2-stability" of boundary, initial-value problems of the form

$$\left.\begin{aligned} [P(D_t) + Q(D_t)\, L(x, D_x)]\, u + N(u) = 0 \ \text{ in } \ [0,\infty)\times\mathscr{D}; \\ (D_t^k u)\,(0, x) = f_k(x), \quad k = 0, \ldots, l-1; \\ B_j(x, D_x)\, u\,|_{\partial\mathscr{D}} = 0, \quad j = 1, \ldots, m. \end{aligned}\right\} \qquad (*)$$

Here $L(x, D_x)$ is an elliptic operator of order $2m$ in the bounded, open domain $\mathscr{D} \subseteq \Re^n$, P and Q are polynomials of degrees p and q respectively,

---

Acknowledgement. This research was supported in part by the Air Force Office of Scientific Research under grant no. AFOSR 693-67 and in part by the National Science Foundation under grant no. NSF-GK 2788.

N is a nonlinearity subject to $N(0) = 0$, and $B_j(x, D_x)$ are boundary operators which cover $\mathcal{D}$ (cf. [3] for details).

Set $l = \max(p, q)$ and assume that the boundary-value problem $(L(x, D_x), B_j(x, D_x))$ is self-adjoint and has the (real, discrete, unbounded) spectrum Λ. Let $\operatorname{Re} s \geqq 0 = \{s \in \mathfrak{C} \mid \operatorname{Re} s \geqq 0\}$. Our main stability result is contained in the

Theorem. *Suppose the following conditions are satisfied:*

C 1: $0 \notin \bigcup_{\lambda \in \Lambda} \{P(s) + Q(s)\lambda \mid \operatorname{Re} s \geqq 0\}$; $Q(s) \neq 0$ *for* $\operatorname{Re} s \geqq 0$;

C 2: $\lim_{|s| \to \infty} \frac{1}{P(s) + Q(s)\lambda} = 0$ *uniformly for* $\lambda \in \Lambda$; $\operatorname{Re} s \geqq 0$;

C 3: *there are numbers* $W, K > 0$ *such that* $\left| \frac{(i\omega)^k}{P(i\omega) + Q(i\omega)\lambda} \right| \leqq K$ *uniformly for* $\lambda \in \Lambda$ *when* $0 \leqq k \leqq l$ *and* $|\omega| \geqq W$.

Suppose also that for some $0 < a < b$,

CC 1: $0 \leqq \int_{\mathcal{D}} [bh(x) - N(h(x))]\,[N(h(x)) - ah(x)]\,dx$ *for all* $h \in L_2(\mathcal{D})$;

CC 2: $-\frac{1}{b} \notin \bigcup_{\lambda \in \Lambda} \left\{ \frac{1}{P(s) + Q(s)\lambda} \,\middle|\, \operatorname{Re} s \geqq 0 \right\}$;

CC 3: *for all* $\lambda \in \Lambda$ *the sets* $\left\{ \frac{1}{P(i\omega) + Q(i\omega)\lambda} \,\middle|\, \omega \in \mathfrak{R} \right\}$ *do not intersect and remain outside of the circle*

$$\mathscr{C} = \left\{ z \in \mathfrak{C} \,\middle|\, \left| z + \frac{1}{2}\left(\frac{1}{a} + \frac{1}{b}\right) \right| \leqq \frac{1}{2}\left(\frac{1}{a} - \frac{1}{b}\right) \right\}.$$

Finally, suppose that for $A > 0$,

B. I. $\sum_{|\alpha| \leqq 2m} \int_{\mathcal{D}} |D_x^\alpha f_k(x)|^2\,dx \leqq A$, $k = 0, \ldots, l - 1$.

Then there is a continuous function $K: [0, \infty) \to [0, \infty)$, $K(0) = 0$, *such that if* u *is any solution of* (*) *which is "locally bounded", i.e.,*

LB 1: $\int_0^a \int_{\mathcal{D}} |D_t^k u|^2\,dx\,dt < \infty$, $k = 0, \ldots, l$; *all* $a > 0$;

LB 2: $\int_0^a \int_{\mathcal{D}} |D_t^k D_x^\alpha u|^2\,dx\,dt < \infty$, $k = 0, \ldots, q$; $|\alpha| \leqq 2m$; *all* $a > 0$;

then u *is in fact "globally bounded", i.e.,*

GB 1: $\int_0^\infty \int_{\mathcal{D}} |D_t^k u|^2\,dx\,dt \leqq K(A)$, $k = 0, \ldots, l$;

GB 2: $\int_0^\infty \int_{\mathcal{D}} |D_t^k D_x^\alpha u|^2\,dx\,dt \leqq K(A)$, $k = 0, \ldots, q$; $|\alpha| \leqq 2m$.

We note that LB 1, LB 2 are automatically true if N satisfies a Lipschitz condition, but the stability result requires no such restriction on N.

The proof of this theorem involves a reduction of (*) to a system:

$$u(t) + \int_0^t \varphi(t-\tau)\, N\big(u(\tau)\big)\, d\tau = u_0(t), \qquad (**)$$

where u and u_0 take values in a suitable Hilbert space H, $N: H \to H$, and for each value of its argument, φ is a bounded linear map of H into itself. For this reduction the assumptions C 1—C 3, as well as the coerciveness theory, are employed. The global boundedness of $\{f_k, k = 0, \ldots, l-1\}$ finds its counterpart in that of u_0, so the assumptions CC 1—CC 3 allow us to bring the results of the extended *circle-criterion* to bear on (**), yielding GB 1 and GB 2.

As an example, consider the system

$$\left.\begin{aligned} u_t - \Delta_x u + N(u) &= 0 \quad \text{in} \quad [0,\infty)\times Q, \\ u(0, x) &= f(x), \\ u &= 0 \quad \text{on} \quad \partial Q, \end{aligned}\right\} \qquad (***)$$

where $\Delta_x = \frac{\partial^2}{\partial x_1^2} + \cdots + \frac{\partial^2}{\partial x_n^2}$, $Q = [0, \pi] \times \overset{n\text{-times}}{\cdots} \times [0, \pi]$, and ∂Q is the boundary of Q. Suppose N is a function as in Fig. 1: Then since $\Lambda = \{k^2 = m_1^2 + \cdots + m_n^2,\ m_i$ any positive integer$\}$ we have the plots of

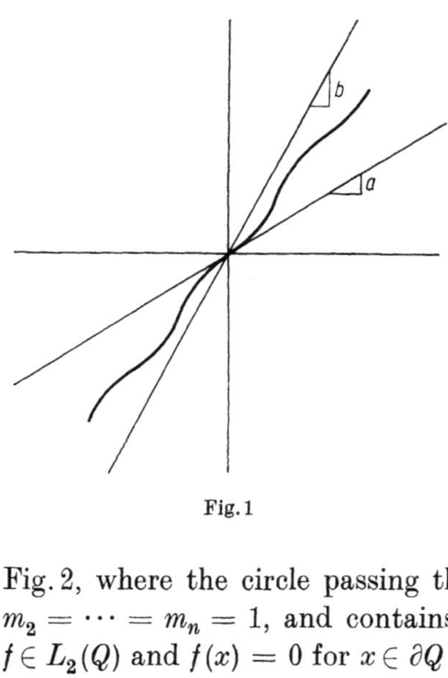

Fig. 1

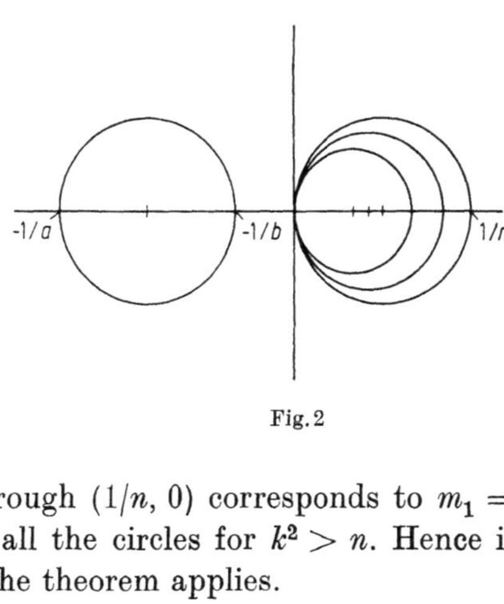

Fig. 2

Fig. 2, where the circle passing through $(1/n, 0)$ corresponds to $m_1 = m_2 = \cdots = m_n = 1$, and contains all the circles for $k^2 > n$. Hence if $f \in L_2(Q)$ and $f(x) = 0$ for $x \in \partial Q$ the theorem applies.

References

1. Zames, G.: On the input-output stability of time-varying non-linear feedback systems, Parts I, II. IEEE Trans. Autom. Control **AC-11**, 228—238, 465—476 (1966).
2. Freedman, M. I., Falb, P., Zames, G.: A Hilbert space stability theory over locally compact abelian groups. SIAM J. Control **7** (1969).
3. Schechter, M.: General boundary value problems for elliptic partial differential equations. Comm. Pure Appl. Math. **12**, 94—106 (1959).

A Note on Imperfection Sensitivity of Thin Plates on a Non-linear Elastic Foundation

By

E. Reissner

University of California, San Diego, La Jolla, Cal., U.S.A.

The considerations which follow are intended as a contribution to the asymptotic theory of the effects of imperfections on the initial post-buckling behavior of non-linear elastic structures, in the sense of KOITER.

We are dealing with a uniform infinite isotropic elastic plate, on a non-linear elastic foundation, and in a state of uniform two-dimensional hydrostatic compression prior to the onset of buckling.

Mathematically, the problem consists in an analysis of certain classes of solutions of the system of differential equations

$$\begin{aligned} D\nabla^4 w + N\nabla^2(w + \overline{w}) + k_1 w - K(w) \\ = (\overline{w} + w)_{,xx} F_{,yy} + (\overline{w} + w)_{,yy} F_{,xx} - 2(\overline{w} + w)_{,xy} F_{,xy}, \end{aligned} \tag{1}$$

$$C\nabla^4 F = 2\overline{w}_{,xy} w_{,xy} - \overline{w}_{,xx} w_{,yy} - \overline{w}_{,yy} w_{,xx} + w^2_{,xy} - w_{,xx} w_{,yy}. \tag{2}$$

To be obtained are asymptotic results showing the effect of the non-linear terms in (1) and (2) in removing indeterminacies and singularities in the behavior of the corresponding solutions of the associated problems governed by the linear differential equation

$$D\nabla^4 w + N\nabla^2(w + \overline{w}) + k_1 w = 0. \tag{3}$$

The solutions of (3) which are the starting point of our considerations are the buckling solutions w_c, which exist for $\overline{w} = 0$, and which are

$$w_c = \sum W_i \cos \alpha_i \frac{x}{L} \cos \beta_i \frac{y}{L}, \quad \alpha_i^2 + \beta_i^2 = 1 \tag{4}$$

Acknowledgement. This work was supported by the Office of Naval Research.

with the amplitude factors W_i arbitrary, with $L = (D/k_1)^{1/4}$ and with $N = N_c = 2(k_1 D)^{1/2}$.

In addition to the indeterminateness of the W_i in (4) we are concerned with the singularities which — upon assuming $\overline{w} = \sum \overline{W}_i \cos \alpha_i x/L \cos \beta_i y/L$ — arise in the solution

$$w = \sum \frac{N \overline{W}_i}{N_c - N} \cos \alpha_i \frac{x}{L} \cos \beta_i \frac{y}{L} \tag{5}$$

of (3) as N increases from below towards N_c.

In assessing the influence of the non-linear terms in (1) and (2) we limit ourselves in what follows to functions K given by

$$K(w) = k_2 h^{-1} w^2 + k_3 h^{-2} w^3, \tag{6}$$

where h is the thickness of the plate and where k_2 and k_3 are positive constants, at most of the order of magnitude of k_1.

To arrive at the results which are wanted, we introduce dimensionless dependent and independent variables in the form

$$\xi = \frac{x}{L}, \quad \eta = \frac{y}{L}, \quad \frac{w}{w_0} = f(\xi, \eta), \quad \frac{\overline{w}}{\overline{w}_0} = \overline{f}(\xi, \eta), \quad \frac{F}{F_0} = g(\xi, \eta). \tag{7}$$

In this the functions f, $\overline{f}$ and g and all their derivatives with respect to ξ and η are to be $O(1)$, at most.

An inspection of the resulting system of equations indicates that we should choose the scale factor F_0 in terms of w_0 as

$$F_0 = C^{-1} w_0^2, \tag{8}$$

whereupon (1) and (2) can be written as

$$\begin{aligned} \nabla^4 f + 2\nabla^2 f + f = {} & 2\left(1 - \frac{N}{N_c}\right) \nabla^2 f + \frac{k_2 w_0}{k_1 h} f^2 + \frac{k_3 w_0^2}{k_1 h^2} f^3 \\ & - 2 \frac{\overline{w}_0}{w_0} \nabla^2 \overline{f} + 2\left(1 - \frac{N}{N_c}\right) \frac{\overline{w}_0}{w_0} \nabla^2 \overline{f} \\ & + \frac{w_0^2}{CD} (f_{,\xi\xi} g_{,\eta\eta} + \cdots) + \frac{\overline{w}_0}{w_0} \frac{w_0^2}{CD} (\overline{f}_{,\xi\xi} g_{,\eta\eta} + \cdots), \end{aligned} \tag{9}$$

$$\nabla^4 g = f_{,\xi\eta}^2 - f_{,\xi\xi} f_{,\eta\eta} + \frac{\overline{w}_0}{w_0} (2 \overline{f}_{,\xi\eta} f_{,\xi\eta} - \overline{f}_{,\xi\xi} f_{,\eta\eta} - \overline{f}_{,\eta\eta} f_{,\xi\xi}). \tag{10}$$

We now consider that $CD = \varrho h^2$ where ϱ is a dimensionless constant of order unity and we have then that (9) and (10) involve the *three* parameters $1 - N/N_c$, w_0/h and $\overline{w}_0/w_0$. We interpret *initial* postbuckling behavior and imperfection sensitivity to be associated with "sufficiently small" values of $1 - N/N_c$, w_0/h and $\overline{w}_0/w_0$.

To obtain information on the behavior of the solutions of (9) and (10) for small values of the three parameters we use an *iterative procedure* and we propose that *initial postbuckling behavior and imperfection sensitivity may be determined from a requirement that there be "no secular terms" in the solution of the second step of the procedure.*

Assuming $1 - N/N_c$, w_0/h and $\overline{w}_0/w_0$ to be small, the first step of the iterative procedure consists in solving the equations

$$\nabla^4 f_1 + 2\nabla^2 f_1 + f_1 = 0, \quad \nabla^4 g_1 = f_{1,\xi\eta}^2 - f_{1,\xi\xi} f_{1,\eta\eta}. \tag{11}$$

In setting up the second step of the procedure, we are led to distinguish the two cases

$$\text{(i)} \quad \frac{k_2}{k_1} = O(1), \qquad \text{(ii)} \quad \frac{k_2}{k_1} \ll 1. \tag{12}$$

For case (i) we have

$$\nabla^4 f_2 + 2\nabla^2 f_2 + f_2 = 2\left(1 - \frac{N}{N_c}\right)\nabla^2 f_1 + \frac{k_2}{k_1}\frac{w_0}{h} f_1^2 - 2\frac{\overline{w}_0}{w_0}\nabla^2 \overline{f}, \tag{13}$$

$$\nabla^2 g_2 = f_{2,\xi\eta}^2 - f_{2,\xi\xi} f_{2,\eta\eta} + \frac{\overline{w}_0}{w_0}(2\overline{f}_{,\xi\eta} f_{1,\xi\eta} - \cdots). \tag{14}$$

For case (ii), with $k_2 = 0$ for simplicity's sake, we have instead of (13) the relation

$$\begin{aligned}\nabla^4 f_2 + 2\nabla^2 f_2 + f_2 = 2\left(1 - \frac{N}{N_c}\right)\nabla^2 f_1 + \frac{k_3}{k_1}\frac{w_0^2}{h^2} f_1^3 \\ - 2\frac{\overline{w}_0}{w_0}\nabla^2 \overline{f} + \frac{1}{\varrho}\frac{w_0^2}{h^2}(f_{1,\xi\xi} g_{1,\eta\eta} + \cdots)\end{aligned} \tag{15}$$

while Eq. (14) remains as it is.

A comparison of (13) with (15) indicates the remarkable fact that while initial postbuckling behavior for the cubic foundation is affected by the midplane stretching of the plate which is associated with deflections of the order of the plate thickness, this is not the case for the quadratic foundation. In this sense the treatment of initial postbuckling behavior for the quadratic foundation may be thought of as a more elementary problem than the corresponding treatment for the cubic foundation.

We illustrate use of the foregoing by considering Eqs. (11) and (13) for the plate on a quadratic foundation with initial imperfection function $\overline{f} = \sum \overline{a}_i \cos \alpha_i \xi \cos \beta_i \eta$. Eq. (11) implies as associated supplementary deflection function $f_1 = \sum a_i \cos \alpha_i \xi \cos \beta_i \eta$. Eq. (13) becomes

$$\begin{aligned}\nabla^4 f_2 + 2\nabla^2 f_2 + f_2 = 2\sum\left[\frac{\overline{w}_0}{w_0}\overline{a}_i - \left(1 - \frac{N}{N_c}\right)a_i\right]\cos \alpha_i \xi \cos \beta_i \eta \\ + \frac{k_2}{k_1}\frac{w_0}{h}\left(\sum a_i \cos \alpha_i \xi \cos \beta_i \eta\right)^2.\end{aligned} \tag{16}$$

In order that the function f_2 be free of secular terms the right-hand side of (16) evidently must not contain any of the terms $\cos \alpha_i \xi \cos \beta_i \eta$ with $\alpha_i^2 + \beta_i^2 = 1$. We find that this precludes the occurence of one-mode solutions for f_1. The simplest multimode solution for f_1 is found to be a two-mode solution. Assuming a_1, a_2, $\overline{a}_1$, $\overline{a}_2$ not equal to zero and all other a_i and $\bar{a}_i$ equal to zero in (16) we find as conditions for the non-appearance of secular terms in f_2 that $\alpha_1 = 0$, $\beta_1 = 1$, $\alpha_2 = \sqrt{3}/2$, $\beta_2 = 1/2$, while at the same time

$$\left(1 - \frac{N}{N_c}\right) a_1 - \frac{1}{8} \frac{k_2}{k_1} \frac{w_0}{h} a_2^2 = \frac{\overline{w}_0}{w_0} \bar{a}_1, \tag{17}$$

$$\left(1 - \frac{N}{N_c}\right) a_2 - \frac{1}{2} \frac{k_2}{k_1} \frac{w_0}{h} a_1 a_2 = \frac{\overline{w}_0}{w_0} \bar{a}_2. \tag{18}$$

Evidently Eqs. (17) and (18) are two simultaneous equations for the quantities $w_0 a_i/h$ in terms of $\overline{w}_0 \bar{a}_i/h$, $1 - N/N_c$ and k_2/k_1.

For an indication of the nature of the further discussion we consider the case $\bar{a}_2 = 0$. We then have $a_2 = 0$ and $(1 - N/N_c)\, w_0 a_1 = \overline{w}_0 \bar{a}_1$, *or* $w_0 a_1 = 2(1 - N/N_c)\, h k_1/k_2$ and $w_0 a_2 = [16(1 - N/N_c)^2 h^2 k_1^2/k_2^2 + 8 \overline{w}_0 \bar{a}_1 k_1/k_2]^{1/2}$. From this we conclude the occurence of a bifurcation stress measure N^* and a bifurcation amplitude $w_0 a_1^*$, determined from the condition that the two distinct expressions for $w_0 a_1$ give equal values,

$$\frac{N^*}{N_c} = 1 - \left(\frac{k_2}{k_1} \frac{\overline{w}_0 \bar{a}_1}{2h}\right)^{1/2}, \qquad \frac{w_0 a_1^*}{h} = \left(2 \frac{k_1}{k_2} \frac{\overline{w}_0 \bar{a}_1}{h}\right)^{1/2}. \tag{19}$$

We note that the above consideration of Eqs. (17) and (18) is analogous to an analysis by Hutchinson [1] of a case of imperfection sensitivity of pressurized spherical shells. A discussion of the relation between the problem of the plate on a quadratic foundation and Hutchinson's version of the spherical shell problem may be found in another place [2], together with a discussion of the three-mode solution of the present problem and of the *one*-mode solution of the plate on a *cubic* foundation.

References

1. Hutchinson, J. W.: J. Appl. Mech. **34**, 49 (1967).
2. Reissner, E.: Studies in Appl. Math. **49**, 45 (1970).

A Stability Criterion for Non-linear Continua

By

P. Villaggio

University of Pisa, Pisa, Italy

1. Introduction

The present paper deals with the dynamical extension of a well known criterion of static stability of a body subjected to conservative loads (PEARSON [1] and HILL [2]) according to the stability analysis of the second method of Liapunov. Since the problem involves partial differential equation, the characterization of the stability depends on a Liapunov functional, instead of a function, as was pointed out by HALE [3], ZUBOV [4] and others, and, consequently, the stability is intended as conservation of a suitable functional "distance" during the motion.

Since the properties of definiteness of the Liapunov functional depend on the constitutive law of the material and the stress and strain of its fundamental state, it is useful to consider an adjunctive term, which makes definite the sign of the Liapunov functional and its time derivative. This device was applied by PARKS [5] in the flutter problem of a two-dimensional panel in a supersonic flow.

By mean of some classical a-priori inequalities connecting the motion with the gradient of displacement, these conditions of definiteness acquire a more expressive formulation, under form of sufficient criteria of stability or instability, whose verification assures a-fortiori the validity of theorems used by MOVCHAN [6].

2. Equations of the motion and Liapunov's functional

Let $\mathscr{B}$ be a body and $\mathscr{B}_R$ a fixed reference configuration where ξ are the positions occupied by the material points X of the body. At the time t the actual configuration $\mathscr{B}_\chi$ is characterized by the functions:

(1) the *motion* χ, with $x = \chi(\xi, t)$ called the *position* at time t of the point X;

(2) the *stress tensor* $T = T^T$;

(3) the *body force* b;

(4) the *boundary data*, that is the displacement χ_0 on a portion $\delta_1\mathscr{B}_\chi$ of the boundary and the surface tractions t_0 in the complementary portion $\delta_2\mathscr{B}_\chi$;

the *initial data*, that is the motion $\chi^\#$ and the velocity field $\dot{\chi}^\#$ at the initial time t_0.

With reference to the configuration $\mathscr{B}_R$ the equations of the motion assume the form:

$$\left.\begin{aligned} \operatorname{div} T_R + \varrho_R b = \varrho_R \ddot{x} \quad &\text{in} \quad \mathscr{B}_R, \\ \chi = \chi_0 \quad \text{on} \quad \delta\mathscr{B}_{1_R}, \quad t_R = t_{0_R} \quad &\text{on} \quad \delta\mathscr{B}_{2_R}, \\ \chi(\xi, t_0) = \chi^\#, \quad \dot{\chi}(\xi, t_0) = \dot{\chi}^\# \quad &\text{in} \quad \mathscr{B}_R, \end{aligned}\right\} \tag{2.1}$$

where T_R is the Piola-Kirchhoff stress tensor related to the Cauchy tensor by the equation (vid. e.g. TRUESDELL and NOLL [7], sec. 48):

$$T_R = J\,T(F^T)^{-1} \tag{2.2}$$

having indicated by F the gradient of χ and J its determinant; ϱ_R is the density in the reference configuration.

If the functions χ are considered as members of a Banach space, whose norm is defined by:

$$\|\chi\|^2 = \frac{1}{2} \int_{\mathscr{B}_R} x x \, dv_R + \frac{1}{2} \int_{\mathscr{B}_R} \varrho_R \dot{x}\dot{x} \, dv_R, \tag{2.3}$$

the state χ is said stable in the sense of Liapunov (cf. HALE [3]) if for all solutions χ^* of the problem (2.1) differing from χ for a perturbation of the initial data such that:

$$\|\chi_0^\# - \overset{*}{\chi}{}_0^\#\| < \delta_{\varepsilon, t_0} \tag{2.4}$$

we have

$$\|\chi - \chi^*\| < \varepsilon \tag{2.5}$$

for all $t \geq t_0$. The definitions of *uniform stability* and of *asymptotical instability* are extended in usual manner (ZUBOV [4]).

The characterization of the stability depends on the sign of a Liapunov functional $V(x, t)$ and its Frechet's time derivative $\dot{V}(x, t)$. If $V(x, t)$ is positive definite, the sign of $\dot{V}(x, t)$ qualifies the *stability*, *asymptotical stability* or *instability*.

We now seek a Liapunov functional of the form $V = V_1 + \lambda V_2$, with:

$$V_1 = \frac{1}{2} \int_{\mathscr{B}_R} \operatorname{tr}(\Delta \boldsymbol{T}_R \Delta \boldsymbol{F}^T)\, dv_R + \frac{1}{2} \int_{\mathscr{B}_R} \varrho_R \Delta \dot{\boldsymbol{x}}\, \Delta \dot{\boldsymbol{x}}\, dv_R, \tag{2.6}$$

and

$$V_2 = \int_{\mathscr{B}_R} \varrho_R \Delta \boldsymbol{x}\, \Delta \dot{\boldsymbol{x}}\, dv_R, \tag{2.7}$$

where $\Delta \boldsymbol{T}_R = \boldsymbol{T}_R^* - \boldsymbol{T}_R$, $\Delta \boldsymbol{F} = \boldsymbol{F}^* - \boldsymbol{F}$, $\Delta \boldsymbol{x} = \boldsymbol{x}^* - \boldsymbol{x}$, $\Delta \dot{\boldsymbol{x}} = \dot{\boldsymbol{x}}^* - \dot{\boldsymbol{x}}$ denote the differences between stresses, deformations and displacement from the perturbed state and the fundamental one.

Using Eqs. (2.1) and the circumstance that the difference solution has homogeneous boundary conditions, we obtain:

$$\dot{V}_1 = \frac{1}{2} \int_{\mathscr{B}_R} \operatorname{tr}(\Delta \dot{\boldsymbol{T}}_R \Delta \boldsymbol{F}^T - \Delta \boldsymbol{T}_R \Delta \dot{\boldsymbol{F}}^T)\, dv_R, \tag{2.8}$$

$$\dot{V}_2 = \int_{\mathscr{B}_R} [\varrho_R \Delta \dot{\boldsymbol{x}}\, \Delta \dot{\boldsymbol{x}} - \operatorname{tr}(\Delta \boldsymbol{T}_R \Delta \boldsymbol{F}^T)]\, dv_R. \tag{2.9}$$

A-priori inequalities. The discussion of the properties of the Liapunov functional is simplified by the following a-priori inequalities:

(a) In the fundamental state $\boldsymbol{\chi}$, the operator (2.1) is *strictly monotonous*, i.e. the inequality holds:

$$\int_{\mathscr{B}_R} \operatorname{tr}(\Delta \boldsymbol{T}_R \Delta \boldsymbol{F}^T)\, dv_R \geq \gamma \int_{\mathscr{B}_R} \operatorname{tr}(\Delta \boldsymbol{F}\, \Delta \boldsymbol{F}^T)\, dv_R \tag{2.10}$$

with γ positive constant. Eq. (2.10) assures a-fortiori the *static superstability* of fundamental state (vid. e.g. TRUESDELL and NOLL [7]).

(b) Between the difference displacements $\Delta \boldsymbol{x}$ and their gradients the so-called Poincaré-Friedrichs inequality lies:

$$\int_{\mathscr{B}_R} \Delta \boldsymbol{x}\, \Delta \boldsymbol{x}\, dv_R \leq \sigma \int_{\mathscr{B}_R} \operatorname{tr}(\Delta \boldsymbol{F}\, \Delta \boldsymbol{F}^T)\, dv_R, \tag{2.11}$$

where σ is a positive constant. Inequality (2.11) is valid both with homogeneous boundary conditions on the displacements (Poincaré inequality, cf. MIKHLIN [8]), and with homogeneous boundary conditions on the tractions on the whole boundary, but in this case we must restrict the class of $\Delta \boldsymbol{x}$ admissible to those, for which:

$$\int_{\mathscr{B}_R} \Delta \boldsymbol{x}\, dv_R = 0, \quad \int_{\mathscr{B}_R} \boldsymbol{p}_x \Delta \boldsymbol{x}\, dv_R = 0 \tag{2.12}$$

being $\boldsymbol{p}$ the radius vector of the variable point in the reference configuration (Friedrichs inequality, cf. MIKHLIN [8]). An identical inequality holds between the velocity $\Delta \dot{\boldsymbol{x}}$ and its gradient $\Delta \dot{\boldsymbol{F}}$.

3. Criteria of stability

From the collection of inequalities (2.10) and (2.11) the following two sufficient criteria of stability and instability can be deduced:

Criterion I (*Stability*). *A sufficient condition for stability is that:*
(a) *the expression*

$$\gamma\, \Delta \boldsymbol{x}\, \Delta \boldsymbol{x} + 2\lambda\, \Delta \boldsymbol{x}\, \Delta \dot{\boldsymbol{x}} + \varrho_R\, \Delta \dot{\boldsymbol{x}}\, \Delta \dot{\boldsymbol{x}},$$

as quadratical form in $\Delta \boldsymbol{x}$, $\Delta \dot{\boldsymbol{x}}$ be positive definite;
(b) *the expression*

$$\operatorname{tr}(-\Delta \dot{\boldsymbol{T}}_R\, \Delta \boldsymbol{F}^T + \Delta \boldsymbol{T}_R\, \Delta \dot{\boldsymbol{F}}^T - 2\varrho_R \lambda \sigma\, \Delta \dot{\boldsymbol{F}}\, \Delta \dot{\boldsymbol{F}}^T + 2\gamma\lambda\, \Delta \boldsymbol{F}\, \Delta \boldsymbol{F}^T),$$

as (generally non-linear) form in $\Delta \boldsymbol{F}$, $\Delta \dot{\boldsymbol{F}}$ be positive semi-definite.

Criterion II (*Instability*). *A sufficient condition for instability is that:*
(a) *the expression.*

$$\gamma\, \Delta \boldsymbol{x}\, \Delta \boldsymbol{x} - 2\lambda\, \Delta \boldsymbol{x}\, \Delta \dot{\boldsymbol{x}} + \varrho_R\, \Delta \dot{\boldsymbol{x}}\, \Delta \dot{\boldsymbol{x}},$$

as quadratical form in $\Delta \boldsymbol{x}$, $\Delta \dot{\boldsymbol{x}}$ be positive definite;
(b) *the expression*

$$\operatorname{tr}(\Delta \dot{\boldsymbol{T}}_R\, \Delta \boldsymbol{F}^T - \Delta \boldsymbol{T}_R\, \Delta \dot{\boldsymbol{F}}^T + 2\varrho_R \lambda \sigma\, \Delta \dot{\boldsymbol{F}}\, \Delta \dot{\boldsymbol{F}}^T + 2\gamma\lambda\, \Delta \boldsymbol{F}\, \Delta \boldsymbol{F}^T),$$

as (generally non-linear) form in $\Delta \boldsymbol{F}$, $\Delta \dot{\boldsymbol{F}}$, be positive definite.

Differential form. The infinitesimal version of the two previous criteria is obtained (TRUESDELL and NOLL [7], sec. 52) posing

$$\boldsymbol{F}^* = (\boldsymbol{1} + \varepsilon \boldsymbol{H})\, \boldsymbol{F}, \tag{3.1}$$

where $\boldsymbol{H}$ is a non-singular tensor and ε a real parameter. Thus if we set

$$f(\varepsilon) = \operatorname{tr}\, \{\boldsymbol{T}_R(\boldsymbol{F}^*)\, (\boldsymbol{H}\boldsymbol{F})^T\},$$

from the mean value theorem it follows, as is known (TRUESDELL and NOLL [7], sec. 52)

$$f(\varepsilon) - f(0) = \varepsilon f'(\bar{\varepsilon}) = \varepsilon \operatorname{tr}\{\boldsymbol{A}(\bar{\boldsymbol{F}})\, [\boldsymbol{H}\boldsymbol{F}]\, (\boldsymbol{H}\boldsymbol{F})^T\}, \tag{3.2}$$

where $0 < \bar{\varepsilon} < \varepsilon$ and

$$\bar{\boldsymbol{F}} = (\boldsymbol{1} + \bar{\varepsilon} \boldsymbol{H})\, \boldsymbol{F},$$

and $\boldsymbol{A}(\bar{\boldsymbol{F}})\, [\boldsymbol{H}\boldsymbol{F}]$ is a linear tensor function of $\boldsymbol{H}\boldsymbol{F}$, being $\boldsymbol{A}(\bar{\boldsymbol{F}})$ the *gradient* of $\boldsymbol{T}_R$.

With analogous considerations, from the position

$$g(\varepsilon) = \operatorname{tr}\{\dot{\boldsymbol{T}}_R^*(\boldsymbol{H}\boldsymbol{F})^T\},$$

where, for the chain rule, it is

$$\dot{T}_R^* = A(F^*)\,[\dot{F}^*],$$

we also obtain

$$g(\varepsilon) - g(0) = \varepsilon g'(\bar{\bar{\varepsilon}}) = \varepsilon\,\mathrm{tr}\{A_\varepsilon(\bar{\bar{F}})\,[\dot{\bar{\bar{F}}}]\,(HF)^T + A(\bar{\bar{F}})\,[\dot{\overline{\overline{HF}}}]\,(HF)^T\},$$

having set

$$\bar{\bar{F}} = (1 + \bar{\bar{\varepsilon}}H)\,F, \quad \text{with} \quad 0 < \bar{\bar{\varepsilon}} < \varepsilon.$$

Now, using the identity

$$A_\varepsilon(\bar{\bar{F}})\,[\dot{\bar{\bar{F}}}] = \dot{A}(\bar{\bar{F}})\,[\overline{\overline{HF}}]\,^1,$$

we can write

$$\varepsilon\,\mathrm{tr}\{\Delta\dot{T}_R(HF)^T\} = \varepsilon^2\mathrm{tr}\{\dot{A}(\bar{\bar{F}})\,[\overline{\overline{HF}}]\,(HF)^T + A(\bar{\bar{F}})\,[\dot{\overline{\overline{HF}}}]\,(HF)^T\}. \tag{3.3}$$

Thus, to the limit, forms (b) of criteria I and II become the following quadratical forms in the pair of tensor ΔF, $\Delta\dot{F}$:

$$\mathrm{tr}\{(2\gamma\lambda\,\Delta F - \dot{A}(F)\,[\Delta F])\,\Delta F^T - A(F)\,[\Delta\dot{F}]\,\Delta F^T + A(F)\,[\Delta F]\,\Delta\dot{F}^T - 2\varrho_R\lambda\sigma\,\Delta\dot{F}\,\Delta\dot{F}^T\}, \tag{3.4}$$

and

$$\mathrm{tr}\{(2\gamma\lambda\,\Delta F + \dot{A}(F)\,[\Delta F])\,\Delta F^T + A(F)\,[\Delta\dot{F}]\,\Delta F^T - A(F)\,[\Delta F]\,\Delta\dot{F}^T - 2\varrho_R\lambda\sigma\,\Delta\dot{F}\,\Delta\dot{F}^T\}, \tag{3.5}$$

where we have set as before

$$\Delta F = \varepsilon HF \quad \text{and} \quad \Delta\dot{F} = \varepsilon\dot{\overline{HF}}.$$

The ordinary algebraical criteria of definiteness for quadratical forms characterize the asymptotical stability or instability of the motion.

4. Examples

The previous criteria are now applied to two classes of materials.

(a) *Hyperelastic materials.* For hyperelastic materials, whose constitutive equation is (vid. e.g. TRUESDELL and NOLL [7], sec. 82):

$$T = \varrho F\sigma_F(F)^T, \tag{4.1}$$

the tensor A is symmetric and the second and third term in (3.4) and (3.5) cancel each other, so that we have stability if the form

$$\mathrm{tr}\{(2\gamma\lambda\,\Delta F - \dot{A}(F)\,[\Delta F])\,\Delta F^T\}$$

[1] In fact we have

$$A_\varepsilon(\bar{\bar{F}})\,[\dot{\bar{\bar{F}}}] = B(\bar{\bar{F}})\,[\overline{\overline{HF}}],$$

where $B(\bar{\bar{F}})$ is the fourth order tensor $A_F(F)\,[\dot{\bar{\bar{F}}}]$, which by chain rule is $\dot{A}(\bar{\bar{F}})$.

is positive semi-definite, and instability if

$$\operatorname{tr}\{(-2\gamma\lambda \varDelta \boldsymbol{F} + \dot{\boldsymbol{A}}(\boldsymbol{F})[\varDelta \boldsymbol{F}])\, \varDelta \boldsymbol{F}^T\}$$

is positive definite.

(b) *Infinitesimal viscous materials.* The material constitutive equation is of the form (cf. COLEMAN and NOLL [9]):

$$\boldsymbol{T} = \boldsymbol{M}(0)\,[\boldsymbol{F}(t)] + \int_0^\infty \dot{\boldsymbol{M}}(s)\,[\boldsymbol{F}(t-s)]\,ds\,, \tag{4.2}$$

where $\boldsymbol{M}[\;]$ is a linear tensor function of $\boldsymbol{F}(t)$. From (4.2), remembering the definition of $\boldsymbol{A}$, we derive

$$\boldsymbol{A} = \boldsymbol{M}(0) \tag{4.3}$$

and so also, deriving (4.2) and integrating by parts, we can write

$$\dot{\boldsymbol{A}} = \dot{\boldsymbol{M}}(0) \tag{4.4}$$

whence, by algebraical considerations on (4.3), (4.4), it is possible to illustrate the cases where the coefficients of viscous response define stability or instability.

References

1. PEARSON, C. E.: Quart. Appl. Math. **14**, 133–144 (1956).
2. HILL, R.: J. Mech. Phys. Solids **5**, 229–241 (1957).
3. HALE, J. K.: Proc. Nat. Acad. Sci. **50**, 942–946 (1963).
4. ZUBOV, V. I.: Methods of A. M. Liapunov and their Application, Groningen: P. Noordhoff 1964.
5. PARKS, P. C.: Int. Symp. on Diff. Eq. and Dyn. Syst., eds. J. K. HALE and J. P. LA SALLE, New York: Academic Press 1967, p. 287–298.
6. MOVCHAN, A. A.: Prikl. Mat. Mekh. **23**, 483–493 (1959).
7. TRUESDELL, C., NOLL, W.: The Non-linear Field Theories of Mechanics, Berlin/Heidelberg/New York: Springer 1965.
8. MIKHLIN, S. G.: Variational Methods in Mathematical Physics, Oxford: Pergamon Press 1964.
9. COLEMAN, B. D., NOLL, W.: Rev. Mod. Phys. **33**, 239–249 (1961).

L'Étude de la Stabilité Élastique et des Efforts du IIe Ordre par la Méthode des Charges Supplémentaires

Par

A. Scarlat

Institut de Construction, Bucarest, Roumanie

La méthode des charges supplémentaires permet une simplification de l'étude de la stabilité élastique et des efforts du IIe ordre des structures et une appréciation plus claire de l'importance des différents facteurs qui influencent l'état de stabilité.

I. Objectifs. Hypothèses

Soit M^0 le moment fléchissant résulté d'un calcul du Ier ordre (dans lequel les équations d'équilibre sont écrites en considérant l'axe nondéformé de la structure) et M le moment fléchissant résulté d'un calcul du IIe ordre (dans lequel les équations d'équilibre sont écrites en considérant l'axe déformé de la structure). On désigne par M^* le «moment supplémentaire»:

$$M^* = M - M^0. \tag{1}$$

Le calcul du IIe ordre peut être effectué par une variante: les équations d'équilibre sont écrites en considérant l'axe nondéformé de la structure mais, en compensation, on introduit à côté des charges réelles, un groupe de «charges supplémentaires». La nature et l'intensité des charges supplémentaires sont déterminées de sorte que les moments qu'elles causent coïncident avec les moments supplémentaires M^*.

Nous nous sommes proposé d'étendre l'emploi des charges supplémentaires — qui ont été jusqu'à présent appliquées seulement à l'étude de la barre droite — à l'étude des structures.

On admet les hypothèses classiques de la théorie de la stabilité élastique: les déplacements sont très petits, de sorte qu'on peut les assimiler aux déplacements infinitésimaux; les contraintes restent dans le domaine de Hooke; les charges varient proportionnellement, de sorte qu'un seul paramètre suffit à caractériser les grandeurs de toutes les forces; les forces qui agissent sur la structure conservent, pendant le processus de déformation, leurs directions initiales; on néglige l'influence de la déformation due à l'effort tranchant T et à l'effort normal N; les efforts normaux résultés d'un calcul du IIe ordre (N) sont presque égaux aux efforts normaux résultés d'un calcul du Ier ordre (N^0), de sorte qu'on peut approximer $N \approx N^0$.

Les efforts normaux N et les efforts tranchants T sont définis par référence à la direction de la tangente, respectivement de la normale, à l'axe nondéformé.

II. L'expression des charges supplémentaires

Soit un élément infinitésimal d'une barre droite déformée sous l'action des charges distribuées d'intensite q (latérales) et q_t (tangentielles); les équations différentielles d'équilibre sont (Fig. 1a):

$$\frac{dT}{dx} = -q; \quad \frac{dN}{dx} = q_t; \quad \frac{dM}{dx} = T + N\frac{dv}{dx} = T + N\varphi. \qquad (2)$$

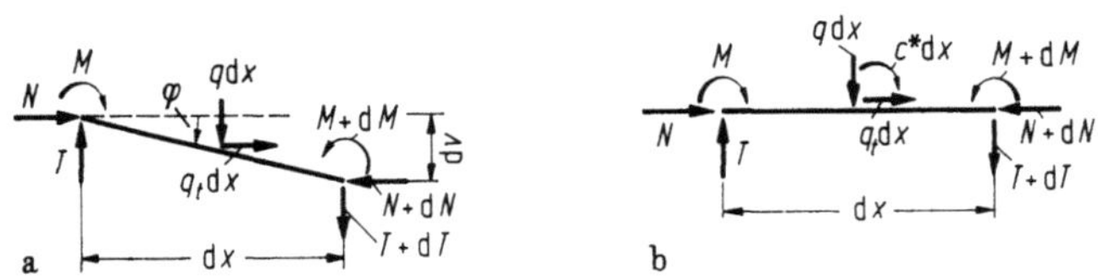

Fig. 1

En considérant le même élément appartenant à une barre droite nondéformée soumise à l'action des charges q, q_t et, aussi, à l'action des charges supplémentaires formées des couples répartis d'intensité c^* (Fig. 1b), on obtient:

$$\frac{dT}{dx} = -q; \quad \frac{dN}{dx} = q_t; \quad \frac{dM}{dx} = T + c^*. \qquad (3)$$

En comparant les deux séries d'équations on remarque que, si l'on choisit l'intensité des couples supplémentaires:

$$c^* = N\varphi, \qquad (4)$$

les équations (2) et (3) coïncident. Par conséquent *le calcul du IIe ordre peut être effectué en considérant l'axe nondéformé, à condition que la structure soit chargée, à côté des charges réelles, par une charge supplémentaire conventionnelle formée des couples répartis d'intensité $c^* = N\varphi$* (voir F. VAN DEN DUNGEN [9]).

Remarque. Souvent il est convenable d'exprimer les conditions d'équilibre par le principe des travaux virtuels. Dans le procédé classique les déplacements virtuels doivent être mesurés par rapport à l'axe déformé (Fig. 2a) et le travail extérieur revêt la forme:

$$L = \Sigma Q \cdot \delta\bar{v} + \Sigma P \cdot \delta\bar{u}.$$

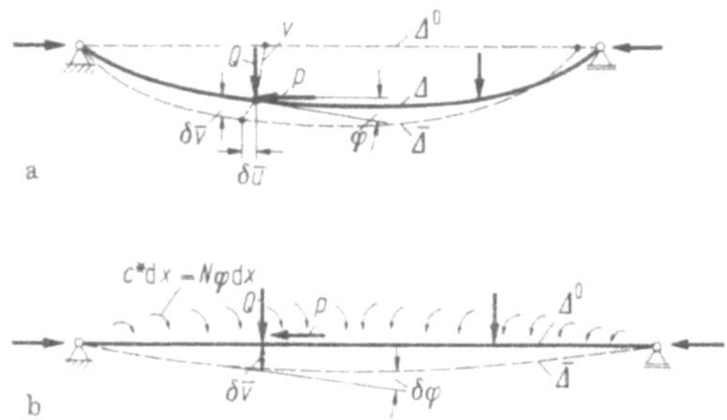

Fig. 2. Δ^0 — l'axe nondéformé, Δ — l'axe d'équilibre élastique, $\bar{\Delta}$ — la courbe des déplacements virtuels.

En considérant la procédé des charges supplémentaires, les déplacements virtuels seront mesurés par rapport à l'axe nondéformé. Le travail L a deux composantes: L^0 — le travail résulté d'un calcul du Ier ordre et L^* — le travail supplémentaire, dû aux couples $c^* = N\varphi$:

$$L = L^0 + L^* = \Sigma Q \cdot \delta\bar{v} + \int N\varphi \cdot \delta\bar{\varphi} \cdot dx,$$

$$L^* = \Sigma P \cdot \delta\bar{u} = \int N\varphi \cdot \delta\bar{\varphi} \cdot dx.$$

III. Cadres à noeuds fixes

Considérons une barre droite ij de section constante, à encastrements rigides, comprimée avec un effort normal N (Fig. 3). La barre

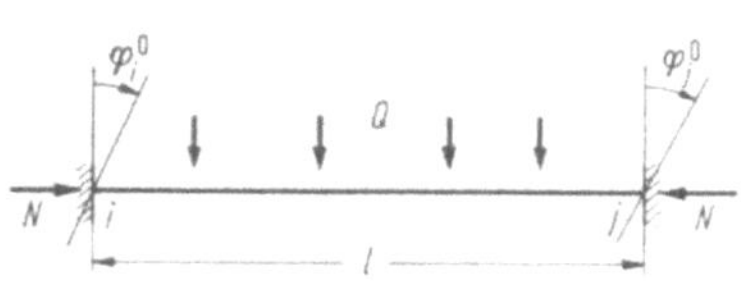

Fig. 3

est soumise à l'action des charges latérales Q et aux angles de rotation φ_i^0 et φ_j^0. Par un calcul du Ier ordre on obtient les moments d'encastrement supplémentaires:

$$\Delta\mathfrak{M}_i = -\frac{1}{30} Nl(4\varphi_i^0 - \varphi_j^0) + \Delta\mathfrak{M}_i^Q;\ \Delta\mathfrak{M}_j = -\frac{1}{30} Nl(4\varphi_j^0 - \varphi_i^0) + \Delta\mathfrak{M}_j^Q, \tag{5}$$

où $\Delta\mathfrak{M}^Q$ désigne le moment d'encastrement dû aux charges latérales; dans le cas d'une charge latérale uniforme d'intensité q, on a:

$$\Delta\mathfrak{M}^Q = \frac{1}{60} \cdot \frac{Nl^2}{EI} \cdot \mathfrak{M}^0;\quad \mathfrak{M}_i^0 = -\mathfrak{M}_j^0 = -\frac{1}{12} ql^2.$$

On détermine, par un calcul du Ier ordre, les moments M^0 et les angles de rotation φ^0 dûs aux charges réelles et puis on calcule les moments d'encastrement $\Delta\mathfrak{M}$. Par une methode itérative quelconque on détermine les moments supplémentaires ΔM.

$$M \approx M^0 + \Delta M.$$

IV. Cadres à noeuds libres de se déplacer

En considérant les barres flexibles mais inextensibles, les translations des noeuds peuvent être caractérisées à l'aide d'un nombre de schémas cynématiques indépendants (a, b, ...) égal au degré de liberté élastique de la structure (Fig. 4). L'angle de rotation de l'axe d'une barre résulte par une combinaison linéaire des angles ψ_a, ψ_b, ...:

$$\psi = \psi_a \cdot \varphi_a + \psi_b \cdot \varphi_b + \cdots, \tag{6}$$

où φ_a, φ_b, ... sont les inconnues du problème.

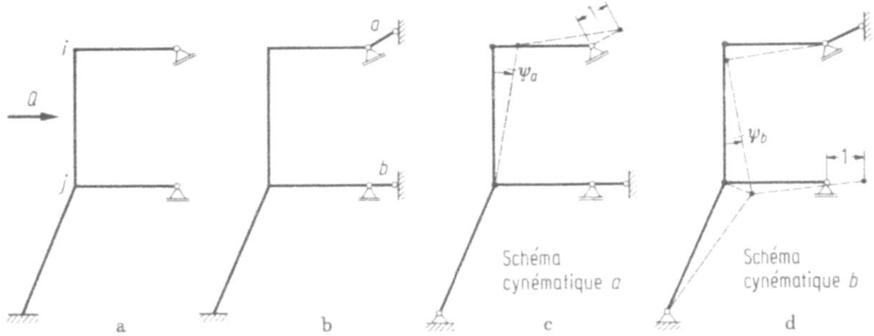

Fig. 4

L'angle de rotation φ d'une section quelconque peut être décomposé en:

$$\varphi = \varphi^F + \psi, \tag{7}$$

où φ^F désigne l'angle de rotation rapporté à l'axe incliné $i'j'$ (Fig. 5). Par conséquent les couples supplémentaires $c^* = N\varphi$ ont deux composantes aussi:

$$c^* = c^F + c^\psi = N\varphi^F + N\psi. \tag{8}$$

L'intensité des couples $c^F = N\varphi^F$ est variable au long de l'axe ij; au contraire, l'intensité des couples $c^\psi = N\psi$ est constante. Dans la Statique des constructions on démontre que les moments d'encastrement dûs aux couples d'intensité constante sont nuls; les couples $c^\psi = N\psi$ peuvent donc être remplacés par les forces concentrées équivalentes

$$Q^* = \pm c^\psi l \cdot \frac{1}{l} = \pm N\psi \tag{9}$$

agissant aux noeuds.

L'influence des forces supplémentaires $Q^* = \pm N\psi$ est beaucoup plus grande que l'influence des couples répartis $c^F = N\varphi^F$. En partant de cette remarque, on peut formuler des procédés approximatifs pour le calcul des charges critiques, procédés qui réclament:

— soit à négliger les couples supplémentaires $c^F = N\varphi^F$;

— soit à les considérer dans une forme simplifiée.

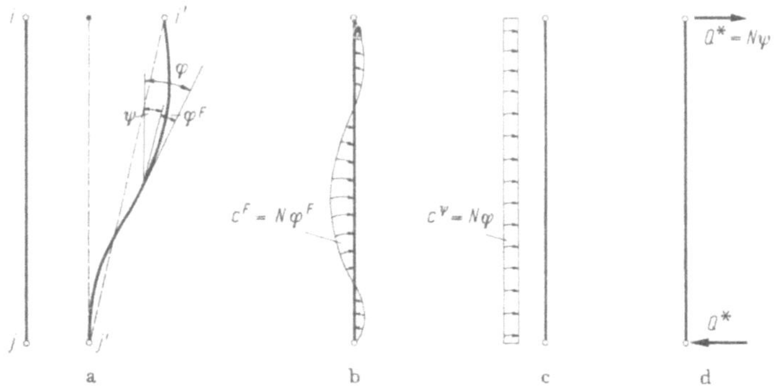

Fig. 5

V. Interprétation matricielle de la méthode

Considérons les barres de la structure divisées en plusieurs segments 1, 2, ..., ayant les longueurs $l_1, l_2, \ldots$, les rigidités $EI_1, EI_2, \ldots$ et les efforts normaux $N_1, N_2, \ldots$ (Fig. 6). On définit les vecteurs:

$$\varphi = \begin{Bmatrix} \varphi_1 \\ \varphi_2 \\ \vdots \end{Bmatrix}; \quad c^* = \begin{Bmatrix} C_1^* \\ C_2^* \\ \vdots \end{Bmatrix} \qquad (\text{où } C_i^* = N_i \varphi_i l_i)$$

et les matrices:

$$\mathscr{L} = \begin{bmatrix} l_1 & & \\ & l_2 & \\ & & \ddots \end{bmatrix}; \quad \mathscr{N} = \begin{bmatrix} N_1 & & \\ & N_2 & \\ & & \ddots \end{bmatrix}; \quad \mathscr{A} = \begin{bmatrix} \alpha_{11} & \alpha_{12} & \cdots \\ \alpha_{21} & \alpha_{22} & \cdots \\ \cdots & \cdots & \cdots \end{bmatrix}; \quad \mathscr{I} = \begin{bmatrix} 1 & & \\ & 1 & \\ & & \ddots \end{bmatrix},$$

où $\mathscr{A}$ représente une «matrice d'influence»[1]; α_{ij} désigne l'angle de rotation φ_i causé par le couple $C_j^* = 1$.

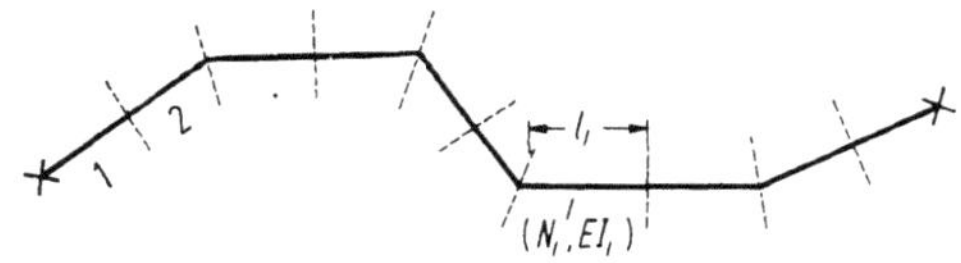

Fig. 6

[1] Les matrices d'influence ont été introduites dans l'étude de la stabilité élastique par A. SMIRNOV [8].

Afin de mettre en évidence l'effort normal critique d'un segment «i» ($N_{i_{cr}}$) on emploie la matrice $\mathcal{N}_i = \frac{1}{N_i} \cdot \mathcal{N}$.

$$c^* = \mathcal{L}\mathcal{N}\varphi = N_i \cdot \mathcal{L}\mathcal{N}_i\varphi; \quad \varphi = \mathcal{A}c^* = N_i \cdot \mathcal{B}\varphi,$$

où $\mathcal{B} = \mathcal{A}\mathcal{L}\mathcal{N}_i$.

$\varphi(N_i\mathcal{B} - \mathcal{I}) = 0$, mais la condition d'équilibre critique impose $\varphi \neq 0$; par conséquent:

$$\mathcal{B} - \lambda\mathcal{I} = 0, \tag{10}$$

où $\lambda = 1/N_i$.

L'équation (10) représente l'équation caractéristique de la matrice $\mathcal{B}$; afin de déterminer $N_{i_{cr}}$ on doit calculer la valeur propre $\lambda_{\max}$ de la matrice $\mathcal{B}$.

Dans le cas où les segments 1, 2, ... coïncident avec les barres du cadre on a $\varphi \approx \psi$ et on trouve un procédé de calcul équivalent du point de vue de l'approximation au procédé fondé sur l'élimination des couples supplémentaires $c^F = N\varphi^F$.

VI. Plaques planes

La surface moyenne déformée ayant un caractère spatial, on doit introduire des couples supplémentaires répartis agissant aux deux directions normales (Fig. 7): les couples d'intensité c^x et les couples d'intensité c^y.

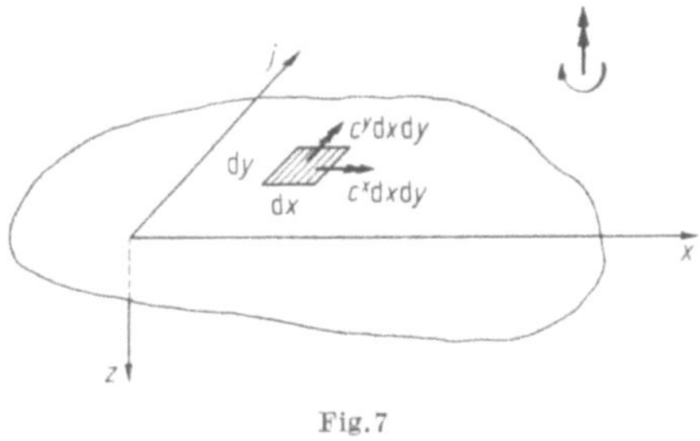

Fig. 7

L'équation différentielle de la surface moyenne déformée par les couples c^x, c^y, résultée d'un calcul du Ier ordre, revêt la forme

$$\frac{\partial^4 w}{\partial x^4} + 2\frac{\partial^4 w}{\partial x^2\,\partial y^2} + \frac{\partial^4 w}{\partial y^4} = \frac{1}{D}\left(\frac{\partial c^y}{\partial x} - \frac{\partial c^x}{\partial y}\right). \tag{11}$$

Afin de déduire les expressions des intensités des couples supplémentaires c^x et c^y on procède de la même manière qu'on a employée dans l'étude des barres droites: on compare les équations différentielles d'équilibre de l'élément $dx \cdot dy$ résultées d'un calcul du IIe ordre avec les mêmes équations résultées d'un calcul du Ier ordre (où on a introduit les couples supplémentaires c^x, c^y):

$$c^x = c^x_{(N)} + c^x_{(H)} = N_y \frac{\partial w}{\partial y} + H \frac{\partial w}{\partial x}\,; \; c^y = c^y_{(N)} + c^y_{(H)} = -N_x \frac{\partial w}{\partial x} - H \frac{\partial w}{\partial y}\,, \tag{12}$$

où N_x, N_y désignent les efforts normaux et H les efforts tangentiels.

En substituant c^x et c^y en (11) on trouve l'équation de la surface moyenne voilée:

$$\frac{\partial^4 w}{\partial x^4} + 2 \frac{\partial^4 w}{\partial x^2 \partial y^2} + \frac{\partial^4 w}{\partial y^4} = -\frac{1}{D}\left(N_x \frac{\partial^2 w}{\partial x^2} + 2H \frac{\partial^2 w}{\partial x \partial y} + N_y \frac{\partial^2 w}{\partial y^2}\right). \tag{13}$$

Afin de calculer les moments du IIe ordre on considère, dans une première étape, la plaque soumise à l'action des charges réelles et on détermine, par un calcul du Ier ordre, les moments M^0 et les flèches w^0; dans une seconde étape on calcule les intensités des couples supplémentaires c^x, c^y et l'on déduit les moments supplémentaires M^* correspondants. En sommant les résultats, on a:

$$M \approx M^0 + M^*.$$

Bibliographie

1. Klöppel, K.: Beitrag zur Berechnung von Stockwerkrahmen auf Stabilität und nach der Spannungstheorie II. Ordnung. Stahlbau **1** (1962).
2. Livesley, R.: Matrix Methods of Structural Analysis, Oxford: Pergamon Press 1964.
3. Pikovski, A.: Statika sterjnevih sistem so sjatimi elementami, Moskva: Fizmatgiz 1961.
4. Sattler, K.: Das Durchbiegungsverfahren zur Lösung von Stabilitätsproblemen. Bautechnik **10/11** (1953).
5. Scarlat, A.: Eine Eigenschaft der auf mittigen Druck und Biegung beanspruchten Stäbe. Bautechnik **5** (1962).
6. Scarlat, A.: Stabilitatea şi calculul de ordinul II al structurilor, Bucureşti: Ed. Tehnică 1969.
7. Scarlat, A.: Stabilitatea structurilor. Probleme speciale, Bucureşti: Ed. Tehnică 1969.
8. Smirnov, A.: Ustoicivosti i kolebania soorujenii, Moskva: Transjeldorizdat 1958.
9. van den Dungen, F.: Cours de technique des vibrations, Bruxelles 1926.

Contributions to Hydrodynamic (In)stability by Use of the Lemma of Nagumo and Westphal

By

E. Adams

University of Karlsruhe, Karlsruhe, and Deutsche Forschungs- und Versuchsanstalt für Luft- und Raumfahrt (DFVLR), Freiburg i. Br., Germany

§ 1. Introduction

A vectorial stationary "basic" velocity field $\boldsymbol{v}_0 = \boldsymbol{v}_0(\boldsymbol{x}) = (u_0, v_0, w_0)$ is considered; here, $\boldsymbol{x}$ is the position vector in a Cartesian basis and u_0, v_0, w_0 denote components of $\boldsymbol{v}_0$ with respect to these axes. Time-dependent vectorial "fluctuations" (= perturbations) $\boldsymbol{v}_1(\boldsymbol{x}, t)$ are superimposed to $\boldsymbol{v}_0$ to yield the perturbed velocity field $\boldsymbol{v} = \boldsymbol{v}_0 + \boldsymbol{v}_1$. Correspondingly, e.g., $T = T_0(\boldsymbol{x}) + T_1(\boldsymbol{x}, t)$ represents a perturbed temperature field. In general, $\boldsymbol{v}_1, T_1, \ldots$ in any flow in nature are induced by some external interference which manifests itself as follows in data changes: (a) transient changes of boundary conditions, e.g., oscillations of a wall, (b) initial conditions $\boldsymbol{v}_1(\boldsymbol{x}, 0), T_1(\boldsymbol{x}, 0), \ldots$ defining perturbations at time $t = 0$, and (c) the presence of perturbation terms in the differential operators. For the majority of flows in nature, expected data changes are known only within intervals bounded by upper and lower "data bounds". By use of the lemma of Nagumo and Westphal[1] on parabolic differential operators as presented in [6], a pair of upper and lower "field bounds" will be determined for each one of the "response functions" $\boldsymbol{v}_1(\boldsymbol{x}, t), T_1(\boldsymbol{x}, t), \ldots$ from a set S which refers to the set D of data changes within given data bounds. The set D is assumed to be non-empty and to consist only of those functions for which res-

[1] The first applications of this lemma to fluid dynamics were published by Görtler [2] and by Nickel [5].

ponse functions exist in $0 < t < \infty$. In order to simplify the mathematical work involved in calculating the field bounds, they will be assumed to depend on t only and to solve suitably defined ordinary initial value problems. A sufficient condition for "stability in the sense of Liapunov" of a basic field such as $\boldsymbol{v}_0(\boldsymbol{x})$ is obtained (a) if there exists a $|\boldsymbol{v}_1(\boldsymbol{x}, t)| \not\equiv 0$ and (b) if the corresponding upper and lower field bounds are stable in the sense of Liapunov which implies in particular their existence and boundedness in $0 < t < \infty$. A sufficient condition for "instability" of a basic field such as $\boldsymbol{v}_0(\boldsymbol{x})$ is obtained (a) if there exists a $|\boldsymbol{v}_1(\boldsymbol{x}, t)| \not\equiv 0$ and (b) if the corresponding upper and lower field bounds possess the same sign and exist in $0 < t < \infty$. As examples for pertinent applications, Liapunov-stable field bounds will be determined in § 2 and in § 3; in § 4, an example of instability will be mentioned.

§ 2. Velocity fluctuations in boundary layers with pressure decrease

Two-dimensional stationary laminar basic boundary layer flows with a favorable pressure gradient are considered in Cartesian coordinates for incompressible, isothermal, and chemically homogeneous fluids:

$$\boldsymbol{v}_0 = \big(u_0(x, y), v_0(x, y), 0\big). \tag{2.1}$$

This type of flow is encountered, e.g., upstream of the maximum cross section $x = L_0$ of an airfoil mounted in a wind tunnel of width $-b < z < +b$. In the following, field bounds of the fluctuation $u_1(\boldsymbol{x}, t)$ of the velocity component u will be determined in the outer region G_e of the boundary layer, where $G_e := \{0 < x < L_0,\ y_i(x) < y < \infty,\ -b < z < b\}$. As was shown experimentally by S. J. Kline in [4], the production of turbulent secondary motion ("spots") in a boundary layer occurs in a narrow vicinity of the wall. In [4], the following correlation of experimental data is given for the frequency F of spot production in "bursts/sec": $F/F_{\text{in}} \approx 1 - K/4.4$ where F_{in} denotes an initial value and $K := (\mu/\varrho\, u_e^2)\,(du_e/dx)$ with u_e the velocity at the outer edge of the boundary layer. Relaminarization occurs for $K > 4.4$, [4]. According to observations reported in [4], only isolated spots of turbulence enter G_e and do not reach the outer edge $y \to \infty$ of the boundary layer if both $y_i(x)$ and K are sufficiently large. Such boundary layer flows will be considered here, i.e.:

$$p(x, y, z, t) = p_0(x) \quad \text{in } G_e \tag{2.2}$$

is valid if the pressure gradient $p_0'(x) < 0$ and if K is sufficiently large.

If x_0 and t_0 denote arbitrary values of x and t in $0 < x_0 \leq L_0$ and

$0 < t_0 < \infty$, a parabolic region $H := H' + R'$ is defined by use of the Cartesian product $H' := G_e x t$ and the set

$$R' := \begin{Bmatrix} x = x_0, & y_i(x) < y < \infty, & -b < z < b, & 0 < t \leq t_0 \\ t = t_0, & 0 < x \leq x_0, & +y_i(x) < y < \infty, & -b < z < b \end{Bmatrix}.$$

It is assumed that the fluctuating boundary layer flows under consideration are governed by the momentum equation in x-direction

$$P_u(\boldsymbol{v}) = \frac{\partial u}{\partial t} + u\frac{\partial u}{\partial x} + v\frac{\partial u}{\partial y} + w\frac{\partial u}{\partial z} + \frac{1}{\varrho}\frac{dp_0}{dx} - \frac{\mu}{\varrho}\frac{\partial^2 u}{\partial y^2} - \frac{\mu}{\varrho}\frac{\partial^2 u}{\partial z^2} = 0 \text{ in } H, \tag{2.3}$$

whose validity for study of stability requires $R(x) \gg L_0$ for the radius of curvature R of the wall. Since $\boldsymbol{v}_1$ possibly changes spatially at the same rates in y-direction and in z-direction, it is not assumed that $\partial^2 u/\partial z^2$ is negligibly small. The functions $\boldsymbol{v}_0$ and $\boldsymbol{v}_1$ are assumed to be classical solutions of the pertinent boundary value problems.

By use of constant data bounds $\overline{\overline{u}}_1(0)$, G_1^*, and G_2^*, the following boundary conditions will be considered because of (2.2):

$$\left.\begin{aligned} u_1(\boldsymbol{x}, t) &\equiv 0 \text{ for } y \to \infty \text{ and } x = 0 \text{ and } z = \pm b; \\ w_1(\boldsymbol{x}, t) &\equiv 0 \text{ for } y \to \infty; \\ \underline{\underline{u}}_1(0) < u_1(x, y, z, 0) &< \overline{\overline{u}}_1(0) = -\underline{\underline{u}}_1(0); \\ \left|u_1\big(x, y_i(x), z, t\big)\right| &\leq G_1^* = \text{const.} > 0 \text{ (problem I)} \\ \text{or } &\leq G_2^* = 0 \text{ (problem II)}. \end{aligned}\right\} \tag{2.4}$$

Problem I refers to disturbances (spots) in G_e which are due to initial perturbations at $t = 0$ and to boundary perturbations on $y = y_i(x)$. In problem II, the perturbations are confined to G_e in order to study the development of spots without interference from the environment of G_e. Because of (2.4) and the continuity equation, $\lim_{y\to\infty} \partial v_1(x, y, z, t)/\partial y = 0$ in H. Therefore it is reasonable to assume that there exists an arbitrarily large positive constant c_v in

$$|\boldsymbol{v}_1(\boldsymbol{x}, t)| \leq c_v = \text{const. in } H. \tag{2.5}$$

A function $\overline{\overline{u}}_1(t, c_v) = -\underline{\underline{u}}_1(t, c_v)$ is defined as solution of the initial value problems

$$\left.\begin{aligned} \overline{Q}(\overline{\overline{u}}_1) &= \overline{\overline{u}}_1' + \overline{\overline{u}}_1 \inf_{\boldsymbol{x}} \frac{\partial u_0}{\partial x} - c_v \sup_{\boldsymbol{x}} \left|\frac{\partial u_0}{\partial y}\right| \\ &= 0 \text{ in } t > 0, \ \overline{\overline{u}}_1(0, c_v) = \overline{\overline{u}}_1(0), \\ \underline{Q}(\underline{\underline{u}}_1) &= \underline{\underline{u}}_1' + \underline{\underline{u}}_1 \inf_{\boldsymbol{x}} \frac{\partial u_0}{\partial x} + c_v \sup_{\boldsymbol{x}} \left|\frac{\partial u_0}{\partial y}\right| \\ &= 0 \text{ in } t > 0, \ \underline{\underline{u}}_1(0, c_v) = \underline{\underline{u}}_1(0). \end{aligned}\right\} \tag{2.6}$$

The perturbation terms with coefficients c_v are due to the fact that $v_1(\boldsymbol{x}, t)$ is not known with the exception of the a priori estimate (2.5). Obviously, $\inf_{\boldsymbol{x}} \partial u_0/\partial x > 0$ and thus $y_i(x) > 0$ are necessary for bounded solutions of (2.6) in $0 < t < \infty$. Because of $p_0' < 0$,

$$\lim_{y\to\infty} \frac{\partial u_0(x, y)}{\partial y} = 0 \text{ and } \lim_{y\to\infty} \frac{\partial u_0(x, y)}{\partial x} = u_e'(x) > 0. \qquad (2.7)$$

For any finite c_v, therefore the boundary $y = y_i(x)$ of G_e can be selected so as to ensure the validity of the relation

$$\left|\lim_{t\to\infty} \overline{\overline{u}}_1(t, c_v)\right| = c_v \sup_{\boldsymbol{x}} \left|\frac{\partial u_0}{\partial y}\right| \Big/ \inf_{\boldsymbol{x}} \frac{\partial u_0}{\partial x} = \varepsilon \ll 1 \qquad (2.8)$$

for the stationary point of (2.6). Because of (2.3) and (2.6),

$$\underline{P_u(u_0 + \underline{\underline{u}}_1, v_0 + v_1, w_1)} = P_u(\boldsymbol{v}_0) + \left[\underline{\underline{u}}_1' + \underline{\underline{u}}_1 \frac{\partial u_0}{\partial x} + v_1 \frac{\partial u_0}{\partial y}\right]$$

$$\leq Q(\underline{\underline{u}}_1) = \underline{P_u(\boldsymbol{v}_0 + \boldsymbol{v}_1)} = 0 = \overline{Q}(\overline{\overline{u}}_1) \qquad (2.9)$$

$$\leq \underline{P_u(u_0 + \overline{\overline{u}}_1, v_0 + v_1, w_1)} = P_u(\boldsymbol{v}_0) + \left[\overline{\overline{u}}_1' + \overline{\overline{u}}_1 \frac{\partial u_0}{\partial x} + v_1 \frac{\partial u_0}{\partial y}\right] \text{ in } H,$$

where $\leq$ and $\geq$ are a consequence of comparing the individual terms in the underlined operators. This explains the choice of terms in (2.6): the positive (negative) terms in $\overline{Q}(\overline{\overline{u}}_1)$, which tend to decrease (increase) $\overline{\overline{u}}_1(t, Q)$, are not larger (smaller) than the corresponding terms in $P_u(u_0 + \overline{\overline{u}}_1, v_0 + v_1, w_1)$. Since $\overline{\overline{u}}_1(t, c_v)$ is a monotonic function of t, functions $\overline{u}_1(t) = -\underline{u}_1(t)$ are defined as follows for any finite fixed c_v:

$$\overline{u}_1(t) := \sup\{\overline{\overline{u}}_1(t, c_v), G_j^*\} =: -\underline{u}_1(t) \text{ in } 0 < t < \infty \text{ for } j = 1 \text{ or } 2. \quad (2.10)$$

A simple and obvious generalization of the lemma of Nagumo and Westphal as given in [6, p. 246—249] is applicable to the functions $\underline{u}_1(t)$, $u_1(\boldsymbol{x}, t)$, and $\overline{u}_1(t)$:

$$\underline{u}_1(t) \leq u_1(\boldsymbol{x}, t) \leq \overline{u}_1(t) \text{ in } H. \qquad (2.11)$$

Even though these field bounds $\underline{u}_1$ and $\overline{u}_1$ do not permit any conclusions on particular perturbations, the validity of (2.8) in case of problem II shows that (a) $|u_1(\boldsymbol{x}, t)|$ becomes negligibly small as $t \to \infty$ if $y_i(x)$ is sufficiently large and (b) branching (= bifurcation) of $\boldsymbol{v}_0(\boldsymbol{x})$ then could yield stationary solutions $u(\boldsymbol{x}) \neq u_0(\boldsymbol{x})$ only within an ε-vicinity of $u_0(\boldsymbol{x})$. In case of problem I, it is seen for spots moving into G_e across $y = y_i(x)$ that $|u_1(\boldsymbol{x}, t)|$ does not increase in G_e.

§ 3. On temperature fluctuations in channel flow

The flow of a chemically homogeneous incompressible fluid with the following properties is considered in the channel G:

$$\left.\begin{aligned}
&G: = \{\boldsymbol{x}: -\infty < x < \infty,\ 0 < y < B,\ -\infty < z < \infty\}\\
&\qquad\qquad\qquad\text{(Cartesian coordinates)},\\
&R_G: = \text{closed boundary of } G,\\
&H:\ = G \times \{t: t > 0\} \text{ (parabolic region)},\\
&R:\ = G \times \{t: t = 0\} + R_G \times \{t: t \geq 0\} \text{ (parabolic boundary)},\\
&T(x, 0, z, t) = T_0(0),\ T(x, B, z, t) = T_0(B) \text{ with } T_0(y) \text{ given},\\
&T_1(\boldsymbol{x}, t) \equiv 0 \text{ on } R_G,\\
&\text{density } \varrho = \text{const., viscosity } \mu = \text{const.},\\
&\text{thermal conductivity } k = a + b \cdot T,\\
&\text{specific heat } c_p = \varrho^{-1}(c + d \cdot T)^{-1}\\
&\text{with positive constants } a,\ -b,\ c,\ d, \text{ and } \frac{d}{dT}(k/\varrho c_p) \leq 0;\\
&\text{heat of dissipation negligible.}
\end{aligned}\right\} \quad (3.1)$$

Because of (3.1), the energy equation takes the form

$$\begin{aligned}
P_T(\boldsymbol{v}, T) = \frac{\partial T_1}{\partial T} + (\boldsymbol{v}_0 + \boldsymbol{v}_1) \cdot \nabla (T_0 + T_1) - \frac{K(T_0 + T\,)}{\varrho c_p(T_0 + T_1)} \nabla \cdot \nabla (T_0 + T_1)\\
- \frac{dK(T_0 + T_1)}{dT} [\nabla (T_0 + T_1)]^2 / \varrho c_p (T_0 + T_1) = 0 \text{ in } H.
\end{aligned} \quad (3.2)$$

The functions $\boldsymbol{v}(\boldsymbol{x}, t)$ and $T(\boldsymbol{x}, t)$ are assumed to be classical solutions of the pertinent boundary value problem with $|\nabla T_1|$ and $|\Delta T_1|$ bounded by arbitrarily large constants.

A laminar stationary "basic flow" in the channel is defined as the following rigorous solution of the pertinent boundary value problem:

$$\left.\begin{aligned}
&p_0(\boldsymbol{x}) = -p_\infty x,\ \ p_\infty = \text{const.} > 0,\ \ v_0(\boldsymbol{x}) = w_0(\boldsymbol{x}) = 0,\\
&u_0(\boldsymbol{x}) = \frac{p_\infty}{2\mu}(B - y)\, y,\ \ P_T(\boldsymbol{v}_0, T_0) = 0,\\
&T_0(\boldsymbol{x}) = \frac{1}{|b|}[a - \sqrt{c_0 + c_1 y}],\\
&\quad c_0 = \text{const.} > 0,\ c_1 = \text{const.} > 0
\end{aligned}\right\} \text{ in } G. \quad (3.3)$$

This basic flow is perturbed by an external interference occurring at time $t = 0$, e.g., by a spark discharge. This causes fluctuations $\boldsymbol{v}_1(\boldsymbol{x}, t)$ and $T_1(\boldsymbol{x}, t)$ in H. Arbitrarily large constant data bounds $|\underline{T}_1(0)| > 0$ and $\overline{T}_1(0) > 0$ are assumed to be known in G for the set

D_0 of continuous functions $T_1(\boldsymbol{x}, 0)$ representing the initial perturbation of T. An arbitrarily large positive constant c_v is assumed to be known in the a priori estimate (2.5) which is valid here because $\boldsymbol{v}_1(\boldsymbol{x}, t)$ is due to initial perturbations at $t = 0$ but is not affected by $T_1(\boldsymbol{x}, t)$ in $t > 0$ since $\mu = \text{const}$. Constants α, β, β^*, and γ are defined as follows by use of the given data $T_0(y)$, $k(T)$, and $c_p(T)$:

$$\left.\begin{aligned} &0 < \alpha \leq -T_0''(y) \frac{d}{dT}\left(\frac{K(T_0)}{\varrho c_p(T_0)}\right) + |b|\, dT_0'^2(y) < \infty, \\ &\frac{d}{dT}\left(\frac{K(T_0)}{\varrho c_p(T_0)}\right) = bc + ad + 2bdT_0, \\ &0 < \beta^* \leq |b|\, dT_0''(y) \leq \beta < \infty,\ 0 \leq c_v\, |T_0'(y)| \leq \gamma < \infty. \end{aligned}\right\} \tag{3.4}$$

The solutions of the initial value problems

$$\left.\begin{aligned} &\underline{Q}(\underline{T}_1) = \underline{T}_1' + \gamma + \alpha \underline{T}_1 + \beta \underline{T}_1^2 = 0, \quad \underline{T}_1(0) < 0, \\ &\overline{Q}(\overline{T}_1) = \overline{T}_1' - \gamma + \alpha \overline{T}_1 + \beta^* \overline{T}_1^2 = 0, \quad \overline{T}_1(0) > 0 \end{aligned}\right\} \tag{3.5}$$

exist in $0 < t < \infty$ and can be expressed in closed-form by use of hyperbolic functions [3, p. 298]. Obviously, $\underline{T}_1(t) \leq 0$ and $\overline{T}_1(t) \geq 0$ in $0 \leq t < \infty$. Whereas the terms with coefficients α, β, and β^* are due to $k(T)$ and $c_p(T)$, the perturbation terms $+\gamma$ and $-\gamma$ in (3.5) are due to the fact that $\boldsymbol{v}_1(\boldsymbol{x}, t)$ is not known with the exception of the a priori estimate (2.5). Because of (3.2), (3.4), and (3.5)

$$\begin{aligned} &P_T(\boldsymbol{v}_0 + \boldsymbol{v}_1, T_0 + \underline{T}_1) \leq \underline{Q}(\underline{T}_1) = P_T(\boldsymbol{v}_0 + \boldsymbol{v}_1, T_0 + T_1) - P_T(\boldsymbol{v}_0, T_0) \\ &= \frac{\partial T_1}{\partial t} + (u_0 + u_1)\frac{\partial T_1}{\partial x} + v_1 \frac{\partial T_1}{\partial y} + v_1 T_0' + w_1 \frac{\partial T_1}{\partial z} - \frac{K(T_0 + T_1)}{\varrho c_p(T_0 + T_1)} \\ &\quad \times \left[\frac{\partial^2 T_1}{\partial x^2} + \frac{\partial^2 T_1}{\partial y^2} + \frac{\partial^2 T_1}{\partial z^2}\right] - \frac{K'(T_0 + T_1)}{\varrho c_p(T_0 + T_1)} \\ &\quad \times \left[\left(\frac{\partial T_1}{\partial x}\right)^2 + \left(\frac{\partial T_1}{\partial y}\right)^2 + \left(\frac{\partial T_1}{\partial z}\right)^2 + 2T_0' \frac{\partial T_1}{\partial y}\right] - T_0''[(bc + ad + 2bdT_0) \\ &\quad \times T_1 + bdT_1^2] - bdT_0'^2 T_1 = 0 = \overline{Q}(\overline{T}_1) \leq P_T(\boldsymbol{v}_0 + \boldsymbol{v}_1, T_0 + \overline{T}_1) \text{ in } H. \end{aligned} \tag{3.6}$$

According to [6, p. 167], the lemma of Nagumo and Westphal yields

$$\underline{T}_1(t) < T_1(\boldsymbol{x}, t) < \overline{T}_1(t) \text{ in } H.^{1} \tag{3.7}$$

If $\underline{T}_1'$ and $\overline{T}_1'$ are ignored, (3.5) yields two real stationary points $\{\underline{T}_{1\infty,1}, \overline{T}_{1\infty}\}$ and $\{\underline{T}_{1\infty,2}, \overline{T}_{1\infty}\}$ which are shown in Fig. 1 and, by use of

[1] Since $\Phi = \Phi(\boldsymbol{v}_0 + \boldsymbol{v}_1) =: \Phi_0(\boldsymbol{v}_0) + \Phi_1(\boldsymbol{v}_0 + \boldsymbol{v}_1) \geq 0$ is valid for the heat of dissipation, the field bound $\underline{T}_1(t)$ accounts for Φ if $+\gamma$ in $\underline{Q}(\underline{T}_1)$ is expressed by $\operatorname{Max}_y c_v\, |T_0'(y)| + \operatorname{Max}_x \Phi_0(v_0)$, where $\Phi_0(\boldsymbol{v}_0)$ is known together with $\boldsymbol{v}_0(\boldsymbol{x})$.

$T^*: = a/|b|$, depend as follows on the data of the problem:

$$\left.\begin{aligned}
&\frac{\alpha}{\beta^*} = \frac{c}{d} + T_0(B), \quad \frac{\alpha}{\beta} = \left[\frac{c}{d} + T_0(B)\right]\left[\frac{T^* - T_0(0)}{T^* - T_0(B)}\right]^3,\\
&\frac{\gamma}{\beta} = \frac{c_v B}{|b|\,d} \frac{[T^* - T_0(0)]^2}{\left[T^* - \frac{1}{2}(T_0(0) + T_0(B))\right][T_0(0) - T_0(B)]},\\
&\frac{\gamma}{\beta^*} = \frac{c_v B}{|b|\,d} \frac{[T^* - T_0(B)]^3}{\left[T^* - \frac{1}{2}(T_0(0) + T_0(B))\right][T_0(0) - T_0(B)][T^* - T_0(0)]}.
\end{aligned}\right\} \quad (3.8)$$

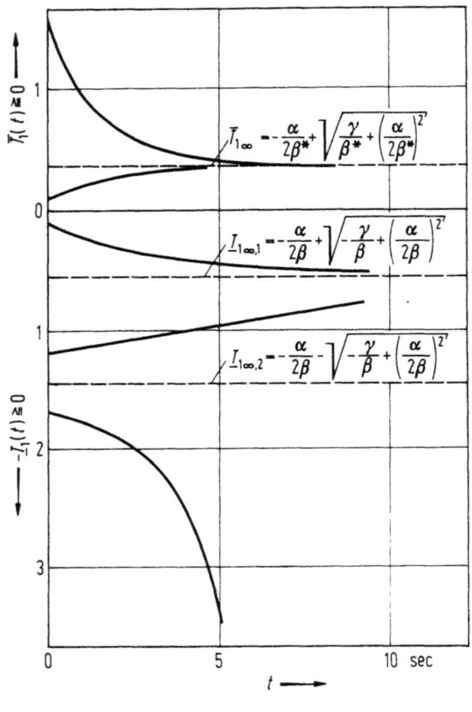

Fig. 1. Bounds for temperature fluctuations in channel flow. Data plotted for $\alpha = 0.5$, $\beta^* = 0.125$, $\beta = 0.25$, $\gamma = 0.2$.

If the numerical values of the material constants a, b, c, and d are selected so as to approximate $k(T)$ and $c_p(T)$ for water in the temperature range $300\,°\mathrm{K} \leq T \leq 600\,°\mathrm{K}$, $|\boldsymbol{v}_1|\, B \leq c_v B$ should not exceed the order of $10^{-8}\,\mathrm{m}^2/\mathrm{sec}$ to yield $\overline{T}_{1\infty} - \underline{T}_{1\infty,1} \leq 1\,°\mathrm{K}$. This result is expected to be rather conservative because of the way $\overline{T}_1$ and $\underline{T}_1$ have been constructed. For any finite c_v in the a priori estimate (2.5), however, it is concluded from (3.8) and Fig. 1 that $\lim_{t\to\infty} |T_1(\boldsymbol{x}, t)| \ll T_0(\boldsymbol{x})$ can be achieved if the numerical values of both $|b|\, d/Bc_v$ and $|b|\, d/ac$

are sufficiently large. This requires that the constitutive functions employed for $k(T)$ and/or $c_p(T)$ possess a sufficiently large magnitude of their derivatives. It is irrelevant whether or not fluids with such constitutive functions exist, which have been selected here for presentation since they yield closed-form solutions for $\overline{T}_1$ and $\underline{T}_1$. Also for more general flows and constitutive functions, $T_1(\boldsymbol{x}, t)$ by use of $\underline{T}_1(t)$ and $\overline{T}_1(t)$ can be restricted to a small vicinity of $T_0(\boldsymbol{x})$ provided $k'(T)$ and/or $c_p'(T)$ possess a sufficiently large magnitude and the appropriate sign in the entire interval $T_0 + \underline{T}_1 \leq T \leq T_0 + \overline{T}_1$.

§ 4. Concluding remarks

Field bounds yield sufficient conditions for stability of fields in the sense of Liapunov or for instability of fields. In the majority of applications, field bounds which depend on t only will tend to ∞ as t increases. This trivial result does not occur if the flow possesses "suitable physical characteristics" which have to be sufficiently strong to avoid conservative field bounds such as for water in § 3. In particular, coefficients in the balances of energy and chemical constituents can depend "suitably" on the thermodynamic state, such as $k(T)$ or $c_p(T)$ in § 3. Correspondingly, the dependency of the chemical production rates on this state will be employed in a forthcoming publication to explain why the chemical state in a wide class of hypersonic flows deviates only very little from the one determined by use of the assumption of thermochemical equilibrium. In many applications, the existence of field bounds in $0 < t < \infty$ requires the restriction of unknown fluctuations by use of (a) arbitrarily large constants such as c_v in (2.5) or (b) empirical data such as the information taken from [4]. Liapunov-stable field bounds which depend on t only can be determined for certain flows with internal radiant heat transfer and for classes of waves which are consistent with [6, p. 144—152].

Field bounds can be used to show instability. E.g., a combustible mixture at $T_0 =$ const. is considered at rest in an isolated container whose walls permit passage of radiation only. The heat of reaction is assumed to be given by $\sqrt{\tau_1}\, f(\ldots)$ where $\tau_1 := T - T_0$ and $f > 0$. Initial perturbations $\tau_1(\boldsymbol{x}, 0) > 0$ are due to the absorption of radiation. A function $\underline{\tau}_1(t)$ is determined as the largest solution of $\underline{\tau}_1' - \sqrt{\underline{\tau}_1} \inf f = \underline{\tau}_1(0) = 0$; $\underline{\tau}_1(t)$ exists in $0 < t < \infty$ and increases. The same is true for an upper bound $\overline{\tau}_1(t)$. If there exists a fluctuation $\tau_1(\boldsymbol{x}, t)$ in $t > 0$, it necessarily branches off T_0, thus implying bifurcation and instability.

References

1. ADAMS, E., SPREUER, H.: Von der Zeit abhängige Störungsschranken in der Fluiddynamik. ZAMM (to be published).
2. GÖRTLER, H.: Über die Lösung nichtlinearer partieller Differentialgleichungen vom Reibungsschichttypus. ZAMM **30**, 265–267 (1950).
3. KAMKE, E.: Differentialgleichungen, Lösungsmethoden und Lösungen, 3. Aufl., Leipzig: Akad. Verlagsges. 1944.
4. KLINE, S. J.: Observed Structure Features in Turbulent and Transitional Boundary Layers. In: Fluid Mechanics of Internal Flow, ed. G. SOVRAN, Amsterdam/London/New York: Elsevier 1967.
5. NICKEL, K.: Einige Eigenschaften von Lösungen der Prandtlschen Grenzschicht-Differentialgleichung. Arch. Rat. Mech. Anal. **2**, 1–31 (1958).
6. WALTER, W.: Differential- und Integral-Ungleichungen, Berlin/Göttingen/Heidelberg/New York: Springer 1964.

Stability Regions of Cellular Fluid Flow

By

F. H. Busse

Max-Planck-Institut für Physik und Astrophysik, Munich, Germany

It is well known that the Navier-Stokes equation of motion for a homogeneous incompressible fluid has a unique stationary solution, if the body forces acting on the fluid and the prescribed values of the velocity at the boundary are sufficiently small. We shall call this solution the primary solution. While the primary solution is a possible solution for all values of the prescribed parameters, it is not unique anymore when the parameters exceed certain critical values. The simplest kind of nonuniqueness is caused by stationary secondary solutions for which the principle of exchange of stability holds, i.e. the secondary solution branching off the primary solution becomes stable in place of the primary solution for values of the relevant parameter beyond the critical branching point. Physically this process corresponds to the appearance of a secondary motion superimposed onto the primary state of the fluid system. The secondary motion is usually characterized by a typical scale which is not shown by the primary state and which is exhibited in the form of a cell-like subdivision of the flow field. It is for this reason that the secondary motion for which the principle of exchange of stability holds has been called cellular flow.

The best known examples of cellular flow are the Taylor vortices between differentially rotating concentric cylinders and the convection in a layer heated from below. These cellular flows have been studied intensively because they differ in many respects from flows which are described by primary solutions of the equations of motion. They can be regarded in fact as the simplest kind of turbulence since it is not unreasonable to define turbulence as the physically realized part among the manifold of all statistically stationary solutions of the equations of motion apart from the primary solution. Although cellular flows are not time-dependent, they show a number of features which are characteristic for turbulent flows in the usual sense. One of these features is the nonuniqueness which we shall consider in the following.

The physically realized cellular flow is, in general, not determined uniquely. In contrast to the case of the primary solution the asymptotic stationary solution depends on the history of the system. Although the physically realizable cellular flows correspond to the fraction of all possible stationary solutions, which is stable with respect to small disturbances, the property of stability does not distinguish a unique preferred cellular flow.

For the discussion of the instabilities which confine the manifold of stable solutions we shall use two-dimensional convection as an example. Two-dimensional convection has the form of rolls which are characterized by a horizontal wave number α. Above the critical value R_c of the Rayleigh number R, which is the relevant parameter of the convection problem, a range of wave numbers α in the neighbourhood of the critical value α_c corresponds to possible stationary solutions. This is indicated qualitatively by the dotted line in Fig. 1. Only a part of this range, however, corresponds to stable solutions which can be realized physically. In order to discuss the various instabilities which restrict the region of stability we consider the dimensionless equations of convection

$$(\nabla^6 - R\,\Delta_2)\,v = \boldsymbol{\delta} v \cdot \nabla\nabla^4 v + \frac{\partial}{\partial t}\nabla^4 v. \tag{1}$$

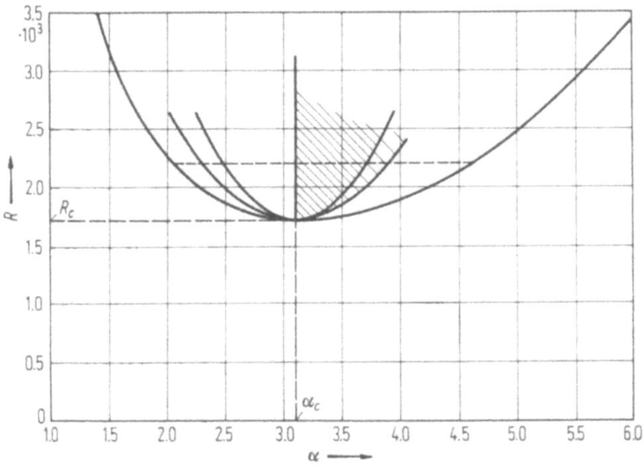

Fig. 1. Stability boundaries corresponding to the criteria (14), (15), and (18), the latter in the case $Pr = \infty$. Cellular convection in the form of rolls with the wave number α is stable in the strongly shaded region for all Prandtl numbers. In the weakly shaded region the stability boundary depends on the Prandtl number.

Δ_2 is the Laplacian with respect to the horizontal x, y-coordinates. The vector

$$\boldsymbol{\delta} v = \left(\frac{\partial^2}{\partial x\,\partial z} v, \quad \frac{\partial^2}{\partial y\,\partial z} v, \quad -\Delta_2 v\right)$$

describes the velocity. For simplicity we have assumed the case of infinite Prandtl number. The explicit results will be given in a form which is valid for arbitrary Prandtl numbers Pr. The boundary conditions at $z = \pm 1/2$ require v, $\nabla^4 v$, and $\frac{\partial^2}{\partial z^2} v$ or $\frac{\partial}{\partial z} v$ to vanish depending on whether stress-free or rigid boundaries are assumed. Further details on the problem can be found, for instance, in Ref. [1] or [2]. For small convection amplitudes ε a solution of Eq. (1) can be obtained in the form of an expansion in powers of ε,

$$v = \varepsilon v^{(1)} + \varepsilon^2 v^{(2)} + \cdots,$$
$$R = R^{(0)} + \varepsilon R^{(1)} + \varepsilon^2 R^{(2)} + \cdots.$$

The lowest order of Eq. (1) admits solutions of the form

$$v^{(1)} = \sum_{n=-N}^{N} c_n v_n \qquad \text{with } c_{-n} = c_n^* \tag{2}$$

and

$$v_n \equiv f(z)\, e^{i\boldsymbol{k}_n \boldsymbol{r}} = v_{-n}^*. \tag{3}$$

The star denotes the complex conjugate. The following normalization conditions will be assumed.

$$\sum_{n=-N}^{N} c_n c_n^* = 1, \quad \langle \Delta_2 v_n^* \nabla^4 v_n \rangle = -1, \tag{4}$$

where the angular brackets indicated the average over the fluid layer. The vectors $\boldsymbol{k}_n$ are arbitrary horizontal vectors with the constant length α. The eigenvalue $R^{(0)}$ reaches its lowest value R_c for $\alpha = \alpha_c$. For small values of $|\alpha - \alpha_c|$

$$R^{(0)} = R_c + \eta(\alpha - \alpha_c)^2 \tag{5}$$

can be assumed. The dependence of the Rayleigh number on the amplitude ε is determined by the solvability conditions for the linear inhomogeneous equations which result from Eq. (1) in the higher orders of ε. It turns out that $R^{(1)}$ vanishes while $R^{(2)}$ is determined by

$$R^{(2)} = +\langle \nabla^4 v^{(1)}, \quad \boldsymbol{\delta} v^{(1)} \cdot \nabla\nabla^4 v^{(2)} + \boldsymbol{\delta} v^{(2)} \cdot \nabla\nabla^4 v^{(1)} \rangle. \tag{6}$$

It was shown in Ref. [1] that all stationary solutions are unstable for sufficiently small amplitudes ε with the exception of the case $N = 1$, corresponding to convection in the form of rolls. In order to determine quantitatively their region of stability we consider the equation for an infinitesimal disturbance $\tilde{v}$ of the stationary solution v,

$$(\nabla^6 - R\,\Delta_2)\,\tilde{v} = \boldsymbol{\delta}\tilde{v}\cdot\nabla\nabla^4 v + \boldsymbol{\delta}v\cdot\nabla\nabla^4\tilde{v} + \sigma\,\nabla^4\tilde{v}. \tag{7}$$

A time dependence of the form $\exp\{\sigma t\}$ has been assumed. Since the stationary solution is given by an expansion in powers of ε an analogous expansion is suggested for $\tilde{v}$, σ,

$$\tilde{v} = \varepsilon\tilde{v}^{(1)} + \varepsilon^2\tilde{v}^{(2)} + \cdots,$$

$$\sigma = \sigma^{(0)} + \varepsilon\sigma^{(1)} + \varepsilon^2\sigma^{(2)} + \cdots.$$

It has to be kept in mind that the parameter ε in this expansion is not related to the amplitude of the disturbance which is left undetermined by Eq. (7). We assume that the horizontal wave numbers of the disturbance as well as the wave number α differ from α_c only by an amount of the order ε. Accordingly $\sigma^{(0)}$ vanishes and $\tilde{v}^{(1)}$ satisfies the same equation as $v^{(1)}$. We can write

$$\tilde{v}^{(1)} = \sum_l \tilde{c}_l \tilde{v}_l, \tag{8}$$

where the functions $\tilde{v}_n$ differ from the functions (3) only in that $\boldsymbol{k}_n$ is replaced by $\tilde{\boldsymbol{k}}_l$ with $|\tilde{\boldsymbol{k}}_l| = \alpha_l \approx \alpha + 0(\varepsilon)$. Because of the symmetry of the problem the solvability condition in the second order of Eq. (7) yields vanishing integrals as in the stationary case with the result $\sigma^{(1)} = 0$. In the third order the solvability conditions are obtained by multiplying the right-hand side of Eq. (7) by $\tilde{\theta}_m^* = \nabla^4\tilde{v}_m^*$, and averaging it,

$$\begin{aligned} &\sigma^{(2)}\,\langle|\tilde{\theta}_m|^2\rangle\,\tilde{c}_m - \eta\varepsilon^{-2}\big((\alpha-\alpha_c)^2 - (\alpha_m-\alpha_c)^2\big)\,\tilde{c}_m - R^{(2)}\,\tilde{c}_m \\ &= -\,\langle\tilde{\theta}_m^*, \boldsymbol{\delta}\tilde{v}^{(1)}\cdot\nabla\theta^{(2)} + \boldsymbol{\delta}\tilde{v}^{(2)}\cdot\nabla\theta^{(1)} + \boldsymbol{\delta}v^{(1)}\cdot\nabla\tilde{\theta}^{(2)} + \boldsymbol{\delta}v^{(2)}\cdot\nabla\tilde{\theta}^{(1)}\rangle. \end{aligned} \tag{9}$$

The second term on the left-hand side gives the contribution owing to the difference between the wave numbers α_m and α which has been neglected in the lowest order of the equation. The horizontal dependence of functions with the upper index 2 is described by the sum of two of the basic $\boldsymbol{k}$-vectors. Non-vanishing contributions from the integrals on the right-hand side of Eq. (9) can be expected accordingly only if the sum of the four $\boldsymbol{k}$-vectors in the argument of the exponential function, which describes the horizontal dependence, vanishes,

$$-\tilde{\boldsymbol{k}}_m + \tilde{\boldsymbol{k}}_l \pm \boldsymbol{k}_1 \pm \boldsymbol{k}_1 = 0. \tag{10}$$

It is convenient to characterize possible disturbances by drawing the $\boldsymbol{k}$-vectors corresponding to relation (10). Three qualitatively different types of disturbances which shall be labelled A, B, and C can be distinguished:

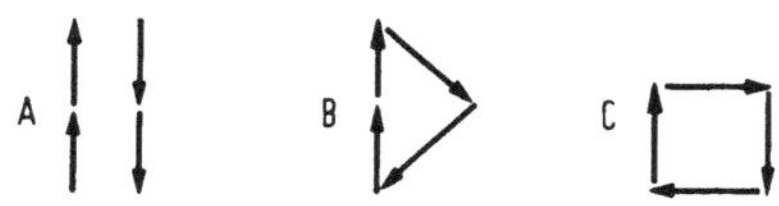

All other kinds of interaction between the $\boldsymbol{k}$-vectors of the disturbance and the vectors $\boldsymbol{k}_1$, $-\boldsymbol{k}_1$ of the roll solution can be regarded as intermediate to those described by the three diagrams. An exact solution of Eq. (7) is

$$\tilde{v} = \boldsymbol{k}_1 \cdot \nabla v \quad \text{with} \quad \sigma = 0. \tag{11}$$

The disturbances of type A and B can be considered as modifications of this exact solution. We shall first discuss the conditions under wich a positive value $\sigma^{(2)}$ can occur in case A. The two equations of the form (9) for the variables $\tilde{c}_l$, $\tilde{c}_m$ are solvable when the characteristic equation for the coefficients vanishes.

$$\det\begin{pmatrix} \sigma^{(2)}\langle|\tilde{\theta}_l|^2\rangle + R^{(2)} - \frac{\eta}{\varepsilon^2}\left((\alpha-\alpha_c)^2 - (\alpha_l-\alpha_c)^2\right) & R^{(2)} \\ R^{(2)} & \sigma^{(2)}\langle|\tilde{\theta}_m|^2\rangle + R^{(2)} - \frac{\eta}{\varepsilon^2}\left((\alpha-\alpha_c)^2 - (\alpha_m-\alpha_c)^2\right) \end{pmatrix} = 0. \tag{12}$$

The terms on the right-hand side of Eqs. (9) have been replaced by their values in the limit $\alpha_m = \alpha_l = \alpha = \alpha_c$. Using the condition $\alpha_m = 2\alpha - \alpha_l$ it is found that the sign of $\sigma^{(2)}$ depends on the sign of

$$\eta\left(3(\alpha-\alpha_c) - (\alpha_l-\alpha_c)\right)(\alpha + \alpha_l - 2\alpha_c) - 2\varepsilon^2 R^{(2)}. \tag{13}$$

The boundary of the region of stability is reached when the expression (13) vanishes with α_l tending towards α. Using $R - R_c - \eta(\alpha-\alpha_c)^2$ in the place of $\varepsilon^2 R^{(2)}$ we can conclude that rolls are unstable for

$$R - R_c - 3\eta(\alpha-\alpha_c)^2 < 0. \tag{14}$$

This stability criterion has been first derived by ECKHAUS [3]. Since it does not depend on the particular form of the nonlinear terms and since the perturbed solution remains two-dimensional, the criterion (14) applies to a variety of cellular motions as, for example, the Taylor vortices. The experimental findings of SNYDER [4] are in agreement with criterion (14), though the observations in the relevant region are not quite accurate enough to confirm quantitatively the theoretical boundary of the stability region. In the case of convection, disturbances of type B and C turn out to be more restrictive, at least for large Prandtl num bers.

The characteristic equation for disturbances of type B is essentially identical to (12) with $\alpha_l = \alpha_m$. It yields

$$\alpha \geq \alpha_c \tag{15}$$

as necessary condition for stability.

Since this stability boundary does not depend on ε, terms of the order $\varepsilon^2(\alpha - \alpha_c)$ have to be considered in the stability equation. Those terms are vanishing in the case of free boundaries because $\Delta_2 v^{(2)}$ is zero. In the case of rigid boundaries the criterion (15) has to be replaced by a criterion of the form

$$R - R_c \gtrless c(\alpha - \alpha_c) \tag{16}$$

depending on whether c is a negative or positive number. Because $|c|$ is a rather large number, however, a criterion of the form (16) is identical with the criterion (15) for all practical purposes.

The characteristic equation for disturbances of type C reduces to the product of the two diagonal elements. For Prandtl numbers not much less than 1, an angle of 90° between the vectors $\boldsymbol{k}_l$ and $\tilde{\boldsymbol{k}}_1$ leads to the maximum value of the growth rate

$$\sigma^{(2)}\langle|\theta_l|^2\rangle = -LR^{(2)} + \eta\varepsilon^{-2}\left((\alpha - \alpha_c)^2 - (\alpha_l - \alpha_c)^2\right) \tag{17}$$

with

$$LR^{(2)} \equiv \frac{1}{2} + L(1) + 2L(0) = \begin{cases} 0.21042 + Pr^{-1}\cdot 0.12686 + Pr^{-2}\cdot 0.07612 \\ 0.17374 + Pr^{-1}\cdot 0.07992 + Pr^{-2}\cdot 0.12927. \end{cases}$$

The numerical values have been taken from Ref. [2] and correspond to free and rigid boundaries in the upper and the lower line, respectively. According to Eq. (17) rolls are unstable for

$$R - R_c - \frac{1+L}{L}\eta(\alpha - \alpha_c)^2 < 0. \tag{18}$$

Since L increases from a value less than 1/2 for $Pr = \infty$ to a value larger than one at small Prandtl numbers, criterion (18) is more restrictive than criterion (14) for larger Prandtl numbers. For $Pr \lesssim 1.1$ in the case of rigid boundaries and for $Pr \lesssim 3.5$ in the case of free boundaries, rolls with $\alpha > \alpha_c$ become unstable for sufficiently low values of $R - R_c$ to disturbances of type A in place of those of type C. This fact is remarkable in so far as the Prandtl number has only a quantitative influence in all other aspects of finite amplitude convection.

The results of the foregoing analysis are summarized in Fig. 1. They have been extended to large Rayleigh numbers in another paper [5] in the case $Pr = \infty$. An analogous discussion for the hexagon solution is given in [6]. The possibility of two-dimensional cellular motions corresponding to a continuous spectrum of wave numbers in place of a single wave number α has been considered by NEWELL and WHITEHEAD [7] and by SEGEL [8].

References

1. SCHLÜTER, A., LORTZ, D., BUSSE, F.: J. Fluid Mech. **23**, 129 (1965).
2. BUSSE, F. H.: J. Fluid Mech. **30**, 625 (1967).
3. ECKHAUS, W.: Studies in Non-Linear Stability Theory, Berlin/Heidelberg/New York: Springer 1965.
4. SNYDER, H. A.: J. Fluid Mech. **35**, 273 (1969).
5. BUSSE, F. H.: J. Math. and Phys. **46**, 140 (1967).
6. BUSSE, F. H.: Dissertation, University Munich 1962.
7. NEWELL, A. C., WHITEHEAD, J. A.: J. Fluid Mech. **38**, 279 (1969).
8. SEGEL, L. A.: J. Fluid Mech. **38**, 203 (1969).

The Stability of Gravity Waves on the Surface of a Flow with Non-uniform Velocity Distribution

By

H. G. M. Velthuizen and L. van Wijngaarden

Twente Institute of Technology, Enschede, Netherlands

1. Introduction

The stability of gravity waves on the free surface of a fluid with nonuniform velocity distribution is affected by two mechanisms. One mechanism is analogous to what J. W. MILES discovered in relation to the generation of gravity waves by wind over the surface. The other mechanism is connected with the momentum transport from the mean flow into the wave and vice versa by viscous stresses. In this contribution the neutral stability problem associated with these mechanisms is investigated.

2. Formulation of the problem

We consider small amplitude gravity waves over a flow with non-uniform velocity distribution $U(y)$.[1] It is convenient to use dimensionless variables. As a characteristic length we will use h, as a characteristic velocity $U(y)$. To satisfy the mass conservation condition we introduce the streamfunction Ψ defined by

$$u = \frac{\partial \Psi}{\partial y}\,; \qquad v = -\frac{\partial \Psi}{\partial x}\,. \tag{2.1}$$

The free surface has the form

$$n = a \exp i\alpha (x - ct)\,. \tag{2.2}$$

[1] For the meaning of the symbols the reader is referred to the nomenclature (at the end of this work).

We assume that the streamfunction has the same x- and t-dependence and therefore write

$$\Psi = \varphi(y) \exp i\alpha(x - ct). \tag{2.3}$$

Substitution of (2.1) and (2.3) into the linearized Navier-Stokes equations results after elimination of the pressure in the Orr-Sommerfeld equation

$$(U - c)(\varphi'' - \alpha^2\varphi) - U''\varphi = \frac{-i}{\alpha R}\{\varphi^{\mathrm{IV}} - 2\alpha^2\varphi + \alpha^4\varphi\}, \tag{2.4}$$

where primes denote differentiation with respect to y. There are two boundary conditions at the bottom and two at the surface. The vanishing of u and v at the bottom requires

$$\varphi(0) = \varphi'(0) = 0. \tag{2.5}$$

At the surface the kinematical condition in linearized form results in

$$\varphi(1) = -a(1 - c). \tag{2.6}$$

It is also easy to show that the second free surface condition which is on the pressure amounts to

$$(1 - c)\varphi'(1) - U'(1)\varphi(1) = -aF_r^{-1}. \tag{2.7}$$

3. Characteristic equation for small values of α

For large values of R the viscous effects that must be taken into account are confined to a thin layer near the wall. Because of the linear character of the Orr-Sommerfeld equation we can construct a solution of the form

$$\varphi = \varphi_{nv} + Bf, \tag{3.1}$$

where φ_{nv} denotes a solution of (2.4) with the right-hand side put equal to zero, which is called the Rayleigh equation. B is a constant and f is a viscous solution, that dies off rapidly as $y \to 1$ and is negligible outside the thin wall layer. So the free surface conditions have to be satisfied by φ_{nv} only. When we impose the bottom conditions (2.5) on (3.1) we obtain

$$\frac{\varphi'_{nv}(0)}{\varphi_{nv}(0)} = \frac{f'(0)}{f(0)}, \tag{3.2}$$

which is called the characteristic equation. The ratio $f'(0)/f(0)$ can be expressed in $\mathcal{F}(z)$ by

$$\frac{f'(0)}{f(0)} = \frac{U'(0)}{c}\,\frac{\mathcal{F}(z)}{1 - \mathcal{F}(z)}. \tag{3.3}$$

The ratio $\varphi'_{nv}(0)/\varphi_{nv}(0)$ is purely inviscid, so we have to solve the Rayleigh equation with the boundary conditions (2.6) and (2.7). For our purposes it is convenient to work with the new independent variable (see MILES [2])

$$\Omega(y) = \frac{\varphi_{nv}}{(U-c)[U'\varphi_{nv} - (U-c)\,\varphi'_{nv}]}, \tag{3.4}$$

which satisfies the Riccati equation

$$\Omega' = \alpha^2 (U-c)^2\,\Omega^2 - (U-c)^{-2}. \tag{3.5}$$

In terms of $\Omega(y)$ and $\mathscr{F}(z)$ the characteristic equation (3.2) takes the form

$$\mathscr{F}(z) = 1 + U'(0)\,c\,\Omega(0). \tag{3.6}$$

For small values of α it is easy to show that

$$\Omega(0) = \int_0^1 \frac{dy}{(U-c)^2} - Fr + O(\alpha^2), \tag{3.7}$$

so the characteristic equation for small values of α is

$$\mathscr{F}(z) = 1 + U'(0)\,c\left\{\int_0^1 \frac{dy}{(U-c)^2} - Fr + O(\alpha^2)\right\}. \tag{3.8}$$

4. Neutral curve

In the neutral case the imaginary part of the wave speed c vanishes. How to cope with the singularity in (3.8), which arises in this case, is treated in LIN [3]. To find the neutral curve we must decompose (3.8) into its real and imaginary parts.

$$\mathscr{F}_r(z) = 1 + U'(0)\,c\left\{\mathrm{Re}\int_0^1 \frac{dy}{(U-c)^2} - Fr\right\}, \tag{4.1}$$

$$\mathscr{F}_i(z) = -\frac{\pi U''(y_c)}{U'(y_c)^3}\,U'(0)\,c, \tag{4.2}$$

y_c is the value of y at the critical layer, where $U = c$. We assume $U''(y) < 0$ and $U'(y) > 0$ in which case the right-hand side of (4.2) is positive because c is positive when a critical layer exists. The argument z has to exceed $z_0 \simeq 2.297$ for $\mathscr{F}_i(z)$ to be positive. In that case $\mathscr{F}_r(z)$ is also positive. The general behaviour of $\mathscr{F}_i(z)$ is such that the neutral curve has a two-branch nature. The form of this curve for a given velocity profile still depends on Fr. So in $a\,(\alpha, R, Fr)$ space we can speak of a neutral surface.

Most of the essential features of the problem can be demonstrated by an example. Let us take for instance

$$U(y) = 1 - e^{-yU'(0)} \quad \text{with} \quad U'(0) \gg 1, \tag{4.3}$$

which in fact represents a whole range of velocity profiles of boundary layer type. Equations (4.1) and (4.2) take the form

$$\mathscr{F}_r(z) + \frac{c}{1-c}\left\{1 - \frac{1}{1-c}\ln\frac{1-c}{c}\right\} = U'(0)\,c\left\{\frac{1}{(1-c)^2} - Fr\right\}, \quad (4.4)$$

$$\mathscr{F}_i(z) = \frac{\pi c}{(1-c)^2}\,. \quad (4.5)$$

From (4.5) it follows that a possible solution c is bounded by a maximum value

$$0 < c < 0.138\,. \quad (4.6)$$

For these values of c the left-hand side of (4.4) is always positive and of order unity, so

$$U'(0)\,c\left\{\frac{1}{(1-c)^2} - Fr\right\} \sim O(1)\,. \quad (4.7)$$

From (4.6) and (4.7) it follows that for a possible solution Fr has to be bounded by a maximum also. We can divide the problem into two classes with different kinds of solutions.

(1) Class I, $1 < Fr < Fr_{\max}$. For these cases c is also bounded by a minimum value as follows from (4.7) and cannot become zero, whatever the value of $U'(0)$.

(2) Class II, $0 < Fr < 1$. Here c can approach zero for very large values of $U'(0)$.

We will examine both cases for very large values of $U'(0)$.

Class I: In this case c becomes equal to $1 - Fr^{-1/2}$ as follows from (4.4). For this value of c equation (4.5) gives two solutions z_1 and z_2, say, with $z_2 > z_1$. For the lower and upper branch of the neutral curve we find respectively

$$R \sim U'(0)^2\, z_1^3 (1 - Fr^{-1/2})^{-3} \cdot \alpha^{-1}, \qquad U'(0) \to \infty, \quad (4.8)$$

$$R \sim U'(0)^2\, z_2^3 (1 - Fr^{-1/2})^{-3} \cdot \alpha^{-1}, \qquad U'(0) \to \infty. \quad (4.9)$$

Class II: From (4.7) it follows that c becomes very small for large values of $U'(0)$. This corresponds with either very large values of z or $z = z_0$ as follows from (4.5). For $z \to z_0$ we have from (4.4)

$$c \sim \frac{\mathscr{F}_r(z_0)}{U'(0)\,(1 - Fr)}\,. \quad (4.10)$$

So the lower branch has the form

$$R \sim z_0^3 U'(0)^5 (1 - Fr)^3 \mathscr{F}_r(z_0)^{-3} \cdot \alpha^{-1}, \qquad U'(0) \to \infty. \quad (4.11)$$

For the upper branch we have to use the asymptotic expressions for $\mathscr{F}_r(z)$ and $\mathscr{F}_i(z)$ as $z \to \infty$.

$$\begin{aligned} \mathscr{F}_r(z) &\sim 1 + \frac{\frac{1}{2}\sqrt{2}}{z^{3/2}} \text{ for } z \to \infty, \\ \mathscr{F}_i(z) &\sim \frac{\frac{1}{2}\sqrt{2}}{z^{3/2}} \text{ for } z \to \infty. \end{aligned} \tag{4.12}$$

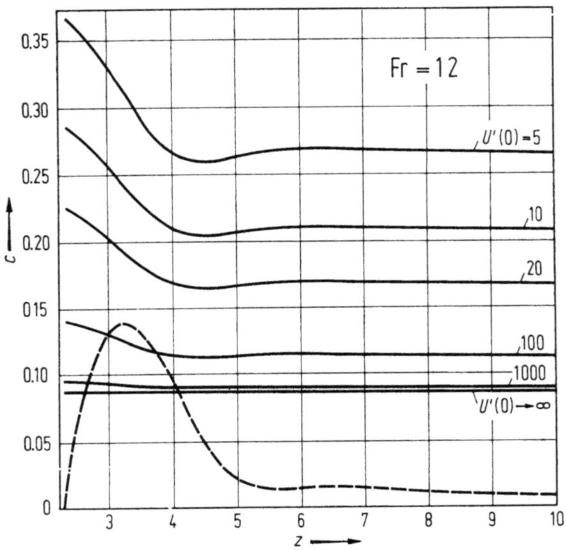

Fig. 1

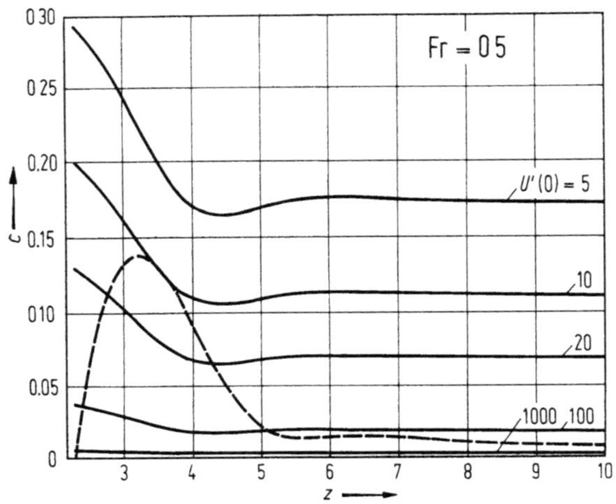

Fig. 2

With these expressions we have from (4.4) and (4.5)

$$c \sim \frac{1}{U'(0)\,(1 - Fr)} \sim \frac{\frac{1}{2}\sqrt{2}}{\pi z^{3/2}} . \tag{4.13}$$

So for the upper branch we get

$$R \sim (2\pi^2)^{-1}\, U'(0)^7\, (1 - Fr)^5\, \alpha^{-1} \quad \text{for} \quad U'(0) \to \infty . \tag{4.14}$$

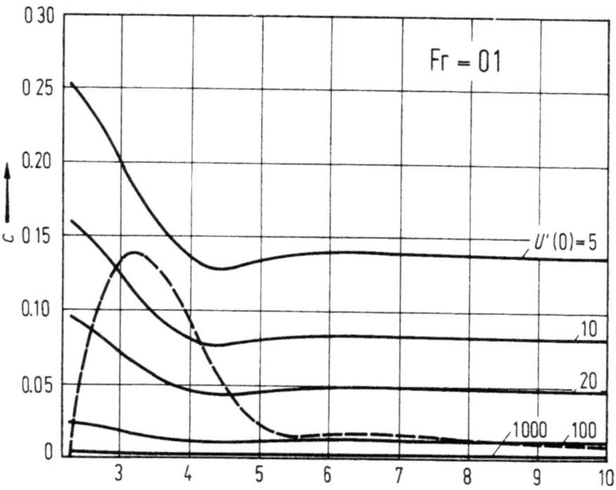

Fig. 3

Figs. 1–3. Graphical representations of (4.4) and (4.5) for different values of *Fr*. The drawn curves follow from (4.4), the dotted one from (4.5).

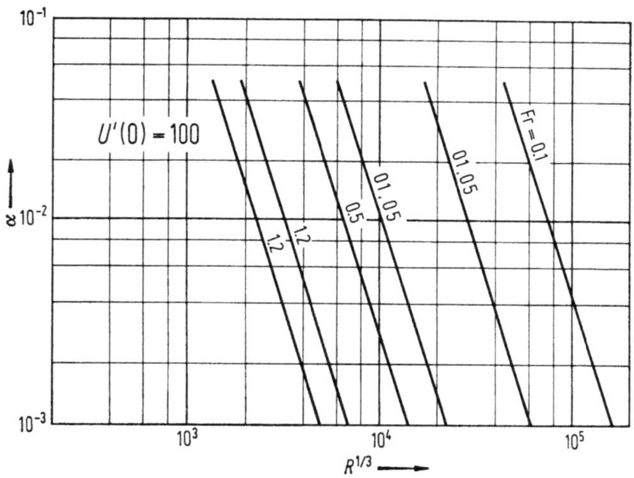

Fig. 4. Asymptotic behaviour of the neutral curve for one value of $U'(0)$ and the different values of *Fr*.

For other values of $U'(0)$ a numerical procedure has to be followed. The typical features of classes I and II are demonstrated in Figs. 1—4. In particular it can clearly be seen that for a neutral wave to exist $U'(0)$ has to increase with increasing Fr.

5. Concluding remarks

From the foregoing section we see that for large values of $U'(0)$ the value of c is almost independent of α. This is in agreement with what we would expect from inviscid wave theory with a homogeneous velocity profile for small α. The values of c for class I are also in agreement with that theory. The class II waves, however, are of a type not included in the potential theory of gravity waves. We can interpret the results in Sec. 4 as follows:

The waves in class I are free surface gravity waves modified by viscous effects, whereas the waves in class II are Tollmien-Schlichting waves, modified by the free surface.

Nomenclature

a	amplitude of the surface wave
c	wave speed
$\mathscr{F}(z)$	modified Tietjens function; for a definition see MILES [1]
Fr	Froude number defined by $\frac{U(h)^2}{gh}$
f	viscous solution of the Orr-Sommerfeld equation
g	gravitational constant
h	depth of the undisturbed fluid
R	Reynolds number defined by $\frac{U(h)\cdot h}{\nu}$
$U(y)$	dimensionless velocity profile of the mean flow
$U'(y)$	first derivative of $U(y)$
u	dimensionless wave induced velocity in x-direction
v	dimensionless wave induced velocity in y-direction
x	dimensionless horizontal coordinate
y	dimensionless vertical coordinate
z	argument of $\mathscr{F}(z)$ defined by $(\alpha R U_0'^{-2})^{1/3} c$
α	dimensionless wave number
η	form of the free surface
ν	kinematic viscosity
φ	amplitude function of the streamfunction
Ψ	streamfunction

References

1. MILES, J. W.: J. Fluid Mech. 8, 593 (1960).
2. MILES, J. W.: J. Fluid Mech. **13**, 427 (1962).
3. LIN, C. C.: The Theory of Hydrodynamic Stability. London/New York: Cambridge Univ. Press 1955.

A Non-linear Investigation of the Stability of Flow between Counter-rotating Cylinders

By

R. C. DiPrima

Rensselaer Polytechnic Institute, Troy, N.Y., U.S.A.

and

R. N. Grannick

University of Chicago, Chicago, Ill., U.S.A.

Consider a viscous, incompressible fluid contained between two infinitely long concentric rotating cylinders with radii R_1 and R_2 ($> R_1$) and angular velocities Ω_1 and Ω_2, respectively. Let (r, θ, z) denote the usual cylindrical coordinates, and let u_r, u_θ, u_z denote the components of velocity in the increasing r, θ, and z directions, respectively. In this discussion we will assume that all variables and parameters are non-dimensional. The Taylor number, T, is defined by $T = -4\Omega_1^2(\mu - \eta^2)d^4/(1 - \eta^2)\nu^2$ where $\mu = \Omega_2/\Omega_1$, $\eta = R_1/R_2 < 1$, $d = R_2 - R_1$, and ν is the kinematic viscosity.

Since the original work of G. I. Taylor [1], extensive experimental and theoretical work has confirmed that there exists a critical Taylor number $T_c(\mu, \eta)$, $\mu < \eta^2$, above which Couette flow is unstable to axisymmetric perturbations. Further, for a wide range of values of μ (certainly $\mu > 0$) this instability yields a steady secondary motion in the form of toroidal vortices (Taylor vortices) spaced regularly along the axis of the cylinders for T slightly greater than $T_c(\mu, \eta)$. This Taylor-vortex flow was first computed by Davey [2] for the cases

---

Acknowledgement. This work was partially supported by the Fluid Mechanics Branch of the Office of Naval Research, Contract N 00014-67-A-0117-0001, and the Army Research Office, Durham, Contract DA-31-124-ARO-D-269.

$\eta = 1/2$, $\mu = 0$; $\eta \to 1$, $\mu = 0$; and $\eta \to 1$, $\mu \to 1$. He followed a procedure proposed by Stuart [3] and Watson [4] which is based on an expansion in an amplitude function $A(t)$ which is order of the square root of the amplification rate of the axisymmetric disturbance.

Recently Krueger, Gross, and DiPrima [5] have shown that for η near 1 and μ sufficiently negative the critical Taylor number occurs for *non-axisymmetric* disturbances. For disturbances proportional to $\exp[i(m\theta + \lambda z)]$ and $\eta = 0.95$ (calculations were also carried out for $\eta \to 1$) they found that the critical Taylor number occurs for the following values of the azimuthal wave number m as a function of μ: $-0.73 < \mu$, $m = 0$; $-0.76 < \mu < -0.73$, $m = 1$; $-0.796 < \mu < -0.76$, $m = 2$; $-0.88 < \mu < -0.796$, $m = 3$; $-1.08 < \mu < -0.88$, $m = 4$; and $? < \mu < -1.08$, $m = 5$. This is depicted in Fig. 1 where the critical value of T for different values of m, T_m, is plotted as a function of $-\mu$. The experimental observations of Snyder [6] are in good qualitative agreement with these predictions. The critical axial wave number, λ_c, was found to be an increasing function of μ for decreasing μ; however, the overall change in λ_c from $\mu \sim -0.75$ to $\mu \sim -1.25$ was only about 10%.

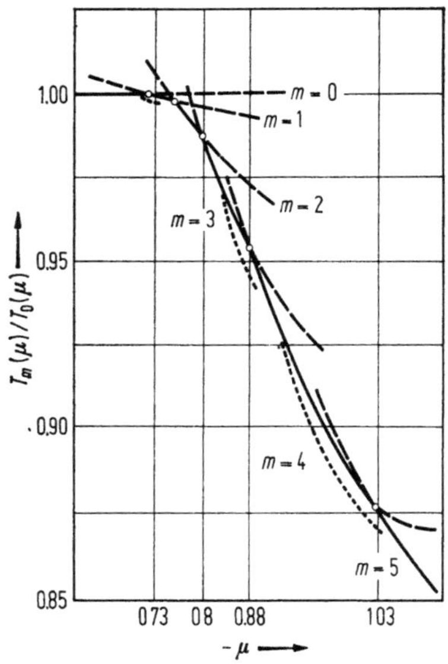

Fig. 1. $T_m(\mu)/T_0(\mu)$ vs. $-\mu$. The critical value of T vs. $-\mu$, ———. Schematic sketch of critical value of T for subcritical instability of Taylor vortex ($m = 0$) and spiral vortex ($m = 3$ and $m = 4$) for a fixed disturbance level, ·····, $\eta = 0.95$.

Within the framework of linear theory, when the critical disturbance is non-axisymmetric it is impossible to distinguish between a wave traveling in both the azimuthal and axial direction and a wave traveling in the azimuthal direction and standing in the axial direction. The experimental observations of Snyder [6] and Coles [7] indicate that the former motion occurs for T slightly greater than T_c. To resolve this indeterminancy theoretically a non-linear model involving the interaction of two disturbances which can lead to either form of motion is considered in the present paper. The analysis is based on the work of STUART [8] and DAVEY, DIPRIMA, and STUART [9]; details of the analysis can be found in GRANNICK [10]. As part of the analysis the development of Taylor-vortex flows for $-0.73 < \mu < 0$ and $\eta = 0.95$ is also considered.

We consider the interaction of two disturbances with the same axial wave number λ, but out of phase by $\pi/2$, and the same azimuthal wave number m. It turns out, after considerable algebra, that the velocity field can be written in the form

$$\begin{aligned} u_\theta(r, \theta, z, t) = {} & V(r) + (B_c h_0 + B_c |B_c|^2 h_1)\, e^{im\theta} \cos \lambda z \\ & + (B_s h_0 + B_s |B_s|^2 l_1) e^{im\theta} \cos \lambda z + (|B_c|^2 + |B_s|)\, F_3 \\ & + (B_c^2 + B_s^2)\, y_1 e^{i2m\theta} + (|B_c|^2 - |B_s|^2)\, m_3 \cos 2\lambda z \\ & + (B_c \tilde{B}_s + \tilde{B}_c B_s)\, m_3 \sin 2\lambda z + (B_c^2 - B_s^2)\, p_1 e^{i2m\theta} \cos 2\lambda z \\ & + 2 B_c B_s p_1 e^{i2m\theta} \sin 2\lambda z + \cdots, \end{aligned} \tag{1}$$

where the amplitudes B_c and B_s are functions of t which satisfy the equations (truncated at cubic order)

$$\begin{aligned} dB_c/dt &= b_0 B_c + b_1 B_c |B_c|^2 + b_2 B_c |B_s|^2 + (b_1 - b_2)\, \tilde{B}_c B_s^2, \\ dB_s/dt &= b_0 B_s + b_1 B_s |B_s|^2 + b_2 B_s |B_c|^2 + (b_1 - b_2)\, \tilde{B}_s B_c^2. \end{aligned} \tag{2}$$

Here $V(r)$ is the basic Couette flow, $h_0, \ldots, p_1$ are functions of r only, a tilda denotes the complex-conjugate, the dots in Eq. (1) denote the complex-conjugates of terms involving θ and higher harmonics which are of cubic order in the amplitudes B_c and B_s. Further b_0, which is a function of T, is the amplification rate according to linear theory of a non-axisymmetric disturbance $h_0(r) \exp[im\theta] \cos \lambda z$ and h_0 is the corresponding eigenfunction; the functions F_3, y_1, m_3, and p_1 can be determined uniquely as solutions of non-homogeneous linear boundary value problems; the Landau constants b_1 and b_2 are determined in the limit $T \to T_c$ so that the non-homogeneous equations for h_1 and l_1 are not ill conditioned. The equations for u_r and u_z are similar. By setting $m = 0$ and $B_s = 0$ the Taylor-vortex mode can be recovered from Eqs. (1) and (2).

The amplitude equations (2) have the following solutions, where subscripts r and i denote the real and imaginary parts respectively, and t_s is an arbitrary time phase.

Spiral vortex:

$$B_c = \beta_e \exp[i\omega(t - t_s)], \quad B_s = \pm i B_c, \quad \beta_e^2 = -b_{0r}/2b_{2r}, \quad \omega = b_{0i} + 2b_{2i}\beta_e^2.$$

Non-axisymmetric simple mode:

$$B_c = \beta_{ce} \exp[i\omega(t - t_s)], \quad B_s = \beta_{se} \exp[i\omega(t - t_s)],$$

$$\beta_{ce}^2 + \beta_{se}^2 = -b_{0r}/b_{1r}, \quad \omega = b_{0i} - b_{0r}b_{1i}/b_{1r}.$$

The spiral vortex gives a wave traveling in both the z and θ directions. It exists if $-b_{0r}/b_{2r} > 0$, and it is stable if $b_{0r} > 0$ and $b_{1r}/b_{2r} > 1$. The non-axisymmetric simple mode gives a wave traveling in the θ direction but standing in the z direction. It exists if $-b_{0r}/b_{1r} > 0$ and is stable if $b_{0r} > 0$ and $b_{2r}/b_{1r} > 1$. It follows from the conditions for stability that when both modes exist only one will be stable.

Calculations were carried out in the "μ intervals" for which $m_c = 3$ and 4. The results for the Landau constants b_1 and b_2 and the corresponding relevant parameters are tabulated below.

$-\mu$	m	λ_m	T_m	b_{1r}	b_{1i}	b_{2r}	b_{2i}
0.80	3 (c)	3.561	13731	−1.186	−0.739	−0.239	5.671
	4	3.588	13942	−2.328	−1.219	−1.158	6.998
1.0	3	3.748	20622	1.567	−1.007	1.716	6.298
	4 (c)	3.680	20068	0.227	−1.644	0.482	7.538

Here T_m is the critical value of T for the given m, and λ_m is the corresponding value of λ. A (c) in the m-column indicates the critical disturbance for the given μ. The Landau constant for Taylor-vortex flow was also computed for $\mu = 0$, -0.125, -0.250, -0.500, and -1.0. Since the variation of this parameter with μ was approximately linear for $\mu = -0.125$ to $\mu = -1.0$, linear interpolation was used to determine the variation of b_{1r} and b_{2r} with μ for $-1.08 < \mu < -0.796$ for $m = 3$ and $m = 4$.

The conclusions that can be drawn are the following.

(1) Axisymmetric disturbances ($m = 0$) are most critical for $\mu > -0.73$. For T slightly greater than T_c Taylor-vortex flow exists and is stable with respect to itself for $\mu > -0.7$. Torque calculations for $\eta = 0.95$, $\mu = 0$ indicate that Davey's results for $\eta \to 1$, $\mu = 0$ are applicable with T_c corrected for gap size. For $-0.73 < \mu < -0.70$ there is a subcritical instability of the Taylor-vortex mode.

(2) For the interval of calculations $-1.08 < \mu < -0.796$, the non-axisymmetric simple mode is *unstable* whenever it exists, and the spiral vortex is *stable* whenever it exists.

(3) For $-0.88 < \mu < -0.796$ the critical value of m is 3. For this range of μ and value of m and T slightly greater than T_c the simple non-axisymmetric mode exists but is unstable. The spiral vortex exists and is stable for T slightly greater than T_c for $-0.83 < \mu < -0.796$; for $-0.88 < \mu < -0.83$ there is a subcritical instability of the spiral vortex.

(4) For $-1.08 < \mu < -0.88$ the critical value of m is 4. For this value of m and for T slightly greater than T_c the simple non-axisymmetric mode exists and for $-0.98 < \mu < -0.88$ but is unstable; and for $-1.08 < \mu < -0.98$ this mode is subcritically unstable. For $-0.94 < \mu < -0.88$ and T slightly greater than T_c the spiral vortex exists and is stable; for $-1.08 < \mu < -0.94$ this mode is subcritically unstable.

(5) While calculations were not carried out for values of μ for which the critical value of m is 1 or 2, results similar to those described in (3) and (4) can be expected.

Notice that at the points $\mu = -0.73$, -0.88, and -1.08 where the critical value of m shifts from 0 to 1, from 3 to 4, and from 4 to 5 respectively the expected mode of the preceding region (larger μ) has become subcritically unstable. Since at the points in question the critical Taylor numbers for the $m = 0$ and $m = 1$ modes, for the $m = 3$ and $m = 4$ modes, ... are the same it is possible, for example, that the spiral vortex mode with $m = 3$ may be observed in intervals on the μ axis for which the spiral vortex flow with $m = 4$ would be expected. This is depicted by the dotted lines in Fig. 1. To determine the subsequent flow resulting from the subcritical instabilities (or even if an equilibrium flow exists) it is necessary to consider 5th order (and possibly higher) terms in the amplitude equations. Inclusion of such terms is also necessary when the Landau constants b_1 and b_2 are near zero. Such an analysis would require a considerable analytical and numerical effort. It is clear, however, from the present results that for $\eta = 0.95$ and μ fixed less than -0.70 the transition with increasing T from Couette flow to a presumably spiral vortex flow may be very complicated. Further, how modes with different azimuthal wave numbers interact, for example the interaction of non-axisymmetric modes with $m = 3$ and $m = 4$ for $\mu < -0.88$, would be of considerable interest. Consideration should also be given to theoretical work in which Ω_2 is held fixed in the opposite direction to that for Ω_1 (rather than μ) and Ω_1 is increased beyond the critical.

References

1. Taylor, G. I.: Phil. Trans. Roy. Soc. London **A 223**, 289–343 (1923).
2. Davey, A.: J. Fluid Mech. **14**, 336–368 (1962).
3. Stuart, J. T.: J. Fluid Mech. **9**, 353–370 (1960).
4. Watson, J.: J. Fluid Mech. **9**, 371–389 (1960).
5. Krueger, E. R., Gross, A., DiPrima, R. C.: J. Fluid Mech. **24**, 521–538 (1966).
6. Snyder, H.: Phys. Fluids **11**, 728–734 (1968).
7. Coles, D.: J. Fluid Mech. **21**, 385–425 (1965).
8. Stuart, J. T.: Adv. in Aero. Sci., Vols. 3–4, London/New York/Paris/Los Angeles: Pergamon 1961, p. 121–142.
9. Davey, A., DiPrima, R. C., Stuart, J. T.: J. Fluid Mech. **31**, 17–52 (1968).
10. Grannick, R. N.: Ph. D. Thesis, Dept. of Math., Renss. Poly. Inst., Troy, N.Y., 1968.

On the Non-linear Stability of Plane Couette Flow

By

T. Ellingsen, B. Gjevik and **E. Palm**

University of Oslo, Oslo, Norway

Plane Couette flow is defined as the flow which takes place between two parallel planes in the case of no pressure gradient in the flow direction. When no disturbance occurs, the velocity will be a linear function of the vertical coordinate z.

The linear stability of this flow has been investigated in several papers. As is well known, it is found that an infinite set of eigenfunctions exists, all of them being stable modes.

Very little is known about the real behaviour of this flow from experiments. The only experiment we know relevant to the present problem is due to REICHARDT [1]. He claims to have found that Couette flow is stable for Reynolds numbers less than about 750 when as characteristic length, H, is used half the distance between the planes and as characteristic velocity, U, half the difference between the velocities of the planes. This indicates that Couette flow must be unstable for finite disturbances. It seems therefore to be an important problem to investigate the stability feature of the non-linear terms, i.e. to examine if these are acting destabilizing and, most important, to examine if the effect of these terms may render the Couette flow unstable.

In the most non-linear problems in hydrodynamics the linearized solution corresponds to an unstable motion, and the effect of the non-linear terms are to stabilize the motion. In the present case, however, the problem is to investigate if the non-linear terms may render the motion unstable, and, eventually, to find the order of magnitude of the necessary amplitude (i.e. the threshold amplitude). We shall in what follows only be concerned with the two-dimensional case.

The vorticity equation may then be written

$$\left(R^{-1}\nabla^2 - z\frac{\partial}{\partial x} - \frac{\partial}{\partial t}\right)\nabla^2\psi = \frac{\partial(\psi, \nabla^2\psi)}{\partial(x, z)} \qquad \left(\nabla^2 \equiv \frac{\partial^2}{\partial x^2} + \frac{\partial^2}{\partial z^2}\right), \tag{1}$$

where ψ denotes the streamfunction.

The boundary conditions are

$$\psi = \psi_z = 0 \qquad \text{for } z = \pm 1. \tag{2}$$

The equation

$$\left(R^{-1}\nabla^2 - z\frac{\partial}{\partial x} - \frac{\partial}{\partial t}\right)\nabla^2\psi = 0 \tag{3}$$

is the corresponding linearized equation discussed by many authors. The normal mode solution of this equation is the Hankel-function of order 1/3.

To find an approximate solution of (1) valid for small amplitudes we apply a method very close to that proposed by STUART [2]. The streamfunction ψ is written as

$$\psi = \sum \psi_n(z, t)\exp\left(in\alpha(x - c_r t)\right), \tag{4}$$

where ψ_1 is the first approximation, $\psi_1 + \psi_2$ the second approximation and so on, and only terms up to the third order are retained. Incorporated in the method is also the "shape assumption"

$$\psi_1 = A(t)\,F(z),$$

where $F(z)$ is an eigenfunction. In our case it seems reasonable for $F(z)$ to choose the least stable eigenfunction. c_r is the corresponding wave velocity and Δc_r, the change of the wave velocity due to non-linear effects, are incorporated in $A(t)$. $A(t)$ satisfies an equation of the Landau type

$$\frac{d\,|A|^2}{dt} = k_1\,|A|^2 + k_2\,|A|^4, \tag{5}$$

where higher order terms are neglected. If the linear solution is exponentially unstable, which is usually true (but not here) k_1 is positive. If, in addition, k_2 is negative, we notice that $|A|$ will tend towards a steady solution. In the problems discussed earlier, a neutral curve exists for the linearized solution. Therefore k_1 may for proper values of the wave number α and Reynolds number R be chosen arbitrarily small. (5) is therefore asymptotically correct.

In our case, however, the linear solution is stable and k_1 is therefore negative. If k_2 is found to be negative, also the considered non-linear terms act stabilizing. On the other hand, if k_2 is found to be positive, they act destabilizing. If this is true, and the initial amplitude is sufficiently large, the non-linear destabilizing effect will overshadow the linear stabilizing tendency. The amplitude just large enough to render the motion unstable is obtained by finding the steady solution of (5). For a given wave number this threshold amplitude depends only on the Reynolds number. This relationship constitutes a non-linear neutral curve, namely the threshold amplitude as a function of the Reynolds number for a given wave number.

True enough, we may say that also in our case k_1 may be chosen arbitrarily small, k_1 being proportional to c_i (c_i the imaginary part of the eigenvalue) which tends towards zero asymptotically as $(\alpha R)^{-1/3}$. It turns out, however, that also k_2 and k_3 are proportional to $(\alpha R)^{-1/3}$, leading to that the stationary value of $|A|$ is not arbitrarily small for any values of the parameters α and R. Since $|A|$ may not be made arbitrarily small, it may be necessary to take into account more terms in (5) in the stationary case. We shall soon return to this question.

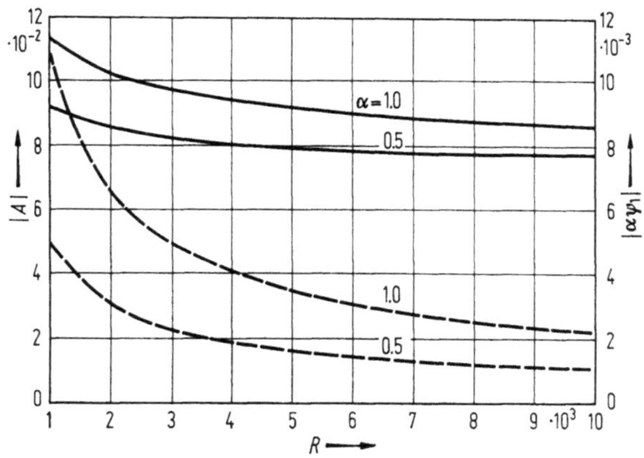

Fig. 1. Amplitudes of the neutral solution as function of R ——; the vorticity amplitude $|A|$ – – –; the maximum vertical velocity $|i\alpha\psi|$.

After some calculation which we shall not discuss here (for nearer details, see Ellingsen, Gjevik and Palm [3]) we find k_2. It turns out that k_2 is negative, i.e. the non-linear terms are acting destabilizing and a threshold amplitude exists. In Fig. 1 the solid lines show the order of the threshold perturbation vorticity. The unit is the vorticity in the basic flow, i.e. U/H. It is noted that the vorticity of the disturbance is about 10% of the vorticity in the basic flow. It is also seen that the necessary disturbance vorticity for large values of R is independent of R. The corresponding horizontal velocity u and vertical velocity w decays as $(\alpha R)^{-1/3}$ and $(\alpha R)^{-2/3}$, respectively. The dotted lines in Fig. 1 show the necessary maximum vertical velocity in the field (occurring near the critical layer) with U as unit. It is seen that for $\alpha = 0.5$ and $\alpha = 1$, the necessary maximum vertical velocity is of the order 10^{-3}.

We therefore obtain that a threshold amplitude curve exists. For amplitudes less than the threshold amplitude, the motion is stable, for larger values the motion is unstable. An important, but difficult question is, however, how good approximation is obtained by truncating the series (5). Important information would have been obtained if we knew the next term $k_3 |A|^6$. From (5) it follows that $|A|^2 = k_1/k_2 \approx 10^{-2}$. If k_3 is of the same order as k_2, obviously $k_3 |A|^6 \ll k_2 |A|^4$. The calculation of k_3 is rather involved and has not been undertaken here. However, one of the typical terms contributing to k_3 is easily computed, namely the contribution due to the change of the mean flow caused by waves of wave number 2α. It is found that for $\alpha = 0.5$ and $(\alpha R) = 10^3$, $k_3 = -118.7$ and for $\alpha R = 10^4$, $k_3 = -130.7$. The ratio $|k_3| |A|^6/|k_2| |A|^4$ is thus about $7.4 \cdot 10^{-2}$ and $8.5 \cdot 10^{-2}$, respectively. As far as this test goes, the approximation involved in (5) is fairly good.

References

1. Reichardt, H.: Z. angew. Math. Mech. **36**, 26 (1956).
2. Stuart, J. T.: J. Fluid Mech. **9**, 353 (1960).
3. Ellingsen, T., Gjevik, B., Palm, E.: J. Fluid Mech. (in print).

Aeroelastic Stability of Plates and Shells: An Innocent's Guide to the Literature

By

E. H. Dowell

Princeton University, Princeton, N.J., U.S.A.

1. Introduction

In the present paper we attempt to provide a guide to the basic literature on the aeroelastic stability of plates and shells or "panel flutter" for the non-expert. In doing so it is hoped that the accomplishments obtained in the field will become available to a wider spectrum of readers and that this will have the twin effects of (i) attracting new workers to the remaining unsolved problems and (ii) indicating how techniques developed for panel flutter may be more widely applied to other physical problems involving the stability or instability of nonlinear, non-conservative systems.

We will attempt to put the problem in both historical and technical perspective. The former because it gives some of the flavor of the way in which such problems are approached and the latter for its direct relevance to the basic objectives of this paper.

For brevity we confine our attention to the literature on the flat plate and do not consider curved plates or shells specifically. In principle, however, the same methods discussed here apply to the latter as well as the former, and hence it is hoped that reader will forgive the all inclusive title of the paper.

2. Physical nature of problem

The basic problem can be stated quite simply. A fluid stream passes over the surface of a thin structure, e.g. plate or shell. See Fig. 1. We

---

Acknowledgement. This work was supported by the National Aeronautics and Space Administration.

seek to determine the stability or instability of the fluid-structural system. In particular we seek to establish the stability or "flutter" boundaries separating stable and unstable motion. If the system is unstable, we further study the nature of the unstable motion, i.e. the nature of the nonlinear limit cycle oscillation. If the system is stable, the structure responds to fluctuations in the fluid stream, generally turbulent boundary pressure fluctuations termed "noise". This stable motion is usually much smaller than the unstable "flutter" oscillations. A schematic of structural response is given in Fig. 2 where we plot plate

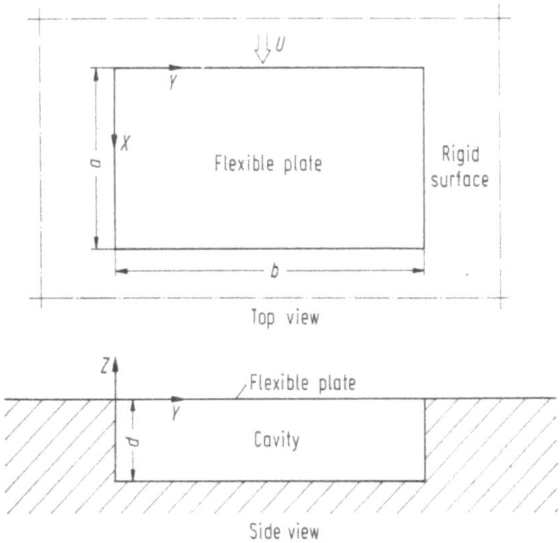

Fig. 1. Plate geometry.

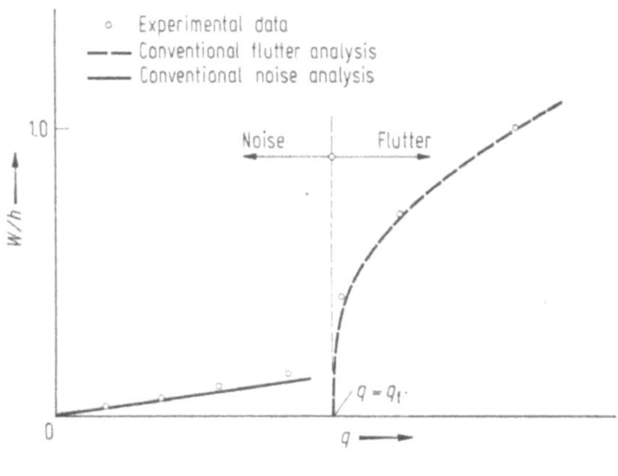

Fig. 2. Schematic of structural response vs. dynamic pressure.

deflection/plate thickness, W/h,[1] vs. the dynamic pressure of the flow, q. Below a certain critical value, $q = q_f$, the structure is stable and responds to "noise". For $q > q_f$, the plate "flutters" with the response increasing much more rapidly with q. Note that for "flutter", $W/h = 0(1)$, indicating that the plate is in the range where geometric structural nonlinearities are important. Physically the nonlinearity results from the tension induced in the plate by its bending deformations. This tension increases with increasing plate deflection and thereby limits the unstable motion to a finite amplitude. A linear model would, of course, predict infinite deflection for all $q > q_f$.

The history of the subject revolves around theoretical and experimental studies made to predict the behavior shown in Fig. 2 for $q \geq q_f$. Also, many "secondary" effects which are important for technological applications in the aerospace field have been studied, such as pressurization, mechanical or thermal in-plane loads, cavity acoustics beneath the structure, structural curvature, etc.

These so-called secondary effects can be quite important in laboratory experimental studies (as well as in practical technological applications) and in the early years often masked the effects of "principal" parameters, such as flow dynamic pressure, Mach number, plate thickness, plate length/width ratio, a/b, and plate boundary conditions.

3. Brief historical sketch

As an introduction to the literature, we present a synopsis of important contributions which retain substantial technical as well as historical interest. More complete references to these contributions are available at the end of the paper. We distinguish between theoretical (Theo.) and experimental (Exp.) papers as well as between linear and nonlinear models or experiments. The essence of each paper is briefly described.

Milestones

1. 1956 Nelson and Cunningham — Theo. — Linear
Two-dimensional plate; full aerodynamics.

2a, b, c, d. 1956 Movchan, 1956 Hedgepeth, 1958 Houbolt — Theo. — Linear
Exact solution for two and three-dimensional plates; piston theory aerodynamics.

3. 1957 Hedgepeth — Theo. — Linear
Galerkin solution for three-dimensional plates; piston theory aerodynamics.

4. 1960 Kordes, Tuovila and Guy — Exp. — Linear
Flutter boundaries for a variety of plate geometries at high Mach number expressed in terms of equivalent isotropic plate. Widely used in design.

[1] For the meaning of the symbols the reader is referred to the nomenclature (at the end of this work).

5. 1961 LOCK and FUNG — Exp. — Linear
Flutter boundaries for plates at low supersonic speeds; attempted correlation with NELSON and CUNNINGHAM [1]. Identified boundary layer as possible source of discrepancy between theory and experiment.
6. 1959—1960 BOLOTIN — Theo. — Nonlinear
Two-mode Galerkin solution for two and three-dimensional plates; piston theory aerodynamics.
7. 1962 KOBAYASHI — Theo. — Nonlinear
Same as BOLOTIN; done independently.
8. 1963 CUNNINGHAM — Theo. — Linear
Three-dimensional plate; full aerodynamics.
9a, b. 1961—1963 GUY, GUY and DIXON — Exp. — Nonlinear
Flutter boundaries for buckled and unbuckled plates at high Mach number; attempted correlation with available linear theory [2, 3].
10. 1963—1965 DOWELL and VOSS — Exp. and Theo. — Linear
Flutter boundaries for a variety of isotropic plates under pressurization; both high and low supersonic Mach numbers. Correlation with available theory [8] satisfactory except at low Mach number for small length/width ratio plates. Substantial improvement in experimental technique.
11. 1965 SHIDELER, DIXON and SHORE — Exp. — Nonlinear
Substantial improvements in experimental technique; same as [9a, b].
12a, b. 1966—1967 DOWELL — Theo. — Nonlinear
Extended BOLOTIN [6], KOBAYASHI [7] to many modes and full aerodynamics.
13. 1968 MUHLSTEIN, GASPERS and RIDDLE — Exp. — Linear
Systematically controlled boundary-layer thickness and demonstrated substantial effect on flutter boundaries at low supersonic Mach number.
14. 1968 DOWELL — Theo. — Linear
Showed that data of MUHLSTEIN et al. [13] when extrapolated to zero boundary-layer thickness, agree well with available theory [8, 10].
15. 1969 VENTRES and DOWELL — Theo. — Nonlinear
Using available theory [12] and extensions thereof showed that effects of pressurization and buckling determined experimentally can be predicted theoretically.

Various authors have provided useful critiques and reviews of the field in considerable more detail than we can or should go into here. We divide these into two categories; the first are somewhat limited in scope but rather detailed critical evaluations; the others essentially cover all of the available literature (up to their date of publication) without going into great detail.

Useful summaries and critiques

16. 1966 DUGUNDJI — Theo. — Linear
Comprehensive treatment ala MOVCHAN, HEDGEPETH and HOUBOLT [2].
17. 1966 DIXON — Theo. and Exp. — Linear
Critique of earlier studies [1—5, 8—10] concerning linear flutter boundaries.
18. 1967 DOWELL — Theo. — Linear and Nonlinear
Summary and critique of state-of-the-art emphasizing nonlinear aspects including [1—3, 6—8, 12].

19. 1969 DOWELL — Exp. and Theo. — Linear and Nonlinear
A complementary study to the above emphasizing experimental technique, results and correlations with theory [1—15].
20. 1969 DOWELL — Theo. and Exp. — Linear and Nonlinear
Discussion of the relationship of "noise" to "flutter" relying primarily on the work of the author and colleagues but also briefly reviewing that of others.

General reviews of the literature and bibliographies

21. 1960 FUNG — A thorough review of the literature to 1960; written for the AGARD Manual on Aeroelasticity.
22. 1961 JOHNS — A review somewhat parallel to that of FUNG [21] to 1961.
23. 1965 JOHNS — An exhaustive bibliography and identification of research programs to 1965.
24. 1969 JOHNS — This will be an up-date of FUNG's article for the AGARD Manual on Aeroelasticity [21].

4. Technical discussion

Theoretical studies

While there has been a voluminous theoretical literature on this problem over the past decade, most analyses can be placed in one of four categories based on the structural and aerodynamic theories employed.

(I) linear structural theory; quasi-steady (inviscid, linear) aerodynamic theory.

(II) linear structural theory; full linearized (inviscid, potential) aerodynamic theory.

(III) nonlinear structural theory; quasi-steady aerodynamic theory.

(IV) nonlinear structural theory; full linearized (inviscid, potential) aerodynamic theory.

Of the above four categories, (I) comprises the great bulk of the literature for the very good reason that it is the simplest. Unfortunately, a type (I) analysis has two major weaknesses:

(a) it does not account for structural nonlinearities; hence it can only determine the flutter boundary and can give no information about the flutter oscillation itself.

(b) the use of quasi-steady aerodynamics neglects the three-dimensionality and unsteadiness (or memory) of the flow; hence it cannot be used too close to a Mach number of one. However, it is in the vicinity of $M = 1$ that flutter is most likely to occur.

A type (II) analysis is intended to remedy (b). It does so in large measure though some have suggested the effect of the boundary layer and transonic flow nonlinearities must also be accounted for near

$M = 1$. This is true under some conditions as will be discussed subsequently. No rational flutter analysis has been made which accounts for the boundary layer or transonic nonlinearities to date. (II) still has weakness (a).

A type (III) analysis remedies (a) but still possesses weakness (b).

A type (IV) analysis remedies both (a) and (b) and represents the most advanced state of the art.

Type (IV) analysis

(IV) contains as special case (I), (II) and (III) by deleting appropriate terms in the governing equations. Therefore for brevity we will be content with a discussion of (IV) and subsequently indicate the simplifications of (I), (II) and (III). The essence of the analysis may be indicated as follows:

(1) Expand the structural deformation in a series of natural, or at least complete, modes (Von Karman plate equations).

(2) Determine the aerodynamic forces for the given modal deformation, e.g., using the techniques of the transform calculus (Acoustic or Wave equation).

(3) Using (1) and (2) in the equations of motion, apply Galerkin's method to arrive at a set of ordinary, nonlinear, integral-differential equations in time for the modal amplitudes.

By omitting the nonlinear terms (due to structural nonlinearities) and the integral terms (which account for the memory and three-dimensionality of the flow) one has a type (I) analysis. By omitting the nonlinear terms only, one has a type (II) analysis and by omitting the integral terms only, one has a type (III) analysis.

Whatever the level of approximation, the set of equations may be solved for the plate motion by a numerical integration with respect of time. If the nonlinear terms are omitted, the oscillating motion will either decay or grow (exponentially) depending on whether the plate is stable or unstable (flutters). If the nonlinear terms are present, the unstable motion will terminate in a finite amplitude or limit cycle oscillation.

Experimental studies

No experimental data have been obtained with the express purpose of investigating the post-flutter (nonlinear) regime. However, in determining the flutter boundary some limited data have been taken in this regime. In Fig. 3 is shown a plot of maximum deflection amplitude (peak to peak) vs. stagnation pressure (dynamic pressure) for a flat plate. Note the plate thickness was 0.025 in. Below the flutter boundary the plate amplitude is small compared to the plate thickness. Above it the plate amplitude is on the order of the plate thickness. This is in qualitative agreement with theory but additional data to larger dynamic pressure are needed for an adequate quantitative comparison.

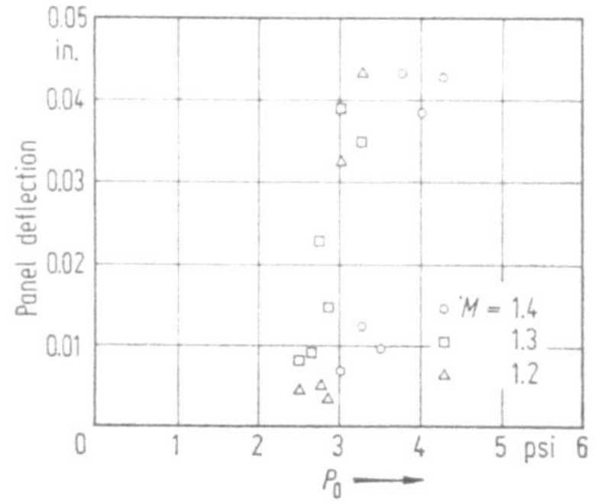

Fig. 3. Plate response.

With data of the type shown in Fig. 3, one may determine (within a certain band of q) the flutter boundary. All of the available quantitative comparisons between theory and experiment are for this boundary. In Fig. 4 are shown such comparisons for plates of three different thicknesses. In dimensional form, we have q_f vs. M. As may be seen, the agreement is quite satisfactory. These data are for $a/b = 0.46$; similar data with similar agreement are presented in [10] for $a/b = 0.73$, 1.37, 2.18, and 4.11. The plates are aluminium and clamped on all edges. The theory is a type (I) analysis.

A smaller amount of data have been obtained in [10] for $M < 2$ which show poorer agreement with theory. Recently MUHLSTEIN, GASPERS and RIDDLE [13] have conducted experiments to determine flutter boundaries of thin, flat plates at low supersonic Mach number, $M = 1.1$ to 1.4. A careful, systematic examination of boundary layer effects on panel flutter has been made. Extrapolating their data to zero boundary-layer thickness, DOWELL [14] has compared their experimental results with theoretical results from [10].

In Fig. 5 the experimental data of [13] and the theory of [10] for a length/width ratio of one-half, $a/b = 0.5$, are presented in terms of (non-dimensional) flutter dynamic pressure λ_f^* vs. Mach number M. In the legend the three numbers designating each plate are length in inches, width in inches and thickness in thousandths of an inch. Experimental data are given for zero boundary-layer thickness (as extrapolated) and the maximum boundary-layer thickness of the tests. Two theoretical curves are given from [10]. One curve is for zero structural damping, and the other for a maximum possible structural damping $g = 0.05$. The plate edges are clamped.

As can be seen, theory and experiment (for zero boundary-layer thickness) are in generally good agreement. Inclusion of structural damping in the theory gives a modest improvement in the agreement.

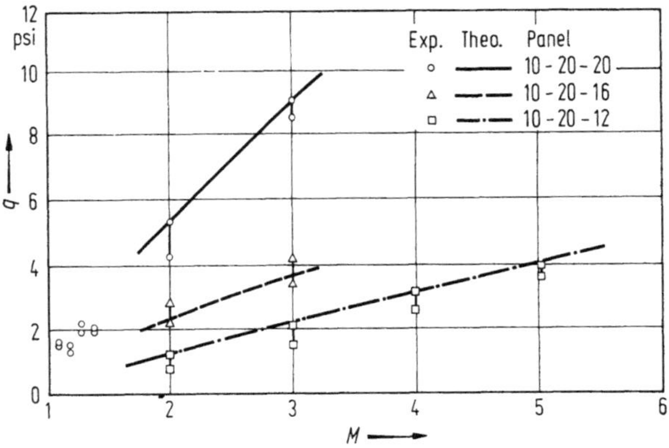

Fig. 4

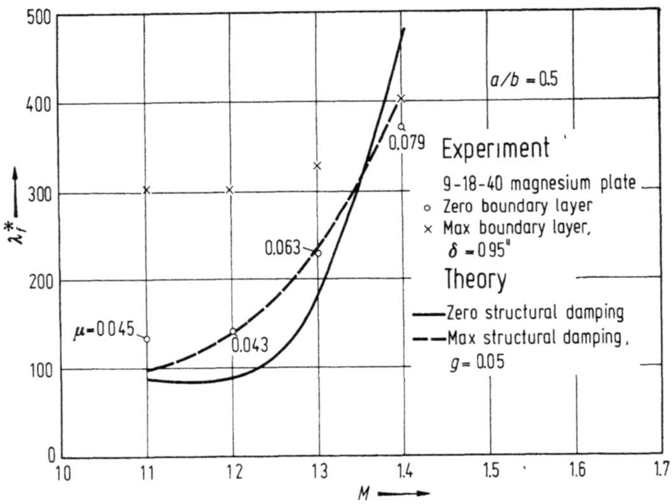

Fig. 5

Figs. 4 and 5. Flutter dynamic pressure vs. Mach number.

Conversely, for nonzero boundary-layer thickness, the experimental data are considerably different from theory particularly at $M \leq 1.3$. Hence, one concludes that much of the difference previously observed between theory and experiment can be attributed to boundary-layer effects.

From the above, one may conclude that the principal needs are for (1) additional experimental data for large a/b and (2) data extending well into the nonlinear regime. Data for large a/b are particularly scarce even for the flutter boundary. Generally speaking, available theoretical tools would appear satisfactory with the exceptions of (1) boundary layer effects and (2) large a/b plates.

Flutter of pressure loaded plates [15]

In this section, the influence of in-plane edge restraint on the flutter behavior of loaded plates will be investigated by determining how a pressure differential alters the stability boundaries of plates with and without in-plane edge restraint. Comparisons will also be made with experimental results for flutter boundaries of pressurized plates reported in [10]. Not all the data for pressurized panels from [10] will be considered. However, the results presented here are typical, and inclusions of the remaining material would not alter the conclusions reached.

The theoretical flutter boundaries calculated for zero and complete edge restraint are shown in Figs. 6 ($a/b = 0.46$) and 7 ($a/b = 2.18$) along with the experimentally obtained flutter boundaries. It should be

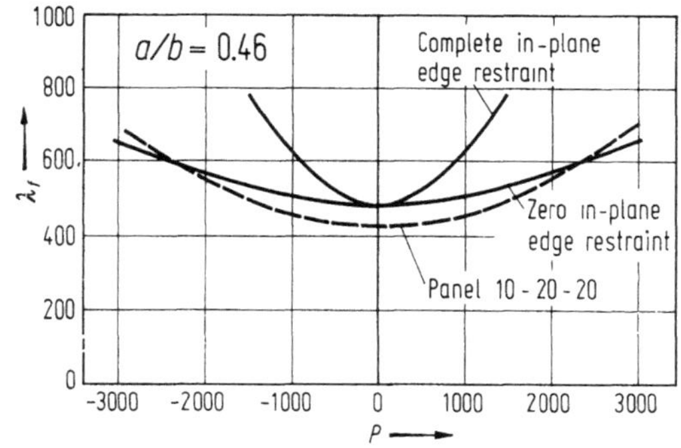

Fig. 6

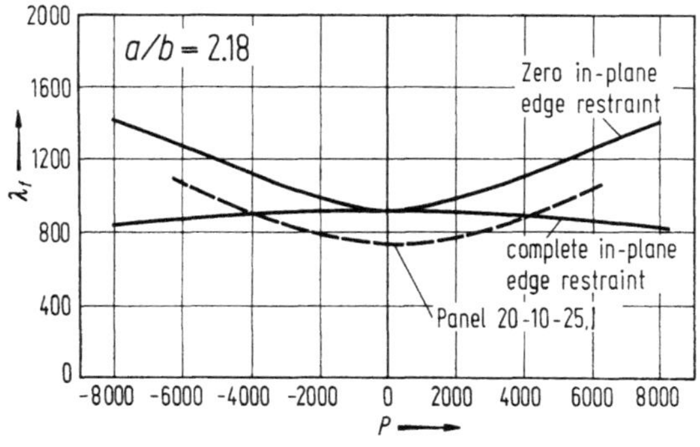

Fig. 7

Figs. 6 and 7. Flutter dynamic pressure vs. static pressure differential.

recognized that the behavior of the plate natural frequencies under a pressure load depends not only on the membrane stresses caused by the load, but on the induced middle surface curvature as well. For restrained plates, the effect of the tension is predominant; but for unrestrained plates, the influence of the curvature must be considered also. It should be recognized that the behavior of the plate natural frequencies under a pressure load depends not only on the membrane stresses caused by the load, but on the induced middle surface curvature as well. For restrained plates, the effect of the tension is predominant; but for unrestrained plates, the influence of the curvature must be considered also.

The best agreement between theory and experiment in Figs. 6 and 7 is obtained by assuming zero edge restraint.

Buckled plates [15]

In-plane compressive loads can cause a plate to buckle, thereby altering its behavior in a uniform airflow. The buckled plate has a curved middle surface and is subjected to mid-plane stresses that depend in a complicated manner on the nature of the in-plane loading, the in-plane boundary conditions, and the external flow. The flutter motion is no longer simple harmonic and may not be perfectly periodic. Moreover, the presence of more than one buckled equilibrium configuration lends some credence to the existence of sustained flutter oscillations below the linear flutter boundary.

Most experimental work has been done on plates buckled by uniform thermal expansion. [11] is the latest in a series of experiments by NASA Langley.

Here comparisons will be made between the type (III) theory and experimental results from [11] for plates with $a/b = 2.9$, buckled by thermal expansion. The testing was done at a Mach number of 3, and at constant dynamic pressure. The plates were heated by the airflow so that the in-plane loading increased with time, exceeding the buckling value before termination of the run. With but few exceptions, flutter occurred prior to buckling, and then ceased at some point subsequent to buckling.

In this sort of test, the in-plane boundary conditions affect not only the large-deflection behavior of the plate, but the nature of the in-plane loading as well. If the edges are completely restrained against in-plane motion, the in-plane stresses $N_x^{(a)}$ and $N_y^{(a)}$ are equal, and

$$\frac{N_x^{(a)} a^2}{D} = \frac{N_y^{(a)} a^2}{D} = 12(1+\nu)\,(a/h)^2\,\alpha\,\Delta T,$$

where ΔT is the uniform temperature differential between the plate and its support. The parameter

$$R_T \equiv 12(1+\nu)\,(a/h)^2\,\alpha\,\Delta T$$

will be used here to describe the thermal loading on the plate, even though it may not be fully restrained.

The assumption of complete in-plane edge restraint produces flutter boundaries in very poor quantitative agreement with experiment. It was thus decided to select values of edge restraint so that buckling would occur for the same R_T experimentally and theoretically. This is equivalent to assuming that the airflow does not affect the static Euler buckling load.

In Fig. 8 two flutter boundaries are shown in the buckled regime: The "linear" or "small disturbance" flutter boundary was computed using a buckled equilibrium configuration as an initial condition for the

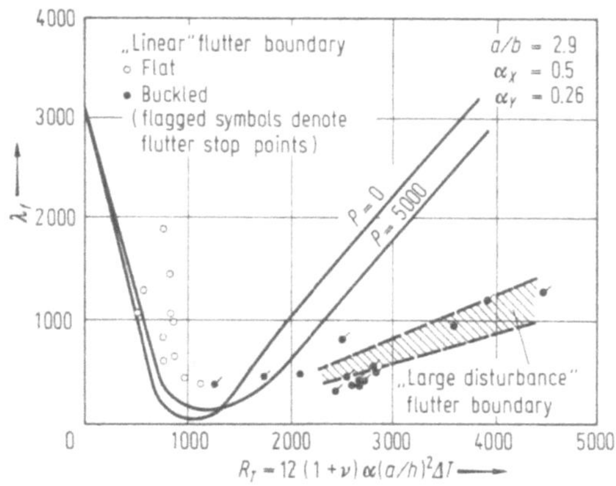

Fig. 8. Flutter dynamic pressure vs. differential temperature.

numerical integration of the modal equations. On the other hand, calculations proceeding from a flat plate initial condition produced sustained flutter-like oscillations below this "linear" flutter boundary. Since the buckled shapes involved deflections on the order of several plate thicknesses, the latter is physically equivalent to applying a large disturbance to the plate. The numerical results obtained suggest that sustained flutter oscillations are possible for λ, R_T combinations above the shaded area in Fig. 8. Note that this "large disturbance" flutter boundary is in excellent agreement with the experimental data, since most of the data in that region are "flutter stop" points. The existence of two "flutter start" points in the same region is not explained by the "large disturbance" flutter boundary, however.

The flutter motion computed from theory agrees qualitatively with descriptions of the behavior observed experimentally. The plate motion is not simple harmonic, nor strictly periodic, but the peak deflections attained are repetitive. The motion is of the traveling wave variety, with two half wavelengths along the length of the plate. The buckled equilibrium shapes computed below the "large disturbance" flutter boundary consisted of two half wavelengths as well. The frequency of flutter (not shown) is in generally good accord with the experimental data for all values of R_T.

Nomenclature

a plate length
b plate width
D plate stiffness
E modulus of elasticity
h plate thickness
$K = \omega \left[\frac{\varrho_m h a^4}{D}\right]^{1/2}$
M Mach number
$Nx^{(a)}$, $Ny^{(a)}$ applied in-plane loads
Δp static pressure differential
$P = \Delta p a^4 / Dh$
$q = \varrho U^2/2$ dynamic pressure
U air velocity
W plate deflection
δ boundary-layer thickness
$\lambda^* = 2qa^3/D$
$\lambda = \lambda^*/M$ or $\lambda^*/(M^2 - 1)^{1/2}$
$\mu = \varrho a/\varrho_m h$
ϱ air density
ϱ_m plate density
ω frequency

Subscripts:

d dynamic
f flutter
p peak

References

1. NELSON, H. C., CUNNINGHAM, H. J.: Theoretical investigation of flutter of two-dimensional flat panels with one surface exposed to supersonic potential flow. NACA Report 1280, 1956.
2a. MOVCHAN, A. A.: On vibrations of a plate moving in a gas. NASA Re 11-22-58 W, 1959.
2b. MOVCHAN, A. A.: On the stability of a panel moving in a gas. NASA Re 11-21-58 W, 1959.
2c. HEDGEPETH, J. M.: On the flutter of panels at high Mach numbers. J. Aero. Sci. **23**, 609 (1956).
2d. HOUBOLT, J. C.: A study of several aerothermoelastic problems of aircraft structures in high speed flight. Doctoral Thesis, ETH Zurich, Mitt. 5, 1958.
3. HEDGEPETH, J. M.: Flutter of rectangular simply supported panels at high supersonic speeds. J. Aero. Sci. **24**, 563—573 (1957).
4. KORDES, E. E., TUOVILA, W. J., GUY, L. D.: Flutter research on skin panels. NASA TN D-451, 1960.
5. LOCK, M. H., FUNG, Y. C.: Comparitive experimental and theoretical studies of the flutter of flat panels in a low supersonic flow. AFOSR TN 670, Guggenheim Aero. Lab., Calif. Inst. Tech., May 1961.
6. BOLOTIN, V. V.: Nonconservative problems of the theory of elastic stability, New York: Macmillan 1963, pp. 274—312.
7. KOBAYASHI, S.: Flutter of simply supported rectangular panels in a supersonic flow — two-dimensional panel flutter. I — Simply supported panel, II — Clamped panel. Trans. Japan Soc. Aero. Space Sci. **5**, 79—118 (1962).

8. Cunningham, H. J.: Flutter analysis of flat rectangular panels based on three-dimensional supersonic unsteady potential flow. NASA TR R-256, Langley Res. Ctr., 1967.
9a. Guy, L. D.: The effects of aerodynamic heating on panel flutter. Symposium Proceedings on Structural Dynamics of High Speed Flight, ACR-62, Vol. 2, Aircraft Industries Association and Office of Naval Research, April 1961.
9b. Guy, L. D., Dixon, S. C.: A critical review of experiment and theory for flutter of aerodynamically heated panels. In: Symposium on Dynamics of Manned Lifting Planetary Entry, New York: John Wiley 1963, pp. 568–595.
10. Dowell, E. H., Voss, H. M.: Experimental and theoretical panel flutter studies in the Mach number range 1.0 to 5.0. ASD-TDR-63-449, USAF, Dec. 1963. Also see AIAA J. **3**, 2292–2304 (1965).
11. Shideler, J. L., Dixon, S. C., Shore, C. P.: Flutter at Mach 3 of thermally stressed panels and comparison with theory for panels with edge rotational restraint. NASA TN D-3498, August 1966.
12a. Dowell, E. H.: Nonlinear oscillations of a fluttering plate I. AIAA J. **4**, 1267–1275 (1966).
12b. Dowell, E. H.: Nonlinear oscillations of a fluttering plate II. AIAA J. **5**, 1856–1862 (1967).
13. Muhlstein, L., Jr., Gaspers, P. A., Riddle, D. W.: An experimental study of the influence of the turbulent boundary layer on panel flutter. NASA TN D-4486, Ames Res. Ctr., 1968.
14. Dowell, E. H.: Theoretical-experimental correlation of plate flutter boundaries at low supersonic speeds. AIAA J. **6**, 1810–1811 (1968).
15. Ventres, C. S., Dowell, E. H.: Influence of in-plane edge support flexibility on the nonlinear flutter of loaded plates. Presented at AIAA Structural Dynamics and Aeroelasticity Specialist Conference, New Orleans, La., April 1969.
16. Dugundji, J.: Theoretical considerations of panel flutter at high supersonic Mach numbers. AIAA J. **4**, 1257–1266 (1966).
17. Dixon, S. C.: Comparison of panel flutter results from approximate aerodynamic theory with results from exact inviscid theory and experiment. NASA TN D-3649, Langley Res. Ctr., 1966.
18. Dowell, E. H.: Nonlinear theory of the flutter of plates and shells. Fluid-Solid Interaction Symposium, ASME Winter Annual Meeting, Pittsburgh, Pa., November 1967.
19. Dowell, E. H.: Panel flutter – a review of the aeroelastic stability of plates and shells. AIAA J. (to be published).
20. Dowell, E. H.: Noise or flutter or both? Presented at the Symposium on Structural Response to Sonic Excitation, British Acoustical Society, Liverpool, England, January 1969.
21. Fung, Y. C.: A summary of the theories and experiments on panel flutter. AFOSR TN 60-224, Guggenheim Aero. Lab., Calif. Inst. Tech., May 1960.
22. Johns, D. J.: The present status of panel flutter. AGARD Advisory Report 484, 1961.
23. Johns, D. J.: A survey on panel flutter. AGARD Advisory Report 1, November 1965.
24. Johns, D. J.: A survey of panel flutter. AGARD Manual on Aeroelasticity (to appear).

Stability of Structures under Stochastic Disturbances

By

S. T. Ariaratnam

University of Waterloo, Waterloo, Ont., Canada

Introduction

This paper examines certain problems in the stability of elastic structures when the external loads fluctuate in a stochastic manner. The study is important for two reasons. First, there exist many examples of loads such as those generated by jet and rocket engines, earthquakes, wind gusts, etc., that cannot be adequately described by deterministic time functions alone, and the stability of structures subject to such loads have to be analysed by statistical methods. Second, even when the loading is principally deterministic, it may be more realistic to investigate the stability of the structure by giving a random perturbation to the loading in addition to the perturbations to the initial state adopted in the conventional approach.

Two classes of problems are studied, namely parametric instability and dynamic snapping. The methods of analysis are indicated only briefly; the details may be found in the references quoted.

Parametric instability

This form of instability commonly occurs in structures that exhibit a bifurcation of equilibrium under static loads, e.g. a column under end thrust. For a structure that is described by a set of n generalised coordinates q_i, the linearised equations of motion may be taken in the

---

Acknowledgement. The author wishes to thank Mr. H. N. Pi for performing the calculations leading to Fig. 4. The research for this paper was supported by the National Research Council of Canada through Grant No. A-1815.

form

$$\ddot{q}_i + \omega_i^2 q_i + \varepsilon \sum_{j=1}^{n} \beta_{ij} \dot{q}_j + f(t) \sum_{j=1}^{n} k_{ij} q_j = 0, \; i = 1, 2, \ldots, n. \tag{1}$$

Here $f(t)$ represents the generalised external load, the coefficients β_{ij}, k_{ij} are constants, and the ω_i are the natural frequencies of the undamped, unloaded structure; $\varepsilon \ll 1$. It may be noted that the load $f(t)$ appears in the equations in the form of a coefficient or parameter. The trivial solution $q_i = 0$ is said to be *stable in the mean* if the mean square response $\langle q_i^2(t) \rangle$ tends to zero as t tends to infinity, $i = 1, 2, \ldots, n$; the angular brackets $\langle \; \rangle$ signify the expectation or ensemble average.

To investigate the stability, Eqs. (1) are first converted to standard form by means of the transformation

$$q_i = a_i \cos \Phi_i, \; \dot{q}_i = -a_i \omega_i \sin \Phi_i, \; \Phi_i = \omega_i t + \varphi_i, \; i = 1, 2, \ldots, n.$$

This procedure gives the set of $2n$ first order equations:

$$\dot{a}_i \omega_i = -\varepsilon \sum_{j=1}^{n} \beta_{ij} a_j \omega_j \sin \Phi_i \sin \Phi_j + f(t) \sum_{j=1}^{n} k_{ij} a_j \sin \Phi_i \cos \Phi_j, \tag{2a}$$

$$a_i \omega_i \dot{\varphi}_i = -\varepsilon \sum_{j=1}^{n} \beta_{ij} a_j \omega_j \cos \Phi_i \sin \Phi_j + f(t) \sum_{j=1}^{n} k_{ij} a_j \cos \Phi_i \cos \Phi_j. \tag{2b}$$

It is now assumed that $f(t)$ is a stationary random process with zero mean and spectral density $\varepsilon S(\omega)$ defined by

$$\varepsilon S(\omega) = \int_{-\infty}^{\infty} \langle f(t) f(t+\tau) \rangle \cos \omega\tau \, d\tau.$$

For small ε, asymptotic solutions of Eqs. (2) may be obtained by the method of averaging. The averaging is done in two stages. First the terms not containing the random excitation $f(t)$ are averaged in accordance with the rule

$$\{F(u, \varphi, t)\}_{\text{av}} = \lim_{T \to \infty} \frac{1}{T} \int_0^T F(u, \varphi, t) \, dt,$$

where the integration is carried out over explicit t. Next the terms containing $f(t)$ are replaced by the sum of the mean value plus the fluctuation about the mean. This is carried out according to the procedure indicated by Stratonovich [1]. After these steps have been carried out, Eq. (2a) becomes

$$\dot{u}_i = -\varepsilon \beta_{ii} u_i + \sum_{j=1}^{n} m_{ij}(u) + \sum_{j=1}^{n} f_{ij}(t), \tag{3}$$

where

$$u_i = a_i^2,$$

$$m_{ij} = \frac{k_{ij}}{\omega_i} \langle a_i a_j f(t) \sin \Phi_i \cos \Phi_j \rangle = \frac{\varepsilon k_{ij}^2 u_j}{4\omega_i^2} S_{ij}^+ + \frac{\varepsilon k_{ij} k_{ji} u_i}{4\omega_i \omega_j} S_{ij}^-,$$

$$f_{ij} = \frac{k_{ij}}{\omega_i} a_i a_j f(t) \sin \Phi_i \cos \Phi_j - m_{ij},$$

$$S_{ij}^+ = S(\omega_i + \omega_j) + S(\omega_i - \omega_j),$$

$$S_{ij}^- = S(\omega_i + \omega_j) - S(\omega_i - \omega_j).$$

The relaxation time of the $u_i(t)$ processes is $0(1/\varepsilon)$. Hence, if the correlation time of the $f(t)$ process is much smaller than this relaxation time, the processes $f_{ij}(t)$ may be replaced by equivalent delta-correlated processes. The response vector $u_i(t)$, $i = 1, 2, \ldots, n$ now approximates a Markov process and the joint probability density $p(u_1, u_2, \ldots, u_n)$ is given by the Fokker-Planck equation

$$\begin{aligned} \frac{\partial p}{\partial t} = &- \sum_{i=1}^{n} \frac{\partial}{\partial u_i} \left[\left\{ -\varepsilon \beta_{ii} u_i + \sum_{j=1}^{n} m_{ij} \right\} p \right] \\ &+ \frac{1}{2} \sum_{i=1}^{n} \sum_{j=1}^{n} \frac{\partial^2}{\partial u_i \, \partial u_i} \left[\left\{ \sum_{k=1}^{n} \sum_{l=1}^{n} \int_{-\infty}^{\infty} \langle f_{ik}(t) \, f_{jl}(t + \tau) \rangle \, d\tau \right\} p \right]. \end{aligned} \tag{4}$$

The differential equations governing the mean square amplitudes $\langle u_i \rangle$ are obtained by multiplying Eq. (4) in turn by each u_i and integrating over all the u_i from 0 to ∞. This procedure gives

$$\begin{aligned} \frac{dM_i}{dt} = &\left[-\varepsilon \beta_{ii} + \frac{\varepsilon k_{ii}}{2\omega_i^2} S(2\omega_i) + \varepsilon \sum_{\substack{j=1 \\ j \neq i}}^{n} \frac{k_{ij} k_{ji}}{4\omega_i \omega_j} S_{ij}^- \right] M_i t \\ &+ \varepsilon \sum_{\substack{j=1 \\ j \neq i}}^{n} \frac{k_{ij}^2}{4\omega_i^2} S_{ij}^+ M_j, \quad i = 1, 2, \ldots, n, \end{aligned} \tag{5}$$

where

$$M_i = \langle u_i \rangle = \langle a_i^2 \rangle.$$

Eq. (5) can be written in the form

$$\frac{d}{dt} M = KM,$$

where M is the column vector with elements $M_i(t)$. The mean square amplitudes $M_i(t)$ remain bounded as t tends to ∞, if all the eigenvalues of the matrix K have negative real parts. The conditions for this may be found by applying the Routh-Hurwitz criterion to the characteristic equation

$$\det(K - \lambda I) = 0.$$

Since $q_i(t) = a_i \cos \Phi_i$, $\langle q_i^2(t) \rangle \leq \langle a_i^2 \rangle$; hence the conditions are sufficient to ensure mean square stability of the responses q_i.

An examination of the terms in the matrix K reveals that only those values of the excitation spectrum at the frequencies $\omega_i \pm \omega_j$ will appear in the stability conditions. Also, to the present first approximation, the coupling of the damping terms has no effect on the stability. In the case of a single degree of freedom system, the stability condition reduces to

$$\beta_{11} > \frac{k_{11}^2}{2\omega_1^2} S(2\omega_1), \tag{6}$$

a result obtained independently by STRATONOVICH and ROMANOVSKI [2] and by GRAEFE [3].

Dynamic snap-through

Snap-through instability is commonly found in the behaviour of thin, shallow, shell-type structures. If the structure may be approximated by a single degree of freedom system, the motion is described by a differential equation of the form

$$\ddot{q} + \beta \dot{q} + \frac{\partial V}{\partial q} = f(t),$$

where $V(q, \lambda)$ represents the potential energy of the system under a static mean load λ, and $f(t)$ denotes the stochastic load fluctuation.

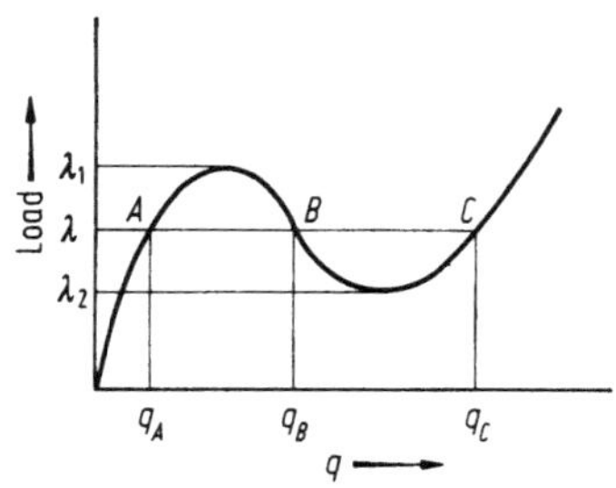

Fig. 1. Static load-deflection curve for structure exhibiting snap-through.

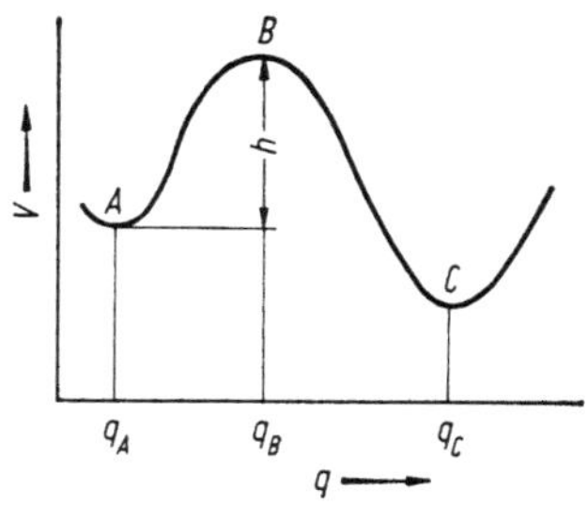

Fig. 2. Potential energy curve.

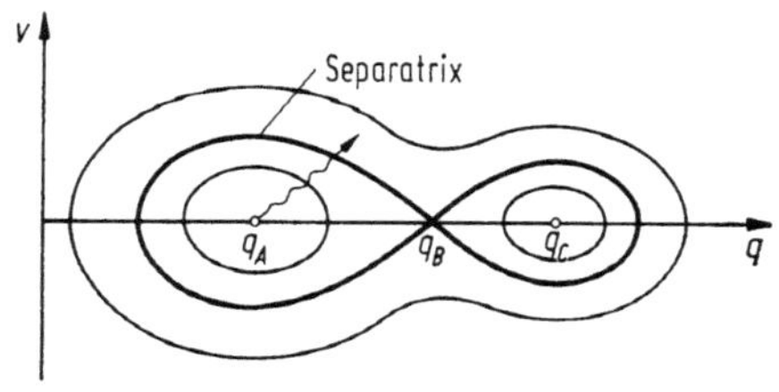

Fig. 3. Trajectories in phase plane.

The static load-deflection behaviour is usually of the form shown in Fig. 1 and the potential energy function $V(q, \lambda)$ for a fixed value of λ is as shown in Fig. 2. For a load λ lying between λ_1 and λ_2, there exist three equilibrium configurations q_A, q_B, and q_C. The configurations q_A and q_C represent stable equilibrium while the configuration q_B is unstable. The motion of the system may be represented by its trajectories in the phase-plane. When both the damping and the stochastic load $f(t)$ are absent, the trajectories corresponding to the free motion are of the form shown in Fig. 3. The heavy curve denotes the separatrix. The trajectories that remain within the left loop of the separatrix correspond to oscillations about the initial equilibrium state q_A; those that lie outside the loop would take the structure towards the equilibrium state q_C and hence to a state of snap-through. It is reasonable therefore to regard the left loop of the separatrix as the boundary between the unsnapped and the snapped states even when both damping and random loads are present.

Let T denote the time it would take a trajectory starting at a point (q, v) within the left loop to reach the boundary for the first time. Then T is a random variable and a problem of interest is the determination of the probability distribution of T. This falls within the class of the so-called "first passage" problems in the theory of probability. If the random excitation is a stationary broad-band process which may be approximated by a delta-correlated process, it is known that the probability distribution of T, $P(t; q, v)$, is governed by the backward Kolmogorov equation with appropriate initial and boundary conditions (see for example Stratonovich [1]). This equation is difficult to solve for the present non-linear problem. However, the expectation or mean value of T is given by another differential equation known as the Pontriagin equation (see Andronov, Pontriagin and Vitt [4]) which can be solved approximately.

Bolotin [5] studied the problem for a long cylindrical shell panel for which the equation of motion is of the form

$$\ddot{q} + 2\zeta\omega\dot{q} + \omega^2[q + \alpha q^2 + \beta q^3 - \lambda] = f(t),$$

where $\alpha < 0$, $\beta > 0$, and employed a Galerkin technique to solve the associated Pontriagin equation

$$\frac{1}{2} S_0 \frac{\partial^2 M}{\partial v^2} + v \frac{\partial M}{\partial q} - [2\zeta\omega v + \omega^2(q + \alpha q^2 + \beta q^3 - \lambda)] \frac{\partial M}{\partial v} = -1 \qquad (7)$$

with boundary condition

$$M(q, v) = 0 \quad \text{on } \Gamma.$$

Here M denotes the mean first passage time of snap-through and Γ the left loop of the separatrix; S_0 is the intensity coefficient of $f(t)$ given by

$$S_0 = \int_{-\infty}^{\infty} \langle f(t)\, f(t+\tau)\rangle\, d\tau .$$

Eq. (7) may also be solved numerically by finite-difference methods. The results of such a computation for a particular example are shown in Fig. 4, along with those of BOLOTIN, in the form of contours of equal

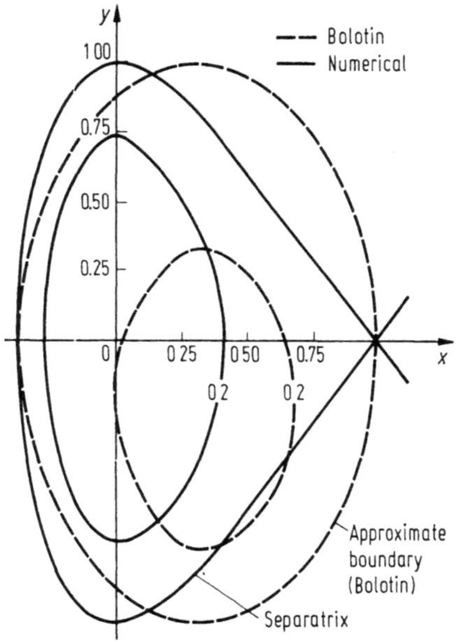

Fig. 4. Contours of equal mean time for snap-through ($M = 0.2$ cycles) (x non-dimensional displacement, y non-dimensional velocity).

mean time for different initial conditions. It appears that BOLOTIN's approach can lead to large errors. In fact, BOLOTIN's solution lies outside the true boundary for the values shown.

Starting from the Kolmogorov equation, it is also possible to derive equations governing the various moments of the distribution of T. In particular, numerical solutions can be obtained for the variance of the first passage time. These results have been calculated for the problem of the cylindrical panel and will be presented in a forthcoming publication.

Conclusions

In this paper, certain typical problems in the stability of structures under stochastic loads have been studied. Due to limitation of space, the methods of analysis have been described only briefly.

In the problem of parametric instability, only linearised equations of motion and broad-band excitation have been considered. It is necessary to extend the investigation to include the case of narrow-band excitation and the effect of non-linear coupling terms in the equations of motion. The problem of dynamic snap-through has been studied using a single degree of freedom model for the structure. The analysis is therefore valid only for very shallow structures whose deformation may be approximated by a single symmetric mode. It must be extended to cover the case of snap-buckling in which the deformation consists principally of more than one mode.

References

1. Stratonovich, R. L.: Topics in the Theory of Random Noise, Vol. I, New York: Gordon and Breach 1963, p. 95 and p. 79—83.
2. Stratonovich, R. L., Romanovski, Y. M.: Non-Linear Transformations of Stochastic Processes, ed. P. I. Kuznetsov, R. L. Stratonovich and V. I. Tikhonov, Oxford: Pergamon 1965, p. 332.
3. Graefe, P. W. U.: Ing.-Arch. **35**, 202 (1966).
4. Andronov, A. A., Pontriagin, L. S., Vitt, A. A.: J. Exp. and Theor. Phys. (in Russian) **3** (1933).
5. Bolotin, V. V.: Proc. Int. Conf. on Dynamic Stability of Structures, ed. G. Herrmann, Oxford/New York: Pergamon 1966, p. 67.

A Theory of Elastic Stability for Perfectly Elastic Materials with Couple-stresses

By

M. F. Beatty

University of Kentucky, Lexington, Ky., U.S.A.

A material that exhibits couple effects is called a *polar* or *Cosserat material*. Although it is possible to consider more general kinds of continua, this note concerns only the class of polar materials for which the angular velocity of a material parcel X is constrained so as to coincide always with the local rigid body angular velocity of the continuum at X. Such a material is called a *constrained, Cosserat continuum*. The objective here is to outline my recent studies of a theory of elastic stability for hyperelastic materials of this type.

In the full text of this work, an exact stability functional for overall conservative systems of forces and couples is derived from the usual energy criterion of stability. When specialized for dead loading, this criterion takes the exact form

$$D_E \equiv \int_{\kappa} \{\operatorname{tr}(\boldsymbol{t}\boldsymbol{D}) + \operatorname{tr}(\boldsymbol{m}_D \boldsymbol{K}^T) - \varrho\, \Delta\varepsilon\}\, dv + \int_{\kappa} \bar{\boldsymbol{\theta}} \cdot \boldsymbol{m}\, da + \int_{\kappa} \bar{\boldsymbol{\theta}} \cdot \boldsymbol{l}\, dm \leq 0 \tag{1}$$

for *all* kinematically admissible virtual deformations $\boldsymbol{u}(X;\tau)$ from the equilibrium configuration $\varkappa$ wherein $\boldsymbol{u}(X;0) = \boldsymbol{0}$. Here $\boldsymbol{t}$ is the Cauchy stress tensor; $\boldsymbol{m}_D$ the deviatoric part of the couple-stress tensor $\boldsymbol{m}$; $\boldsymbol{l}$ the body couple per unit mass m with density ϱ; $\Delta = \hat{\varepsilon}(X, \boldsymbol{F}^*, \nabla\boldsymbol{F}^*) - \hat{\varepsilon}(X, \boldsymbol{F}, \nabla\boldsymbol{F})$ the increase in the stored energy per unit mass in response to the deformation from $\varkappa \to \varkappa^*$ where $\boldsymbol{F}^*(X;\tau) = \nabla\boldsymbol{x}^*(X;\tau)$ is the deformation tensor and $\nabla\boldsymbol{F}^*(X;\tau)$ its material gradient relative to an arbitrary reference configuration $\varkappa_0$. We have $\boldsymbol{F}(X) = \boldsymbol{F}^*(X;0)$

Acknowledgement. The research reported here was supported by a grant from the U.S. National Science Foundation.

and $\nabla F(X) = \nabla F^*(X; 0)$ in $\varkappa$ with $\nabla = \partial/\partial X$, X denoting the place of the material point X in $\varkappa_0$. Its place in $\varkappa$ and $\varkappa^*$ being $x(X)$ and $x^*(X;\tau)$, respectively. Also, $D = H_S$ is the symmetric part of $H = \bar{\nabla} u$, $K = \bar{\nabla}\theta$ defines the traceless curvature-twist tensor, $\theta(X;\tau)$ is the *mean rotation vector*:

$$\theta(X;\tau) \equiv \frac{1}{2}\bar{\nabla}\times u(X;\tau) = \theta^*(X;\tau) - \bar{\theta}(X;\tau) \tag{2}$$

and, by definition, the axial vector

$$\bar{\theta}(X;\tau) \equiv -\int_0^\tau (\dot{H}H^*)_X \, d\tau, \qquad H^* = \bar{\nabla} u; \tag{3}$$

$$\theta^*(X;\tau) = \int_0^\tau \frac{1}{2}\tilde{\nabla}\times \dot{u}\, d\tau \tag{4}$$

being the *virtual angular displacement vector* of the particle X at τ. Here $\tilde{\nabla} \equiv \partial/\partial x^*$, $\bar{\nabla} \equiv \partial/\partial x$ and a superimposed dot means $\partial/\partial\tau$.

This structure includes as a special case the theory of elastic stability for simple, hyperelastic continua developed by Pearson in 1955. Moreover, for infinitesimal deformations from a natural state $\varkappa_0$ of vanishing stress, couple-stress and deformation, (1) reduces to the familiar requirement that the total strain energy of the linearized theory of elastic materials with couple-stress effects be non-negative. Nevertheless, the presence of the last two terms in (1) seems unusual. Apparently, these reflect the strong influence of nonlinear rotational contributions to the work done by all dead load couples. Moreover, it is also evident from (3) that these effects in general do not disappear in any linearization that retains terms of second order in products of H and $\bar{\nabla} H$; therefore, in general, these "extra-rotational" contributions to the work done by all dead load couples will occur in both the exact and the linearized dead load stability functionals.

It is important to observe that the inequality (1) is required to hold for *all* kinematically admissible virtual deformations $u(X;\tau)$. We may exclude at once and without loss all constant rigid body translations; we see easily that (1) vanishes for these. If $D_E = 0$ for some $\dot{u} \neq$ const., but is non-positive for all u in a suitable class C, the equilibrium configuration is called *neutrally stable with respect to C*.

Now in any traction problem the competing deformations in the criterion of stability may be considered arbitrary. Therefore, in particular, we may choose a virtual displacement which corresponds to an arbitrary rigid rotation about a fixed axis. For this choice, it is easy to construct examples to show that frequently (1) can be rendered either positive, negative or zero as we may wish. Consequently, in every traction problem the body must in general be ruled inherently unstable if rigid rotations are admitted as contenders for the criterion of stability. Rather than cast out all rigid motions, which is not necessary, I attempt to characterize those rotations with respect to which the equilibrium configuration is not unstable by examining the consequences of restricting the class of *all* virtual motions so as to satisfy two constraints, called the zero moment condition and the zero work condition.

If all virtual motions are restricted so as to preserve equilibrium of moments in $\boldsymbol{\varkappa}^*$, then $\boldsymbol{u}$ must satisfy the following *zero moment condition*

$$\boldsymbol{\nu}^*(\tau) \equiv \boldsymbol{\nu} + \int_{\varkappa} \boldsymbol{u} \times \boldsymbol{t}\, da + \int_{\varkappa} \boldsymbol{u} \times \boldsymbol{b}\, dm = \boldsymbol{0}. \tag{5}$$

Here $\boldsymbol{\nu}^*(0) = \boldsymbol{\nu} = \boldsymbol{0}$ in $\boldsymbol{\varkappa}$ and $\boldsymbol{b}$ is the body force. Several theorems relating this requirement to rigid deformations and the stability criterion have been found; the main result is this:

Let $\boldsymbol{A} = \int_{\varkappa} \boldsymbol{x} \otimes \boldsymbol{t}\, da + \int_{\varkappa} \boldsymbol{x} \otimes \boldsymbol{b}\, dm$. *Then, if there exists no axis of equilibrium orthogonal to the resultant* $\boldsymbol{\gamma}$ *of dead load couples, the zero moment condition is satisfied by a rotation through a straight angle about every principal direction of* $\boldsymbol{A}$ *which is orthogonal to* $\boldsymbol{\gamma}$, *but no others; if there is an axis of equilibrium orthogonal to* $\boldsymbol{\gamma}$, *every rotation about that line and a rotation through a straight angle about every other real principal direction of* $\boldsymbol{A}$ *which is perpendicular to* $\boldsymbol{\gamma}$ *satisfies the zero moment condition. Moreover, the equilibrium configuration is always neutrally stable with respect to rotations of arbitrary magnitude about an axis of equilibrium orthogonal to* $\boldsymbol{\gamma}$.

Although the zero moment condition alone characterizes almost entirely the class of rigid rotations with respect to which the equilibrium is neutral, there remain some exceptional cases in which it is possible that the zero moment condition may be satisfied by a certain rotation relative to which the equilibrium state is still unstable. This seemingly unhappy situation could be eliminated if we had available some criterion that admits only those rigid rotations about an axis normal to $\boldsymbol{\gamma}$; then the only exceptional cases would be those corresponding to certain rotations through a straight angle as already noted.

The significance of the aforementioned orthogonality requirement may be made clear by an illustration. Suppose that the body is in equilibrium subject to any systems of dead load forces and couples. It is plain that a rotation of magnitude $\theta = 2\pi$ about any direction $\boldsymbol{h}$ returns the body to its initial configuration and the zero moment condition is trivially satisfied; yet (1) yields $D_E = 2\pi \boldsymbol{h} \cdot \boldsymbol{\gamma}$. On the other hand, a rotation of the same magnitude but about $-\boldsymbol{h}$ has the same effect, but (1) now yields $D_E = -2\pi \boldsymbol{h} \cdot \boldsymbol{\gamma}$. Thus, since in general

$\boldsymbol{\gamma} \neq \mathbf{0}$, the equilibrium configuration must be considered inherently unstable unless $\boldsymbol{h}$ is restricted to satisfy the orthogonality requirement $\boldsymbol{h} \cdot \boldsymbol{\gamma} = 0$. Rather than simply impose this condition, I prefer to study the effect of restricting further the class of virtual deformations striving for the criterion of stability by demanding that the class of *all* virtual motions $\boldsymbol{u}$ be such that the work done by all dead load couples in a rotation $\boldsymbol{\theta}^*$ be equal to the work done by the same couples in the corresponding mean rotation $\boldsymbol{\theta}$. Then $\boldsymbol{u}$ must satisfy the following *zero work condition*:

$$\int_{\kappa} \bar{\boldsymbol{\theta}} \cdot d\hat{\boldsymbol{\gamma}} = \int_{\kappa} \boldsymbol{\theta}^* \cdot d\hat{\boldsymbol{\gamma}} - \int_{\kappa} \boldsymbol{\theta} \cdot d\hat{\boldsymbol{\gamma}} = 0, \tag{6}$$

with $d\hat{\boldsymbol{\gamma}} = \boldsymbol{m}\, d\boldsymbol{a} + \boldsymbol{l}\, dm$. The stability criterion (1) now becomes

$$D_{EW} \equiv \int_{\kappa} \{\mathrm{tr}(\boldsymbol{tD}) + \mathrm{tr}(\boldsymbol{m}_D \boldsymbol{K}^T) - \varrho\, \Delta\varepsilon\}\, dv \leq 0 \tag{7}$$

for *all* kinematically admissible virtual deformations that satisfy (6).

In particular, for a rigid rotation $\theta \in (0, \pi]$, the zero work condition demands

$$\int_{\kappa} \bar{\boldsymbol{\theta}} \cdot d\hat{\boldsymbol{\gamma}} = (\theta - \sin\theta)\, \boldsymbol{h} \cdot \boldsymbol{\gamma} = 0. \tag{8}$$

Thus when $\boldsymbol{\gamma} \neq \mathbf{0}$ it is plain from (8) that in order that the zero work constraint be satisfied for any non-trivial rotation, it is requisite and sufficient that the axis of rotation be perpendicular to the resultant of dead load couples. Indeed, it was the "extra-rotational" contributions (8) of all dead load couples that led to the situation illustrated above. Therefore, *if there exists an axis of equilibrium orthogonal to* $\boldsymbol{\gamma}$, *the zero moment condition and the zero work condition are simultaneously satisfied,* $D_{EW} = 0$ *and the equilibrium configuration is neutrally stable with respect to this class of rigid rotations, no other being allowed.* As consequence of the analysis of stability under rigid deformations we are led to a theory of elastic stability for perfectly elastic materials with couple-stresses which resembles more closely that for simple, hyperelastic continua than does the criterion mentioned earlier. Nevertheless, the full physical interpretation and significance of the zero work condition imposed here remains to be found.

This paper concerns only the basic structure of the theory; but more can be done. Specifically, uniqueness theorems in the theory of small deformations superimposed on a (possibly) finitely deformed equilibrium configuration of a hyperelastic body subject to dead load forces and couples can be proved in much the same way as for simple, hyperelastic continua. And, similarly, the problem of small free vibrations superimposed on a finitely deformed equilibrium configuration can be investigated with respect to uniqueness of solutions for the superimposed free motion. Normal mode functions may be introduced, and orthogonality of the normal functions and reality of normal frequencies investigated in relation to stability of the underlying static equilibrium configuration. The physical implications of these results, however, remain to be explored.

Instability of Bars with Stress-dependent Properties

By

B. A. Boley

Cornell University, Ithaca, N.Y., U.S.A.

Abstract. A classification is first given of the various types of problems of bending and buckling of rods whose properties depend on the level and on the distribution of the normal stresses acting on the cross-section. The possibility that this stress-dependence is not the same at all points of the cross-section is included. Four different cases are distinguished, examples of two of which are treated in detail. These are the inelastic buckling of rods and the buckling of rods composed of a bar attached to a buckled thin sheet. In the former case a clarification is presented for the preferred direction of buckling of asymmetric sections, in agreement with experimental results. In the latter, the conventional analysis, employing WAGNER's reduced effective width of sheet, is shown to be incorrect, and the proper procedure for the calculation of buckling loads is described. It is found that three main loading procedures can be defined, each leading to a different buckling load. The significant physical quantity distinguishing these three cases is the relative position of the applied axial load, relative to the moving centroid of the bar-sheet combination. The theory provides an explanation for the experimentally observed preferred direction of buckling.

1. Introduction

Problems in which the cross-sectional properties of an axially loaded rod depend on the magnitude of the axial normal stress arise in a variety of important physical situations; of these, the best known example is probably that of the buckling of a rod stressed beyond the elastic limit [1, 2, 3]. The essential feature in this case is the fact that the loading and unloading behaviors are not the same, but this is by no means the only way in which the stress dependence can arise. It is the purpose of this paper to present a general discussion of problems in which some type of stress-dependence arises, to provide a detailed analysis for some

---

Acknowledgement. This work was partially supported by the Office of Naval Research.

important special cases, and to point out some errors in some conventional analyses. It will be shown that in many cases a preferred direction of buckling exists, in agreement with experimental observations.

It is convenient to begin by classifying all possible cases on the basis of the nature of the stress-dependence of the bar behavior. To this end, two criteria were selected: is the stress-dependence law the same for loading and for unloading ? is it the same at all points of the cross-section ? Four classes of problems may thus be distinguished, according to whether the stress-dependence is (1) the same for loading and unloading and at all points within a cross-section; (2) different for loading and unloading and still the same at all points; (3) the same for loading and unloading but not the same at all points; (4) different for loading and unloading and at various points.

Problems of class 1 present no difficulty, and in fact bars of this type will buckle at the tangent-modulus load. Problems of class 2 are discussed in the next section of this paper, and those of class 3 in the section following. Problems of class 4 are not treated in the present paper.

2. Inelastic buckling of asymmetric columns

The stability of initially straight columns stressed beyond the elastic limit, and the significance of the reduced-modulus load P_r, the tangent-modulus load P_t, and the Euler load P_E, are, as has been mentioned, quite well understood. Briefly, buckling (i.e. a point of bifurcation in a load-deflection plot) can occur at any load P_{cr} intermediate, to P_t and P_E, provided it is accompanied by a change in load ΔP. Actually, however, buckling will occur somewhere in the restricted range $P_t \leq P_{\mathrm{cr}} \leq P_r$, and one is therefore justified in practice [4, 5] in choosing the safe value $P_{\mathrm{cr}} = P_t$.

What has received comparatively little attention in the literature, however, is the fact that, for columns whose cross-section has only one axis of symmetry[1], a preferred direction of buckling exists for all loads P_{cr} except $P_{\mathrm{cr}} = P_t$ and $P_{\mathrm{cr}} = P_E$. To examine this question from a theoretical point of view, consider for example the simplest cross-section, namely the two-flange beam of Fig. 1, for which (with E_{eff} the effective modulus corresponding to a change in load ΔP):

$$P_{\mathrm{cr}} = \pi^2 E_{\mathrm{eff}} I / L^2, \quad E_t \leq E_{\mathrm{eff}} \leq E_r = \frac{E E_t (A_1 + A_2)}{E_t A_1 + E A_2} \tag{1}$$

[1] This does not apply to cross-section with no axis of symmetry, since they can only buckle by a combination of flexure and torsion. Only bending in the plane of symmetry is being considered in the present analysis.

so that

$$\frac{E_{\text{eff}}}{E} = \frac{E_t(A_1 + A_2)(1 + K)}{E_t A_1 + E A_2 + K(A_1 + A_2) E} = \frac{\eta(1+\alpha)(1+K)}{\eta + \alpha + K(1+\alpha)}; \qquad \eta = \frac{E_t}{E}; \qquad \alpha = \frac{A_2}{A_1}. \tag{2}$$

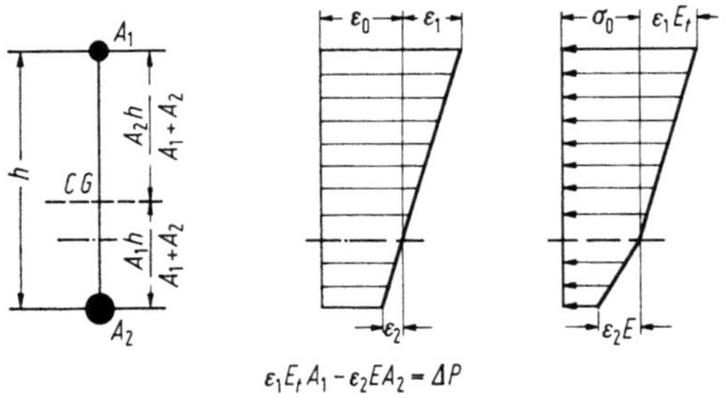

Fig. 1

The change in load ΔP is represented by the quantity $K = \frac{\Delta P}{(A_1 + A_2) E \varepsilon_2}$; $-\frac{\alpha}{1+\alpha} \leq K \leq \infty$ where the three special cases of $K = \infty$, $K = 0$, and $K = -\alpha(1+\alpha)$ correspond respectively to $E_{\text{eff}} = E_t$, $E_{\text{eff}} = E_r$, and $E_{\text{eff}} = E$.

Now α is smaller or larger than unity, depending on whether the smaller or the larger of the flanges is on the convex side. Calculate $(E'_{\text{eff}} - E''_{\text{eff}})$, where E'_{eff} corresponds to α' and E''_{eff} to α'' $(\alpha'' \leq \alpha')$, or

$$\frac{E'_{\text{eff}}}{E} - \frac{E''_{\text{eff}}}{E} = \frac{\eta(1+K)(1-\eta)(\alpha'' - \alpha')}{[\alpha' + \eta + (1+\alpha')K][\alpha'' + \eta + (1+\alpha'')K]} \leq 0. \tag{3}$$

Thus, the higher the value of α, the lower E_{eff}, and hence a lower buckling load results from $\alpha > 1$ than from $\alpha < 1$, provided $E_{\text{eff}} \neq E, E_t$. In other words, there is no preferred direction of buckling if $P_{\text{cr}} = P_t$; but in the range $P_t \leq P_{\text{cr}} < P_r$, one is justified in concluding that buckling will always occur when the smaller of the two flanges is on the concave side of the buckled shape, in agreement with available experimental work [6, 7, 8].

3. The behavior of a bar-sheet combination

Consider a thin sheet reinforced by rods (stringers) and loaded in its plane (Fig. 2). As the load increases each sheet panel will buckle; the load may then be increased further, but because of its buckled condition, the sheet will no longer be fully effective, and in fact, its effectiveness will depend on the stress it carries. The concept of an "effective width" b_e of sheet was introduced by VON KÁRMÁN [9] to account for this reduced load carrying capacity, so that

$$b_e = b_e(\sigma); \qquad 0 < b_e \leq b; \qquad \frac{db_e}{d\sigma} < 0. \tag{4}$$

For more explicit expressions see [9, 10, 11].

The loading of the sheet-stringer combination can continue until the stringers themselves buckle; because of the very small resistance to bending of the buckled central portion of the sheet panels, however, it is permissible to calculate independently the buckling load of each stringer with its accompanying effective width of sheet. We are therefore faced with the problem of the buckling of a composite bar, one part of whose cross-section varies in size according to the stress it carries, while the remainder does not; this is therefore clearly a problem of class 3 as defined in the introduction.

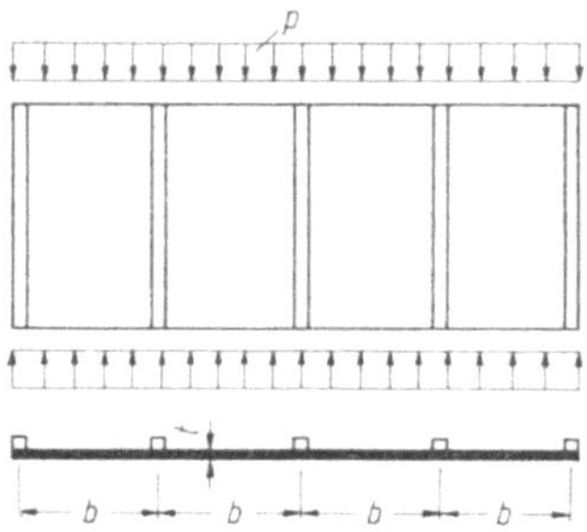

Fig. 2

It was pointed out by WAGNER [12, 13, 14] that the buckling load of this type of bar cannot be obtained simply by employing in the Euler formula the section properties and effective width prevailing at the buckling load, but that a reduced effective width b_e' should be used, defined as follows:

$$b_e' = \frac{d(b_e\sigma)}{d\sigma} = b_e + \sigma\frac{db_e}{d\sigma} < b_e. \tag{5}$$

It will now be shown that, although a reduction in the buckling load does indeed sometimes occur, it is not that predicted by the above theory, and is in fact due to an essentially different physical effect.

Consider the composite rod shown in Fig. 3, in which both sheet and stringer are of the same material, and let the origin be fixed at the

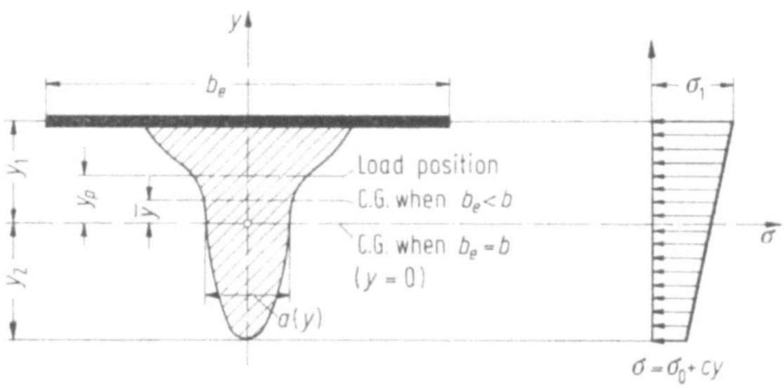

Fig. 3

original centroid. Thus (all integrations always go from the lower limit $-y_2$ to the upper limit y_1):

$$\int y a(y)\, dy + bty_1 = 0, \qquad \bar{y} = -(b - b_e)\, ty_1/A \leq 0, \tag{6}$$

where A is the total area of the composite rod. Under the usual assumptions of strength-of-materials we have

$$\sigma = \sigma_0 + E \frac{d^2v}{dx^2} y, \tag{7}$$

$$\frac{d^2v}{dx^2} + \frac{P}{EI}(v - y_p + \bar{y}) = 0, \tag{8}$$

where v is the deflection, $I(b_e)$ is the moment of inertia of the entire composite rod, y_p specifies the position of the applied load, and

$$\sigma_1 = \sigma_1(b_e, v) = \frac{P}{A} + \frac{P}{I}(y_1 - \bar{y})(y_p - \bar{y} - v). \tag{9}$$

Eqs. (8) and (4) are a set of coupled non-linear equations for the two unknowns v and b_e, which have to be solved under appropriate end conditions. We note the trivial solution $y_p = \bar{y}$, $v = 0$. To find some further information we linearize the problem by setting

$$v(x) = V(x) + v'(x), \quad v' \ll V \tag{10}$$

and, retaining only linear terms in v', we obtain

$$\frac{d^2v'}{dx^2} + \frac{P}{EI}\left[1 + \frac{\partial(\bar{y} - y_p)}{\partial v} - \frac{(v - y_p + \bar{y})}{I}\frac{\partial I}{\partial v}\right]_{v=V} v'$$
$$= -\left[\frac{d^2v}{dx^2} + \frac{P}{EI}(V - y_p + \bar{y})\right]_{v=V}. \tag{11}$$

It is convenient to note here that $\partial\bar{y}/\partial v > 0$ and $\partial I/\partial v > 0$.

We now consider three important special cases of Eq. (11). *Case* I: The position of the load is constrained to coincide at all times with that of the local centroid, so that $y_p \equiv \bar{y}$, also during the buckling process. Clearly here the buckling load is $P_{cr\cdot\mathrm{I}} = \pi^2 EI/L^2$ where I is calculated on the basis of the effective width prevailing when $\sigma = P_{cr\cdot\mathrm{I}}/A$. There is therefore no reduction in the buckling load beyond that implicit in the use of the local properties.

The conditions imposed above are difficult to realize in practice, because buckling occurs too rapidly to enable one to shift the load position simultaneously with the occurrence of the bending deflection. A physically more meaningful sequence of events is therefore that of *Case* II: The position of the load is constrained to coincide with that of the local centroid at all times before buckling, but during the buckling process the load position remains unchanged. Here the buckling load is $P_{cr\cdot\mathrm{II}} = \dfrac{\pi^2 EI}{L^2\left(1 + \dfrac{\partial\bar{y}}{\partial v}\right)} < P_{cr\cdot\mathrm{I}}$.

In the present case it has thus been seen that a reduction in the buckling load does indeed occur. A different situation arises if bending is allowed to take place at loads below the critical, giving rise to *Case* III: The load position is stationary at all times or $y_p \equiv 0$ so that $v \not\equiv 0$ for all positive loads. We are therefore dealing here with a beam-column, and there is no possibility of a typical neutral equilibrium condition as in the two previous cases. We may note, however, that the maximum load will be $P_{cr\cdot \mathrm{III}} = \dfrac{\pi^2 EI}{L^2\left(1 + \dfrac{\partial \bar{y}}{\partial v} - \dfrac{y}{I}\dfrac{\partial I}{\partial v}\right)} < P_{cr\cdot \mathrm{II}}$.
The load $P_{cr\cdot \mathrm{III}}$ is to be regarded as an upper limit to the load at which substantial deflections may be expected to occur. Note, in this connection, that the bending which takes place in this case occurs with the sheet on the concave side, i.e. $v' < 0$. It may be concluded that, since actual conditions usually involve a stationary load near the original centroid, there is a tendency for the bar to buckle away from the sheet, and indeed it is a common observation in the aircraft industry that is the most prevalent occurrence.

We note that, whether a reduction due to the changing properties is included or not, it is not due to a use of a reduced effective width, but is associated primarily with the change of the centroidal position. It is again the latter quantity that determines the preferred direction of buckling.

References

1. Bleich, F., Bleich, H. B.: The Buckling Strength of Metal Structures, New York: McGraw-Hill 1952.
2. Timoshenko, S., Gere, J. M.: Theory of Elastic Stability, 2nd ed., New York: McGraw-Hill 1961.
3. Gerard, G.: Introduction to Structural Stability Theory, New York: McGraw-Hill 1962.
4. Templin, R. L., Sturn, R. G., Hartmann, E. C., Holt, M.: Aluminum Research Lab., Tech. Paper No. 1, 1938.
5. Osgood, W. R., Holt, M.: N.A.C.A. Tech. Rep. 656, 1939.
6. Osgood, W. R.: U.S. Bureau of Standards, J. of Research, Vol. 9, Research Paper No. 492, October 1932.
7. Johnston, B. G.: J. of the Eng. Mech. Division, Proc. A.S.C.E., December 1964.
8. Hariri, Rafi: Doct. Dissert., Univ. of Michigan, Ann Arbor, Mich., June 1967.
9. von Kármán, Th., Sechler, E. E., Donnell, L. H.: Trans. A.S.M.E. **54**, 53 (1932).
10. Marguerre, K.: Luftfahrtforschung **14**, 121 (1937).
11. Boley, Sara R.: J. Aero. Sci. **22**, 336 (1955).
12. Lahde, R., Wagner, H.: N.A.C.A. Tech. Memo. 814, 1936.
13. Hoff, N. J.: J. Roy. Aero. Soc., April 1938.
14. Boley, B. A.: J. Aero. Sci. **17**, 356 (1950).

Trace Effects in Stability

By

H. Ziegler

Eidgenössische Technische Hochschule, Zurich, Switzerland

Zusammenfassung. Es wird anhand von Beispielen (darunter elektrische, Massen- und Wärmeströme) gezeigt, daß selbst kleinste Spuren gewisser Effekte (Masse, Reibung, Kapazität, Selbstinduktion usw.) das Stabilitätsverhalten entscheidend beeinflussen können. Das ist bei Untersuchungen, welche sich auf Modelle stützen, dadurch zu berücksichtigen, daß nur „annehmbare" Modelle verwendet werden.

1. Introduction

Success and reliability of many scientific investigations depend on the skill of the investigator to distinguish between essential and inessential facts, i.e., between quantities that must be taken into account and such as may be discarded. The problem becomes particularly interesting whenever (as in the case of Prandtl's boundary layer theory) an apparently small effect proves to be essential.

Stability problems are especially sensitive with respect to small effects. It is well known that, in certain cases, the stability of a mechanical system is changed if arbitrarily small masses or damping forces are added or neglected. It will be shown here that the situation is similar with respect to traces of capacities or inductances in closed circuits of electrical, mass, or heat flow.

The phenomena to be discussed are of a physical rather than of a mathematical nature. They do not arise from the linearization of an originally nonlinear system nor from the reduction of its degree of freedom as, e.g., the replacement of a continuous system by a discrete model. The problem outlined above may turn up in any type of system

---

Acknowledgement. The author wishes to thank Professor Ch. Wehrli and Mr. P. Niederer for their stimulating criticism of this work.

and is merely concerned with the question whether certain apparently negligible physical effects should be accounted for or may be neglected.

By means of a series of typical examples which, for mere convenience, will be treated as linear, we are going to demonstrate that, in principle, even slight traces of physically possible effects should be taken into account. If this rule is neglected, the most sophisticated mathematics may supply entirely unrealistic results.

It is obvious that this rule has certain consequences for stability investigations based on simplified models. It will be shown that one should confine oneself to a certain class of models which will be denoted as acceptable.

2. Traces of mass

Let us start with a few examples dealing with simple mechanical systems. It sometimes happens that the stability of a mechanical system is affected by trace of mass. The effect occurs if a slight change in the mass distribution is sufficient to suspend the positive definite character of the kinetic energy.

A typical example is the compound pendulum of Fig. 1, which may be considered to be a simplified model of an Euler column built in and acted upon by a dead load at the free end. In this model the elasticity of the actual column is represented by restoring moments acting in the hinges; the mass appears concentrated in a few points as shown in Fig. 1, and damping is neglected. The problem is conservative and nongyroscopic (the systems treated here are classified according to [1] and [2]); it may be solved by the kinetic approach, which consists in the investigation of small oscillations in the vicinity of the equilibrium position. The analysis has been carried out in [3], p. 67. It turns out that, for any $\varepsilon > 0$ and hence also for $\varepsilon \to 0$, the system is stable in the upright equilibrium position for

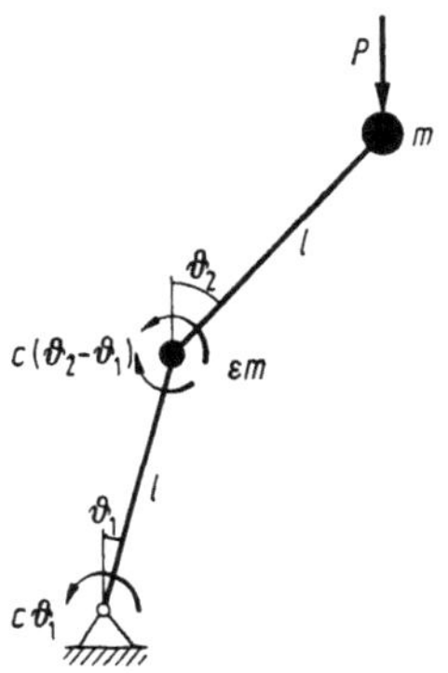

Fig. 1. Model of an Euler column.

$$P \leqq 0.382\,\frac{c}{l} = P_1 \tag{2.1}$$

and unstable for $P > P_1$. However, for $\varepsilon = 0$ the analysis yields a second stable domain

$$2.5\,\frac{c}{l} < P < 2.618\,\frac{c}{l}\,. \tag{2.2}$$

In the interval (2.2) the stability of the pendulum is thus highly sensitive to traces of mass in the hinge connecting the two rigid elements. The reason is evident: for $\varepsilon = 0$ the kinetic energy of the system is not positive definite. Thus the actual problem is unduly simplified by concentration of the entire mass at the upper end.

3. Traces of friction

Small frictional forces acting in a given system are sometimes disregarded. It may happen, however, that the stability of such a system is decisively affected by traces of friction. If, e.g., the system is stabilized by gyroscopic forces, it is apt to be destabilized by damping. For linear systems this has been shown by Lord KELVIN and TAIT [4]; a more general result is given in [3], p. 62.

The effect just mentioned occurs in a rotating shaft equipped with a disc. If the principal flexural rigidities of the shaft are equal, experiments show that there is a single critical angular velocity, ω_1. Provided that damping is neglected, this result is readily confirmed by an analysis of a simple model ([3], p. 16), which shows that the shaft is stabilized for $\omega > \omega_1$ by the gyroscopic Coriolis forces. If internal (material) damping is taken into account, a similar analysis ([3], p. 64) demonstrates that the model is unstable for $\omega > \omega_1$, regardless of the amount of damping. By inclusion of external damping (air drag) one finally is led back to the result ([3], p. 100) that the domain $\omega > \omega_1$, at least in the vicinity of ω_1, is stable.

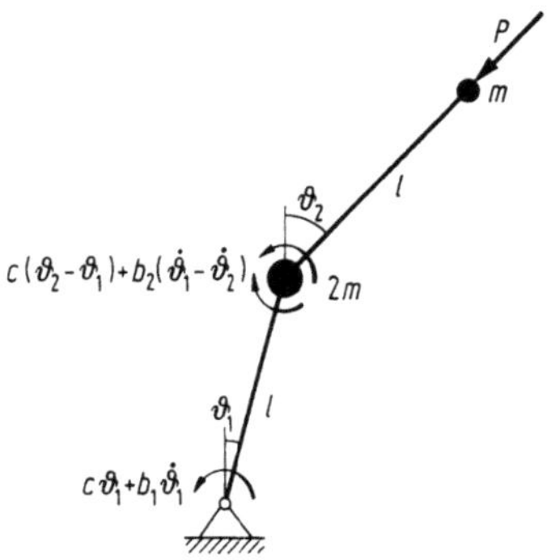

Fig. 2. Model of a column loaded tangentially.

It happens also in systems of the circulatory type that the stability behaviour is sensitive to traces of friction. A simple example which has been discussed by the author [5] and by HERRMANN and JONG [6, 7] is the compound pendulum of Fig. 2 with visco-elastic hinges, loaded by a follower force at the free end. In the absence of friction ($b_1 = b_2 = 0$) the critical load is found to be $P_1 = 2.086\ c/l$. If there is friction in the upper hinge alone ($b_1 = 0$, $b_2 > 0$), one finds $P_1 = 0.333\ c/l$. If the upper hinge alone is free of friction ($b_1 > 0$, $b_2 = 0$), the critical load is $P_1 = 2\ c/l$. Still other values are obtained if friction is present in both hinges. For $b_1 = b_2 \to 0$, e.g., one obtains $P_1 = 1.464\ c/l$.

4. Closed circuits

Effects of the kind considered here arise sometimes in stability problems of a more general type [8]. Mechanical, electrical, and chemical engineers are often confronted with the question whether a process which may be interpreted as a stationary flow in a closed circuit is stable. The most simple circuit of this type consists of a propelling device (e.g., a pump, an electric generator, or a heat source) feeding a consumer (a pipe system, an electric resistance, or a heat reservoir). The stationary process in question is distinguished by a constant current (mass flow, electric current, or heat flux), and the problem arises whether the process is stable with respect to small perturbations.

As an example, let us consider the electric circuit of Fig. 3, consisting of a generator A and a resistance B which may be composed of an ordinary resistance and an arc in series. In the stationary case the current I is constant with respect to time and place. Provided the resistance of the connections between A and B can be disregarded, U_1 is the potential at the entrance of the generator and at the exit of the resistance, and a similar statement holds for U_2. The generator is characterized by a certain relation between the potential difference, $V = U_2 - U_1$, and the current I. The corresponding curve f in Fig. 4 shows a small

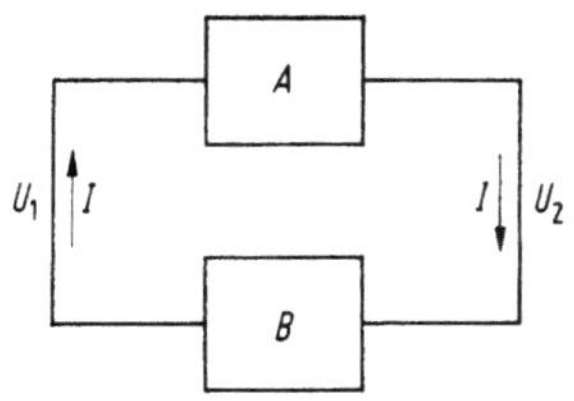

Fig. 3. Simple closed circuit.

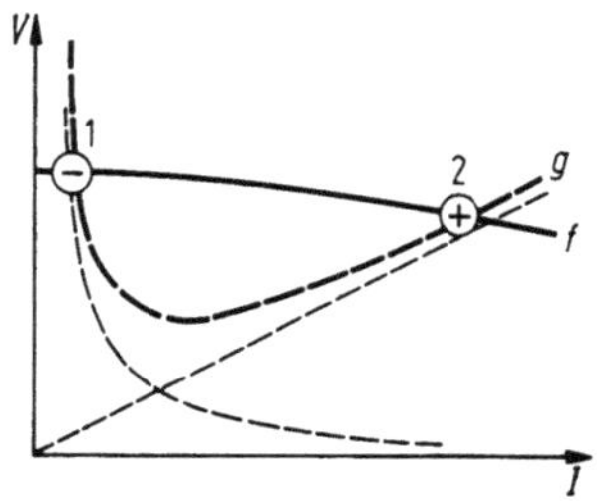

Fig. 4. Characteristics of an electric circuit consisting of generator, ordinary resistance and arc.

decrease of V with I. A second relation, represented by the curve g, characterizes the consumer. For small values of I the curve g is determined by the decreasing characteristic of the arc, while the linear characteristic of the normal resistance prevails for increasing I. The points of intersection of the characteristics f and g represent the two possible stationary regimes. It is known from practice that point 1 corresponds to an unstable regime (denoted by a minus-sign) while 2 is stable (plus-sign).

Figs. 5—7 show three examples of mass flow [9]. In Fig. 5 the curve f, connecting the flow M with the pressure difference p, is the characteristic of a centrifugal pump feeding a reservoir with the characteristic g. In a similar manner, Figs. 6 and 7 represent the characteristics of a sieve tray and of an evaporator tube.

Figs. 8 and 9 finally show to instances of heat flow [9]. The characteristics of Fig. 8, connecting the flow Q with the temperature T, are those of an electrically heated wire and its surroundings, consisting in a more or less vaporized liquid, while Fig. 9 represents a chemical reactor.

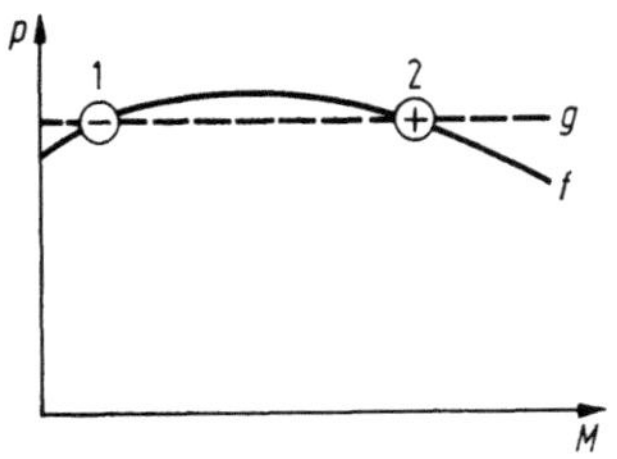

Fig. 5. Characteristics of a centrifugal pump feeding a reservoir.

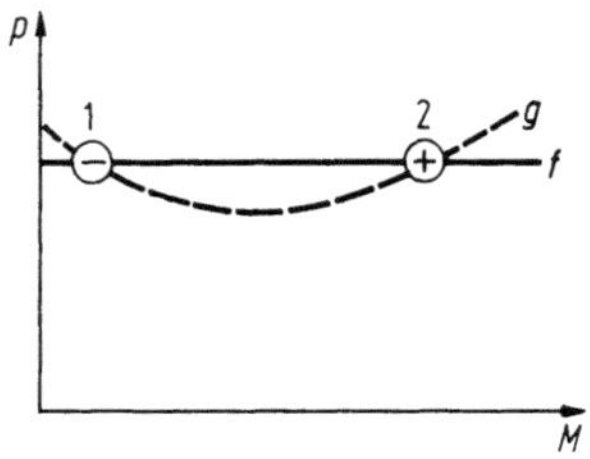

Fig. 6. Characteristics of a sieve tray.

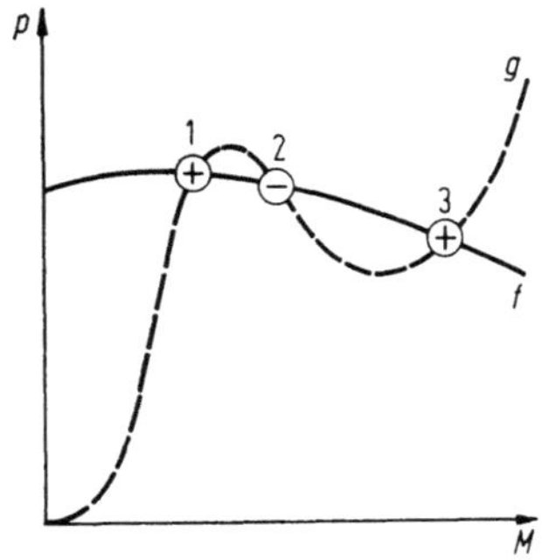

Fig. 7. Characteristics of an evaporator tube.

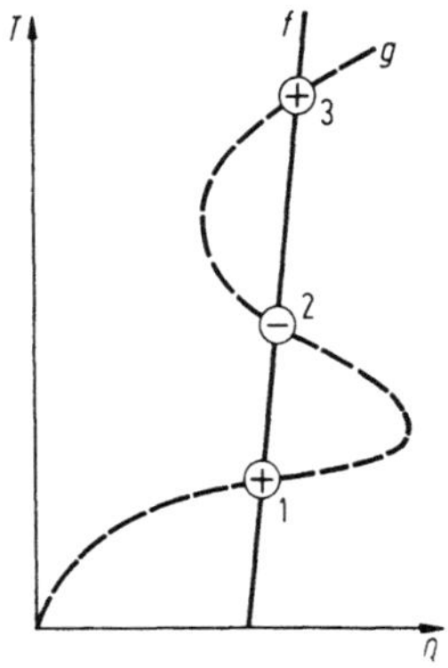

Fig. 8. Characteristics of a heating wire.

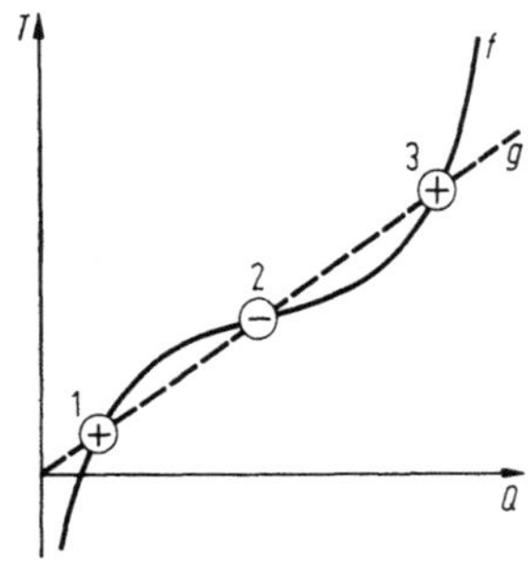

Fig. 9. Characteristics of a chemical reactor.

5. Quasistatic considerations

In each one of the Figs. 4—9 the potential difference (V, p, T) is plotted against the current (I, M, Q). Every point of intersection between the characteristics f and g corresponds to a stationary process. One is tempted to believe that an inspection of the local inclinations of the characteristics is sufficient to decide whether the process is stable or not. However, comparing the various cases one does not find any simple rule.

An argument widely used but suspect since it is quasistatic may be illustrated in the case of point 2 in Fig. 4: If, on account of a casual disturbance, the electric potential at the entrance of the generator A is slightly increased beyond its stationary value U_1, while U_2 remains unaltered, the potential difference V at the generator is momentarily too small. According to Fig. 4 the current consequently increase, and so does the potential difference at the resistance B. Thus the potential at the entrance of A decreases, and one concludes that the stationary regime corresponding to point 2 is stable.

Apart from the fact that this kind of argument is not applicable to cases where (as in Figs. 5 and 6) one of the two characteristics is parallel to the flow axis, it is easily disproved by Fig. 8: here an analogous consideration leads to the conclusion that the actually unstable regime corresponding to point 2 ought to be stable.

It would be easy to accumulate more evidence against this quasistatic reasoning. However, it is unnecessary since we know [1, 2] that, even in Mechanics, the kinetic approach alone is generally reliable.

6. The kinetic approach

Taking once more the electric circuit as an example, we note that the elements represented in Fig. 3 do not admit any deviation from the stationary regime. Such deviations are possible only because the actual circuit contains capacities and inductances. We thus replace Fig. 3 by the circuit of Fig. 10 which, in addition to the former elements, contains a condenser of capacity $C > 0$ and a pair of coils of self-inductances $L_A \geqq 0$, $L_B \geqq 0$, respectively. Incidentally, additional self-inductances on the left-hand side of the circuit would not alter the performance.

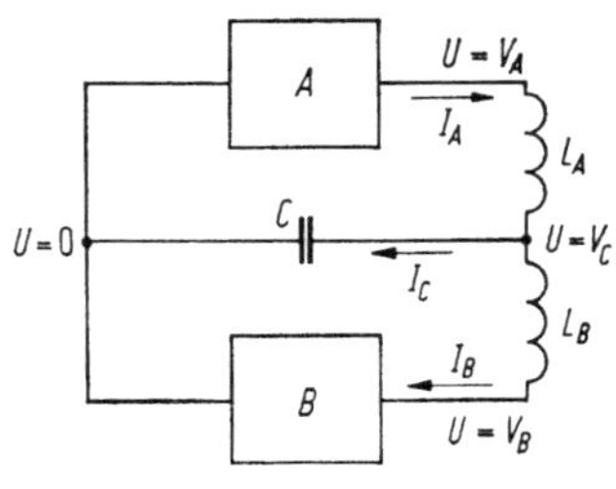

Fig. 10. Circuit containing a capacity and two self-inductances.

Denoting the potential differences along the elements A, B, C by V_A, V_B, V_C, and the currents passing through these elements by I_A, I_B, I_C, we represent the characteristics of A and B by the functions

$$V_A = f(I_A), \qquad V_B = g(I_B). \tag{6.1}$$

The balance of currents in the branch-point on the right-hand side requires

$$I_A = I_B + I_C. \tag{6.2}$$

The self-inductances are characterized by

$$L_A \dot{I}_A = V_A - V_C, \qquad L_B \dot{I}_B = V_C - V_B, \tag{6.3}$$

the capacity by

$$C \dot{V}_C = I_C = I_A - I_B, \tag{6.4}$$

where use has been made of (6.2).

Any stationary state is represented by relations of the form

$$\begin{aligned} I_A &= I_B = I_0, \qquad I_C = 0, \\ V_A &= V_B = V_C = V_0, \end{aligned} \tag{6.5}$$

where I_0 and V_0 are constant. A slight disturbance may be described by the small quantities

$$\begin{aligned} I_A - I_0 &= \varepsilon_A, \quad I_B - I_0 = \varepsilon_B, \quad I_C = \varepsilon_C, \\ V_A - V_0 &= \nu_A, \quad V_B - V_0 = \nu_B, \quad V_C - V_0 = \nu_C. \end{aligned} \tag{6.6}$$

Expanding the first of the two equations (6.1) into a power series, we have

$$V_A - V_0 = f'(I_0)(I_A - I_0) + \cdots. \tag{6.7}$$

Denoting by f' the inclination $f'(I_0)$ of the characteristic f in the vicinity of the stationary regime I_0, expanding the second equation (6.1) in the same way and making use of (6.6), we obtain

$$\nu_A = f' \varepsilon_A, \qquad \nu_B = g' \varepsilon_B \tag{6.8}$$

by way of first approximations. Equations (6.3) and (6.4) yield

$$\begin{aligned} L_A \dot{\varepsilon}_A &= \nu_A - \nu_C, \qquad L_B \dot{\varepsilon}_B = \nu_C - \nu_B, \\ C \dot{\nu}_C &= \varepsilon_A - \varepsilon_B. \end{aligned} \tag{6.9}$$

Eliminating ν_A, ν_B and ν_C from (6.8) and (6.9) we finally obtain the differential equations

$$\begin{aligned} L_A \ddot{\varepsilon}_A - f' \dot{\varepsilon}_A + \frac{1}{C}(\varepsilon_A - \varepsilon_B) &= 0, \\ L_B \ddot{\varepsilon}_B + g' \dot{\varepsilon}_B - \frac{1}{C}(\varepsilon_A - \varepsilon_B) &= 0 \end{aligned} \tag{6.10}$$

for ε_A and ε_B.

The system (6.10) can be integrated by means of exponential functions

$$\varepsilon_A = A e^{rt}, \qquad \varepsilon_B = B e^{rt}. \tag{6.11}$$

Inserting (6.11) in (6.10) we obtain

$$\begin{aligned} \left(L_A r^2 - f' r + \frac{1}{C}\right) A - \frac{1}{C} B &= 0, \\ -\frac{1}{C} A + \left(L_B r^2 + g' r + \frac{1}{C}\right) B &= 0. \end{aligned} \tag{6.12}$$

The characteristic equation becomes

$$\left(L_A r^2 - f' r + \frac{1}{C}\right)\left(L_B r^2 + g' r + \frac{1}{C}\right) - \frac{1}{C^2} = 0 \tag{6.13}$$

or

$$L_A L_B r^4 + (L_A g' - L_B f')\, r^3 + \left(\frac{L_A + L_B}{C} - f' g'\right) r^2 + \frac{g' - f'}{C} r = 0. \tag{6.14}$$

The roots of (6.14) determine the stability of the stationary regime (6.5).

To a certain degree the circuit of Fig. 10 is typical also for the processes illustrated by Figs. 5—9. In all of these cases there exists a capacity C, supplied by the elasticity of the walls or the medium itself in case of mass flow and by the heat capacity of the system in case of heat flow. Furthermore, if the medium is material, its inertia must be accounted for. As a glance at (6.10) shows, this is accomplished by assuming that L_A and L_B are positive. In the case of heat flow, however, there is no inertia and hence $L_A = L_B = 0$. Since this difference proves to be essential, we have to distinguish between the two cases.

7. Circuits without inertia

The value $r = 0$ is always a root of the characteristic equation (6.14), whether the inductances be zero or not. If once for all we exclude the case $f' = g'$ of characteristics equally inclined in the point representing the stationary regime in question, $r = 0$ is a simple root, and the corresponding fundamental solution (6.11) consists in constant values of ε_A and ε_B which, according to (6.10), must be equal. On account of (6.9) also the corresponding values of ν_A, ν_B, and ν_C are constant and equal, and (6.8) requires $\varepsilon_A = \varepsilon_B = 0$. Thus the root $r = 0$ does not imply an instability of the linear system.

The left-hand side of the characteristic equation (6.14) can now be divided by r. Restricting ourselves to systems without inertia, we further set $L_A = L_B = 0$ along with $C > 0$. Since the case $f' = g'$ has been excluded, we obtain a single root

$$r = \frac{1}{C}\left(\frac{1}{f'} - \frac{1}{g'}\right) \qquad (f' \neq 0, \quad g' \neq 0) \tag{7.1}$$

or none (apart from the harmless one $r = 0$) in case $f' = 0$ or/and $g' = 0$. With (7.1), equations (6.11) yield the sufficient stability condition

$$\frac{1}{g'} - \frac{1}{f'} > 0. \tag{7.2}$$

It follows that, in this case, the inclinations of the characteristics alone determine the stability of a stationary regime.

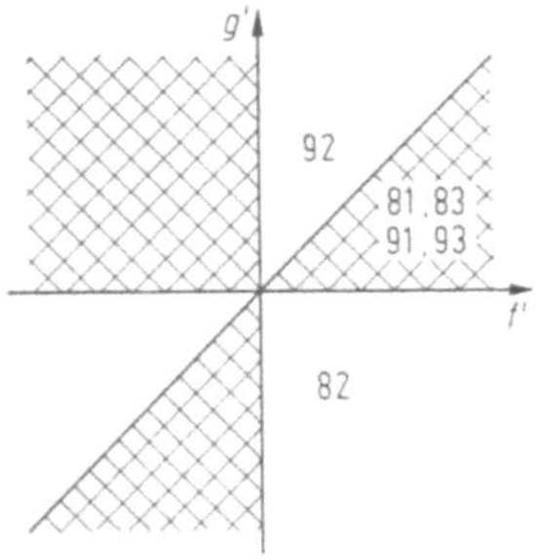

Fig. 11. Stability pattern of the circuit of Fig. 10 with $C > 0$ and $L_A = L_B = 0$.

Fig. 11 shows the stable and unstable domains in the plane f', g'. Here and in the sequel the stable domains are distinguished by cross-hatching from the unstable ones which are left open. The boundaries $f' = 0$ and $g' = 0$ belong to the stable domains while the oblique boundary represents the case $f' = g'$ which has been excluded and requires special attention.

It is easy to see that Fig. 11 confirms the experimental results given in Figs. 8 and 9 for the stability in two instances of heat flow. The numerals in Fig. 11 indicate the various cases: the first digit refers to the figure representing the particular circuit; the second one indicates the stationary regime in question. Thus the numeral 91, e.g., refers to the first regime in Fig. 9.

8. Circuits with inertia

Assuming that $L_A > 0$ and $L_B > 0$ we obtain from (6.14) the characteristic equation

$$p_0 r^3 + p_1 r^2 + p_2 r + p_3 = 0 \tag{8.1}$$

with the coefficients

$$\begin{aligned} p_0 &= L_A L_B > 0, \qquad & p_2 &= \frac{L_A + L_B}{C} - f'g', \\ p_1 &= L_A g' - L_B f', \qquad & p_3 &= \frac{g' - f'}{C}. \end{aligned} \tag{8.2}$$

The characteristic equation, which now is of the third degree, can be treated by means of the stability criteria of Routh [10] and Hurwitz [11]. According to these a stationary regime is stable if, for $p_0 > 0$, the test functions p_1, $p_1 p_2 - p_0 p_3$, and p_3 are positive. If at least one of them is negative, the regime is unstable, and in the limiting case where at least one test function is zero while the other ones are positive a more detailed investigation is necessary. Such a case has been treated in Sec. 7.

Computing the test functions by means of (8.2) and making use of the fact that $C > 0$, one obtains the sufficient stability conditions

$$\left.\begin{aligned} & L_A g' - L_B f' > 0, \\ & \frac{L_A^2}{C} g' - \frac{L_B^2}{C} f' - (L_A g' - L_B f') f' g' > 0, \\ & g' - f' > 0. \end{aligned}\right\} \tag{8.3}$$

These three inequalities take the place of the single relation (7.2) valid in the case of a circuit without inertia. With their help the stability of the stationary regimes can be completely discussed as long as limiting cases are excluded. However, this discussion cannot be rendered as general as in Sec. 7 since the conditions (8.3) now also contain C and even L_A and L_B, i.e., quantities that are not always known with sufficient accuracy. In any case it is to be expected that the stability of a given regime does not alone depend on the local inclinations f', g' of the characteristics but also on the parameters C, L_A, L_B or, more precisely, as (8.3) shows, on the quotients L_A/C and L_B/C.

In the plane f', g' (Fig. 12) the inequalities (8.3) can be interpreted as regions with borders partly dependent on L_A/C and L_B/C. The third one of these regions lies above and to the left of the straight line bisecting the angles of the first and the third quadrants. The first one likewise is a half-plane confined by a straight line through the origin and containing the entire second quadrant. The second region is generally bounded by a third degree curve.

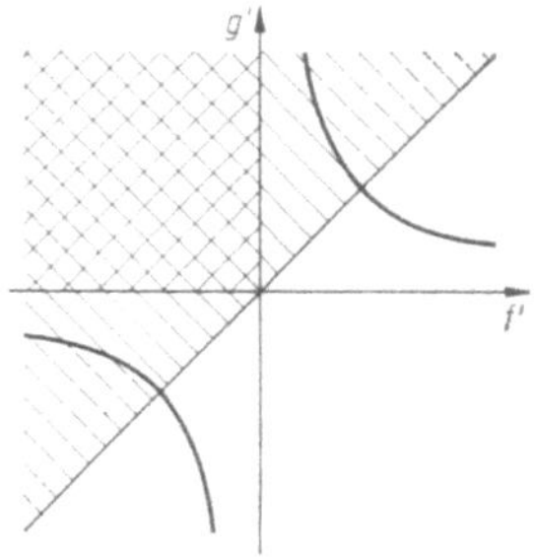

Fig. 12. Stability pattern for $C > 0$ and $L_A = L_B > 0$.

Particularly simple is the case $L_A = L_B > 0$. Here the two straight border-lines coincide and the curved one degenerates into a hyperbola lying in the first and the third quadrant and having the axes f', g' as asymptotes. In Fig. 12 the region that is stable for any choice of $L_A = L_B > 0$ is cross-hatched and the one that is always unstable is open. Simple hatching indicates stability for the special choice of $L_A = L_B$ represented by the hyperbola in Fig. 12, whereas dotted hatching indicates instability for this particular choice.

The geometric interpretation given above of the inequalities (8.3) shows that, in general, none of them is a consequence of the other ones. Thus the conditions (8.3) cannot be further reduced. In spite of this it is possible to state sufficient conditions for stability or instability, depending only on the inclinations of the characteristics. If, e.g.,

$g' > 0$ and $f' \leqq 0$, the three conditions (8.3) are satisfied. This is also true if $g' = 0$ and $f' < 0$. On the other hand, the third condition is violated if $g' < f'$. Thus in Fig. 13 the cross-hatched domain (including its boundary) is always stable while the open region is always unstable. In the region indicated by simple hatching no general result can be obtained from (8.3); it is to be expected that in this whole region stability depends on the values of L_A/C and L_B/C.

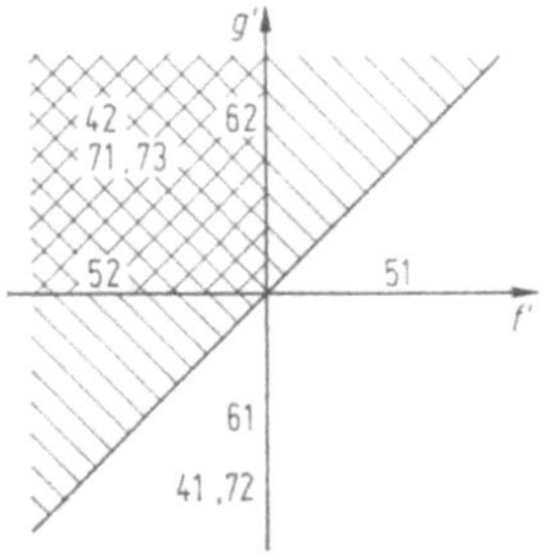

Fig. 13. Stability pattern for $C > 0$, $L_A > 0$, and $L_B > 0$.

Fig. 13 confirms all of the experimental results given in Figs. 4—7 for electric and material flows. The numerals in Fig. 13 refer to the various cases and regimes as explained at the end of Sec. 7. For a more detailed discussion of these circuits, see [8].

9. Traces of inductances and capacities

We are not interested in the particular properties of one or the other circuit but rather in the general response of a system to the addition or deletion of small effects.

Comparing Figs. 11 and 13 we note a marked difference of the stability patterns in the first and third quadrants. Fig. 11 is obviously not obtained from Fig. 13 by letting L_A and L_B tend to zero since this limiting process does not influence the regions that are stable or unstable whatever the values of L_A and L_B. Thus the circuit without inertia does not appear as a limiting case of the one with inertia. On the contrary: arbitrarily small traces of inertia affect the behaviour of the circuit decisively in comparison with the one without inertia. The only property the two circuits have in common is that they are both

$$\begin{aligned} &\text{stable for} \quad && g' > 0, \quad f' < 0, \\ &\text{unstable for} \quad && g' < 0, \quad f' > 0. \end{aligned} \tag{9.1}$$

The situation is by no means improved if only one of the two inductances is assumed to be zero.

For $L_B = 0$ the characteristic equation (6.14) reduces to

$$L_A g' r^2 + \left(\frac{L_A}{C} - f'g'\right) r + \frac{g' - f'}{C} = 0. \tag{9.2}$$

If $g' = 0$, the only stability condition is $f' < 0$. For $g' \neq 0$ the two conditions $1 - f'/g' > 0$ and $L_A/g' - Cf' \geqq 0$ are obtained. The corresponding stability domains are given in Fig. 14 and are bounded by the same lines as those of Fig. 12. The region that is stable for any choice of L_A is again cross-hatched while the one that is always unstable is open. Simple hatching indicates stability for the special choice of L_A/C represented by the hyperbola in Fig. 14, whereas dotted hatching indicates instability for this particular case. Although the border-lines are the same in Figs. 12 and 14, the stability patterns differ in the third quadrant.

For $L_A = 0$ the characteristic equation (6.14) becomes

$$-L_B f' r^2 + \left(\frac{L_B}{C} - f'g'\right) r + \frac{g' - f'}{C} = 0. \tag{9.3}$$

If $f' = 0$, the only stability condition is $g' > 0$. For $f' \neq 0$ the two conditions $1 - g'/f' > 0$ and $Cg' - L_B/f' \geqq 0$ are obtained. These inequalities yield the stability pattern of Fig. 15, which differs from Fig. 12 in the first quadrant and from Fig. 14 also in the third.

Let us finally assume that $C = 0$ while at least one of the two inductances is different from zero. Going back to (6.9) and (6.8) we readily obtain the special form

$$(L_A + L_B)\, r + g' - f' = 0 \tag{9.4}$$

of the characteristic equation (6.14). The stability condition is $f' < g'$. It provides the stability pattern of Fig. 16, which is entirely independent

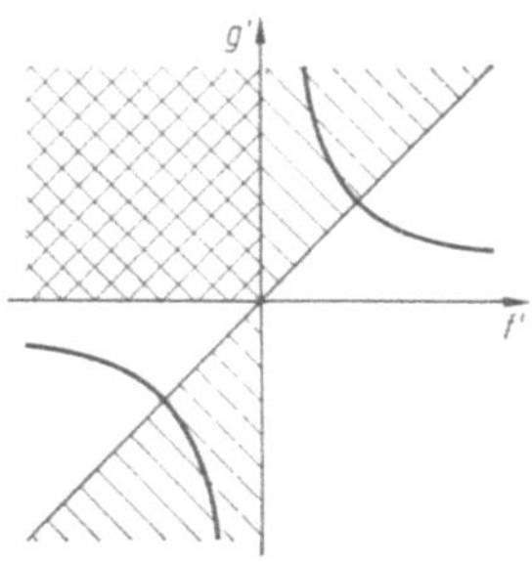

Fig. 14. Stability pattern for $C > 0$, $L_A > 0$, and $L_B = 0$.

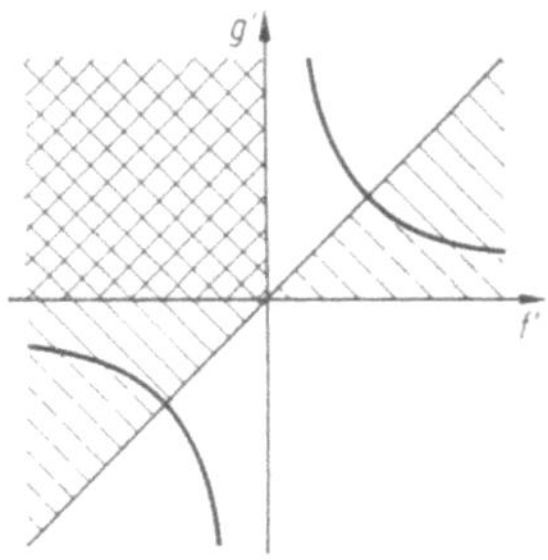

Fig. 15. Stability pattern for $C > 0$, $L_A = 0$, and $L_B > 0$.

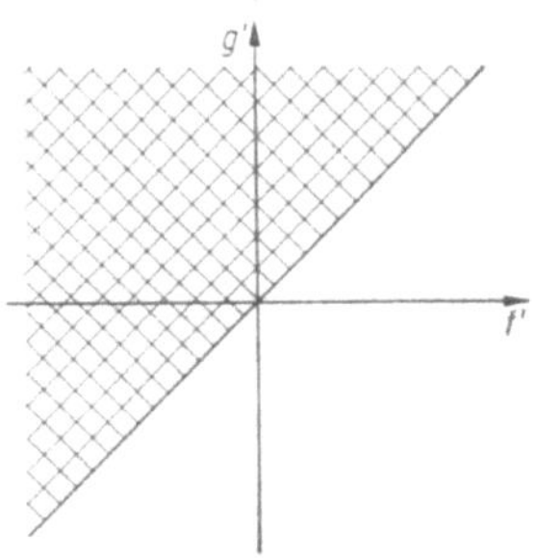

Fig. 16. Stability pattern for $C = 0$ and L_A or/and L_B positive.

of the values of L_A and L_B provided that one of them at least is different from zero. Although the patterns of Figs. 13 and 16 look somewhat similar, the last one cannot be considered a limiting case for $C \to 0$ of the former one. If, e.g., a point in Fig. 13 satisfies the last inequality (8.3) but not the first one, it represents an unstable regime for arbitrary small values of $C > 0$ but a stable one for $C = 0$.

The difference between the patterns of Figs. 11—16 is rather surprising. However, we note one common property; all these patterns comply with the sufficient conditions (9.1). This last observation is easily explained by an analysis of the differential equations (6.10) which shows that in any case the circuit is a purely dissipative system with positive definite energies, and that the dissipation is positive for $g' > 0$, $f' < 0$ and negative for $g' < 0$, $f' > 0$.

10. A rule concerning traces

The various cases compared in Secs. 7—9 show that the simple circuit of Fig. 3 is at least as sensitive to small inductances and capacities as the mechanical examples of Secs. 2 and 3 were with respect to traces of mass and friction.

In the case of the compound pendulum of Fig. 2 Herrmann and Jong [7] have analyzed the discontinuous behaviour in regard to traces of friction, and they have shown how the discontinuity can be explained by means of a concept called "degree of instability". There is no doubt that a similar reasoning can be applied to all the other cases treated here. All the same the fact remains that in general the answer to a given stability problem depends on the inclusion or deletion of small effects.

Bodies without mass, or hinges without friction do not exist. If an elastic body oscillates, internal and external friction are always present. In an electric circuit there are always traces of inductances and capacities; in a material flow inertia and damping effects cannot be excluded. Since even traces of such effects have been shown to be essential, we state the following

Rule 1: *In stability problems even small effects that are physically possible must be taken into account.*

It is true that, in certain cases and with respect to certain effects, this rule may be disregarded. It has been proved in [3] that complete damping does not affect the stability of a linear mechanical system of the nongyroscopic conservative type. In such systems it is therefore allowed to neglect traces of damping. Moreover, it is usually possible to exclude certain effects by physical reasons. Whereas, e.g., any one of the circuits discussed in connection with Fig. 3 has a certain capacity C, the inductances L_A and L_B represent inertia effects and hence are absent in cases of heat flow. Here the stability pattern must be the one of Fig. 11, and comparison with Fig. 13 shows that this is confirmed by the experimental results transferred from Figs. 8 and 9.

On the other hand, inertia effects cannot be disregarded if the circuit of Fig. 3 represents a material flow or an electric current. Comparing Figs. 13, 14, and 15 we note however that it makes quite a difference whether we represent the inertia of the circuit by one or the other or both of the inductances. The experimental facts transferred from Figs. 4—7 to Fig. 13 do not decide in favor of any one of the three possibilities. Since inertia is actually present all along the circuit observation of the rule given above stipulates that we take $L_A > 0$ and $L_B > 0$, accepting thus the stability pattern of Fig. 13.

11. Acceptable and inacceptable models

The last remark as well as the desire to continue the process leading from Fig. 3 to Fig. 10 raise a final and rather important problem.

Figs. 1—3 and 10 represent systems with merely a few degrees of freedom. They may be considered as models of more complex and even of continuous systems. Thus the compound pendulum of Fig. 1 has been used as a model of a loaded column; similarly, Fig. 10 may be interpreted as a model of a circuit with infinitely many degrees of freedom. One may expect to improve the results by using models which have more degrees of freedom and hence are better approximations of the real system. Thus the number of hinged members of which the pendulum of Fig. 1 consists may be increased. In a similar manner, the circuit of Fig. 10 may be rendered more complex by addition of elements like condensers and coils. However, in view of the differences encountered between Figs. 11—16 there is some danger that these refinements might result in further abrupt chances of the stability pattern. It is here where the concept of "acceptable models" becomes important, for it turns out that such abrupt changes can be avoided by a careful choice of the models.

In [1, 2, 3] a classification of mechanical systems has been given, and it has been shown that the behaviour of the various classes with respect to stability is different. This classification can be extended in order to cover the problems considered here, and it may be used as a guide in the choice of appropriate models.

The systems considered here are governed by second order differential equations the number of which increases with every refinement of the model. Thus, equations (6.10) characterized the linearized model of

Fig. 10. The second order terms represent the inertia of the system; they may or may not be derivable from a *kinetic energy* T. The first order terms represent damping and gyroscopic influences according to whether they are or are not derivable from a *dissipation function* D. The remaining terms correspond to restoring influences which may or may not possess a *potential energy* V. Each one of the functions T, D, and V may be *positive* or *negative definite*, *semidefinite* or *indefinite* in their principal arguments.

We have confined ourselves to the linear case. Here the differential equations have the form

$$\sum_{k=1}^{n} (m_{ik}\ddot{\varepsilon}_k + g_{ik}\dot{\varepsilon}_k + c_{ik}\varepsilon_k) = 0, \; (i = 1, 2, \ldots, n), \tag{11.1}$$

where (m_{ik}), (g_{ik}), and (c_{ik}) are constant matrices. The qualities just enumerated manifest themselves in the properties of the three matrices: they may be *symmetric*, *antimetric*, or *asymmetric*, and their symmetric parts, which represent, in turn, the functions T, D, and V, may be *positive* or *negative definite*, *semidefinite* or *indefinite*.

If a simple system is to serve as a model of a more elaborate one, it is necessary that there exists a certain *similarity* between the two systems. This similarity cannot be generally defined in exact terms, but is easily established in any single case. If, in addition, the given system and the model have *the same properties as far as the functions* T, D, *and* V, i.e., the matrices (m_{ik}), (g_{ik}), (c_{ik}) *are concerned*, the model shall be called *acceptable*.

To demonstrate this, let us return to the compound pendulum of Fig. 1. If $\varepsilon \neq 0$, it is an acceptable model of the Euler column. In fact, the kinetic energy of either system exists and is positive definite; damping and gyroscopic forces are absent, and a total potential energy of the load and the restoring forces exists and is positive definite for small values of P. For $\varepsilon = 0$ the model becomes inadmissible since its kinetic energy is merely positive semidefinite. This accounts for the discontinuity discussed in Sec. 2.

There are other instances. Let us assume, e.g., that Fig. 10 serves as a model of a mass flow originating from a centrifugal pump A and passing through an evaporator tube B. Since every element of the real tube is elastic, the capacity is actually distributed over the whole circuit, and a similar statement holds for the inertia of the flowing medium. In Fig. 10 the actual circuit is considerably simplified. However, the model is acceptable since it contains all of the essential features of the real circuit, and it may therefore be expected that the stability pattern of Fig. 13 provides a fair picture of the actual situation. In contradistinction, the model becomes unacceptable if we set $L_A = 0$

or $L_B = 0$ since these additional simplifications create subcircuits that are free of inertia. The function T from which the second order terms in (6.10) may be derived, becomes positive semidefinite, and this accounts for the anomalies in the third quadrant of Fig. 14 and in the first one of Fig. 15.

Similar circuits of a more complicated structure may be studied by means of more elaborate models. To render them acceptable, it will be necessary, as a rule, to exclude negative capacities. In mass or electric currents, care must be taken that each one of the closed subcircuits contains a positive inductance while, in the case of heat flow, all inductances must be zero. It may even be necessary to equip the subcircuits with positive resistances.

It is obvious that the problem of acceptable models needs more detailed attention. On the strength of the foregoing examples, however, it seems safe to establish the following

Rule 2: *If, in stability, models are used, one should confine oneself to acceptable ones.*

References

1. Ziegler, H.: Linear elastic stability. Z. angew. Math. Phys. **4**, 89, 167 (1953).
2. Ziegler, H.: On the concept of elastic stability. Adv. in Appl. Mech. **4**, 351 (1956).
3. Ziegler, H.: Principles of structural stability, Waltham, Mass.: Blaisdell 1968.
4. Lord Kelvin (Sir William Thomson), Tait, P. G.: Treatise on natural philosophy; reproduced under the title "Principles of mechanics and dynamics" Part I, New York: Dover 1962, p. 391.
5. Ziegler, H.: Die Stabilitätskriterien der Elastomechanik. Ing.-Arch. **20**, 49 (1952).
6. Herrmann, G., Jong, I. C.: On nonconservative stability problems of elastic systems with slight damping. J. Appl. Mech. **33**, 125 (1966).
7. Herrmann, G., Jong, I. C.: On the destabilizing effect of damping in nonconservative elastic systems. J. Appl. Mech. **32**, 592 (1965).
8. Grassmann, P., Ziegler, H.: Zur Stabilität von Strömungen in geschlossenen Kreisen. Chem. Ing. Techn. (in print).
9. Grassmann, P.: Physikalische Grundlagen der Chemie-Ingenieur-Technik, Aarau/Frankfurt a. M.: Sauerländer 1961.
10. Routh, E. J.: Advanced Rigid Dynamics, New York: Dover 1933, p. 210.
11. Hurwitz, A.: Über die Bedingungen, unter welchen eine Gleichung nur Wurzeln mit negativen reellen Teilen besitzt. Math. Ann. **46**, 273 (1895).

A Stability Study of Continuous Systems under Parametric Excitation via Liapunov's Direct Method

By

C. S. Hsu

University of California, Berkeley, Cal., U.S.A.

and

T. H. Lee

University of California, Berkeley, Cal., U.S.A.
and Lockheed Aircraft Corp., Burbank, Cal., U.S.A.

Continuous systems under parametric excitation are governed by partial differential equations with time dependent coefficients, [1]. In this paper the stability of certain systems of this type is investigated. Here the direct method of Liapunov, [2], is used. The analysis follows in spirit the work of KOZIN [3], CAUGHEY and GRAY [4], and INFANTE [5] on dynamical systems with finite degrees of freedom subjected to random excitation, and that of WANG [6] on continuous systems. Our interest is, however, on deterministic systems. A key innovation here is believed to be the use of time dependent Liapunov functionals to obtain frequency dependent criteria.

1. Preliminaries

Let a dynamical system be defined on a spatial k-dimensional and Euclidean domain D with its boundary C. Let $x = (x_1, \ldots, x_k)$ be the coordinate variable and let T denote the time interval $[0, \infty)$. More-

Acknowledgement. The results presented here were obtained in the course of research supported by a grant from the U.S. National Science Foundation.

over, let the system be governed by

$$\dot{u}(t, x) = L(t)\, u(t, x) \quad \text{for} \quad x \in D,\, t \in T, \tag{1}$$

where $u(t, x)$ is the state vector with $2N$ components satisfying certain boundary conditions

$$Gu(t, x) = 0 \quad \text{for} \quad x \in C \tag{2}$$

and appropriate initial conditions. In (1) $L(t)$ is a $2N \times 2N$ matrix whose elements are linear partial differential operators in spatial variables with time dependent coefficients. For this system $u(t, x) \equiv 0$ is an equilibrium state, and the stability of this state is to be studied.

Let H be a real Hilbert space with elements being all the vectors having continuous functions defined in D as their $2N$ components. The space H has an inner product defined for each pair $y, z \in H$ as

$$\langle y, z \rangle = \int_D y'(t, x)\, z(t, x)\, dD, \tag{3}$$

where the prime denotes the transpose. We define the state space S of the system to be a subspace of H with elements being all the $2N$-vectors $u(t, x)$ satisfying the boundary conditions (2) and other necessary smoothness conditions. A transformation A is said to be an operator in H with domain $H_1 \subset H$ if it relates to each element $y \in H_1$ an element $Ay \in H$.

Metric for stability measure. On S we introduce a metric which for each pair $u, v \in S$ is defined as

$$\varrho^2(u, v) = \langle A_0 z, A_0 z \rangle + \langle A_1 z, A_1 z \rangle + \cdots + \langle A z_n, A z_n \rangle, \tag{4}$$

where $z = u - v$, $A_0 = I =$ identity operator, and $A_1, A_2, \ldots, A_n$ are linear differential operators in H with domain in S. This metric will be used as a measure of closeness between states in the stability analysis. Obviously, the metric $\varrho(u, 0)$ may be written as

$$\begin{aligned}
\varrho^2(u, 0) &= \langle A_0 u, A_0 u \rangle + \langle A_1 u, A_1 u \rangle + \cdots + \langle A_n u, A_n u \rangle \\
&= \langle u, u \rangle + \langle u, A_1^* A_1 u \rangle + \cdots + \langle u, A_n^* A_n u \rangle + \text{boundary terms} \\
&= \langle u, Mu \rangle + \text{boundary terms}.
\end{aligned} \tag{5}$$

In going from the first line to the second, we use integration by parts. $(A)^*$ of any operator A means the adjoint of A. If the boundary terms vanish then one has

$$\varrho^2(u, 0) = \langle u, Mu \rangle. \tag{6}$$

This is true for systems treated in this paper.

2. Liapunov functionals and stability criteria

Let us introduce a functional $V(u)$ defined as

$$V(u) = \langle u, B(t)\, u \rangle, \tag{7}$$

where $u \in S$ and $B(t)$ is a symmetric operator $H \to H$. Unlike most of the previous work on continuous systems we shall allow the possibility of $B(t)$ being time dependent. The property that $V(u)$ is positive definite with respect to ϱ is established if one can show that

$$V(u) \geqq \alpha \varrho^2(u, 0) \quad \text{for some} \quad \alpha > 0. \tag{8}$$

When (6) is used for $\varrho(u, 0)$, then the condition (8) becomes

$$\langle u, (B - \alpha M)\, u\rangle \geqq 0 \quad \text{for some} \quad \alpha > 0. \tag{9}$$

To establish the stability criterion we follow a general idea used by Infante for discrete systems [5]. The time derivative of $V(u)$ along a trajectory is

$$\dot{V}(u) = \langle Lu, Bu\rangle + \langle u, BLu\rangle + \langle u, \dot{B}u\rangle = \langle u, N(t)\, u\rangle, \tag{10}$$

where

$$N(t) = L^*B + BL + \dot{B} \tag{11}$$

and where we have imposed a restriction $\langle Lu, Bu\rangle = \langle u, L^*Bu\rangle$. We also assume that $N(t)$ so obtained is symmetrical. Next, consider along the trajectory the following quotient

$$Q(t) = \frac{\dot{V}(u)}{V(u)} = \frac{\langle u, N(t)\, u\rangle}{\langle u, B(t)\, u\rangle}. \tag{12}$$

From the extreme properties of Rayleigh's quotient

$$Q(t) = \frac{\dot{V}(u)}{V(u)} \leqq \lambda_{\max}(t) \quad \text{for all} \quad t \in T, \tag{13}$$

where $\lambda_{\max}(t)$ is the maximum eigenvalue of the associated eigenvalue problem,

$$N(t)\, y = \lambda B(t)\, y, \qquad y \in S. \tag{14}$$

Let us assume that $B(t)$ is so chosen that $B^{-1}N$ is a compact operator and that $\lambda_{\max}(t)$ is bounded. In that case, one has from (13) after integration

$$V\big(u(t)\big) \leqq V\big(u(0)\big) \exp\left[\frac{1}{t}\int_0^t \lambda_{\max}(\tau)\, d\tau\right] t. \tag{15}$$

Let

$$E\{\lambda_{\max}(t)\} = \lim_{t\to\infty} \frac{1}{t}\int_0^t \lambda_{\max}(\tau)\, d\tau. \tag{16}$$

One sees then that if $E\{\lambda_{\max}(t)\} \leqq -\varepsilon$ for some $\varepsilon > 0$, then $V(u) \to 0$ as $t \to \infty$. Furthermore, on account of (8), $\varrho(u, 0) \to 0$ as well, as $t \to \infty$. Consequently, the trivial solution is asymptotically stable.

Asymptotic stability theorem: *If a functional $V(u)$ exists so that (i) $V(u)$ is positive definite with respect to $\varrho(u, 0)$, (ii) $N(t)$ is symmetric,*

(iii) (14) *defines an eigenvalue problem, and* (iv) $\lambda_{\max}(t)$ *of* (14) *is bounded, then the trivial solution* $u(t, x) \equiv 0$ *is asymptotically stable with respect to* ϱ *if*

$$E\{\lambda_{\max}(t)\} \leqq -\varepsilon \quad \textit{for some} \quad \varepsilon > 0. \tag{17}$$

Thus, the determination of the stability criterion entails selecting a class of operator $B(t)$, solving an eigenvalue problem, obtaining from $\lambda_{\max}(t)$ the desired stability criterion, and finally sharpening the criterion by optimizing $B(t)$ within the given class. We remark here that in solving the eigenvalue problem the variable t is regarded strictly as a parameter and not as an independent variable.

3. Application to supported columns

Let $w(t, x)$ be the displacement. Then the equation of motion in non-dimensional form may be written as

$$\frac{\partial^2 w}{\partial t^2} + \beta \frac{\partial w}{\partial t} + \frac{\partial^4 w}{\partial x^4} + f(t) \frac{\partial^2 w}{\partial x^2} = 0. \tag{18}$$

Let $v(t, x)$ be the velocity. Then in the form of (1),

$$u(t, x) = \begin{bmatrix} w(t, x) \\ v(t, x) \end{bmatrix}, \qquad L(t) = \begin{bmatrix} 0 & 1 \\ -\frac{\partial^4}{\partial x^4} - f(t)\frac{\partial^2}{\partial x^2} & -\beta \end{bmatrix}. \tag{19}$$

The boundary conditions are

$$u(t, 0) = u(t, 1) = 0, \qquad \frac{\partial^2 u(t, 0)}{\partial x^2} = \frac{\partial^2 u(t, 1)}{\partial x^2} = 0. \tag{20}$$

For this problem we take

$$\varrho^2(u, 0) = \int_0^1 \left[\left(\frac{\partial^2 w}{\partial x^2}\right)^2 + \left(\frac{\partial w}{\partial x}\right)^2 + w^2 + v^2\right] dx \tag{21}$$

and hence,

$$M = \begin{bmatrix} \frac{\partial^4}{\partial x^4} - \frac{\partial^2}{\partial x^2} + 1 & 0 \\ 0 & 1 \end{bmatrix}. \tag{22}$$

For a specific example of application we shall take

$$f(t) = \left(\frac{P_0}{P_{cr}}\pi^2\right) \cos \Omega t, \tag{23}$$

where P_0 is the amplitude of the applied pulsating load and P_{cr} the Euler critical load. For the matrix $B(t)$ we take

$$B(t) = \begin{bmatrix} \frac{\partial^4}{\partial x^4} + g(t)\frac{\partial^2}{\partial x^2} + \alpha_2 + \frac{\beta^2}{4} & \frac{\beta}{2} \\ \frac{\beta}{2} & 1 \end{bmatrix}. \tag{24}$$

(A) Functional with constant coefficients. In this case we take $g(t) = 0$ and $\alpha_2 = \beta^2/4$. Thus,

$$V(u) = \int_0^1 \left[\left(\frac{\partial^2 w}{\partial x^2}\right)^2 + \frac{\beta^2}{2} w^2 + \beta w v + v^2\right] dx. \tag{25}$$

After solving the associated eigenvalue problem, we obtain

$$\lambda_{\max}(t) = -\beta + \operatorname*{Max}_n \left(\frac{[2n^2\pi^2 f(t) + \beta^2]^2}{4n^4\pi^4 + \beta^2}\right)^{1/2}. \tag{26}$$

This $\lambda_{\max}(t)$ takes on different values of n at different instants of time. The relation depends to a large extent upon the sign of $f(t)$. The stability criterion (17) with this $\lambda_{\max}$ has been computed and is shown in Fig. 1. A similar result has also been obtained independently by INFANTE and PLAUT [7].

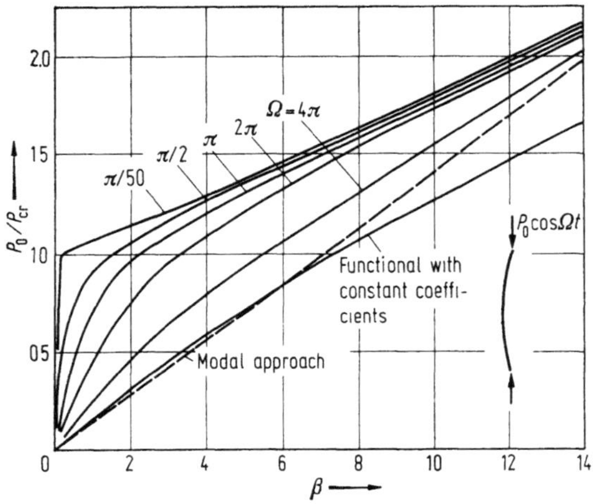

Fig. 1. Stability criteria for simply supported columns under harmonic axial loadings.

(B) Modal approach. One may also use the modal approach by reducing the partial differential equation first to a system of uncoupled ordinary differential equations and then apply the stability criterion derived in [5]. This leads to

$$E\{f^2(t)\} \leqq \beta^2 - \varepsilon \quad \text{for some} \quad \varepsilon > 0 \tag{27}$$

or

$$(P_0 P_{cr}) < 2^{1/2}\beta/\pi^2. \tag{28}$$

This result is also shown in Fig. 1.

(C) Functional with time dependent coefficients. Both of the above results (26) and (28) establish criteria on the amplitude of excitation as functions of the damping coefficient. The effects of the excitation frequency could not, however, be ascertained. Moreover, the criteria are too overly conservative for cases with small damping and small excitation frequencies. With an aim to improve on these results we take a Liapunov functional with time dependent coefficients by using $B(t)$ of (24). Thus,

$$V(u) = \int_0^1 \left[\left(\frac{\partial^2 w}{\partial x^2}\right)^2 - g(t)\left(\frac{\partial w}{\partial x}\right)^2 + \left(\alpha_2 + \frac{\beta^2}{4}\right) w^2 + \beta w v + v^2\right] dx. \quad (29)$$

The condition that $V(u)$ be positive definite requires

$$g(t) < \pi^2 - \varepsilon_1 \text{ for some } \varepsilon_1, \quad (30)$$

$$\alpha_2 > -\varepsilon_1 \pi^2. \quad (31)$$

The $\lambda_{\max}$ of the associated eigenvalue problem is found to be

$$\lambda_{\max} = \underset{n}{\mathrm{Max}} \left\{ -\beta - \frac{\dot g(t)}{2n^2\pi^2 D_2} + \left[\left(\frac{\dot g(t)}{2n^2\pi^2 D_2}\right)^2 + \frac{1}{D_2}\left(f(t) - g(t) + \frac{4\alpha_2 + \beta^2}{4n^2\pi^2}\right)^2\right]^{1/2} \right\}, \quad (32)$$

where

$$D_2 = 1 - \frac{g(t)}{n^2\pi^2} + \frac{\alpha_2}{n^4\pi^4}. \quad (33)$$

For harmonic excitation, $f(t)$ is given by (23). For $g(t)$ we choose

$$g(t) = \pi^2 R_g \cos c\Omega t, \quad (34)$$

where c is a positive constant which may take on values 1, 2, ... or 1/2, 1/3, ... It is seen that $g(t)$ is also harmonic but with the amplitude and frequency yet undetermined. Of course, apart from the restriction (31), α_2 is also free. The most desirable choices of α_2, c and R_g were made by computation. In most cases the sharpest condition was found with $c = 1$ and $\alpha_2 = \beta^2/4$. With c and α_2 fixed at these values the optimum values of R_g can be determined. The stability criteria thus obtained for various values of Ω are shown in Fig. 1. These curves give us new criteria on the excitation amplitude which depend upon the damping coefficient as well as the excitation frequency. The criteria are quite sharp for small damping coefficients and small excitation frequencies. For example, for very small damping and very small excitation frequencies the stability criterion for the amplitude of a harmonic axial loading approaches the Euler critical load. A study has also been made by allowing α_2 in (24) also to be time dependent. The improvement so far obtained is, however, not appreciable.

4. Application to clamped columns

The case of a clamped column subjected to periodic axial loading has also been studied. This is a much more significant case in that the differential equation is not separable by the modal approach. Because of the space limitation, however, the analysis will not given here.

References

1. Bolotin, V. V.: The Dynamic Stability of Elastic Systems, San Francisco: Holden-Day 1964.
2. Zubov, V. I.: Method of A. M. Lyapunov and their Application, Leningrad 1957; English translation, Groningen: P. Noordhoff 1964.
3. Kozin, F.: J. Math. and Phys. **42**, 59 (1963).
4. Caughey, T. K., Gray, A. H., Jr.: J. Appl. Mech. **32**, 365 (1965).
5. Infante, E. F.: J. Appl. Mech. **35**, 7 (1968).
6. Wang, P. K. C.: J. Appl. Mech. **33**, 182 (1966).
7. Infante, E. F., Plaut, R. H.: AIAA J. **7**, 766 (1969).

Probleme nichtlinearer Operatoren bei Untersuchung der Stabilität dünner Platten und Schalen

Von

R. Kodnár

Slowakische Akademie der Wissenschaften, Bratislava, Č.S.S.R.

Es sei Ω ein begrenzter Bereich in E_2 mit der Lipschitzschen Grenze. Nehmen wir das kartesische Produkt

$$W = W_2^{(2)}(\Omega) \times W_2^{(1)}(\Omega) \times W_2^{(1)}(\Omega)$$

und definieren wir für $\boldsymbol{u} = (w, u, v)$, wo $w \in W_2^{(2)}(\Omega)$, $u \in W_2^{(1)}(\Omega)$, $v \in W_2^{(1)}(\Omega)$ die Norm

$$\|\boldsymbol{u}\|_W^2 = \|w\|_{W_2^{(2)}}^2 + \|u\|_{W_2^{(1)}}^2 + \|v\|_{W_2^{(1)}}^2.$$

Wir setzen

$$V = W_2^{(2)}(\Omega) \times W_2^{(1)}(\Omega) \times W_2^{(1)}(\Omega).$$

Es sei P_1 der Raum sämtlicher Polynome höchstens ersten Grades. Bezeichnen wir mit P jenen Unterraum des Raumes $P_1 \times P_1 \times P_1$, der die Basis $(0, 1, 0)$, $(0, 0, 1)$, $(0, y, x)$ hat. Wir bezeichnen mit V/P den Raum der Klassen $\tilde{\boldsymbol{u}}$ solcher Funktionen $\boldsymbol{u} \in V$, daß $\boldsymbol{u}, \boldsymbol{v} \in \tilde{\boldsymbol{u}} \equiv \boldsymbol{u} - \boldsymbol{v} \in P$.

V/P ist der Hilbertsche Raum und daher reflexiv.

Es seien weiter die Funktionen $q \in L_2(\Omega)$, $h_1 \in L_2(\partial\Omega)$, $h_2 \in L_2(\partial\Omega)$ gegeben.

Wir untersuchen die Existenz der schwachen Lösung des Systems der Differentialgleichungen der Theorie der schlanken Wände in den Verschiebungen

$$\left.\begin{aligned} &\frac{D}{h}\Delta^2 w = \frac{\partial^2 w}{\partial x^2}\sigma_x + \frac{\partial^2 w}{\partial y^2}\sigma_y + 2\,\frac{\partial^2 w}{\partial x\,\partial y}\tau + \frac{q}{h}, \\ &\frac{\partial \sigma_x}{\partial x} + \frac{\partial \tau}{\partial y} = 0, \\ &\frac{\partial \tau}{\partial x} + \frac{\partial \sigma_y}{\partial y} = 0 \end{aligned}\right\} \quad (1)$$

bei Randbedingungen

$$w = \frac{\partial w}{\partial n} = 0, \quad \sigma_x n_x + \tau n_y = h_1, \quad \tau n_x + \sigma_y n_y = h_2. \tag{2}$$

n ist die Außennormale zum $\partial\Omega$; n_x, n_y ihre Komponenten. Wir fordern die Erfüllung der Gleichungen (1) im Sinne der Distributionen.

Das Element $\boldsymbol{\alpha} = (w, u, v) \in V/P$ ist die schwache Lösung des gegebenen Randwertproblems, wenn für jedes Element $\boldsymbol{\beta} = (\tilde{w}, \tilde{u}, \tilde{v}) \in V/P$ gilt

$$\begin{aligned} A(\boldsymbol{\alpha})\,\boldsymbol{\beta} = &\int_\Omega \frac{D}{h}\left(\frac{\partial^2 w}{\partial x^2}\frac{\partial^2 \tilde{w}}{\partial x^2} + 2\,\frac{\partial^2 w}{\partial x\,\partial y}\,\frac{\partial^2 \tilde{w}}{\partial x\,\partial y} + \frac{\partial^2 w}{\partial y^2}\frac{\partial^2 \tilde{w}}{\partial y^2}\right) d\Omega \\ &+ \int_\Omega \left(\frac{\partial w}{\partial x}\,\frac{\partial \tilde{w}}{\partial x}\,\sigma_x + \frac{\partial w}{\partial y}\,\frac{\partial \tilde{w}}{\partial y}\,\sigma_y + \frac{\partial w}{\partial x}\,\frac{\partial \tilde{w}}{\partial y}\,\tau + \frac{\partial w}{\partial y}\,\frac{\partial \tilde{w}}{\partial x}\,\tau\right) d\Omega \\ &+ \int_\Omega \left(\sigma_x\,\frac{\partial \tilde{u}}{\partial x} + \tau\,\frac{\partial \tilde{u}}{\partial y} + \tau\,\frac{\partial \tilde{v}}{\partial x} + \sigma_y\,\frac{\partial \tilde{v}}{\partial y}\right) d\Omega - \frac{1}{h}\int_\Omega q\tilde{w}\,d\Omega \\ &- \int_{\partial\Omega} h_1 \tilde{u}\,dS - \int_{\partial\Omega} h_2 \tilde{v}\,dS \equiv 0. \end{aligned} \tag{3}$$

Der so definierte Operator $A(\boldsymbol{\alpha})$ ist potentiell. Es existiert also das Funktional $f(\boldsymbol{\alpha})$, für welches $\operatorname{grad} f(\boldsymbol{\alpha}) = A(\boldsymbol{\alpha})$, $\boldsymbol{\alpha} \in V/P$ gilt. Aus dem Ausdruck (3), der für jedes $\boldsymbol{\beta} \in V/P$ gelten soll, geht also hervor, daß es bei der Untersuchung der Existenz einer Lösung um die Untersuchung der kritischen Punkte des Funktionals $f(\boldsymbol{\alpha})$ auf V/P geht. Durch Berechnung wird festgestellt, daß das Potential des Operators $A(\boldsymbol{\alpha})$ das Funktional

$$\begin{aligned} f(\boldsymbol{\alpha}) = &\frac{D}{2h}\int_\Omega \left[\left(\frac{\partial^2 w}{\partial x^2}\right)^2 + 2\left(\frac{\partial^2 w}{\partial x\,\partial y}\right)^2 + \left(\frac{\partial^2 w}{\partial y^2}\right)^2\right] d\Omega \\ &+ \frac{E}{2(1-\mu^2)}\int_\Omega \left[\mu\left(\frac{\partial u}{\partial x} + \frac{\partial v}{\partial y}\right)^2 + (1-\mu)\left\langle\left(\frac{\partial u}{\partial x}\right)^2 + \left(\frac{\partial v}{\partial y}\right)^2\right\rangle\right. \\ &\left. + \frac{1-\mu}{2}\left(\frac{\partial u}{\partial y} + \frac{\partial v}{\partial x}\right)^2\right] d\Omega \\ &+ \frac{E}{8(1-\mu^2)}\int_\Omega \left[\left(\frac{\partial w}{\partial x}\right)^2 + \left(\frac{\partial w}{\partial y}\right)^2\right] d\Omega \\ &+ \frac{E}{2(1-\mu^2)}\int_\Omega \left[\frac{\partial u}{\partial x}\left(\frac{\partial w}{\partial x}\right)^2 + \frac{\partial v}{\partial y}\left(\frac{\partial w}{\partial y}\right)^2 + \mu\,\frac{\partial v}{\partial y}\left(\frac{\partial w}{\partial x}\right)^2 + \mu\,\frac{\partial u}{\partial x}\left(\frac{\partial w}{\partial y}\right)^2\right] d\Omega \\ &+ \int_\Omega \frac{E}{2(1+\mu)}\left(\frac{\partial u}{\partial y} + \frac{\partial v}{\partial x}\right)\frac{\partial w}{\partial x}\,\frac{\partial w}{\partial y}\,d\Omega - \frac{1}{h}\int_\Omega q w\,d\Omega \\ &- \int_{\partial\Omega} h_1 u\,dS - \int_{\partial\Omega} h_2 v\,dS \end{aligned} \tag{4}$$

ist. Es läßt sich nachweisen, daß das Funktional auf V/P von unten schwach halbstetig ist. Setzen wir weiter voraus, daß h_1, h_2 senkrecht zu den entsprechenden Polynomen ersten Grades sind. Es gelten Sätze, welche das Existenzproblem für kleine h_1, h_2 vollkommen lösen (NEČAS, PORACKÁ und KODNÁR [1]).

Wenn wir das Existenzproblem der Lösung für flache Schalen bei Randbedingungen (2) untersuchen, dann wird die Beweisführung ähnlich sein wie für dünne Platten.

Nehmen wir den konkreten Fall, eine rechtwinkelige Wand mit Seitenlängen a, b. Setzen wir voraus, daß der Anfang des Koordinatensystems sich an der linken unteren Ecke der Wand befindet, also $\Omega \equiv \langle 0, a\rangle \times \langle 0, b\rangle$. Weiter wollen wir voraussetzen, daß der Rand gelenkig gelagert ist. Das Grundsystem der Differentialgleichungen nehmen wir in der Form

$$\begin{aligned} Gw &\equiv \frac{D}{h}\Delta\Delta w - \frac{\partial^2\varphi}{\partial y^2}\frac{\partial^2 w}{\partial x^2} - \frac{\partial^2\varphi}{\partial x^2}\frac{\partial^2 w}{\partial y^2} + 2\frac{\partial^2\varphi}{\partial x\,\partial y}\frac{\partial^2 w}{\partial x\,\partial y} - \frac{q}{h} = 0, \\ \Delta\Delta\varphi &= E\left[\left(\frac{\partial^2 w}{\partial x\,\partial y}\right)^2 - \frac{\partial^2 w}{\partial x^2}\frac{\partial^2 w}{\partial y^2}\right]. \end{aligned} \tag{5}$$

Wählen wir ein System von Koordinatenfunktionen $\{\varphi_n\}_{n=1}^{\infty}$ und lösen wir das Gleichungssystem (5) nach der Art von Papkowitsch. Es gilt (KODNÁR [2])

Satz 1. *Die Folge der Ritzschen Annäherungen ist eine minimalisierende.*

Satz 2. *Es sei Gâteaux's Differential des Operators G positiv definit in $L_2(\Omega)$. Dann konvergiert die minimalisierende Folge in $L_2(\Omega)$.*

Die Konvergenzbedingungen der minimalisierenden Folge können verstärkt werden.

Definition 1. *$H_0 \subset H$ sind Hilbert-Räume. Es sei jedes lineare Funktional $(x, y) \in H^*$, $x \in H$, $y \in H_0$ ein lineares Funktional über H_0. Es sei $A \in (H_0 \to H)$ (ansonsten ein beliebiger Operator). Den Operator A werden wir fast positiv definit zu H_0 nennen, falls $(Ax, x)_H \geq \gamma^2 \|x\|_{H_0}^2$ für alle $x \in H_0$ ist.*

Satz 3. *Es sei Gâteauxs Differential des Operators G fast positiv definit in der Norm $W_2^{(2)}$. Dann konvergiert die minimalisierende Folge für das Funktional (5) in $W_2^{(2)}$.*

Die positive Definitheit und die fast positive Definitheit kann bei Ausschluß des Problems der Spannungskonzentration bewiesen werden. Außerdem gilt (KODNÁR [3, 4])

Satz 4. *Unser System der Ritzschen oder Galerkinschen algebraischen Gleichungen hat mindestens eine Lösung.*

Satz 5. *Das Problem der Bifurkationspunkte für Gleichungen der schlanken Wand bei Randbedingungen*

$$w\big|_{\partial\Omega} = \frac{\partial w}{\partial n}\bigg|_{\partial\Omega} = 0 \quad \textit{oder} \quad w\big|_{\partial\Omega} = \frac{\partial^2 w}{\partial n^2}\bigg|_{\partial\Omega} = 0$$

und die eine Spannungskonzentration für φ *ausschließen, kann linearisiert werden, und jeder Eigenwert der Fréchet-Ableitung in* θ *ist ein Bifurkationspunkt, welchem ein stetiger Zweig der Eigenfunktionen der Ausgangsaufgabe entspricht.*

Satz 6. *Außerdem seien die Randbedingungen für* φ *von einer solchen Form, daß der Operator*

$$G_1\eta = \frac{d}{d\varepsilon}(\bar{\varphi}_{xx})_{\varepsilon=0}\, k_2 + \frac{d}{d\varepsilon}(\bar{\varphi}_{yy})_{\varepsilon=0}\, k_1$$

symmetrisch in $L_2(\Omega)$ *ist.* $\bar{\varphi}$ *ist die Lösung der Gleichung*

$$\Delta\Delta\bar{\varphi} = E\left[(\varepsilon\eta_{xy})^2 + \varepsilon^2\eta_{xx}\eta_{yy} - \varepsilon k_2\eta_{xx} - \varepsilon k_1\eta_{yy}\right].$$

Dann besteht die Möglichkeit, das Problem der Bifurkationspunkte für Gleichungen flacher Schalen zu linearisieren. Jeder Eigenwert der linearisierten Aufgabe ist ein Bifurkationspunkt. Dem Bifurkationspunkt entspricht der stetige Zweig der Eigenfunktionen der nichtlinearen Ausgangsaufgabe.

Satz 7. *Die linearisierte Aufgabe hat unendlich viele Eigenwerte, und die entsprechenden Eigenfunktionen bilden ein System, ganz im Sinne der energetischen Normen ihrer Operatoren.*

Zeigen wir ein solches Beispiel: Untersuchen wir die gelenkig gelagerte schlanke Wand im Seitenverhältnis $a/b = 1$ und setzen wir voraus, daß bei Belastung die gegenseitige Verschiebung der Ränder nach der linearen Abhängigkeit verläuft. Die Wand sei durch eine Kombination der Biegung und des Druckes gespannt. Zur Berechnung der Bifurkationspunkte können wir die Variationsmethode von RITZ verwenden. Wenn wir eine genügende Zahl der Glieder der Reihenfolge wählen, können wir die Bifurkationspunkte sehr genau berechnen. Bei einer kleinen Gliederanzahl müssen diese entsprechend gewählt werden. Es müssen solche Basisglieder gewählt werden, welche die Lösungsformen am meisten beeinflussen, welche entstehen sollen. Wenn wir Glieder bei den Koeffizienten w_{11}, w_{12}, w_{13} wählen, bekommen wir die Lösung $\lambda_1 = 7{,}8123$; $\lambda_2 = 22{,}2222$; $\lambda_3 = 58{,}8235$. Der Wert λ_1 ist der genaue Wert beim Entstehen des ersten Zweiges der stabilen nichttrivialen Lösung. λ_2 ist der genaue Wert beim Entstehen des zweiten Zweiges. Der Wert λ_3 ist auch verhältnismäßig genau. Wenn wir die Glieder bei den Koeffizienten w_{31}, w_{32}, w_{33} wählen, erhalten wir die Lösungen $\lambda_1 = 19{,}9644$; $\lambda_2 = 40{,}2890$; $\lambda_3 = 105{,}3742$. Wenn wir die Glieder bei den Koeffizienten w_{12}, w_{13} wählen, erhalten wir die Lösungen $\lambda_1 = 47{,}74$; $\lambda_2 = 246{,}80$. Wir sehen also, daß wir durch die Wahl der Basisglieder die Form der Lösungen vorschreiben, welche sie erst bei einem gewissen Parameterwert aufweisen können. Die Werte der Bifurkationspunkte sind daher abhängig von der Möglichkeit, welche nichttrivialen Formen die Lösung bei der gegebenen Wahl der Glieder der Koordinatenfolge haben kann. Auch bei Voraussetzung der Stabilität aller Schritte der numerischen Lösung können die geeigneten Basisglieder und die Anzahl für ein entsprechend genaues Feststellen der Entstehungspunkte der Zweige nichttrivialer stabiler Lösungen nur intuitiv geschätzt werden.

Kehren wir zurück zum Funktional (4). Nehmen wir das konkrete lineare Funktional $\psi_1 \in H \subset V/P$, $\|\psi_1\| = 1$, für welches die Aufgabe (für $\lambda\psi_1$; $\lambda \in \langle c, d\rangle \subset E_1$) eine Lösung hat. $\psi = f + f_6 + f_7 + f_8$.

Es sei $\mathfrak{M}^{(i)}$ der Zweig der nichttrivialen Lösungen, entsprechend dem i-ten Bifurkationspunkt λ_i. Es sei $\lambda > \lambda_i$, $\boldsymbol{u}_0 = \boldsymbol{u}(\lambda\psi_1)$.

Definition 2. *Wir werden sagen, daß die Lösung $\boldsymbol{u}_0$ auf dem gegebenen Zweig fortschreitet, wenn*

$$\forall_{\varepsilon>0}\ \exists_{\delta>0,\, \boldsymbol{u}[(\lambda>\delta)\psi_1]\in\mathfrak{M}^{(i)}} \left(\boldsymbol{u}(\lambda\psi_1 - \delta\psi_1) \in \{\boldsymbol{u};\ \|\boldsymbol{u} - u_0\| < \varepsilon\}\right).$$

Definition 3.

$$\lambda_{\mathfrak{M}^{(i)}}(\psi_1) = \sup\{\lambda;\ \boldsymbol{u}(\lambda\psi_1) \in \mathfrak{M}^{(i)}\}.$$

Satz 8. *Es sei* $\lambda > \lambda_{\mathfrak{M}^{(i)}}(\psi_1)$. *Dann existiert entweder keine stabile Lösung, oder sie gehört in den Zweig* $\mathfrak{M}^{(k)}$, $k \neq i$.

Dieser Satz zusammen mit dem Satz über die Bifurkationspunkte löst das Problem der möglichen Formen der nichttrivialen Lösungen und deren Verschwinden.

Nehmen wir nun eine allgemeine Aufgabe. Es sei $H \subset V/P$ und ψ_2 das lineare begrenzte Funktional auf H.

Definition 4.

$$M_s = \left\{\psi_2 \in H^*;\ \liminf_{\|\boldsymbol{u}\|\to\infty} \left(\psi(\boldsymbol{u}) - (\boldsymbol{u}, \psi_2)\right) = +\infty\right\},$$

$$M_i = \left\{\psi_2 \in H^*;\ \liminf_{\|\boldsymbol{u}\|\to\infty} \left(\psi(\boldsymbol{u}) - (\boldsymbol{u}, \psi_2)\right) \neq \pm\infty\right\},$$

$$M_l = \left\{\psi_2 \in H^*;\ \liminf_{\|\boldsymbol{u}\|\to\infty} \left(\psi(\boldsymbol{u}) - (\boldsymbol{u}, \psi_2)\right) = -\infty\right\}.$$

Für $\psi_2 = (0, 0, q)$ existiert laut vorangehendem Satz [1] das absolute Minimum des Funktionals $\psi(\boldsymbol{u}) - (\boldsymbol{u}, q)$, also die Lösung einer bestimmten Randwertaufgabe. Also $M_s \neq \emptyset$.

Richtig ist

Satz 9. *Die Menge M_s ist konvex. Es sei* $\psi_2 \in H$, $\|\psi_2\| = 1$.

Definition 5. $\lambda_H(\psi_2) = \sup\{\lambda;\ \lambda\psi_2 \in M_s\}$.

Es gilt

Satz 10. $\lambda_H(\psi_2) > 0$.

Satz 11. *Es sei* $\lambda > \lambda_H(\psi_2) \Rightarrow \lambda\psi_2 \in M_l$.

Hiermit ist die Aufgabe der Existenz für verschiedene Typen der Randbedingungen und rechten Seiten gelöst.

Literatur

1. NEČAS, J., PORACKÁ, Z., KODNÁR, R.: Mat. čas. (1970).
2. KODNÁR, R.: Stav. čas. **16**, 400 (1968).
3. KODNÁR, R.: Stav. čas. **16**, 641 (1968).
4. KODNÁR, R.: Stav. čas. **17**, 201 (1969).

Some Applications of Liapunov Functionals

By

P. C. Parks

University of Warwick, Coventry, U.K.

1. Introduction

The idea of extending the Liapunov function technique to examine stability problems of partial differential equations was first proposed by Zubov [1] and Movchan [2] in 1959, but until recently there have been few applications to physical problems. A useful bibliography is that of Wang [3], and a theoretical background using Hilbert and Sobolev spaces has been given by Buis, Vogt et al. [4]. In the present paper some specific applications are reviewed, and a useful method for constructing Liapunov functionals is suggested.

2. Metric spaces and Liapunov functionals

The second method of Liapunov, as applied to the discussion of the asymptotic stability of an equilibrium point in the state or phase system described by ordinary differential equations, is essentially a method of showing that the distance ϱ from the equilibrium point of a point moving along a trajectory of the disturbed system tends to zero as time increases. Although the point is described by an n-vector the stability argument is entirely in terms of scalar quantities — the Liapunov function V, its rate of change $\dot{V}$ and the distance ϱ.

When partial differential equations are considered it is still possible to consider scalar measures of the system disturbed from its equilibrium state. These measures must obey simple rules for metric space norms such as the triangle inequality described in books [5] on functional analysis.

A Liapunov function V or functional such that

$$\alpha\varrho^2 \leq V \leq \beta\varrho^2, \dot{V} \leq -\gamma\varrho^2,$$
$$\alpha, \beta, \gamma > 0$$

proves uniform asymptotic stability in the large, of the equilibrium point or state. Weaker forms of stability are possible, [6].

3. Applications

3.1 Panel flutter [7]

Oscillations (x, t) of a two-dimensional uniform panel with pinned edges at $x = 0$ and $x = 1$ under the influence of a supersonic air stream in the x direction are governed by the non-dimensional equation

$$d\frac{\partial^4 z}{\partial x^4} + \mu\frac{\partial^2 z}{\partial t^2} - f\frac{\partial^2 z}{\partial x^2} + M\frac{\partial z}{\partial x} + \frac{\partial z}{\partial t} = 0$$

with boundary conditions $z = \partial^2 z/\partial t^2 = 0$ at $x = 0$, $x = 1$.

A stability criterion in terms of M is sought. This is a well known problem with an extensive literature [8] but is treated here using the Liapunov functional technique.

A suitable measure of the panel disturbance from its equilibrium position, in which $z(x, t) = 0$, $0 \leq x \leq 1$, is

$$\varrho = \left\{\int\limits_{x=0}^{1} z^2 + \left(\frac{\partial z}{\partial x}\right)^2 + \left(\frac{\partial^2 z}{\partial x^2}\right)^2 + \left(\frac{\partial z}{\partial t}\right)^2 dx\right\}^{1/2}.$$

A Liapunov functional, V, was obtained in [7] by considering twice the total energy of the vibrating panel and modifying this so as to obtain a sign definite form for dV/dt. This gave

$$V = \int\limits_{x=0}^{1} \mu\left(\frac{\partial z}{\partial t}\right)^2 + f\left(\frac{\partial z}{\partial x}\right)^2 + d\left(\frac{\partial^2 z}{\partial x^2}\right)^2 + \lambda\left(z^2 + 2\mu z\frac{\partial z}{\partial t}\right) dx$$

and

$$\frac{dV}{dt} = \int\limits_{x=0}^{1} - 2\left(\frac{\partial z}{\partial t}\right)^2 - 2M\frac{\partial z}{\partial t}\cdot\frac{\partial z}{\partial x} + \lambda\left(2\mu\left(\frac{\partial z}{\partial t}\right)^2 - 2f\left(\frac{\partial z}{\partial x}\right)^2 - 2d\left(\frac{\partial^2 z}{\partial x^2}\right)^2\right) dx.$$

An optimum value of λ is $\lambda = 1/2\mu$, when, making use of the Rayleigh inequality that $\int\limits_{x=0}^{1}\left(\frac{\partial^2 z}{\partial x^2}\right)^2 dx \geq \pi^2 \int\limits_{x=0}^{1}\left(\frac{\partial z}{\partial x}\right)^2 dx$, two stability criteria are obtained:

$$f + \pi^2 d > 0, \qquad M^2 < \frac{(f + \pi^2 d)}{\mu}.$$

The first is precisely the Euler buckling criterion for the panel under compression; for panels of large chord the second condition is essentially a well-known wave-making criterion — that is the air speed must be less than the speed of any flexural waves in the panel.

If these two criteria are satisfied then V and dV/dt may be shown [7] to have bounds in terms of ϱ^2 as in Sec. 2 above. The condition on M, which is sufficient but not necessary for stability, is compared with the traditional Galerkin result in Fig. 1.

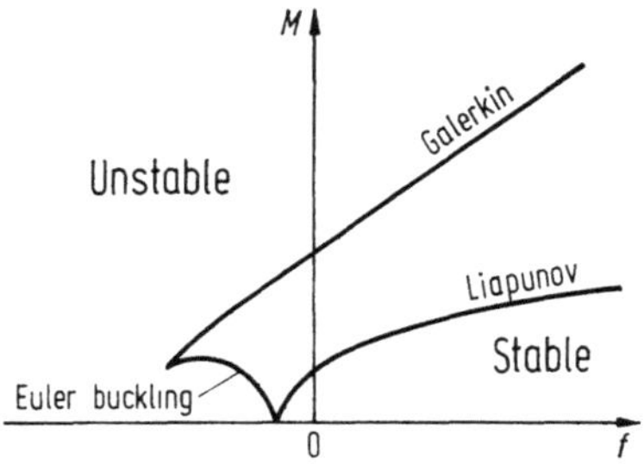

Fig. 1. Panel flutter: comparison of stability criteria.

3.2 Couette flow between concentric rotating cylinders [9]

A transformed version of the Navier-Stokes and continuity equations for a perturbed Couette flow of a viscous incompressible fluid between two concentric rotating cylinders may be written in conventional notation

$$\frac{\partial u}{\partial t} + u\,\frac{\partial u}{\partial r} + w\,\frac{\partial u}{\partial z} - \frac{v^2}{r} - \frac{T^{1/2}}{\varkappa^{1/2} r^2}(1 - \varkappa r^2)\,v = -\,\frac{\partial p}{\partial v} + \Delta u - \frac{u}{r^2},$$

$$\frac{\partial v}{\partial t} + u\,\frac{\partial v}{\partial r} + w\,\frac{\partial v}{\partial z} + \frac{uv}{r} - T^{1/2}\varkappa^{1/2} u = \Delta v - \frac{v}{r^2},$$

$$\frac{\partial w}{\partial t} + u\,\frac{\partial w}{\partial r} + w\,\frac{\partial w}{\partial z} = -\,\frac{\partial p}{\partial z} + \Delta w,$$

$$\frac{1}{r}\,\frac{\partial}{\partial r}(ru) + \frac{\partial w}{\partial z} = 0$$

with boundary condition $u = v = w = 0$ for $r = \eta$ and $r = 1$, and assuming rotational symmetry. For a given η a condition in terms of T and $\varkappa$ is sought.

A suitable measure of the disturbance is

$$\varrho = \left\{\int\limits_z \int\limits_r (u^2 + v^2 + w^2)\,dr\,dz\right\}^{1/2}$$

and a Liapunov functional is

$$V = \frac{1}{2}\int\limits_z \int\limits_r r\,(u^2 + v^2 + w^2)\,dr\,dz,$$

for which, after some integrations by parts,

$$\frac{dV}{dt} = \int\limits_z \int\limits_r \left[\frac{T^{1/2}}{\varkappa^{1/2}}\left(\frac{uv}{r}\right) - \left\{r\left(\frac{\partial v}{\partial r}\right)^2 - rv\,\frac{\partial^2 v}{\partial z^2} + \frac{v^2}{r} + r\left(\frac{\partial u}{\partial v}\right)^2 \right.\right.$$
$$\left.\left. -\,ru\,\frac{\partial^2 u}{\partial z^2} + \frac{u^2}{r} + r\left(\frac{\partial w}{\partial r}\right)^2 + r\left(\frac{\partial w}{\partial z}\right)^2\right\}\right] dr\,dz.$$

If we now assume, as customary, that the perturbations are periodic in z and are zero at intervals of length π/k in z we may obtain, integrating by parts and using the Rayleigh inequality,

$$\frac{dV}{dt} \leq \int\limits_z \int\limits_r \frac{T^{1/2}}{\varkappa^{1/2}} \frac{uv}{r} - \left\{ r \left(\frac{\partial v}{\partial r}\right)^2 + k^2 r v^2 + r \left(\frac{\partial u}{\partial r}\right)^2 + \frac{v^2}{r} \right.$$
$$\left. + k^2 r u^2 + r \left(\frac{\partial w}{\partial r}\right)^2 + r \left(\frac{\partial w}{\partial z}\right)^2 \right\} dr\, dz.$$

Using the continuity equation the terms in w may be exchanged for terms in u so that

$$\frac{dV}{dt} \leq \int\limits_z \int\limits_r \frac{T^{1/2}}{\varkappa^{1/2}} \frac{uv}{r} - \left\{ r \left(\frac{dv}{dr}\right)^2 + k^2 r v^2 + \frac{v^2}{r} + 2r \left(\frac{\partial u}{\partial r}\right)^2 + k^2 r u^2 + \frac{2u^2}{r} \right.$$
$$\left. + \frac{1}{k^2} \left(\frac{3}{r} \left(\frac{\partial u}{\partial r}\right)^2 - \frac{3u^2}{r^3} + r \left(\frac{\partial^2 u}{\partial r^2}\right)^2 \right) \right\} dr\, dz.$$

Use of the calculus of variations to obtain inequalities between $(\partial v/\partial r)^2$ and v^2 terms and between $(\partial u/\partial r)^2$, $(\partial^2 u/\partial r^2)^2$ and u^2 terms leads to an inequality of the form

$$\frac{dV}{dt} \leq \int\limits_z \int\limits_r \frac{T^{1/2}}{\varkappa^{1/2}} \frac{uv}{r} - \left\{ \left[\frac{A}{r} + k^2 r + \frac{1}{r} \right] v^2 \right.$$
$$\left. + \left[\frac{2A}{r} + k^2 r + \frac{2}{r} + \frac{B}{k^2 r^3} \right] u^2 \right\} dr\, dz.$$

We then obtain a sufficient stability criterion that

$$T < 4\varkappa [A + k^2 r^2 + 1] \left[2A + k^2 r + 2 + \frac{B}{k^2 r^3} \right]$$

for all $k > 0$ and all r, $\eta \leq r \leq 1$.

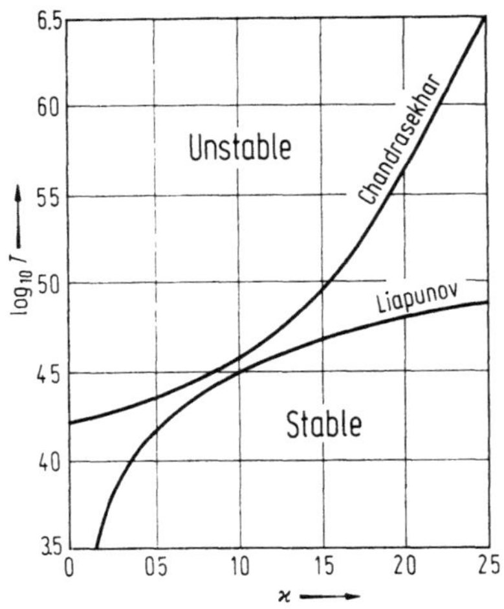

Fig. 2. Couette flow: comparison of stability criteria.

For $\eta = 1/2$ the resulting criterion is compared with the necessary and sufficient criterion arising from the linearised theory of CHANDRASEKHAR [10] in Fig. 2. Here, however, the non-linear terms in the Navier-Stokes equations have been retained in the calculation of dV/dt.

3.3 Feedback control of distributed parameter systems [11]

Consider the temperature control problem of Fig. 3. It is desired to control the heat input at $x = 0$ in such a way that the temperature $T(x, t)$ in the uniform insulated bar tends to the demanded temperature T_i everywhere. The temperature is measured at $x = c$ where $0 \le c \le l$, and this is fed back to form an error signal for closed loop control.

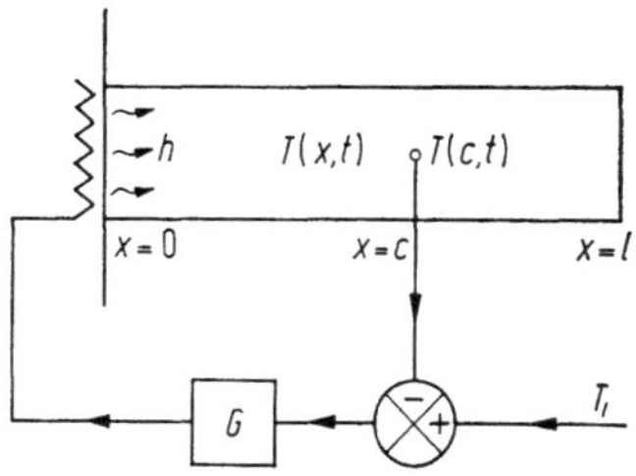

Fig. 3. Temperature control loop.

The heat equation is

$$\frac{\partial T}{\partial t} = k \frac{\partial^2 T}{\partial x^2}$$

with boundary conditions $h = -K\, \partial T/\partial x$ at $x = 0$, $\partial T/\partial x = 0$ at $x = l$. The Liapunov functional

$$V = \frac{1}{2} \varrho^2 = \frac{1}{2} \int\limits_{x=0}^{l} (T - T_i)^2\, dx$$

yields

$$\frac{dV}{dt} = \frac{k}{K} (T - T_i)|_{x=0}\, h - k \int\limits_0^l \left(\frac{\partial T}{\partial x}\right)^2 dx.$$

Let us consider making $h = G(T_i - T)_{x=c}$, when

$$\frac{dV}{dt} = -\frac{Gk}{K} (T - T_i)|_{x=0} (T - T_i)|_{x=c} - k \int\limits_{x=0}^{l} \left(\frac{\partial T}{\partial x}\right)^2 dx.$$

Splitting the integral into two parts and using the Schwarz inequality

$$\left\{\int\limits_0^c 1 \cdot \frac{\partial T}{\partial x}\, dx\right\}^2 \le \int\limits_0^c 1 \cdot dx \int\limits_0^c \left(\frac{\partial T}{\partial x}\right)^2 dx$$

we obtain

$$\frac{dV}{dt} \le -k \int\limits_c^l \left(\frac{\partial T}{\partial x}\right)^2 dx - \frac{Gk}{K} (T - T_i)|_{x=0} (T - T_i)|_{x=c}$$

$$- \frac{k}{c} \{(T - T_i)|_{x=0} - (T - T_i)|_{x=c}\}^2.$$

Consideration of the quadratic form in $(T - T_i)|_{x=0}$ and $(T - T_i)|_{x=c}$ leads to a sufficient stability condition on the gain G that $G < 4K/c$. Fig. 4 compares this gain with a calculated (and simulated) critical gain.

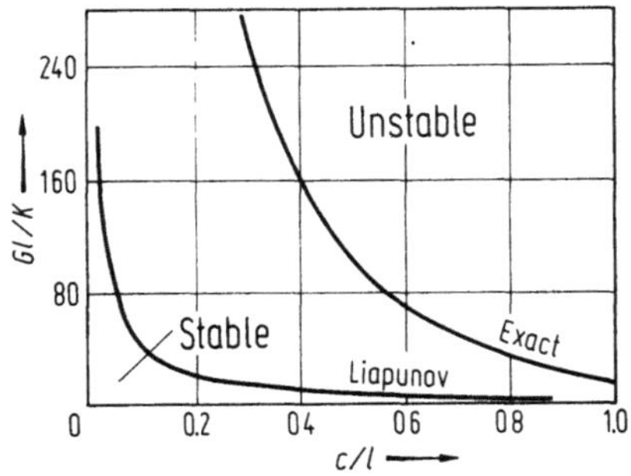

Fig. 4. Temperature control: comparison of stability criteria.

The Liapunov technique would also work for a non-uniform bar, when the conventional Laplace transform method fails. In [11] PARKS and PRITCHARD consider also feedback controls of flexible bodies involving the wave equation.

4. Construction of functionals

An outstanding problem is to find methods of constructing Liapunov functionals. One method which works well in a number of cases is as follows. If the partial differential equation is written $L(u) = 0$ where L is an operator in $\partial/\partial t$ and also space derivatives, then we differentiate L partially with respect to the symbol $\partial/\partial t$ to form a new operator N. We then consider the product $L(u)\,N(u)$ and its integral with respect to space and time. The terms in this product are put into two groups, the first being a group of terms which may be integrated explicitly with respect to t. The Liapunov functional is taken to be the first group of terms. Since $L(u)\,N(u) = 0$ the second group of terms (without integration with respect to time) is $-dV/dt$.

Consider for example the panel flutter equation of Sec. 3.1 above

$$L \equiv d\,\frac{\partial^4}{\partial x^4} + \mu\,\frac{\partial^2}{\partial t^2} - f\,\frac{\partial^2}{\partial x^2} + M\,\frac{\partial}{\partial x} + \frac{\partial}{\partial t},$$

then

$$N \equiv 2\mu\,\frac{\partial}{\partial t} + 1,$$

and

$$\begin{aligned}\int_x\int_t L(z)\,N(z)\,dt\,dx = {} & \int_x \mu^2\left(\frac{\partial z}{\partial t}\right)^2 + \frac{1}{2}z^2 + \mu z\,\frac{\partial z}{\partial t} + \mu f\left(\frac{\partial z}{\partial x}\right)^2 + \mu d\left(\frac{\partial^2 z}{\partial x^2}\right)^2 dx \\ & + \int_x\int_t 2\mu M\,\frac{\partial z}{\partial x}\,\frac{\partial z}{\partial t} + d\,\frac{\partial^4 z}{\partial x^4}\,z - f\,\frac{\partial^2 z}{\partial x^2}\,z + Mz\,\frac{\partial z}{\partial x} \\ & + \mu\left(\frac{\partial z}{\partial t}\right)^2 dt\,dx.\end{aligned}$$

Taking

$$V = \int_x \mu^2\left(\frac{\partial z}{\partial t}\right)^2 + \mu f\left(\frac{\partial z}{\partial x}\right)^2 + \mu d\left(\frac{\partial^2 z}{\partial x^2}\right)^2 + \frac{1}{2}z^2 + \mu z\,\frac{\partial z}{\partial t}\,dx$$

we have

$$\frac{dV}{dt} = -\int_x \mu\left(\frac{\partial z}{\partial t}\right)^2 + 2\mu M \frac{\partial z}{\partial x}\frac{\partial z}{\partial t} + f\left(\frac{\partial z}{\partial x}\right)^2 + d\left(\frac{\partial^2 z}{\partial x^2}\right)^2 dx,$$

carrying integration by parts with respect to x. It will be noted that V is μ times the optimised functional developed in Sec. 3.1 above, but obtained here in a quite different way.

The present method can be adapted for ordinary differential equations in phase space, by dropping the space integration and differentiation and replacing $\partial/\partial t$ above by d/dt. In the case of an n-th order linear differential equation with constant coefficients the technique leads to a Liapunov function V having the Hermite matrix [12] in the phase space quadratic form of $-dV/dt$. This is an unexpected and interesting result.

References

1. ZUBOV, V. I.: The Methods of Liapunov and their Application, Leningrad 1959.
2. MOVCHAN, A. A.: Prikl. Mat. Mekh. **23**, 483 (1959).
3. WANG, P. K. C.: Int. J. Control **7**, 101 (1968).
4. BUIS, G. R., VOGT, W. G., et al.: Lyapunov stability for partial differential equations. NASA CR-1100, 1968.
5. KOLMOGOROV, A. N., FOMIN, S. V.: Functional Analysis, Vol. I, Rochester, N.Y.: Graylock Press 1957.
6. KALMAN, R. E., BERTRAM, J. E.: Trans. A.S.M.E. **D 82**, 371 (1960).
7. PARKS, P. C.: Differential Equations and Dynamical Systems, ed. J. P. LASALLE, New York: Academic Press 1967, p. 287.
8. JOHNS, D. J.: A survey on panel flutter. AGARD Advisory Report No. 1, 1965.
9. PRITCHARD, A. J.: J. Inst. Maths. Applics. **4**, 78 (1968).
10. CHANDRASEKHAR, S.: Hydrodynamic and Hydromagnetic Stability, Oxford: Clarendon Press 1961.
11. PARKS, P. C., PRITCHARD, A. J.: Proc. 4th IFAC Congress, Warsaw 1969, Paper 20.5.
12. PARKS, P. C.: IEEE Trans. Autom. Control **AC-11**, 334 (1966).

On the Place of Energy Methods in a Global Theory of Hydrodynamic Stability

By

D. D. Joseph

Imperial College of Science and Technology, London, U.K.

The point of departure for the global theory to be described is the system of the nonlinear Boussinesq equations (1, 2) governing the disturbance of some given motion. For simplicity, let $(\boldsymbol{U}, T, \Gamma)$ be a basic steady velocity, temperature and concentration (say, salt) field which satisfies the Boussinesq equations in a bounded domain (or in a period cell) and such that these fields take on prescribed values on $\partial\mathscr{V}$ (or periodic values on appropriate parts of $\partial\mathscr{V}$). Let $(\boldsymbol{u}, \theta, \gamma)$ be disturbances of the given motion which are induced at time zero as initial conditions. Subsequently [30]

$$(\boldsymbol{u}, \theta, \gamma)/\partial\mathscr{V} = 0, \qquad \operatorname{div} \boldsymbol{u} = 0/\mathscr{V}, \tag{1a, b}$$

$$\frac{d\boldsymbol{u}}{dt} + R(\boldsymbol{u} \cdot \nabla)\, \boldsymbol{U} + (\boldsymbol{u} \cdot \nabla)\, \boldsymbol{u} = -\nabla P - (\mathscr{R}\theta - \mathscr{C}\gamma)\, \boldsymbol{\eta} + \Delta \boldsymbol{u}, \tag{2a}$$

$$\mathscr{P}_T \frac{d\theta}{dt} + \mathscr{P}_T\, \boldsymbol{u} \cdot \nabla\theta + \mathscr{R}\boldsymbol{u} \cdot \boldsymbol{\eta}_T = \Delta\theta, \tag{2b}$$

and

$$\mathscr{P}_\Gamma \frac{d\gamma}{dt} + \mathscr{P}_\Gamma \boldsymbol{u} \cdot \nabla\gamma + \mathscr{C}\boldsymbol{u} \cdot \boldsymbol{\eta}_\Gamma = \Delta\gamma, \tag{2c}$$

where

$$\frac{d}{dt} = \frac{\partial}{\partial t} + R\boldsymbol{U} \cdot \nabla.$$

Here, the vector fields $\boldsymbol{\eta}$, $\boldsymbol{\eta}_T$, $\boldsymbol{\eta}_\Gamma$ are respectively, the normalized gravity, temperature and salt gradient vectors for the basic motion. $\mathscr{P}_T$ and $\mathscr{P}_\Gamma$ are Prandtl numbers for heat and salt and R, $\mathscr{R}$ and $\mathscr{C}$ are the Reynolds number and square root of the Rayleigh numbers for heat and salt, respectively. Instead of three stability parameters R, $\mathscr{R}$ and $\mathscr{C}$ we work with one polar parameter ϱ introduced by the transformation

$$R = \varrho \cos a, \qquad \mathscr{R} = \varrho \sin a \sin b, \qquad \mathscr{C} = \varrho \sin a \cos b .$$

The polar angles a and b are preassigned and we solve a stability problem on a fixed ray from the origin.

The stability of the basic motion (U, T, Γ) is judged by the behaviour of the disturbance "energy"

$$E(t, \lambda_T, \lambda_\Gamma) = \tfrac{1}{2} \langle |\boldsymbol{u}|^2 + \mathscr{P}_T \lambda_T \theta^2 + \lambda_\Gamma \mathscr{P}_\Gamma \gamma^2 \rangle,$$

where λ_T and λ_Γ are any preassigned positive numbers and the angle bracket denotes integration over $\mathscr{V}$. A *stable motion* we define as one for which $E(t)/E(0) \to 0$ as $t \to \infty$ (asymptotic stability in the mean). An *unstable motion* is not stable. A motion is *globally stable*[1] if it is stable for all bounded initial energies $E(0) < \infty$. A value ϱ_G is a *global stability limit* if we have global stability for all $\varrho < \varrho_G$. A motion is *linearly stable* if stable for sufficiently small $E(0)$. A value ϱ_L is a *linear stability limit* if we have linear stability for all $\varrho < \varrho_L$. A *sublinearly unstable motion* is unstable but linearly stable, that is, unstable for $\varrho_L > \varrho \geq \varrho_G$.

Since sublinear instabilities are commonplace in fluid mechanics we cannot expect the stability criterion of linear theory to have predictive value. But a motion which is unstable in the linearized theory is unstable so that the linear theory of *instability* has a place in a global theory, that is, we have instability if $\varrho \geq \varrho_L$.

For sufficient conditions for stability we use the energy method. The method starts from the evolution equation

$$\frac{dE}{dt} = \varrho I - D, \tag{3}$$

where

$$D = \langle |\nabla \boldsymbol{u}|^2 + \lambda_T |\nabla \theta|^2 + \lambda_\Gamma |\nabla \gamma|^2 \rangle$$

is the (stabilizing) dissipation and

$$I = TM + TH + TS$$

is the (destabilizing) energy source defined by

$$\begin{Bmatrix} TM \\ TH \\ TS \end{Bmatrix} = - \left\langle \begin{Bmatrix} \boldsymbol{u} \cdot \nabla \boldsymbol{U} \cdot \boldsymbol{u} \cos a \\ (\boldsymbol{\eta} + \lambda_T \boldsymbol{\eta}_T) \cdot \boldsymbol{u} \theta \sin a \sin b \\ (-\boldsymbol{\eta} + \lambda_\Gamma \boldsymbol{\eta}_\Gamma) \cdot \boldsymbol{u} \gamma \sin a \cos b \end{Bmatrix} \right\rangle .$$

Equation (3) is mere restatement of the nonlinear problem (1, 2) and its general solution is beyond our ability. But if we consider (3) for a class of smooth functions which satisfy (1) and contain solutions of (1, 2) as a subset then on this class of kinematically admissible func-

[1] Globally stable motions are necessarily unique. For nonunique solutions one needs a more restricted stability concept. For example, in the global theory the long rod broadside down is *unstable* because there is a perturbation which stands the rod end up.

tions (called $\mathscr{k}$) one can establish that there exists positive numbers ϱ_E and ξ^2 such that

$$\frac{I}{D} \leq \frac{1}{\varrho_E(\lambda_T, \lambda_\Gamma)} = \max_{\mathscr{k}} \frac{I}{D}, \qquad \frac{E}{D} \leq \frac{1}{\xi^2} = \max_{\mathscr{k}} \frac{E}{D}. \tag{4, 5}$$

If $\varrho < \varrho_E$, then

$$\frac{D - \varrho I}{E} \geq \sigma = \min_{\mathscr{k}} \frac{D - \varrho I}{E} \geq \xi^2 \left(1 - \frac{\varrho}{\varrho_E}\right), \tag{6, 7}$$

and we can integrate the inequality which arises from (3) and (6) to find a central estimate of energy theory

$$E(t) \leq E(0) \exp(-\sigma t). \tag{8}$$

The value of ξ^2 in (7) is not smaller than π^2 in a domain $\mathscr{V}$ which can be contained in a layer of height one [24, 28, 31, 34].

When $\varrho < \varrho_E$ *every* kinematically admissible disturbance will decay so that global stability with exponential decay is assured. When $\varrho > \varrho_E$, however, one can find an initial disturbance whose energy will increase initially. For example, the vector that wins the competition (4) makes E increase. This vector is called the most persistent disturbance. The motion is called *strongly stable* if every disturbance of it decays from the start. Motions which are stable but not strongly stable can have persistent disturbances which make E increase for a time. The energy limit defines a necessary and sufficient condition for strong stability [1].

---

[1] Energy methods are generally associated with the work of REYNOLDS [26] and ORR [23]. REYNOLDS derived the evolution equation (for the Navier-Stokes system) and used it to form stability criteria from guessed approximations to hydrodynamic solutions. ORR was the first to see that the proper procedure for drawing deductions from Reynold's integral was a variational method leading to eigenvalue problem for the variational equations. ORR did not prove that his criterion leads to strong stability, cf. Eq. (8), nor did he realise, as THOMAS [32] later proved, that these integrals could be estimated to establish the existence of strong stability in bounded domains. Despite ORR's insight in establishing a correct mathematical procedure for deducing stability limits from Reynolds' integrals I am not aware of a single stability limit calculated correctly by him. He has given incorrect values for plane Couette flow [12], for Hagen-Poiseuille flow [16] and HARRISON, on the urging of ORR, gave the incorrect limit for the stability of Couette flow between cylinders.

The modern version of energy theory dates from the work of SERRIN [28] which brings together the ideas of THOMAS and HOPF [10] on the existence of stability and ORR's variational method. SERRIN showed how to obtain an explicit criterion for strong stability of Navier-Stokes solutions by solving a variational problem for (essentially) the decay constant ξ^2, cf. Eq. (8). The proof that ORR's criterion implies strong stability I gave jointly with SERRIN [12]. For the existence of stability and for the important problem of the rate of decay of stable disturbances estimates of the value of the decay constant ξ^2 is the central problem and has been treated by SERRIN [28], VELTE [34], PAYNE and WEINBERGER [24] and SORGER [31].

The relevance of these stability definitions can be made clear by application to the problem of stability of Poiseuille flow. Theoretical and experimental results for this problem are shown in Fig. 1. The bottom line represents results of energy analysis [16]. Actually only the points on the curve are stability limits and the values of N give the azimuthal periodicity of the most persistent disturbance. Below the energy line the Poiseuille flow is strongly stable. The line ϱ_G gives the trend of experimental results for transition in cylindrical annuli from nine different authors. The top curve gives the result of a numerical integration of the linearized disturbance equations but only for axially symmetric disturbances [21]. The variation of the linear limit with the radius ratio η seems to have nothing common with experiments. On the other hand experimental and energy limits are alike with regard to their variation with η. For pipe flow ($\eta = 0$) the first (spiral) mode azimuthal periodicity of the most persistent disturbance is in qualitative agreement with the experimental observations of Fox, Lessen and Bhat [6]. The integers on the energy line give the variation of azimuthal periodicity with radius ratio $N(\eta)$ and these are offered as hypotheses for experiment.

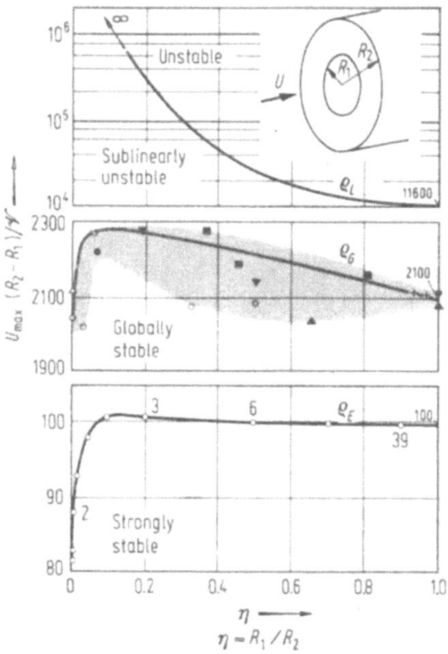

Fig. 1. Stability and instability limits for Poiseuille flow. The linear line ϱ_L is taken from a numerical calculation [21] for axisymmetric disturbance. The energy line ϱ_E [16] is also an outcome of numerical calculation, but over unrestricted periodic disturbances. The integers on ϱ_E indicate the azimuthal periodicity of the most unstable disturbance. The shaded band contains experimentally observed transition points compiled by R. Hanks [8] and the line called ϱ_G is an interpolation formula for the data. The line ϱ_G should roughly correspond to the as yet theoretically unknown global stability limit.

It is relevant to our considerations, but I cannot develop the idea here, that the linear limit collapses onto the energy line if the pipe is made to rotate ever faster about its symmetry axis [16, 25, 38].

For purposes of discussion let us assume that ϱ_G is the global stability limit. We want to know if the distinction made between "globally stable" and "strongly stable" is physically relevant. The experiments of Davies and White [4] and Lindgren [19, 20], in particular, but also of CAROTHERS [2], GRINDLEY and GIBSON [7] and NAUMANN [22] all suggest the strongly affirmative answer: In the region of strong stability decay of even very large norm disturbances is so rapid that the disturbance cannot be convected far downstream before it has decayed away. About their channel flow experiments (see Fig. 2) DAVIES and WHITE stress that there is:

"... a distinct deviation from true viscous flow if initial disturbing factors are present, and the influence of such disturbing factors does not disappear entirely until a second well-defined limit is reached, which has a value of about one tenth of the lower critical number" (global stability limit).

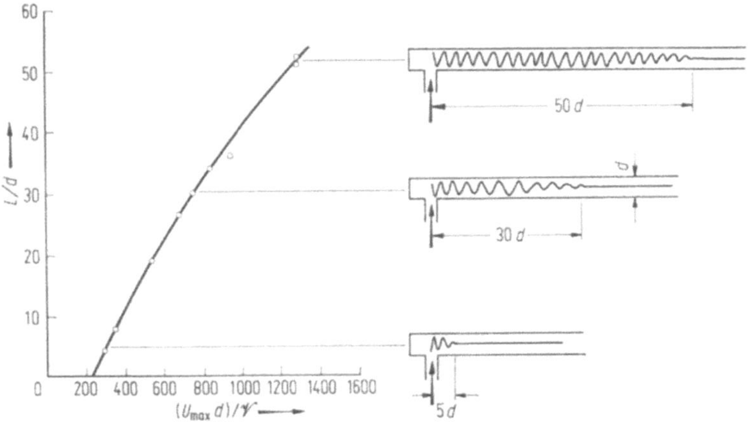

Fig. 2. Length of channel over which an entrance disturbance persists (after DAVIES and WHITE [4]). The experiment apparatus is a rectangular channel of large aspect ratio. The flow is disturbed at the entrance by turning a corner (as shown). The $R = 210$ intercept is an interpolation of the data and the true curve could be expected to pass through the origin. It is clear that the decay must become very rapid in the neighbourhood of $R = 200$. LINDGREN [19] says about this nearly identical result for round tubes that "... below a Reynolds number R of the order of 200 no disturbances were observed in the flow even quite near the tube inlet, however, strong disturbances were agitated in the entrance flow."

For the Boussinesq equations the problem of finding limits for stability is enriched by the presence of the numbers λ_T and λ_Γ. For fixed λ_T, λ_Γ the estimate (8) reduces the stability problem to a standard maximum problem (5) (this is equivalent to an eigenvalue problem for Euler's variational equations). Suppose (5) is solved. Then we choose λ_T and λ_Γ obtain the largest region of stability

$$\varrho_E = \max_{\lambda_T, \lambda_\Gamma} \varrho_E(\lambda_T, \lambda_\Gamma). \tag{9}$$

I call this the problem of the "optimum stability boundary". It can be resolved as an ordinary maximum problem. One finds that

$$\lambda_T \langle \boldsymbol{\eta}_T \cdot \boldsymbol{u}\theta \rangle = \langle \boldsymbol{\eta} \cdot \boldsymbol{u}\theta \rangle, \quad \lambda_\Gamma \langle \boldsymbol{\eta}_\Gamma \cdot \boldsymbol{u}\gamma \rangle = -\langle \boldsymbol{\eta} \cdot \boldsymbol{u}\gamma \rangle. \tag{10a, b}$$

The problem of the optimum stability boundary is relevant for Bénard convection [11, 12], convection in variable gravity spherical shells [15], convection with heat sources [14, 17], unsteady convection [29], thermohaline convection [30, 13], convection in porous materials [37], and for surface tension driven convection [3]. The heat source result is representative and is briefly reviewed below.

Consider the stability of the conduction solution in a homogeneous fluid layer ($a = b = \pi/2$, $\gamma = TM = TS = 0$) heated from below ($\boldsymbol{\eta} = -\boldsymbol{k}$, $\boldsymbol{k}$ is a unit vector in the direction of increasing z) with constant heat sources of intensity ζ ($\boldsymbol{\eta}_T = -(1 + \zeta z)\,\boldsymbol{k}$). We consider periodic disturbances of the conduction solution and integrations are relative to the period rectangle. Also, for simplicity, we consider free strips on which

$$w = \boldsymbol{u} \cdot \boldsymbol{k} = \boldsymbol{k} \times \partial \boldsymbol{u}/\partial z = 0, \qquad z = \pm 1/2. \tag{11}$$

The results hold under much more general circumstances. For the energy limit we maximize $I/D = TH/D = \langle (1 + \lambda_T \zeta z)\, w\theta \rangle / D$ and find the best value $\lambda_T = \langle w\theta \rangle / \langle (1 + \zeta z)\, w\theta \rangle$. When $\zeta = 0$, $\lambda_T = 1$ and the maximum problem defines the classical linear problem for Bénard convection so that $\varrho_E = \varrho_L$ and no sublinear instabilities exist when $\zeta = 0$ [12]. For $\zeta \neq 0$ [14] there is a band of values which increases in size with the magnitude of ζ in which sublinear solutions are possible (Fig. 3). Sublinear steady convection (in hexagons) can be rigorously demonstrated to exist[1] when $\zeta \neq 0$ but the proof requires a sufficiently small norm E and source intensity ζ. The only available global result is that given in Fig. 3.

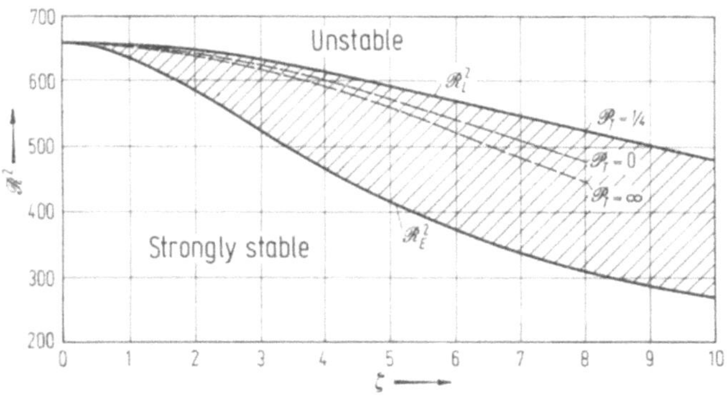

Fig. 3. Regions of stability and instability for a fluid layer with free surfaces heated from below and internally [17]. The principle of exchange of stability holds for the free surface problem even for $\varrho \neq 0$. Both the linear and energy limits are obtained from numerical integration. In the shaded region sublinear solutions of the Boussinesq equations cannot be excluded and at least one family is known to exist [5, 18] for sufficiently small norms and ζ, and is reported as observed [18].

[1] The proof is constructed using the method of Busse [1] and requires the use of a double series in parameters which Paul Fife and I [5] have proved converges. For the heat source problem the relevant calculation has been given, but incorrectly, by Krishnamurti [18] who also reports observing sublinear hexagons in experiments.

A similar situation holds for the salty fluid layer. Here we consider constant gradients of salt and temperature. Then, if the fluid is heated below ($\boldsymbol{\eta}_T = -\boldsymbol{k}$) and salty above ($\boldsymbol{\eta}_\Gamma = \boldsymbol{k}$) we find from (10) that $\lambda_T = \lambda_\Gamma = 1$ and again it is possible to completely exclude sublinear instabilities. But if the fluid is destabilized by heating and stabilized by salt[1] ($\boldsymbol{\eta}_\Gamma = \boldsymbol{\eta}_T = -\boldsymbol{k}$) then the linear limit can occur as either steady or oscillating convection [27, 35] and sublinear instabilities are possible, depending on the values $\mathscr{P}_\Gamma$ and $\mathscr{P}_T$. If one consults (10) one finds an acceptable value $\lambda_T = 1$ but also the unacceptable value $\lambda_\Gamma = -1$. It is not hard to show [30] that for this problem the maximum for (5) is taken on for $\gamma \neq 0$ (but $\theta \not\equiv 0$, $\boldsymbol{u} \neq 0$) so that (10b) holds but does not give the best value for λ_Γ. The criterion $\mathscr{R}^2 < 27\pi^4/4$ which guarantees linear stability when $\mathscr{C} = 0$ also holds globally when the salt field is *stabilizing* and $\mathscr{C} \neq 0$. This seems obvious enough from physical point of view but it is not at all obvious that any criterion better than $\mathscr{R}^2 \leq 27\pi^4/4$ will suffice for *strong stability*. There is, in fact, no better criterion which is independent of $\mathscr{P}_T$ and $\mathscr{P}_\Gamma$ because it is easy to show [30] that if $\mathscr{P}_T$ is fixed and $\mathscr{P}_\Gamma \to \infty$ then $27\pi^4/4$ is attained as the *linear limit*. In this way one concludes that $\mathscr{R}^2 \leq 27\pi^4/4$ is a necessary as well as sufficient condition for the stability of the conduction-diffusion solution when $\mathscr{P}_\Gamma \to \infty$.

It is clear then that if we are to have a global stability criterion which is better than $\mathscr{R}^2 < 27\pi^4/4$ it cannot be independent of $\mathscr{P}_\Gamma$ and $\mathscr{P}_T$. It is possible to have such a criterion in the class steady solutions of (1, 2). The criterion which we find below[2] does depend on $\mathscr{P}_T$ and $\mathscr{P}_\Gamma$ but not, as in (1, 2) on their separate values, but only on their ratio $\mathscr{P}_\Gamma/\mathscr{P}_T = \tau$.

Consider the steady disturbance equation (1, 2) for the constant gradient conduction-diffusion solution. Every solution of this problem is such that

$$\frac{1}{\varrho} = \frac{I}{D} \leq \max_{\mathscr{h}} \frac{I}{D}, \tag{12}$$

---

[1] This configuration models a "solar Pond". The "Pond" is a contained fluid layer which is both heated and salted below, so that the upper fluid layers thermally insulate the lower. Like the Dead Sea, the pond is washed by fresh water at its free surface and salted at its bottom, ensuring the existence of a stabilizing salt gradient in the vertical. The dark bottom of the pond is an effective absorber of radiative energy of the sun, which has the effect of heating the pond from below. Without the stable salt gradient, the limit of heating that could be achieved in this way is determined by the stability condition for the onset of convective motions [30].

[2] The central equation (13) which makes the improved criterion possible is valid for any bounded domain provided only that $\boldsymbol{u} \cdot \boldsymbol{N} = 0/\partial\mathscr{V}$ and div $\boldsymbol{u} = 0$. The theory is given here only for zero boundary values (but see [13]).

where we have put $\lambda_T = \lambda_\Gamma = 1$. Moreover, for each such solution since div $\boldsymbol{u} = 0$ and $\boldsymbol{u} \cdot \boldsymbol{k} = 0$, $z = \pm 1/2$, we have

$$0 = \langle \nabla \cdot (\boldsymbol{u}\theta\gamma) \rangle = \langle \theta (\boldsymbol{u} \cdot \nabla) \gamma + \gamma (\boldsymbol{u} \cdot \nabla) \theta \rangle. \tag{13}$$

Now, using (2b, c) we can write (13) as

$$\mathscr{I}(\varrho) = (1 + \tau) \langle \nabla\theta \cdot \nabla\gamma \rangle - \varrho \cos b \langle \theta w \rangle - \tau\varrho \sin b \langle \gamma w \rangle = 0. \tag{14}$$

If we require that functions competing for the maximum (12) in $\mathscr{k}$ also satisfy (14) then if $\cos b \neq 0$ (14) will not allow the previous unconstrained maximilizing function for which $\gamma \equiv 0$. It is also clear from (12) that values ϱ such that

$$\frac{1}{\varrho} > \max_{\mathscr{k}'} \frac{I}{D}$$

could not be associated with steady solutions of (1, 2).

Theorem: *The steady-conduction-diffusion solution, that is, the null solution of* (1, 2) *is unique provided only that*

$$\frac{1}{\varrho} > \max_{\mathscr{k}} \frac{I}{D} \equiv \frac{1}{\hat{\varrho}}, \tag{15}$$

where $\mathscr{k}'$ is the class kinematically admissible functions which also satisfy the side constraint $\mathscr{I}(\varrho) = 0$.

To find $1/\varrho$ we use variational calculus and require that $-\delta\{(I + 2\mu\mathscr{I})/D\} = 0$ where μ is a Lagrange multiplier. For the layer heated and salty below one finds that

$$\Delta \boldsymbol{u} + \varrho\{(\theta + \mu\tau\gamma)\sin b + \mu\theta\cos b\}\boldsymbol{k} + \nabla p = 0, \tag{16a}$$

$$\Delta\theta + \varrho(\sin b + \mu\cos b)\, w + \mu(1 + \tau)\Delta\gamma = 0, \tag{16b}$$

$$\Delta\gamma + \tau\varrho\mu \sin b w + \mu(1 + \tau)\Delta\theta = 0 \tag{16c}$$

for doubly periodic functions (in X, Y) such that div $\boldsymbol{u} = 0/\mathscr{V}$ and $\theta = \gamma = 0$, $z = \pm 1/2$. Equations (16b) and (16c) can be rearranged as

$$\Delta\theta + (\varrho\Phi w)/\chi = 0 \text{ and } \Delta(\Psi\theta + \Phi\gamma) = 0, \tag{17a, b}$$

where $\chi = 1 - \mu^2(1 + \tau)^2$, $\Phi = \mu\cos b + [1 - \mu^2\tau(1 + \tau)]\sin b$ and $\Psi = \mu \sin b + \mu^2 \cos b(1 + \tau)$. The boundary condition, (17b) and potential theory show that $\Psi\theta + \Phi\gamma = 0$ at each point in $\mathscr{V}$.

Let Λ^2 be the minimum eigenvalue for the problem $\Delta^3 w - \Lambda^2(\partial^2/\partial X^2 + \partial^2/\partial Y^2)\, w = 0$ for w doubly-periodic (in X, Y) satisfying rigid or free boundary conditions at $z = \pm 1/2$. By reducting the system (16a) and (17a, b) to this problem we show that

$$\left(\frac{\Lambda}{\varrho \sin b}\right)^2 \chi = (1 + 2\mu\cot^2 b + \mu^2\cot^2 b + \mu^2\tau^2) - 2\mu^2\tau(1 + \tau)(1 + \mu\cot b). \tag{18}$$

To complete the problem we select the multiplier μ to satisfy the condition $\mathscr{I} = 0$. This is equivalent to the equation

$$\frac{-\cot b}{1 + \cot^2 b} = \frac{\mu}{h(\mu^2, \tau)}, \tag{19}$$

where

$$h(\mu^2, \tau) = \mu^2 + \left(1 - \mu^2\tau\left(\frac{1}{2} + \tau\right)\right)^2 + \mu^4(1 + 2\tau^2 + 11\tau/4).$$

To find (19) set $\gamma = -\theta\Psi/\Phi$ in $\mathscr{I}$, then use (17a) to eliminate $\langle w\theta\rangle$ and rearrange.

The solution of the relatively simple algebraic problem (18, 19) is given in [13]. The results for free surfaces ($\Lambda^2 = 27\pi^4/4$) are these: when $\tau = 0$ the linear and constrained energy results coincide. There are therefore no steady sublinear solutions in either of the limits ($\mathscr{P}_\Gamma \to \infty$) or ($\mathscr{P}_T \to \infty$, $\mathscr{P}_\Gamma \to 0$). For other values of $\mathscr{P}_T$ and $\mathscr{P}_\Gamma$ a band of possible sublinear solutions cannot be excluded. For very large $\mathscr{C}$ the result guarantees uniqueness when $\mathscr{R}^4 < \frac{27}{4}\pi^4\mathscr{C}^2/\tau^2$. We have represented the result for the value $\tau = 100$ in Fig. 4. The estimates hold relative to periodic disturbances but are otherwise unrestricted. One might expect that the sublinear solutions, like the hexagons mentioned earlier and the sublinear turbulence in pipes, have a three-dimensional structure. The small scale sublinear turbulent motions observed by TURNER and STOMMEL [33] in their experiment on

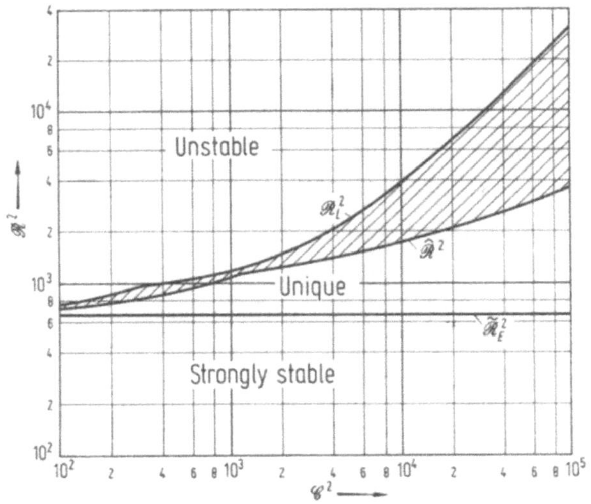

Fig. 4. Stability and uniqueness limits for the conduction-diffusion solution with a stabilizing solute ($\tau = \sqrt{10}$, $\mathscr{P}_T = 100$) gradient. For small $\mathscr{C}^2$ the linear limit $\mathscr{R}_L^2$ first occurs as exchange of stability and then, after the kink, as oscillations. The uniqueness limit $\hat{\mathscr{R}}^2$ also has a kink. The shaded region is open to steady sublinear solutions.

thermohaline convection in an open box are three-dimensional. On the other hand, two-dimensional, steady sublinear solutions have been calculated theoretically and by different (approximate) methods by SANI [27] and VERONIS [35, 36]. For these reasons it seems likely that sublinear solutions do exist and are perhaps very deep.

In closing, we note that the sense in which the uniqueness criterion of the theorem also implies stability has yet to be delineated. Moreover, though the result for $\tau = 0$ and $\tau = \infty$ cannot be improved it is likely that a better criterion can be found for intermediate values by an optimal selection of the coupling constants λ_T and λ_Γ.

Note added in proof: The problem mentioned in the last paragraph of this paper is solved in [39].

References

1. BUSSE, F.: Dissertation, Munich 1962. Cf. also The stability of finite amplitude cellular convection and its relation to an extremum principle. J. Fluid Mech. **4**, 225 (1968).
2. CAROTHERS, S.: Portland experiments on the flow of oil in tubes. Proc. Roy. Soc. **87**, 154 (1912).
3. DAVIS, S.: Surface tension driven convection by the method of energy. J. Fluid Mech. **39**, 347—359 (1969).
4. DAVIES, S. J., WHITE, C. M.: An experimental study of the flow of water in pipe of rectangular section. Proc. Roy. Soc. **119**, 92 (1928).
5. FIFE, P., JOSEPH, D.: Existence of convective solutions of the generalized Benard problem which are analytic in their norm. Arch. Rat. Mech. Anal. **33**, 116 (1969).
6. FOX, J., LESSEN, M., BHAT, W.: Experimental investigation of the stability of Hagen-Poiseuille flow. Phys. Fluids **11**, 1 (1968).
7. GRINDLEY, J., GIBSON, A.: On the frictional resistances to the flow of air through a pipe. Proc. Roy. Soc. **80**, 114 (1907/1908).
8. HANKS, R.: The laminar-turbulent transition for flow in a pipe, concentric annuli, and parallel plates. A.I.Ch.E.J. **9**, 45 (1963).
9. HARRISON, W.: On the stability of the steady motion of viscous liquid contained between two rotating circular cylinders. Trans. Camb. Phil. Soc. **22**, 455 (1921).
10. HOPF, E.: On nonlinear partial differential equations. Lecture series of the Symposium on partial differential equations. University of California 1955, pp. 7—11.
11. JOSEPH, D.: On the stability of the Boussinesq equations. Arch. Rat. Mech. Anal. **20**, 59 (1965).
12. JOSEPH, D.: Nonlinear stability of the Boussinesq equations by the method of energy. Arch. Rat. Mech. Anal. **22**, 163 (1966).
13. JOSEPH, D.: Uniqueness criteria for the conduction-diffusion of the Boussinesq equations. Arch. Rat. Mech. Anal. **35**, 169—177 (1969).
14. JOSEPH, D., SHIR, C. C.: Subcritical convective instability. Part 1: Fluid layers heated from below and internally. J. Fluid Mech. **26**, 753—768 (1966).
15. JOSEPH, D., CARMI, S.: Part 2: Spherical shells. J. Fluid Mech. **26**, 769 (1966).
16. JOSEPH, D., CARMI, S.: Stability of Poiseuille flow in pipes, annuli and channels. Quart. Appl. Math. **26**, 575—599 (1969).

17. Joseph, D., Goldstein, R., Graham, D.: Subcritical instability and exchange of stability in a horizontal fluid layer. Phys. Fluids **11**, 903 (1968).
18. Krishnamurti, R.: Finite amplitude convection with changing mean temperature. Part I: Theory. J. Fluid Mech. **33**, 445 (1968).
19. Lindgren, R.: The transition process and other phenomena in viscous flow. Arkiv för Fysik **12**, 1—169 (1957).
20. Lindgren, R.: Liquid flow in tubes. Part I: The transition process under highly disturbed entrance flow conditions. Arkiv för Fysik **15**, 97 (1959).
21. Mott, J., Joseph, D.: Stability of parallel flow between concentric cylinders. Phys. Fluids **11**, 2065 (1968).
22. Naumann, A.: Experimentelle Untersuchungen über die Entstehung der turbulenten Rohrströmung. Forsch. Geb. Ingw. **2**, 85 (1931).
23. Orr, W. McF.: The stability or instability of steady motions of a liquid. Part II: A viscous liquid. Proc. Roy. Irish Acad. **A 27**, 69 (1907).
24. Payne, L., Weinberger, H.: An exact stability bound for Navier-Stokes flow in a sphere. Nonlinear problems, ed. R. E. Langer, University of Wisconsin Press 1963.
25. Pedley, T.: On the stability of viscous flow in a rapidly rotating pipe. J. Fluid Mech. **35**, 97 (1969).
26. Reynolds, O.: On the dynamical theory of incompressible viscous fluids and the determination of the criterion. Phil. Trans. Roy. Soc. London **A 186**, 123 (1895).
27. Sani, R.: On finite amplitude roll cell disturbances in a fluid layer subjected to heat and mass transfer. A.I.Ch.E.J. **11**, 971 (1965).
28. Serrin, J.: On the stability of viscous fluid motions. Arch. Rat. Mech. Anal. **3**, 1 (1959).
29. Shir, C. C.: Ph. D. thesis, University of Minnesota, 1967.
30. Shir, C. C., Joseph, D.: Convective instability in a temperature and concentration field. Arch. Rat. Mech. Anal. **30**, 38 (1968).
31. Sorger, P.: Über ein Variationsproblem aus der nichtlinearen Stabilitätstheorie zäher inkompressibler Strömungen. Z. angew. Math. Phys. **17**, 201 (1966).
32. Thomas, T. Y.: On the uniform convergence of the solutions of the Navier-Stokes equations. Proc. Nat. Acad. Sci. U.S.A. **29**, 243 (1943).
33. Turner, J., Stommel, A.: A new case of convection in the presence of combined vertical salinity and temperature gradient. Proc. Nat. Acad. Sci. U.S.A. **52**, 49—53 (1964).
34. Velte, W.: Über ein Stabilitätskriterium der Hydrodynamik. Arch. Rat. Mech. Anal. **9**, 9 (1962).
35. Veronis, G.: On finite amplitude instability in thermohaline convection. J. Marine Res. **23**, 1 (1965).
36. Veronis, G.: Effect of a stabilizing gradient of solute on thermal convection. J. Fluid Mech. **34**, 315 (1968).
37. Westbrook, D.: The stability of convective flow in a porous medium. Phys. Fluids **12**, 1547—1551 (1969).
38. Joseph, D. D., Munson, B.: Global stability of spiral flow. J. Fluid Mech. (in press).
39. Joseph, D. D.: Global stability of the conduction-diffusion solution. Arch. Rat. Mech. Anal. **36**, 285—292 (1970).

Interfacial Instability between Fluids and Granular Beds

By

M. H. Gradowczyk

Comisión Nacional de Energía Atómica, Buenos Aires, Argentinia

1. Introduction

When a fluid flows over an erodible non-cohesive bed, the interfacial boundary surface may become unstable. Some of the most striking characteristics of this instability, for the case of a free-surface flow, are shown in Fig. 1. The different bed-form domains can be classified according to the flow régime into:

(α) Lower régime[1] ($F < F_f < 1$); ripples ($kH > 1$), "incipient" flat bed, ripples on dunes, dunes ($kH < 1$); all these waves move downstream.

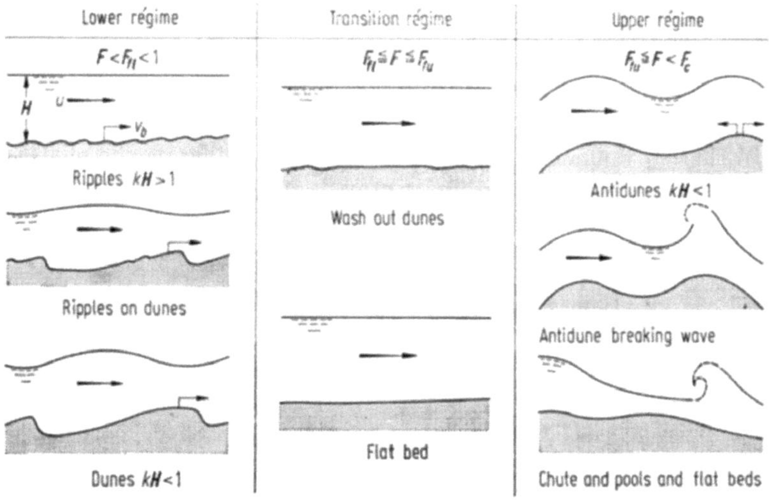

Fig. 1. Different bed régimes in one-dimensional open-channel flow.

[1] Symbols are defined at end of work.

(β) Transition régime ($F_{fl} \leq F \leq F_{fu}$): wash-out dunes and flat beds.

(γ) Upper régime ($F_{fu} < F < F_c$, $F_c > 1$): flat beds, antidunes moving upstream and downstream, stationary waves[1] and chute and pools.

Different attempts[2] have been made in order to explain the origin of that complex interaction between two-phase media:

(1) quasi-steady[3] shallow water theory: EXNER [1], REYNOLDS [2], RAUDIVIKI [3].

(2) Linear (unsteady) shallow-water theory: GRADOWCZYK [4].

(3) Quasisteady linear potential theory: ANDERSON [5], KENNEDY [6, 7], REYNOLDS [2].

(4) Laminar-turbulent model: DUNGAN SMITH [8].

(5) Rotational model with varying density: ENGELUND [9].

In models 1 to 4 the motion of bed particles is characterized by the bed-material discharge q_b — an empirical formula related to flow variables —, which measures the amount of convected material transported by the flow mostly near the bed. The effect of varying flow density along the water depth, which is produced by the mixture of fluid + suspended sediment, has been taken into account in model 5.

These models give only a partial explanation of the interfacial instability and even their results are sometimes contradicting. For example, DUNGAN SMITH's analysis, valid for small Froude numbers (régime α), assures that all bed perturbations will lead to unstable bed forms; model 2 and the results of this work predicts neutral bed forms whereas model 5 shows that only a flat bed is possible in the same region.

In this work model 3 will be extended by considering the unsteadiness of the flow and the non-linear growth of bed perturbations will be discussed from the point of view of kinematic waves.

2. Bed waves as neutral infinitesimal waves

Consider the steady and uniform flow of Fig. 2. The possibility of having a system of neutral coupled bed and surface waves of the form

$$\eta = E^* \sin k(x - vt), \quad \xi = A^* \sin k(x - vt) \qquad (1, 2)$$

[1] All waves of régime γ are relatively short, i.e. $kH < 1$.

[2] Unfortunately the author was unable to trace eastern references.

[3] In a quasi-steady (unsteady) model the unsteadiness of the fluid flow due to the bed perturbation is (is not) disregarded in comparison with the temporal changes of the bed.

will be investigated. These perturbations can be studied by introduction of a potential $\varphi(x, y, t)$. If η/H, $\xi/H \ll 1$, φ satisfies

$$\nabla^2\varphi = 0, \qquad 0 < y < -h, \tag{3}$$

$$\xi_t + U\xi_x - \varphi_y = 0, \quad \varphi_t + U\varphi_x + g\xi = 0 \quad \text{for } y = 0, \tag{4, 5}$$

$$\eta_t + U\eta_x - \varphi_y = 0 \qquad \text{for } y = -H. \tag{6}$$

In order to make the problem determinate, an equation that gives the balance of convected bed material, will be introduced:

$$\eta_t + q_{bx} = 0 \text{ for } y = -H, \tag{7'}$$

where q_b is usually given by an empirical formula of the type

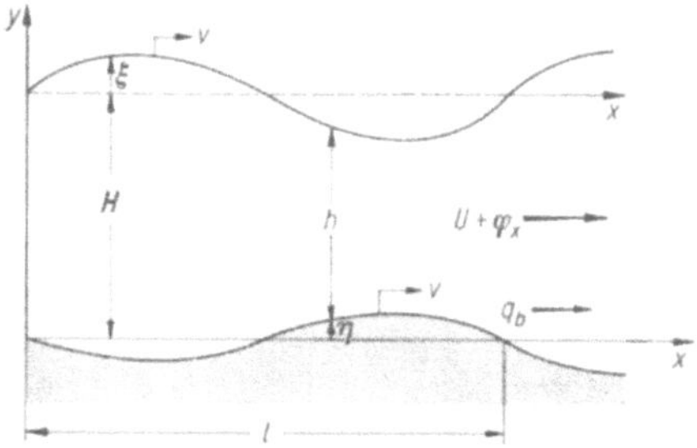

Fig. 2. Symbols in potential theory.

$$q_b = f(U + \varphi_x; h; \tau_0) \quad \text{for} \quad y = -H, \tag{8}$$

where $h = H + \xi - \eta$. The linearized version of (7′) reads

$$\eta_t + q_1\varphi_{xx} + q_2(\xi_x - \eta_x) = 0, \tag{7}$$

where $q_1 = \partial f/\partial U$, $q_2 = \partial f/\partial H$ are computed at the undisturbed flow. Let us assume that the solution of Eqs. (3—7) is $\varphi = \psi(y) \cos k(x - vt)$, then

$$E^* = \left[1 - \left(\frac{F_f}{F}\right)^2\right] A^* \cosh kH, \; F_f^2 = \frac{\tanh kH}{kH(1 - \bar{v}^2)}, \tag{9, 10}$$

which in turn yield

$$\begin{aligned} &F^2\{\bar{v}^3(1 + NM) - \bar{v}^2[2 + (3N + Q)M]\} \\ &+ \bar{v}\left\{F^2[1 + (3N + 2Q)M] - \frac{\tanh kH}{kH} + \bar{q}_1 M\right\} \\ &+ M\left[\bar{q}_1 - (N + Q)F^2 - \bar{q}_2\frac{\tanh kH}{kH}\right] = 0, \end{aligned} \tag{11}$$

as the dispersion equation for the dimensionless wave velocity $\bar{v} = v/U$. Its smallest root is

$$\bar{v}_3 = \bar{v}_b = \frac{kH\bar{q}_1 - \bar{q}_2 \tanh kH - (N + Q)F^2}{\tanh kH - kHF^2} M + O(M^2), \tag{12}$$

which gives the velocity of the bed wave; the remaining ones give the velocity of surface (gravity) waves moving over erodible beds. It is clear that possible bed-form régimes will depend on F, kH and M. Since $M \ll 1$, it will be considered only their dependence on F and kH, M being considered as a parameter. Taking $\bar{q}_2 = 0$, an assumption only valid for one-dimensional wave propagation, one observes from

(12) that for $F^2 = \coth kH/kH = F_1^2$, say, bed an surface waves are stationary. Therefore it is conjectured [2] that F_1 is the limit of possible bed forms. Experimental evidence gathered by GUY et al. [10] is in agreement with that conjecture, as shown in Fig. 3.

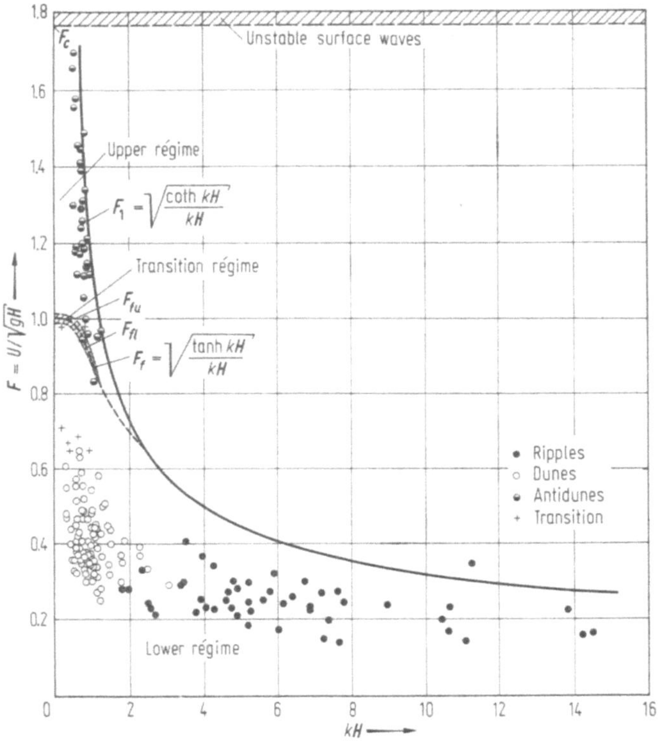

Fig. 3. Bed-form domains in one-dimensional flow for a given value of the erosion number M.

When $F < F_f$, the model gives neutral infinitesimal waves, which are out of phase with the surface wave, that move in the direction of the flow. These waves resemble ripples of régime (α). When the velocity of the flow is increased, the bed velocity $\bar{v}_b$ increases. In the neighbourhood of F_f, $\bar{v}_b = \{[1 - (\tanh kH)^2]\,\bar{q}_1 M/2\}^{1/2} + O(M)$ and the bed amplitude will tend to zero[1]. Hence the transition régime can be determined for values $|E^*/A^*| = \mu \ll 1$ as

$$F_{fl} \leq F \leq F_{fu} \tag{13}$$

where

$$F_{fu,l}^2 = F_f^2 \Big/ \left(1 \mp \frac{\mu}{\cosh kH}\right).$$

[1] Prof. C. C. MEI of Massachusetts Institute of Technology, in an unpublished work, has shown that the singularity for $F = F_f$ can be removed by considering the non-linear terms neglected in (4, 5, 6).

A similar discussion to the one given in [4] shows that in the region of the diagram (F^2, kH) limited by F_{fu} and F_1 antidunes moving upstream, downstream and stationary antidunes are possible bed forms. It is to be noticed that when $F \geq F_c$ (Gradowczyk [4]), the free-surface may become unstable and discontinuous water waves as chutes and pools may appear. Thus, F_c gives an upper limit to the diagram of Fig. 1. An important feature of our analysis is the fact that the main features of previous work [4] have been extended to small bed forms like ripples. The analysis shows that proper consideration of *all* time derivatives in Eqs. (4, 5, 6) leads to a reasonable explanation of some features of bed-form régimes, without the use of Kennedy's [6] fase shift δ.

It may be of interest to compare our results with those derived by Engelund, who used the Boussinesq approximation for the fluid and a diffusion-type equation for the concentration of suspended particles. In his first model, Engelund obtained that antidunes are unstable bed forms, i.e. their amplitude grows indefinitely in time. His antidune region practically coincides with the upper régime of Fig. 3. He also found that for the lower curve ($F \approx F_f$) surface waves give stable flat bed configurations and for the upper curve ($F \approx F_1$) the bed and surface wave are in phase and stationary, which is in agreement with potential theory. In a second model that takes shear stresses into account, Engelund showed that neutral (stable) antidunes may exist inbetween curves very close to F_f and F_1. The two models gave flat beds in the lower régime, since bed waves are damped in time.

The fact that the curves of neutral instability of [9] are in close agreement with potential theory indicates that regions of stability are practically independent of the suspended particles carried by the flow.

3. Non-linear growth of bed perturbations

Experimental work has shown that typical non-linear phenomena as the growth of a perturbation into a fully developed dune or the passage from ripples to dunes occurs even when $F \ll 1$. If dynamical effects are altogether neglected, it turns out that $h + e = d =$ const., and $uh = q =$ const. The only equation that needs to be considered is (7′), which can be written as

$$e_t + v_b e_x = 0, \tag{14}$$

where $v_b(e) = (\partial q_b/\partial e)_{x=\text{const.}} = (2 + n)_{rq} M/(d - e)(1 - H)$ when q_b is expressed by Meyer-Peter-Müller formula [4]. This rather simple model, first proposed by Exner [1], governs the propagation of non-linear bed disturbances in one-dimensional channel flow. Thus, a bed perturbation propagates like a kinematic wave (Lighthill and Whitham [11]).

Let us consider the propagation of a sinusoidal perturbation $e(x, t_0) = E \sin kx$; then the solution e is $E \sin k[x - v_b(e)\, t]$. Since v_b is a monotonically increasing function of e, for increasing values of e the wavelets spread out in the backward part of the perturbation. In the forward part e decreases and the wavelets will cross for a sufficiently long time. Since the crossing of two wavelets indicate that the wave is multivalued, this indeterminancy can be overcome by the insertion of shocks. The minimum time t_s for which the first crossing will occur is $t_0 - [\max\,(\partial v_b/\partial x)_{t_0}]^{-1}$ and shocks must be introduced after this time. Fig. 4 shows the evolution of the sinusoidal perturbation; the

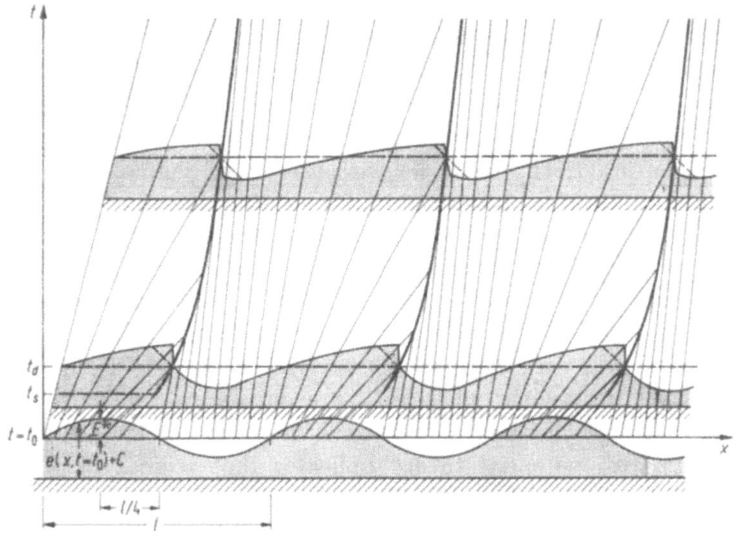

Fig. 4. Growth of a sinusoidal perturbation into a fully developed bed form.

shocks drawn with a thick line have been computed using a graphical procedure given by Lighthill [12]. The resulting bed forms[1] resemble ripples or dunes of régime (α). The instant t_d for which the semi-period wave develops into a fully developed wave is equal to $t_d = [v_b(e = E) - v_b(0)]/4$. If $v_b(E) \gg v_b(0)$, then $t_d = (1 - \boldsymbol{H})\,(d - e)\,[4(2 + n)\, r q_b]^{-1}$; i.e. the dune growth time increases (decreases) when the bed discharge decreases (increases) and the remaining parameters are held constant. This estimate is compared in Table 1 with actual observations of the dune-growth phenomenon due to Raichlen and Kennedy [13]. Since the wavelength l is unknown, relative values t_d/t_{dc} are compared, where t_{dc} corresponds to run 6—10.

[1] Due to the lack of cohesion of the bed particles the wave front must have an angle $\theta < 90°$ with the horizontal, as shown in Figs. 4 and 5.

Table 1

Run	t_d/t_{dc}	
	measured	computed
6–10	0.147	0.100
6–11	1	1
8–5	0.470	0.423

These bed waves will continue to spread out and change continuously their form and the velocity of the shocks also diminishes. Since the relation $v_{\text{shock}} = [q_b]/[e]$ must hold along the shock and $v_{\text{shock}} = \beta(t)\, v_b$ $(\beta(t) < 1)$, the dune depth $[e] \propto h(1 - \boldsymbol{H})/[(2 + n)\, r\beta]$. This estimate compares well with the data of Ref. [13], and is similar to an empirical formula proposed by Knoroz [14].

A different kinematic bed wave that appears when a flat erodible bed flushes with a rigid bed has been discussed theoretically and experimentally by Gradowczyk, Maggiolo and Raggi [15].

If the bed perturbation is not strictly periodic but modulated in the x-direction as shown in Fig. 5, the application of the kinematic method

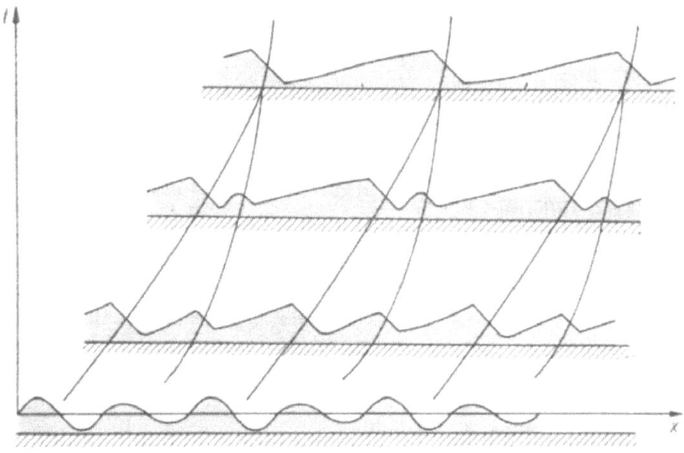

Fig. 5. Growth of a specially modulated bed perturbation.

shows that the faster bed waves that develop earlier shocks will eventually overtake the slower waves and consequently will increase their lengths. This is a rough explanation of why short waves as ripples grow into longer bed forms: dunes.

Nomenclature

$\boldsymbol{F} = U/\sqrt{g\boldsymbol{H}}$	Froude number of undisturbed flow
$U, \boldsymbol{H}$	mean velocity and mean depth of undisturbed flow
$k = 2\pi/l$	wavenumber
l	wavelength
q_b	bed-material discharge
τ_0	threshold shear stress of granular material
τ	shear stress transmitted by the fluid to the bed
$\bar{q}_1 = Q_b^{-1} q_1$	
$\bar{q}_2 = Q_b^{-1} q_2$	
Q_b	bed discharge of undisturbed flow
$\boldsymbol{M} = Q_b/(U\boldsymbol{H})$	erosion number
$N = k\boldsymbol{H}\bar{q}_1 \tanh k\boldsymbol{H}$	
$Q = \bar{q}_2(1 - \cosh k\boldsymbol{H})/\cosh k\boldsymbol{H}$	
$\boldsymbol{H} = \lvert\tau\rvert/\tau_0$	bed shear number
n, r	positive exponents
t_0	initial time

References

1. Exner, F.: Sitzungsberichte Oest. Akad. der Wiss. Abt. II, **134**, 165 (1925).
2. Reynolds, A. J.: J. Fluid Mech. **22**, 113 (1965).
3. Raudiviki: J. Fluid Mech. **26**, 507 (1966).
4. Gradowczyk, M. H.: J. Fluid Mech. **33**, 93 (1968).
5. Anderson, A. G.: Proc. 3rd West. Conf. on Fluid Mech., University of Minnesota 1953, p. 379.
6. Kennedy, J. F.: J. Fluid Mech. **16**, 521 (1963).
7. Kennedy, J. F.: J. Geophys. Res. **69**, 1517 (1964).
8. Dungan Smith, J.: Stability of a sand bed subjected to a shear flow of low Froude number. Unpublished manuscript, Dept. of Oceanography, University of Washington.
9. Engelund, F.: On the formation of standing waves and antidunes in alluvial channels. Unpublished manuscript, Hydraulic Laboratory, Tech. University of Denmark.
10. Guy, H. P., Simons, D. B., Richardson, E. V.: U.S. Geol. Survey, Prof. Paper 462-I, 1966.
11. Lighthill, M. J., Whitham, G. B.: Proc. Roy. Soc. **A 229**, 281 (1955).
12. Lighthill, M. J.: Survey in Mechanics, eds. G. K. Batchelor and R. M. Davies, London: Cambridge Univ. Press 1956, p. 250.
13. Raichlen, F., Kennedy, J. F.: Proc. XI. Congress of I.A.H.R., Leningrad 1965, paper 3.7.
14. Knoroz, V. Z.: Trudy Laboratori i Plantin, Gidrouzlov. **62**, 75 (1959).
15. Gradowczyk, M. H., Maggiolo, O. J., Raggi, R.: Proc. XII. Congress of I.A.H.R., Fort Collins, **3**, 314 (1967).

Unstable Flows in Two Dimensions: Comparison of Laboratory Experiments with Numerical Simulation

By

N. J. Zabusky and **G. S. Deem**

Bell Telephone Laboratories, Whippany, N.J., U.S.A.

I. Introduction

In recent years analysts have expended much effort to characterize the development of unstable two-dimensional flows described by the Navier-Stokes equations. Sato and Kuriki (SK) [1] and Mattingly [2], have performed laboratory experiments to characterize the development of unstable two-dimensional wakes. The gap between analysis and laboratory measurements is still large because of the usual difficulty of solving nonlinear dynamical problems.

This paper presents a brief review of the results of a series of numerical finite difference calculations with the two-dimensional incompressible Navier-Stokes equations (initial boundary-value problems) [3]. We have simulated the SK wake experiment and obtained results that have a large number of features in common with the experiment, including:

(1) development of a double row of elliptical vortices;

(2) mean flow evolution (profile spread) including "backflow";

(3) good first harmonic "cross-stream" amplitude and phase evolution;

(4) the effects of Reynolds number, Re; computations with $\nu = 0$ or small show little qualitative differences and demonstrate that the measured results of the SK experiment are inviscid phenomena.

These computational results are being studied in the analytical/computational/experimental/synergetic (ACES) [4] working mode. They are providing us with a deeper insight of the nonlinear dynamical processes involved.

II. Equations, boundary and initial conditions

A. Equations

We solve the two-dimensional Navier-Stokes equations

$$\partial_t \boldsymbol{u} + \boldsymbol{u} \cdot \nabla \boldsymbol{u} = -\nabla p + \nu \nabla^2 \boldsymbol{u}, \tag{2.1}$$

$$\nabla \cdot \boldsymbol{u} = 0, \tag{2.2}$$

$$\boldsymbol{u} = u(x, y, t)\, \boldsymbol{e}_x + v(x, y, t)\, \boldsymbol{e}_y, \tag{2.3}$$

on a rectangle with periodic boundary conditions in the x (downstream) and y (cross-stream) directions. We have made inviscid calculations for (2.1) with $\nu = 0$ (described below) and finite Reynolds number calculations with Re $= 750$ (see [3]).

The numerical scheme used is second-order; it employs an energy conserving algorithm for the convective terms ($\boldsymbol{u} \cdot \nabla \boldsymbol{u}$) and a DuFort-Frankel algorithm for the viscous terms. The pressure is calculated by demanding that (2.2) be identically satisfied in finite difference form. The resulting discrete Poisson equation is solved using an algorithm of Hockney.

B. Boundary and initial conditions

Fig. 1 shows the flat plate wake experiment we are simulating. We *assume* that the flow is confined to a rectangle and impose periodic

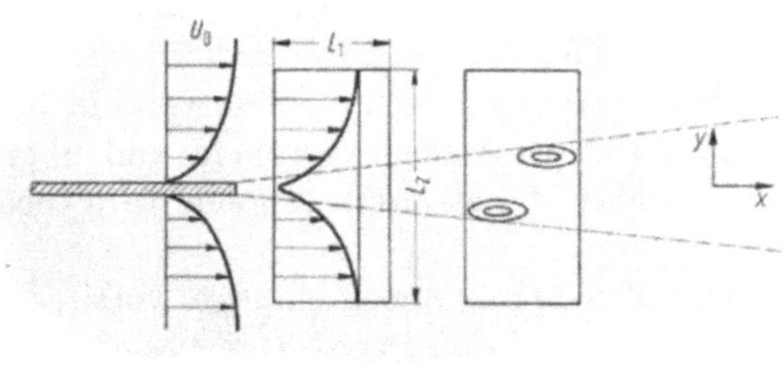

Fig. 1. Schematic of two-dimensional flat plate wake evolution.

boundary conditions in x and y. The cross-stream periodicity is a reasonable approximation since the flow is observed, *a posteriori*, to remain laterally confined. The downstream periodicity condition is more difficult to justify. This condition inhibits the coupling of energy to downstream wavelengths larger than the box length. This is a good approximation to laboratory reality if: the linearly most unstable wavelength is small compared to the box length; initially, there is little energy in the long wavelengths; strong subharmonic coupling does not exist (as in, for example, a separated shear layer). The last two conditions are properties of the SK experiment. We make a Taylor, or frozen flow identification of experimental downstream distance with the computation time t to compare our results with experiment.

For the initial velocity field we take a parallel flow $U(y)$ with perturbation $\tilde{\boldsymbol{u}}$

$$\boldsymbol{u}(x, y, 0) = \boldsymbol{e}_x U(y) + \tilde{\boldsymbol{u}}(x, y, 0), \tag{2.4}$$

where U is either a Bickley or a Gaussian profile

$$U(y) = U_0 \operatorname{sech}^2(y/\Delta^2) \tag{2.5}$$

or

$$U(y) = U \exp(-y^2/\Delta^2). \tag{2.6}$$

The initial perturbation $\tilde{\boldsymbol{u}}(x, y, 0)$ is formed from eigenfunctions of the Rayleigh equation

$$[d^2/dy^2 - \alpha^2 - (d^2U/dy^2)/(U - c)]\, \varphi = 0, \tag{2.7}$$

where

$$\tilde{\boldsymbol{u}} = \boldsymbol{e}_x\, \partial_y \psi - \boldsymbol{e}_y\, \partial_x \psi, \tag{2.8}$$

$$\psi = \operatorname{Re}\,\{\varphi(y) \exp[i\alpha(x - ct)]\}. \tag{2.9}$$

Table 1 summarizes the parameters and results of two computations using (2.5) and (2.6).

Table 1. *Parameters and results of computation*

Profile	Lattice size	Box dimensions $(L_1 \times L_2)$	Cross-stream lattice intervals in Δ	$L_2/2$	ν	U_0
Bickley (2.5)	64×64	20.94×10	6.4	10	0	1.0
Gaussian (2.6)	128×128	8.38×16	12.8	10	0	0.692

Profile	Number of (equally excited) modes used initially	Fraction of total energy in initial perturbation $\tilde{\boldsymbol{u}}$	Exponentiation time t_E of most unstable mode
Bickley (2.5)	5	0.0148	6.21
Gaussian (2.6)	1	0.0404	9.32

Profile	Computation time step	Run duration time steps	Run duration (t_F/t_E)
Bickley (2.5)	0.05	600	4.83
Gaussian (2.6)	0.05	1200	6.44

III. Results

Initially the results of both computations agree well with linear theory. Exponentiation times $t_E = (\alpha c_i)^{-1}$ and phase velocities c_r agree with eigenvalues $c = c_r + ic_i$ of the Rayleigh equation.

After an exponentiation time t_E of the most unstable mode we observe higher harmonic generation and the instability "saturates" with the formation of a stable vortex structure. Fig. 2 shows computer print-outs of the $2D$ flow fields, with superposed heavy lines to bring out salient features. The Bickley results at $t/t_E = 0$ and 4.83 are given

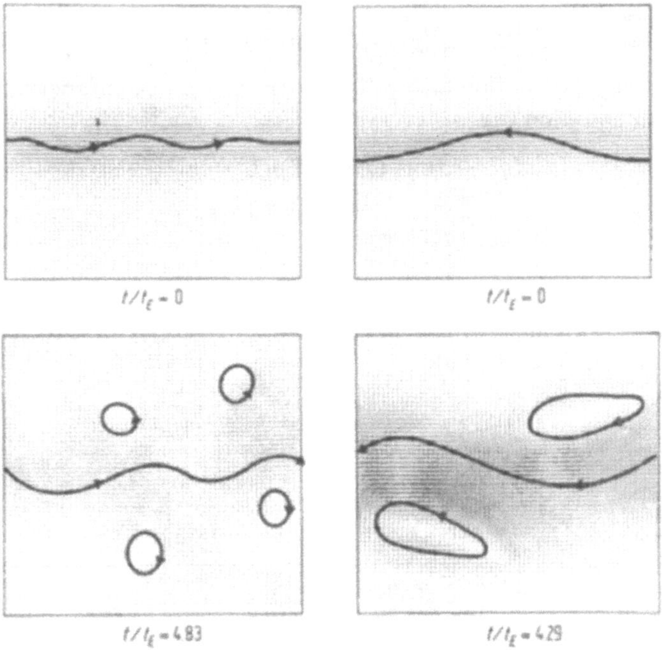

Fig. 2. Velocity fields for unstable shear profiles with $\nu = 0$.
Left: Bickley profile, Eq. (2.5), excited by 5 equal modes. Mode 3 is most unstable.
Right: Gaussian profile, Eq. (2.6), excited by its most unstable mode.

on the left and those for the Gaussian at $t/t_E = 0$ and 4.29 are given on the right. For the latter we see an initial sinusoidal disturbance "roll up" into a pair of *elliptical vortices* near the center of the flow region. Note that the backflow or overshoot observed by SK is simply the result of vortex circulation at the lateral flow extremities. The 5-mode Bickley profile quickly feeds energy into the third (most unstable) mode but after a time, $t/t_E = 4.83$ energy has been coupled into the lower modes and one sees only four distinct vortices. Note that the flow has spread to the boundaries.

Fig. 3 shows cross-stream distributions of downstream averaged properties of the Gaussian wake. These simulate time-averaged measurements of SK. Fig. 3a shows the spread in time of the mean downstream velocity and the development of an overshoot (compare SK, Fig. 22). Fig. 3b shows the corresponding first harmonic energy grow,

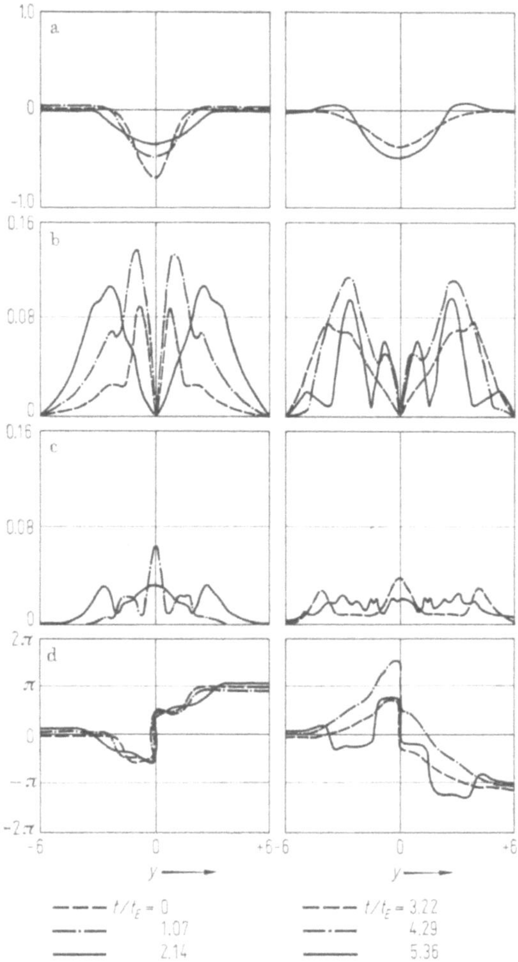

Fig. 3. Cross-stream features for one-mode Gaussian profile, $\nu = 0$.
a) Mean longitudinal velocity; b) Mean fundamental amplitude; c) Mean second harmonic energy; d) Fundamental phase.

spread and *develop a null* off axis (compare SK, Fig. 18). Fig. 3c shows the second harmonic energy (compare SK, Fig. 20). Fig. 3d shows the phase of the fundamental (compare SK, Fig. 19). These results are in *excellent agreement* with experiment.

The detailed properties of this evolution are explained by *a slow "nutation" of the elliptical vortex with respect to the horizontal.* This nutation frequency f_N is a new result which depends only weakly on viscosity. On the left of Fig. 4 we show pressure contours at $t/t_E = 0$ and 6.44. On the right we show the time evolution of a low pressure region to bring out the nutation frequency. The pitch angle varies

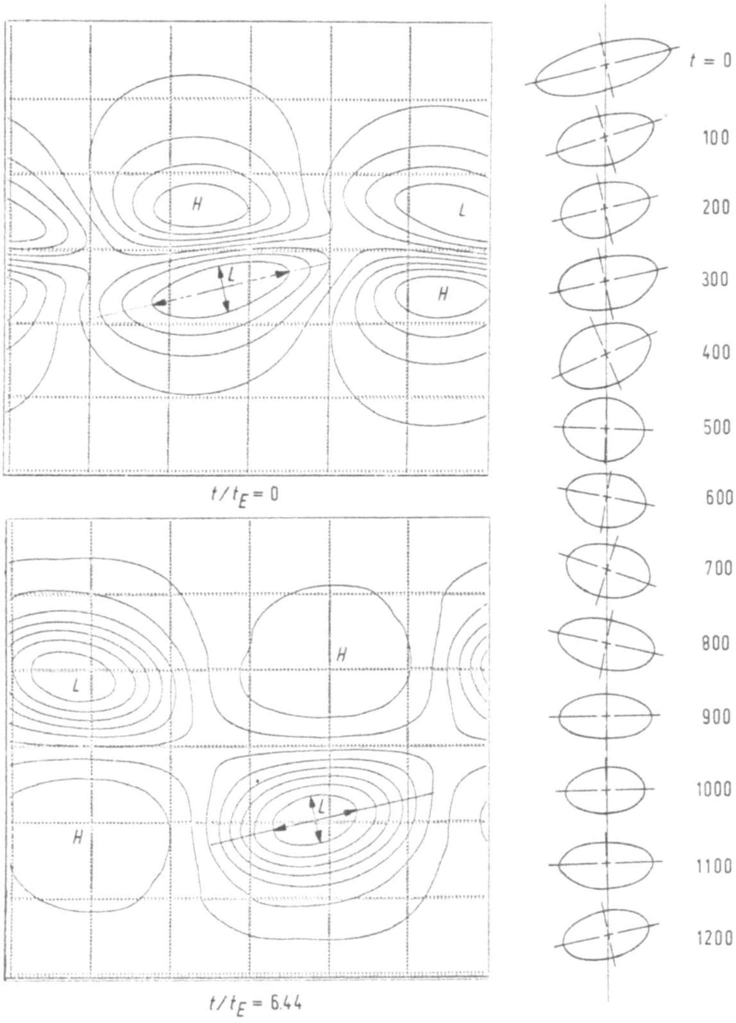

Fig. 4. Pressure contours for the one-mode Gaussian profile, $\nu = 0$. (Nine levels, eight contour lines.) Left, the initial condition and the result after 1200 computation steps ($t/t_E = 6.44$). Right, the innermost pressure contours surrounding pressure lows in the bottom half of the flow. The tilt of the axis with the horizontal shows a longtime nutation. The numbers at right represent computational time steps.

from $\approx -25°$ at $t/t_E = 0$ to $+25°$ at $t/t_E = 3.76$ and finally back to $-25°$ at $t/t_E = 6.44$. We estimate that

$$f_N t_E \approx 5.8. \tag{2.10}$$

IV. Conclusions

High Reynolds number (Re > 750) unstable parallel flows in two dimensions give rise to vortex states which inhibit the instability. The new state achieved undergoes slow quasiperiodic motion associated with a *nutating vortex* — the result undoubtedly of an interaction between the mean flow and the vortex states.

References

1. Sato, H., Kuriki, K.: J. Fluid Mech. **11**, 321 (1961).
2. Mattingly, Jr., G. E.: The Stability of a Two-Dimensional Incompressible Wake, Princeton University, Department of Aerospace and Mechanical Sciences, Report No. 858, December 1968.
3. In works to be published we will describe the numerical methods used including the effects of boundary conditions, lattice interval and aliasing. We will also give extensive details on the runs described above with $\nu = 0$ as well as a longer run with Re $= 750$. We have also studied the evolution of "turbulent" initial conditions, that is where the velocity field is composed of modes with a power law dependence and random phase.
4. The interaction of the analytic and computational mode of working has already been described by the author for another nonlinear continuous problem, the Korteweg-deVries equation. See Zabusky, N. J.: Nonlinear Partial Differential Equations, ed. W. Ames, New York: Academic Press 1967, p. 223. Also see M. D. Kruskal's paper, this meeting.

On the Effect of Sidewalls in Cellular Convection

By

L. A. Segel

Rensselaer Polytechnic Institute, Troy, N.Y., U.S.A.

Much progress in understanding the stability of continuous systems has come from intensive investigation of the prototype situations of an infinite layer of fluid heated from below (Bénard problem) and of fluid contained between concentric infinite rotating cylinders (Taylor problem) (see the survey by SEGEL [1]). Implicit in these investigations of unbounded regions of fluid is the assumption that the results are meaningful for sufficiently broad convecting layers or sufficiently long rotating cylinder apparatuses. The present paper reports on work which begins the tasks of verifying this assumption and estimating the effects of bounding walls. The approach is to introduce slow spatial modulation of the solutions derived from the nonlinear stability theory of the unbounded region. Although illustrated on the Bénard problem, it should be clear that the analysis is of general applicability. The reader should consult the paper by NEWELL in this volume for a report on closely related work and should refer to a paper of the author [2] for a number of details of the basic analysis which are omitted here.

Consider a horizontal fluid layer of depth d, heated from below. Let us choose cartesian coordinates so that the layer is bounded above and below by the planes $z = 1$ and $z = 0$. (All variables are non-dimensional. The length, velocity, and time (t) scales are d, $\varkappa/d$, and $d^2/\varkappa$, where $\varkappa$ is the thermal conductivity.) The velocity components are denoted by (u, v, w). We assume that convection is in the form of *rolls* ($v = 0$) in line with experimental evidence due to KOSCHMIEDER [3] and we employ the Boussinesq equations. We impose "free" boundary

Acknowledgement. This work was largely supported by the Office of Naval Research (Mechanics Branch) and partially by the Army Research Office (Durham).

conditions at the horizontal boundaries $z = 0$ and $z = 1$, and no-slip conditions at the vertical boundaries $x = 0$, $x = L$, $y = 0$, $y = S$. We require that the temperature perturbation vanish everywhere on the boundary.

For the unbounded region, the standard methods of nonlinear stability theory lead to a solution of the form

$$w = 2\varepsilon\,[A^{(r)}(\tau)\cos \pi\alpha x - A^{(i)}(\tau)\sin \pi\alpha x]\sin \pi z + \varepsilon^2 w_2 + \cdots, \tag{1}$$

or, with $A \equiv A^{(r)} + iA^{(i)}$,

$$w = \varepsilon\,[A\exp(i\pi\alpha x) + \bar{A}\exp(-i\pi\alpha x)]\sin \pi z + \varepsilon^2 w_2 + \cdots. \tag{2}$$

There are similar expansions for u and the temperature perturbation T (see Ref. [1]). The amplitude A satisfies

$$dA/d\tau = A - \beta A^2\bar{A}, \tag{3}$$

where

$$\tau \equiv 2(\alpha^2 + 1)\,\pi^2\varepsilon^2\gamma t,\ \varepsilon^2 \equiv [\mathcal{R} - \mathcal{R}_c(\alpha)]/\mathcal{R}_c(\alpha),\ \gamma \equiv (1/2)\,\mathcal{P}(1 + \mathcal{P})^{-1}. \tag{4a, b, c}$$

$\mathcal{R}$ is the Rayleigh number (dimensionless temperature difference across the layer) and $\mathcal{R} = \mathcal{R}_c(\alpha)$ is the neutral curve according to linear stability theory. $\mathcal{P}$ is the Prandtl number.

For a given value of the wavenumber α, nonuniqueness ensues for the unbounded steady problem when and only when $\mathcal{R}$ exceeds $\mathcal{R}_c(\alpha)$. [$\mathcal{R}_c(\alpha)$ has a minimum of $\mathcal{R}_c \equiv 27\pi^4/4$ when $\alpha^2 = \alpha_c^2 = 1/2$.] We assume that ε is small so that our analysis deals with the portion of parameter space near the bifurcation point for the unbounded problem.

The presence of ε in (4a) shows that A is a slowly varying function of time in the solution for the unbounded region. We attempt to satisfy the boundary conditions required when rigid vertical walls enclose the layer by means of a solution which has the same form as (2) except that $A = A(\tau, X, Y)$, where $X = \mu(\varepsilon)$, $xY = \eta(\varepsilon)\,y$, and $\mu(\varepsilon)$ and $\eta(\varepsilon)$ approach zero with ε. A detailed calculation shows that all equations and boundary conditions are satisfied if we choose $\mu = \varepsilon$, $\eta = \varepsilon^{1/2}$, and if

$$A_\tau = A + \delta_1 A_{XX} - 2i\,\delta_{12} A_{XYY} - \delta_2 A_{YYYY} - \beta A^2\bar{A}. \tag{5}$$

$$\text{On } X = 0, \Lambda\colon A = 0. \quad \text{On } Y = 0, \Sigma\colon A = A_{YY} = 0. \tag{6}$$

Here $\Lambda = \varepsilon L$, $\Sigma = \varepsilon^{1/2} S$ and the δ's, nominally functions of α, can be considered constants because $\alpha - \alpha_c$ must be $O(\varepsilon^2)$ to avoid extra terms in (5).

It is a fairly straightforward matter to perform a linear stability analysis of the motionless bounded layer by ignoring the nonlinear term in (5). The analysis shows that disturbances of wavenumber α_b will grow when the Rayleigh number first exceeds $\mathscr{R}_b$ where

$$\alpha_b^2 = \frac{1}{2} - \frac{5}{3} L^{-2} - \frac{1}{2} S^{-4} + \cdots, \ \mathscr{R}_b = \mathscr{R}_c \left[1 + \frac{8}{3} L^{-2} + \cdots \right]. \qquad (7)$$

Since $\mathscr{R}_b$ is independent of S the instability should set in as rolls parallel to the shorter side of the rectangular layer, to make L as long as possible. (Recall that L is the length of the side parallel to the roll axes.) Our results are in accord with the experimental evidence of KOSCHMIEDER [3] and the Galerkin calculations of DAVIS [4].

In studying finite amplitude solutions we limit ourselves for simplicity to the case where Y variations are sufficiently slow so that terms involving Y-derivatives drop out of (5). The solutions which vanish at $x = 0$ and $x = L$ (i.e. at $X = 0$ and $X = \Lambda$) are

$$A_i = a_i \, sn \, b_i X \text{ where } a_i^2 = 2\beta^{-1} k_i^2 (1 + k_i^2)^{-1}, b_i^{-2} = \delta_1 (1 + k_i^2). \qquad (8)$$

The boundary conditions require that k_i, the possible moduli of sn, satisfy $b(k_i)\,\Lambda = 2iK(k_i)$, where $4K$ is the period of sn, $i = 1, 2, 3, \ldots N$. N is the largest integer in $\Lambda/(\pi\delta_1^{1/2})$. A_i possesses $i - 1$ interior zeros.

Let us examine the stability of these solutions. Writing

$$A(\tau, X) = A_i(X) + \exp(\sigma_i \tau) \, D_i(\xi_i), \quad \xi_i \equiv b_i X,$$

and linearizing, we find from (5) that D_i satisfies

$$d^2 D_i/d\xi_i^2 + D_i[H_i - 6k_i^2 \, sn^2 \, \xi_i] = 0; \ D_i(0) = D_i(2iK) = 0. \qquad (9)$$

In the Lame equation (9)

$$H_i \equiv (1 + k_i^2)(1 - \sigma_i) \text{ so } \sigma_i = 1 - H_i(1 + k_i^2)^{-1}. \qquad (10)$$

A solution will be stable if $\sigma_i < 0$ for the lowest eigenvalue H_{i1} of (9). When $i = 1$ reference to ARSCOTT [5] shows that $sn\xi_1 \, dn \, \xi_1$ is the appropriate eigenfunction (with no interior zeros). The corresponding eigenvalue is $1 + 4k_1^2$ so A_1 is stable. When Λ just exceeds $i\pi \, \delta_1^{1/2}$, A_i is virtually sinusoidal and $k_i \approx 0$. In this case the stability of A_i will therefore be governed by

$$d^2 D_i/d\xi_i^2 + H_i D_i = 0, \ D_i(0) = D_i(i\pi) = 0, \text{ so } H_i = i^{-2}$$

and the D_i are unstable when their modulus k_i is small. When k_i is at the other end of its possible range ($k_i \approx 1$) the solutions $A_i(x)$ for $i > 1$ behave like $\beta^{-1} \tanh \xi_i$ near their widely spaced interior zeros. The demonstration by NEWELL and WHITEHEAD [6] that such solutions are unstable, coupled with our verification of instability for small k_i ($i > 1$), confirms the conjecture of Ref. [2] that of the possible solutions $A_i(X)$ only $A_1(X)$ is stable.

Our conclusion is that slow spatial modulation of finite amplitude rolls "fits" them into a rectangular container. For a sufficiently large container solutions are possible which vanish at interior "imaginary walls" but these are unstable. The stable solution has the uniform amplitude $\beta^{-1/2}$ (characteristic of the solution for the unbounded layer) except within a distance $(2\delta_1)^{1/2}$ of the walls where the amplitude decreases to zero.

References

1. SEGEL, L.: Non-equilibrium Thermodynamics, Variational Techniques and Stability, eds. R. DONNELLY, R. HERMAN, and I. PRIGOGINE. Chicago: Univ. of Chicago Press 1966, Chapter 10, p. 165–197.
2. SEGEL, L.: J. Fluid Mech. **38**, 203–224 (1969).
3. KOSCHMIEDER, L.: Beitr. z. Phys. Atm. **39**, 1–11 (1966).
4. DAVIS, S.: J. Fluid Mech. **30**, 465–478 (1967).
5. ARSCOTT, F.: Periodic Differential Equations, New York: Macmillan 1964, p. 205.
6. NEWELL, A., WHITEHEAD, J.: J. Fluid Mech. **38**, 279–303 (1969).

The Stability of Cellular Branching Solutions of the Navier-Stokes Equations

By

K. Kirchgässner

University of Bochum, Germany

and

H. Lange

University of Freiburg i. Br., Germany

1. Introduction

In 1958 J. T. Stuart [1] introduced an energy balance method for the study of the growth of Taylor vortices for supercritical Reynolds numbers where linear stability theory does not apply anymore. Besides the classical energy method leading to sufficient stability limits (cf. [2]), this was the first systematic attempt toward an understanding of nonlinear phenomena in hydrodynamic stability theory.

In 1962 A. Davey [3] published a somewhat modified version of Stuart's method which seemed to be especially well fitted for a rigorous mathematical justification in the case of the full Navier-Stokes equations. Recently, Davey, DiPrima, and Stuart [4] were able to apply this method to the study of wavy toroidal vortices in the Taylor model.

The Stuart-Davey method proved to be so powerful in the study of nonlinear cellular stability problems that its justification would put many interesting results on a safe basis. The paper presented here displays the results of our attempts in this direction, and shows which requirements are necessary at the present stage of knowledge for the applicability of the Stuart-Davey method.

2. The Stuart-Davey method

We consider a bounded domain Ω in $\mathbb{R}^3$ with sufficiently regular boundary, or (for the study of periodic solutions) a cylindrical region $\Omega' \times \mathbb{R}^1$ (Ω' bounded in $\mathbb{R}^2$). Let u^*, p^* be a stationary solution (ev. periodic) of the Navier-Stokes boundary value system with prescribed non-zero boundary values on $\partial\Omega$, $\partial\Omega'$ respectively. For a pair u, p of solutions of the Navier-Stokes initial boundary value problem we set $u = u^* + v$, $p = p^* + q$ and obtain for v, q the following system:

$$v_t - \Delta v + \lambda\big((u^*\nabla)\, v + (v\nabla)\, u^*\big) + (v\nabla)\, v = -\lambda\, \nabla q,$$
$$\nabla \cdot v = 0,\; v_{|\partial\Omega} = 0,\; v_{t=0} = v_0. \tag{1}$$

Here λ denotes the Reynolds number. We are investigating the solutions of (1) in a neighbourhood of a simple eigenvalue $\lambda_1 > 0$ of the corresponding stationary linearized problem:

$$-\Delta w + \lambda\big((u^*\nabla)\, w + (w\,\nabla)\, u^*\big) = -\lambda\, \nabla\tilde{q},\; \nabla \cdot w = 0,\; w_{|\partial\Omega} = 0. \tag{2}$$

Let φ be the eigenfunction corresponding to λ_1, and $\lambda \in (\lambda_1, \lambda_1 + \delta)$, $\delta > 0$. Solutions of (1) are sought in the form

$$v(x, t) = \sum_{k=1}^{\infty} A^k(t)\, v_k(x, t), \qquad v_1(x) = \varphi(x), \tag{3}$$

where the *amplitude function* $A(t)$ satisfies a differential equation of the form $\dot{A}(t) = \sum_{k=1}^{\infty} a_k A^k(t)$. The coefficients $a_2, a_3, \ldots$ are unknown and are to be determined by comparison of equal order terms in $A(t)$; a_1 is the *amplification rate* of the linear stability theory. For special problems an a priori knowledge of some a_k's can be obtained; e.g. for the Taylor problem we have: $a_{2n} = 0$ $(n = 1, 2, \ldots)$ (cf. [3]).

3. The existence of strong solutions

It is obvious that the initial value problem (1) can in general not be solved by functions representable as in (3). But a slight modification of (3) will be sufficient to ensure existence of "strong" solutions for any given time interval if certain conditions are met. The solutions can be calculated in a quite analogous way as the expansion (3).

To treat the problem mathematically we use a Hilbert space approach. We project the problem (1) onto the subspace N of all solenoidal $\mathbb{R}^3$-vector functions in $L_2(\Omega)$; N is defined as the closure in the L_2-norm of the set $C_{0,\sigma}^{\infty}(\Omega)$ of all arbitrarily often differentiable and divergence-free vector functions with compact support in Ω. We norm N with the L_2-norm $|\cdot| := \|\cdot\|_{L_2}$; the orthogonal complement of N in $L_2(\Omega)$ is the space of all functions which are gradients of scalar functions (cf. [5]). By N_1 we denote the closure of $C^\infty_{0,\sigma}(\Omega)$ in the Dirichlet norm

$$\|v\|^2 = \sum_{i,k=1}^{3} \int_\Omega \left|\frac{\partial v_i}{\partial x_k}\right|^2 dx .$$

Let $N_2 := N_1 \cap W_2^2(\Omega)$, where W_2^2 denotes the usual Sobolev space. The problem (1) can be written in the form

$$v_t + A(\lambda)\, v + \lambda R(v, v) = 0, \qquad v_{t=0} = v_0, \tag{4}$$

where $A(\lambda) := A + \lambda M(u^*)$, $A := -P_0 \Delta$; P_0 is the orthogonal projection of L_2 onto N. The linear operator $M(u^*)$ is defined by $M(u^*)\, v := R(u^*, v) + R(v, u^*)$ where we have set $R(u, v) := P_0(u \nabla)\, v$. It can be shown that $A(\lambda)$ is a closed linear operator with domain of definition $D(A(\lambda)) = D(A) = N_2$ (cf. [6]). The following estimates are a main tool of our analysis (for a proof cf. [6], $u^* \in N_2$): $|M(u^*)\, v| \leq c_1 \|v\|$; $|R(u, v)| = c_2 \|u\| \cdot |Av|$.

Definition: *A function v is called a* ***strong solution*** *of* (1) *in a given time interval* $[0, T]$ *if it satisfies equation* (4) *a.e. with $v_0 \in N_1$ and if $v \in C([0, T]; N_1) \cap L_2([0, T]; N_2)$, $v_t \in L_2([0, T]; N)$.*

Prodi [6] has proven the existence of strong solutions of a system as (4) for small $\|v_0\|$ if the spectrum $\sigma(A(\lambda))$ of $A(\lambda)$ lies in the right half plane $\mathbb{C}^+$ of the complex plane $\mathbb{C}$. We make a similar assumption:

$$\sigma(A(\lambda)) \subset \mathbb{C}^+ \ (\lambda \in [0, \lambda_1));\ \sigma(A(\lambda_1)) \subset \mathbb{C}^+ \cup \{0\},$$ [1]

where zero is to be a simple eigenvalue of $A(\lambda)$ with corresponding eigenfunction φ, $\|\varphi\| = 1$. Let P denote the projector onto the null-space of $A(\lambda_1)$, $Q := I - P$; and let $\tilde{A}(\lambda)$ be the restriction of $A(\lambda)$ to $Q(N) \cap D(A)$.

We are looking for solutions of (4) in the form:

$$v(x, t) = \varepsilon z(t)\, \varphi(x) + \sum_{k=2}^{\infty} \varepsilon^k v_k(x, t) = \varepsilon z \varphi + \tilde{v}, \tag{5}$$

where $Pv = \varepsilon z \varphi$, $Qv = \tilde{v}$; ε is a real constant. If we insert the representation (5) into (4) we get the following system:

$$\begin{gathered} \dot{z} - \sigma z + \frac{\lambda}{\varepsilon}\big(PR(\varepsilon z\varphi + \tilde{v}), \varphi\big) = 0,\ z(0) = z_0 \\ \big(\sigma = (\lambda - \lambda_1) \cdot (A\varphi, \varphi)/\lambda_1\big), \end{gathered} \tag{6}$$

$$v_{k,t} + \tilde{A}(\lambda)\, v_k + \lambda \sum_{j=1}^{k-1} QR(v_j, v_{k-j}) = 0,\ v_k(0) = v_k^0\ (v_1 = z\varphi;\ k \geq 2). \tag{7}$$

[1] This assumption is called the "*Principle of Exchange of Stabilities*" (P.E.S.).

Theorem 1. *Let* $||v_k^0|| \leq a/k^2$ *and* $|z(t)|^2 + \int_0^T |z(s)|^2 \, ds \leq \delta^2$ $(t \in [0, T])$, $|\varepsilon| < 1$, *a and* δ *sufficiently small constants. Then* $\tilde{v}(x, t) = \sum_{k=2}^{\infty} \varepsilon^k v_k(x, t)$, *where the* v_k*'s are uniquely determined by* (7) *is a strong solution of*

$$\tilde{v}_t + \tilde{A}(\lambda)\, \tilde{v} + \lambda Q R(\varepsilon z \varphi + \tilde{v}) = 0, \; \tilde{v}(x, 0) = \sum_{k=2}^{\infty} \varepsilon^k v_k^0.$$

We can write the differential equation (6) for the amplitude function $z(t)$ in the form

$$\dot{z} - \sigma z + a_2 \varepsilon z^2 + a_3 \varepsilon^2 z^3 + r(t) = 0, \; z(0) = z_0 \tag{8}$$

with $a_2 = \lambda(PR(\varphi, \varphi), \varphi)$, $a_3 = \lambda(PR(\varphi, w) + PR(w, \varphi), \varphi)$ where w is uniquely determined by the equation $(\tilde{A}(\lambda) + 2\sigma)\, w = \lambda Q R(\varphi, \varphi)$; $r(t)$ is a measurable function which has an estimate of the form

$$\int_0^T |r(s)|^2 \, ds \leq c_3 \varepsilon^6 \, \delta^2 \big(||v_2(0) - z^2(0)\, w||^2 + c_4 K(\varepsilon, \delta)\big),$$

where $K(\varepsilon, \delta) := \varepsilon^2 \delta^6$ if $a_2 \neq 0$, $:= \varepsilon^4 \delta^4$ if $a_2 = 0$, $a_3 \neq 0$.

Theorem 2. *Let either* $a_2 > 0$ *or (if* $a_2 = 0$) $a_3 > 0$; *let* $\tau = \lambda - \lambda_1$ $(\tau > 0)$ *and* $||v_0||$ *be sufficiently small. Then there exists in* $[0, \infty)$ *a strong solution of* (1) *in the form* (5) *where* z *and the* v_k*'s are determined by* (6) *and* (7) *resp.; moreover the* ***amplitude function*** $\varepsilon z(t)$ *may be estimated by the solution* $\zeta(t)$ *of the differential equation* $\dot{\zeta} - \sigma\zeta + a_k \zeta^k = 0$ $(k = 2, 3;\ \zeta(0) = \varepsilon z_0)$ *by*

$$|\varepsilon z(t) - \zeta(t)|^2 + \int_0^t |\varepsilon z(s) - \zeta(s)|^2 \, ds \leq c_4 \varepsilon^4 \tau^{2/(k-1)}. \tag{9}$$

4. Conclusions

In the interesting cases of cellular instability as in the Taylor or the Benard problem we have always $a_2 = 0$, $a_3 > 0$ (cf. [7], [8]; the cases $a_2 < 0$ and $a_2 = 0$, $a_3 < 0$ can be excluded since the P.E.S. holds). Our main assumption, the P.E.S., has not yet been proven for the Taylor problem in general, but for the limit case of small gaps, and for the Benard problem the P.E.S. is known to be valid. We have shown that for these cases a global strong solution of the system (1) exists which can be represented in the form (5), and this is our modified version of the Stuart-Davey method. Moreover, since $\lim_{t \to \infty} \zeta(t) = (\sigma/a_k)^{1/(k-1)}$ we can conclude that the amplitude function $\varepsilon z(t)$ remains bounded within an interval about $(\sigma/a_k)^{1/(k-1)}$ with constant radius $\varepsilon^4/\tau^{2/(k-1)}$.

References

1. Stuart, J. T.: J. Fluid Mech. **4**, 1 (1958).
2. Serrin, J.: Arch. Rat. Mech. Anal. **3**, 1 (1959).
3. Davey, A.: J. Fluid Mech. **14**, 336 (1962).
4. Davey, A., DiPrima, R. C., Stuart, J. T.: J. Fluid Mech. **31**, 17 (1968).
5. Ladyshenskaja, O. A.: Funktionalanalytische Untersuchungen der Navier-Stokesschen Gleichungen, Berlin: Akademie-Verlag 1965.
6. Prodi, G.: Rend. Sem. Mat. Univ. Padova **32**, 374 (1962).
7. Yudovich, V. I.: J. Appl. Math. Mech. (PMM) **31**, 103 (1967).
8. Kirchgässner, K., Sorger, P.: Quart. J. Mech. Appl. Math. **22**, 183 (1969).

On Equilibrium States and Periodic Vibrations of Thin Nonlinear Elastic Systems

By

M. S. Berger

University of Minnesota, Minneapolis, Minn., U.S.A.

In this article we hope to demonstrate the applicability of relatively new mathematical advances in the calculus of variations and nonlinear elliptic partial differential equations to concrete problems in nonlinear elasticity. We shall discuss two examples of the following general class of problems: Given an elastic body B acted on by forces λF and described by the potential energy function $V(\lambda)$, determine in a *mathematically rigorous* manner

(i) the totality of equilibrium states of B for all possible values of λ and classify these states,

(ii) the (in)stability of each state by both the energy criterion and the Liapunov criterion,

(iii) the periodic motions in the neighbourhood of a given equilibrium state.

The immense mathematical difficulties in carrying out such a program for even the simplest systems described by partial differential equations has long been recognized. One need only recall the many mathematical studies concerning the existence and stability of equilibrium forms of rotating fluid masses by Poincaré, Liapunov and many others. Nonetheless for certain important problems new mathematical techniques are available, and when combined with computer studies yield many interesting results.

Before proceeding, let us survey the typical difficulties to be surmounted:

(1) *Nonlinearity*. The desired equilibrium states are defined by systems of nonlinear elliptic partial differential equations together with appropriate boundary conditions.

(2) *Asymmetry*. Even though the initial geometry of B may be symmetrical, it has been observed that the stable deformed states are often asymmetric, thus the partial differential equations involved may not be reduced to ordinary ones.

(3) *Nonuniqueness*. For fixed λ the physical system often admits many distinct equilibrium states.

(4) *Higher order effects*. In comparing $V(\lambda)$ with λ fixed, for two near equilibrium states terms of the third and fourth orders in a small parameter must be taken into account.

(5) *Global effects*. As a function of λ, the size of the stable equilibrium state does not often depend continuously on λ.

With these difficulties in mind, we now turn to the common features of the new mathematical techniques to be described:

(1) *Hilbert space formulation*. By considering spaces of functions with generalized derivatives (i.e. Sobolev space) we combine seemingly complicated systems of partial differential equations and boundary conditions into a much simpler operator equation in an appropriate Hilbert space, thus bringing the mathematical structures in the problem to the force.

(2) *L_p regularity theory for linear elliptic differential equations and Sobolev inequalities*. Results on the smoothness of elliptic equations insure that solutions of the operator equation coincide with the classically desired pointwise smooth solutions of the differential equations and boundary conditions. The associated inequalities enable one to give reasonable estimates for the solutions and their derivatives.

(3) *New variational principles*. In terms of the Hilbert space formulation we may formulate new variational principles for equilibrium states as well as apply new results in the theory of critical points.

(4) *Manifolds in Hilbert space*. Nonquadratic constraints in the above variational principles give rise to manifolds in Hilbert space. A geometric study of these manifolds yields new results.

(5) *Bifurcation theory for nonlinear eigenvalue problems* (Berger [1]). This relatively new theory enables one to rigorously compare nonlinear and linearized problems near eigenvalues and is especially interesting for variational problems near a degenerate eigenvalue of a linear problem.

Let us consider the following examples.

Example 1. Buckling of a thin elastic plate of arbitrary shape under general boundary conditions (BERGER [2], BERGER and FIFE [3]).

(a) *Equations* (in non-dimensional form):

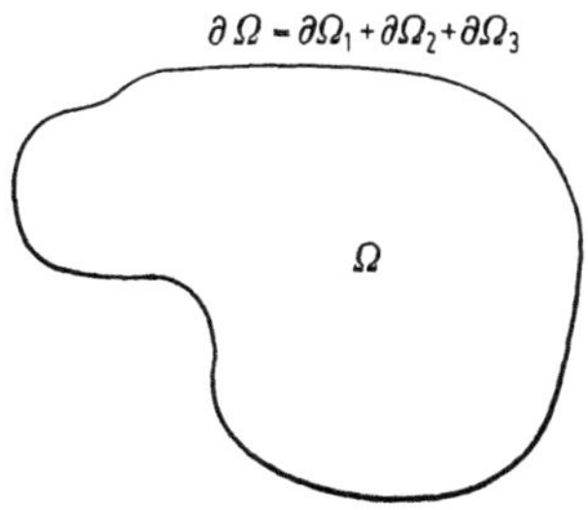

$$\Delta^2 f = -[w, w],$$

$$\Delta^2 w = \lambda [F_0, w] + [f, w],$$

$$f = f_x = f_y = 0 \text{ on } \partial\Omega,$$

$$w = w_x = w_y = 0 \text{ on } \partial\Omega_1 \text{ (clamped)},$$

$$w = B_1 w = 0 \text{ on } \partial\Omega_2 \text{ (simply supported)},$$

$$B_1 w = B_2 w = 0 \text{ on } \partial\Omega_3 \text{ (free)},$$

where

$$B_1 w = w_{nn} + \sigma(w_{ss} - K w_n),$$

$$B_2 w = \Delta w_n + (1 - \sigma)\, w_{ssn},$$

$$[f, g] = f_{xx} g_{yy} + f_{yy} g_{xx} - 2 f_{xy} g_{xy}.$$

$K = K(s)$ denotes the curvature of $\partial\Omega$, and σ is Poisson's ratio.

(b) *Operator equation in Hilbert space* H (H is a closed linear subspace of the Sobolev space $W_{2,2}(\Omega)$):

$$f = -C(w, w),$$

$$w = \lambda C(F_0, w) + C(f, w)$$

or

$$w + Cw = \lambda L w,$$

where

$$C\big(C(w, w), w\big) \equiv Cw \text{ is a map } H \to H.$$

(c) *Linearized equation*: $w = \lambda L w$, where L is a compact, self-adjoint, linear map of $H \to H$.

(d) *Variational principles*:

(1) Find the critical points in H of the potential energy difference functional (fixed λ) $I(w) = ||w||^2 + \frac{1}{2} ||C(w, w)||^2 - \lambda (Lw, w)$.

(2) Find the critical points of the functional (Lw, w) in H subject to the constraint $||w||^2 + \frac{1}{2} ||C(w, w)||^2 = R$, a constant.

(e) *Alternate expression for potential energy difference*:

$$I(w) = -\frac{1}{2} ||C(w, w)||^2.$$

(f) *Mathematical results*:

I. Results concerning the totality of equilibrium states:

(1) *Bifurcation results.* Let λ_n be an eigenvalue of multiplicity p of the linear problem, then for $\lambda > \lambda_n$ at least p one-parameter families of solutions bifurcate from the planar state.

(2) *Global results.* Each of the one-parameter families mentioned above can be continued into the nonlinear range (i.e. for large R).

(3) *Computation results.*

(a) Variational principle for branch bifurcating from λ_1. Maximize (Lw, w) subject to the constraint $||w||^2 + \frac{1}{2}||C(w, w)||^2 = R$, a constant varying from 0 to ∞.

(b) Assume λ_i is a simple eigenvalue, for $\lambda \doteqdot \lambda_i$, and ε sufficiently small

$$u(\lambda) = \varepsilon u_i + O(|\varepsilon|^2) = \lambda_i + O(|\varepsilon|^2),$$

$$I\big(u(\lambda)\big) = -\frac{1}{2}\varepsilon^4 ||C(u_i, u_i)||^2 + O(|\varepsilon|^5).$$

(Similar expressions exist for eigenvalues of higher multiplicity.)

(4) *A priori estimates.* $\left\{\sup_\Omega |D^\alpha u| + \sup_\Omega |D^\alpha f|\right\}^2 \leq \lambda^2 \operatorname{const}(\Omega) \int_\Omega F_0^2$.

II. Results concerning stability:

(5) For any given λ, an equilibrium state minimizing the potential energy always exists.

(6) *Exchange of stabilities.* For $0 \leq \lambda \leq \lambda_1$, this state is planar. For $\lambda > \lambda_1$ this state is nonplanar and an "exchange of stabilities" definitely occurs at $\lambda = \lambda_1$.

(7) *Liapunov stability.* The planar state is stable in the sense of Liapunov for $\lambda < \lambda_1$ and unstable for $\lambda > \lambda_1$.

Some simple proofs

(i) *A "stable" equilibrium state always exists for all λ.*

Proof:

$$I(w) = ||w||^2 + \frac{1}{2}||C(w, w)||^2 - \lambda(C(w, F_0), w)$$

$$= ||w||^2 + \frac{1}{2}||C(w, w)||^2 - \lambda(C(w, w), F_0) \text{ (symmetry of operator)}$$

$$= ||w||^2 + \frac{1}{2}||C(w, w)||^2 - \lambda||C(w, w)|| \, ||F_0|| \text{ (Cauchy inequality)}$$

$$= ||w||^2 + \left(\frac{1}{2} - \frac{\lambda}{2\varepsilon}\right) C(w, w)^2 - \frac{1}{2}\lambda\varepsilon \, ||F_0||^2.$$

Choosing $\lambda = \varepsilon$,

$$I(w) \geq ||w||^2 - \frac{1}{2}\lambda^2 ||F_0||^2. \qquad (*)$$

Now we can find a minimizing sequence w_n, with $||w_n|| \leq M$ such that $I(w_n) \to \inf_H I(w)$. By choosing a weakly convergent subsequence of w_n with limit w, and using *lower-semicontinuity* of $I(w)$ with respect to weak convergence we show $I(w) = \inf_H I(w)$. By L_p regularity theory, w is a *smooth* function and satisfies equations pointwise.

(ii) *A priori estimate* (easy case) ($\alpha = 0$). First note that for a solution, $I(w) = -\frac{1}{2}||C(w, w)||^2$. Hence by (*):

$$\frac{1}{2}\lambda^2 F_0^2 = ||w||^2 + \frac{1}{2}||C(w, w)||^2 = ||w||^2 + \frac{1}{2}||f||^2$$

$$= K_s\left\{\sup_\Omega |w| + \frac{1}{2}\sup_\Omega |f|\right\} \text{ (by Sobolev's inequality).}$$

Since $\int F_0^2 = K_\Omega ||F_0||^2$ where K_Ω is a constant depending only on Ω, the desired estimate (**) below is immediate

$$\sup_\Omega |w|^2 + \sup_\Omega |f|^2 \leq K\lambda^2 \int_\Omega F_0^2. \qquad (**)$$

(iii) *The "stable" equilibrium state is planar when* $\lambda < \lambda_1$ *and non-planar when* $\lambda > \lambda_1$. The proof is immediate by the results of (i) above and the fact that the second variation of $I(w)$ at $w = 0$ is $\delta^2 I(0, v) = ||v||^2 - \lambda(Lv, v)$.

(iv) *Liapunov stability of planar state of* $\lambda < \lambda_1$, *instability for* $\lambda > \lambda_1$. A general argument due apparently to YUDOVICH shows that for Liapunov stability we must prove $\delta^2 I(0, v) \geq k\, ||v||^2$ for all $v \in H$ where k is a constant independent of v, while for instability $\delta^2 I(0, \tilde{v}) \geq -k\, ||\tilde{v}||^2$ for some $\tilde{v} \in H$. Thus the proof is immediate by (iii) above since for $\lambda < \lambda_1$, the operator $I - \lambda L$ is invertible in H, while for $\lambda > \lambda_1$, we may choose $\tilde{v} = u_1$ (an eigenfunction associated with λ_1).

Example 2. Buckling of a thin shallow elastic cap under uniform pressure p ($\equiv \lambda$) (BERGER [4]). (Edge conditions chosen so that spherical shape is a solution of equations for all p.)

(a) *Equations* (in non-dimensional form):

$$\left.\begin{aligned} \frac{1}{\mu^2}\Delta^2 f &= -\frac{1}{2}[w, w] - \Delta w \\ \frac{1}{\mu^2}\Delta^2 w &= -p\,\Delta w + [f, w] + \Delta f \end{aligned}\right\} \text{ in } \Omega,$$

$$\left.\begin{aligned} w &= w_N = 0 \\ f &= f_N = 0 \end{aligned}\right\} \text{ on } \partial\Omega,$$

where $W = p\bar{w} + w$, $F = p\bar{f} + f$, μ is a geometric parameter, Ω is a circle with radius 1, and N is normal to $\partial\Omega$.

(b) *Operator equation in* H:

$$w + \mu^4 \left\{ \frac{1}{2} C(w) + C(w, L_1 w) + \frac{1}{2} L_1 C(w, w) + L_1^2 w \right\} = \lambda L w .$$

(c) *Linearized equation*:

$$w + \mu^4 L_1^2 w = \lambda L w .$$

(d) *Variational principles*:

(1) Find the critical points of $I(w) = ||w||^2 + \mu^4 \left\| \frac{1}{2} C(w, w) + L_1 w \right\|^2 - \lambda (Lw, w)$.

(2) Find the critical points of (Lw, w) on the hypersurface $||w||^2 + \mu^4 \left\| \frac{1}{2} C(w, w) + L_1 w \right\|^2 = R$, a constant.

(e) *Alternate expression for potential energy difference*:

$$I(w) = -\frac{1}{2} \mu^4 \left\{ \left(L_1 w, C(w, w)\right) + \frac{1}{2} ||C(w, w)||^2 \right\}.$$

(f) *Mathematical results*:

I. Results concerning equilibrium configurations:

(1) *Bifurcation results.* Nontrivial equilibrium states bifurcate from the spherical state at all eigenvalues λ_n of the linearized problem.

(2) For sufficiently small μ, the smallest eigenvalue λ_1 is simple and the associated eigenfunction is radially symmetric and so is the bifurcating branch of equilibrium states. As μ increases, λ_1 is no longer simple and the associated eigenfunctions are not radially symmetric, so that the bifurcating branch is also not symmetric.

(3) *Global results.* The equilibrium state bifurcating from λ_1 can be continued for large values of R, by the formula maximize (Lw, w) subject to the constraint

$$||w||^2 + \left\| \frac{1}{2} C(w, w) + L_1 w \right\|^2 = R .$$

(4) *Computation results.*

(a) For $\lambda \doteqdot \lambda_i$ (a simple symmetric eigenvalue)

$$\begin{aligned} u(\lambda) &= \varepsilon u_i + O(|\varepsilon|^2), \\ \lambda &= \lambda_i + \varepsilon \lambda_{i,1} + O(|\varepsilon|^2) \quad (\lambda_{i,1} \neq 0), \\ I\left(u(\lambda)\right) &= \varepsilon^2 f_{i,1} + O(|\varepsilon|^4) \qquad (\beta_{i,1} \neq 0). \end{aligned}$$

(b) Near a double eigenvalue (asymmetric)

$$\begin{aligned} u(\lambda) &= \varepsilon [\alpha_1 u_1 + \alpha_2 u_2] + O(|\varepsilon|^2), \\ \lambda &= \lambda_i + O(|\varepsilon|^2), \\ I\left(u(\lambda)\right) &= O(|\varepsilon|^4). \end{aligned}$$

II. Results concerning stability:

(5) For any given λ, an equilibrium state minimizing the potential energy always exists.

(6) The spherical state is a relative minimum for $\lambda \leq \lambda_1$, and is always unstable for $\lambda > \lambda_1$.

(7) The spherical state is an absolute minimum only for $\lambda \leq \varphi_1$, where φ_1 is the smallest eigenvalue of the "equivalent plate" problem values of μ, $w = \varphi L w$.

(8) For a large range of values of μ, the spherical state is not the absolute minimum for small and negative values of $(\lambda - \lambda_1)$. Thus instability must set in for such values of λ. (However for very large values of μ, this need not be the case.)

Finally we turn to problems concerning *periodic vibrations* in the vicinity of a given equilibrium state. In order to discuss this problem, let us consider "*nonlinear*" normal modes of a clamped but flexible elastic plate. Mathematically stated we wish to find (for a given domain Ω) solutions $(w(x, t), f(x, t))$ periodic in t for the system:

$$\left.\begin{array}{l} \left.\begin{array}{l} \Delta^2 f = -[w, w] \\ -\omega_{tt} = \Delta^2 \omega - [f, w] \end{array}\right\} \text{in } \Omega, \\ \omega = \omega_n = f = f_n = 0 \quad \text{on } \partial\Omega, \\ \omega_{t=0|\,\partial\Omega} \ \text{(appropriately chosen)}. \end{array}\right\} \tag{1}$$

To discuss this system we approximate by a Hamiltonian system with a finite number of degrees of freedom N. Indeed writing a typical solution $\omega = \sum \omega_i(x)\, q_i(t)$ where ω_i are orthonormalized eigenfunctions of Δ^2 over Ω with associated eigenvalues λ_i, the resulting truncated system can be written (in terms of the operator C earlier introduced):

$$\ddot{q}_{i,N} + \lambda_i^2 q_{i,N} + \int C\left(\sum_{i=1}^{N} q_i \omega_i\right) \omega_i = 0 \qquad (i = 1, 2, \ldots, N). \tag{2}$$

Theorem. *For each N, system* (2) *has N distinct one-parameter families of $T_i(R)$-periodic solutions $q_i(R)$ ($i = 1, 2, \ldots, N$) such that $T_i(R) \to \frac{2\pi}{\lambda_i}$ as $R \to 0$ where*

$$2\pi R = T_i(R) \int_0^{T_i(R)} \sum_{i=1}^{N} \dot{q}_i^2(t)\, dt.$$

Outline of proof. Set $t = \sigma s$ and consider 2π-periodic solutions of the autonomous Hamiltonian system

$$\frac{d^2 q_{i,N}}{ds^2} + \sigma^2 \left[\lambda_i^2 q_{i,N} + \int_\Omega C\left(\sum_{i=1}^{N} q_{i,N} \omega_i\right) \omega_i\right] = 0.$$

The 2π-periodic solutions can be regarded as critical points of the functional

$$G(q_{1,N}, \ldots, q_{N,N}) = \int_0^{2\pi} \sum_{i=1}^{N} \lambda_i^2 q_{i,N}^2 + \frac{1}{2} \int \left\{\sum_{i=1}^{N} q_{i,N}\omega_i\right\} C \left(\sum_{i=1}^{N} q_{i,N}\omega_i\right)$$

over the class of odd 2π-periodic N-vector functions $(q_{1,N}, \ldots, q_{N,N})$ subject to the constraint $||q||^2 = R$, where

$$||q||^2 = \int_0^{2\pi} \sum_{i=1}^{N} \dot{q}_{i,N}^2 \quad \text{and} \quad R \text{ a constant.}$$

Now apply the *theory of critical points* in a *Hilbert space* of generalized functions to this problem.

Remark: As $N \to \infty$, the families $q_i(R)$ converge to a periodic solution of (1) by virtue of the Liapunov stability of the associated equilibrium state.

References

1. Berger, M. S.: A bifurcation theory for nonlinear elliptic partial differential equations. In: Bifurcation Theory, ed. J. Keller, New York: Benjamin 1969.
2. Berger, M. S.: On von Karman's equations (I). Comm. Pure Appl. Math. **20**, 687–719 (1967).
3. Berger, M. S., Fife, P.: On von Karman's equations (II). Comm. Pure Appl. Math. **21**, 227–241 (1968).
4. Berger, M. S.: (forthcoming paper).

On Buckling and Instability of Plastic Structural Models

By

G. Augusti

University of Naples, Naples, Italy
(on leave at Brown University, Providence, R.I., U.S.A., 1968/1969)

Introduction

This paper presents an introductory discussion of the phenomena of buckling and instability of plastic structures, making use of elementary *models* with a limited number of degrees of freedom and of the associated *yield profiles* (*yield loci*). It is in fact felt that yield profiles have been unjustly neglected in this field (one exception worth remembering is a so far little-noticed paper by Boyce [1]), while they have been an essential tool in the development of the general theory of plastic structures and continua, where their use made it possible to overcome many of the mathematical difficulties arising from the non-uniform constitutive laws.

Throughout this paper, it will be assumed that the intensity of the loads depends on a single parameter P, whose value does not vary because of the deformation of the structure, and can be increased very slowly (*quasi-static, dead* loading). The terms *buckling* and *bifurcation of equilibrium* will be used as synonyms, and the meaning of the word *stability* will be discussed in detail. Velocity-dependent forces and rate effects in the constitutive laws will not be considered.

1. Plastic Buckling

In a buckling problem, it is possible (in the *generalized stress* space appropriate to the continuum under consideration) to distinguish at each point a *primary stress*, say N, and *secondary stresses*, say Q_i, that are zero if the structure has not buckled (Fig. 1). In the usual situation of structural symmetry, the value of N is proportional to P, and independent of the associated deformation λ; while the equilibrium condi-

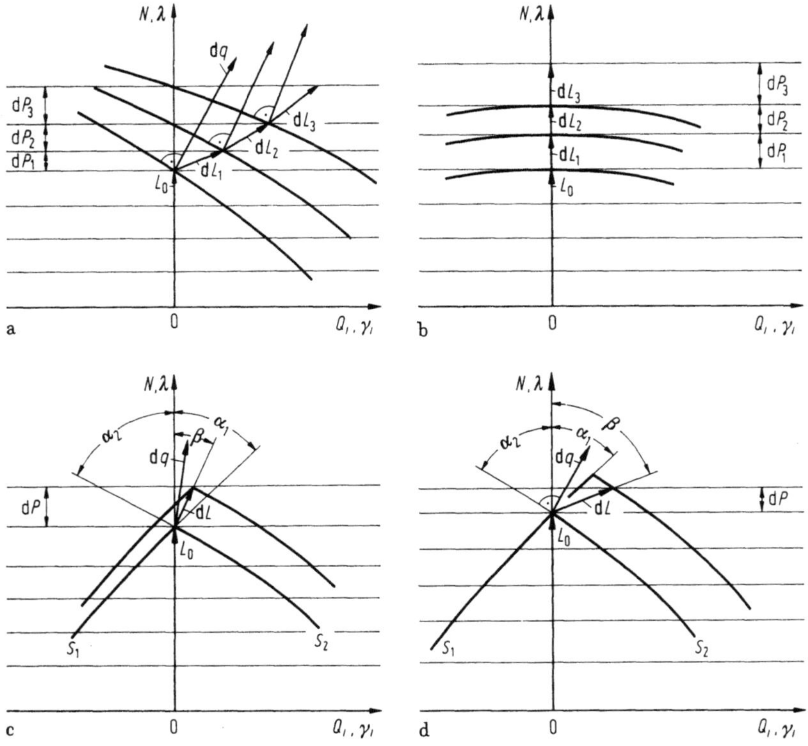

Fig. 1. Loading point engaging the yield profile.

tions relating the secondary stresses to the associated deformations γ_i and to the value of P can be linearized in the vicinity of the N-axis

$$N = bP,$$

$$Q_i = a_{ij}\gamma_j P. \tag{1}$$

When the terminal point of the stress vector $\overline{OL}$ (N, Q_i) is internal to a closed, convex surface called *yield profile*, only *elastic* deformations take place; when L is on the yield profile and tends to move towards its exterior, *plastic* deformations occur while the profile itself moves, and possibly deforms, with L according to some strain-hardening law that will not be discussed here. If the generalized strain (λ, γ_i) space is superimposed on the (N, Q_i) space, the plastic strain increment vector $\boldsymbol{dq}$ is normal to the yield profile in L; its magnitude is determined by the stress increment vector $\boldsymbol{dL}$ and the strain-hardening law.

Suppose that, when P is increased quasi-statically, the stress point L first reaches the yield profile while still on the N-axis (i.e., before elastic buckling occurs, Fig. 1). If a yield profile is smooth in the first contact point L_0 and so remains for further increases of stress, the normality rule, the equilibrium equations (1) and the strain-hardening law determine uniquely $\boldsymbol{dq}$ and $\boldsymbol{dL}$ for any given increment of P (Fig. 1a). In these conditions, there is no *truly plastic* bifurcation of equilibrium: the response of the structure to deformations orthogonal to the plastic strain path is elastic, and the plastic deformations can influence the elastic behavior (and possible buckling) only through geometry changes. In the particular but important case of structural *and* strength symmetry in which the yield surface is orthogonal to the N-axis (Fig. 1b), stress and strain vectors never deviate from their initial *principal* direction (unless, of course, *elastic* buckling occurs).

Assume now that L_0 is in a *corner* (*edge*) of the yield profile, determined by the intersection of two smooth surfaces S_1 and S_2 (Fig. 1c, d): the extension to the case of a *vertex* in which three or more surfaces intersect is immediate. If the stress increment vector $\boldsymbol{dL}$ is comprised within the solid angle $\alpha_1 + \alpha_2$ formed by the tangent planes to S_1 and S_2 (*total loading*, Fig. 1c), there is still a one-to-one relationship between $\boldsymbol{dL}$ and $\boldsymbol{dq}$. On the contrary, whenever $\boldsymbol{dL}$ is not internal to $\alpha_1 + \alpha_2$, $\boldsymbol{dq}$ takes either of two fixed directions, determined by the normality rule (Fig. 1d): this lack of uniqueness in the stress-strain relationship may allow a bifurcation of equilibrium. In other words, the presence of a sharp *corner* in the yield profile (either already in the *virgin* profile, or generated during strain-hardening) is necessary for *truly plastic* buckling, which occurs when the equilibrium conditions allow the stress point L to *disengage* itself from the corner of the yield profile in at least one point of the structure. Graphically (assuming that the two-dimensional section of the yield profile in Fig. 1c, d is orthogonal to the edge in L_0) the angle β between the N-axis and $\boldsymbol{dL}$ at the onset of buckling must satisfy either

$$\beta \geq \alpha_1 \quad \text{or} \quad \beta \geq \alpha_2. \tag{2}$$

It is well known that the plastic buckling load P_B is defined only within a certain interval, say $[P_T, P_U]$, and all known examples have shown that at the smallest buckling load, P_T, the relevant $\boldsymbol{dL}$ is initially tangent to either S_1 or S_2, i.e. either $\beta = \alpha_1$ or $\beta = \alpha_2$.

This fact can be justified as follows. Consider, for simplicity, only one secondary stress Q, related to the associated strain γ and the load P by the linearized equilibrium equation

$$Q = a\gamma P. \tag{3}$$

At the onset of buckling, let

$$P = P_B, \quad Q = 0, \quad \gamma = 0, \quad dN = C_N\, d\lambda. \tag{4}$$

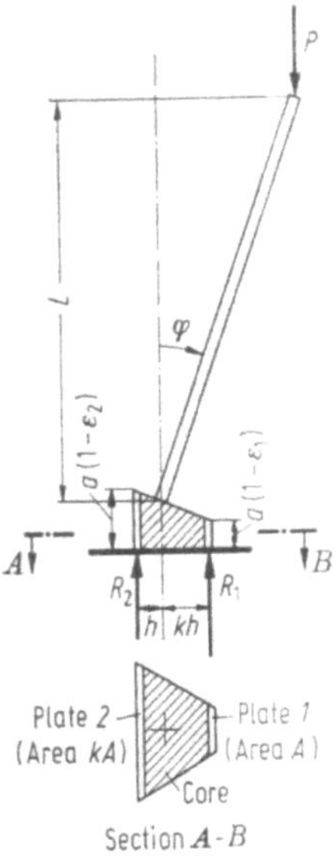

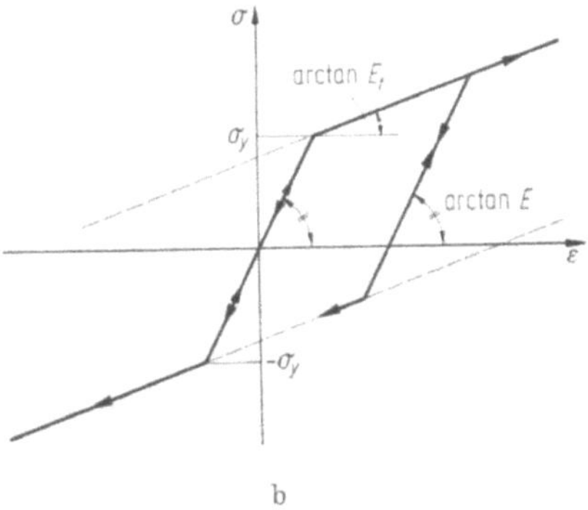

Fig. 2. Single-cell strut model and stress-strain relationship.

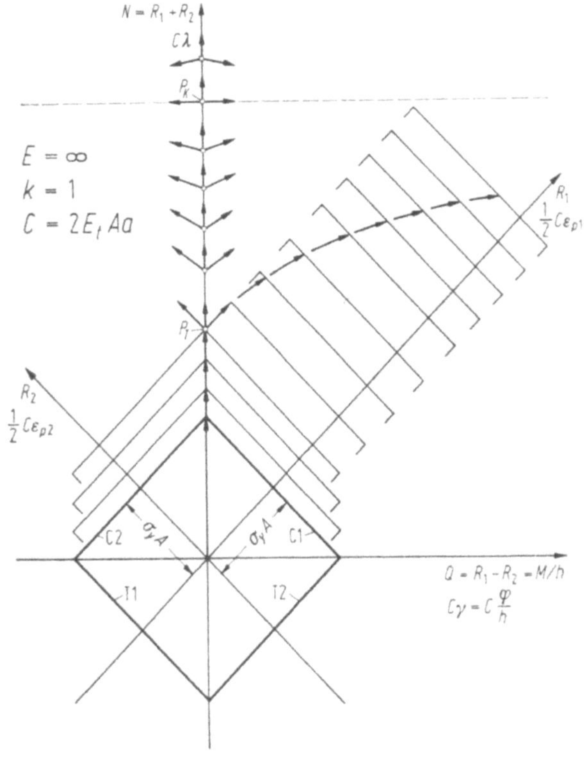

Fig. 3. Symmetric single-cell strut. Yield profile; stress paths.

On either side of the N-axis ($i = 1, 2$), Eqs. (2—4) yield

$$\tan\beta = \left|\frac{dQ}{dN}\right| = \frac{aP_B}{C_N}\left|\frac{d\gamma}{d\lambda}\right| \geq \tan\alpha_i; \quad \text{i.e.} \quad P_B \geq \frac{C_N}{a}\left|\frac{d\lambda}{d\gamma}\right| \tan\alpha_i. \tag{5}$$

Since for most stress-strain laws C_N is a non-increasing function of N, and $|d\lambda/d\gamma|$ is fixed by the normality rule, the right-hand side of the last inequality (5) is a non-increasing function of P_B: thence, $P_T = \min P_B$ corresponds to the equality sign, q.e.d.

The previous considerations are most simply exemplified on a compressed strut (Fig. 2a), in which the deformation is *lumped* in a *cell* constituted by a soft *care* and two *plates* whose stress-strain relationship is shown in Fig. 2b[1]. Indicating by R_1 and R_2 the (compressive) forces in the plates, and by

$$N = R_1 + R_2 = P \quad \text{and} \quad Qh = kR_1 - R_2 \tag{6}$$

the axial load and bending moment in the cell, the yield profile can be immediately derived both in $R_1 - R_2$ and in $N - Q$ coordinates as shown in Figs. 3 and 4, respectively for plates equal ($k = 1$) and different ($k > 1$). Each side of the profile corresponds to yielding of one plate,

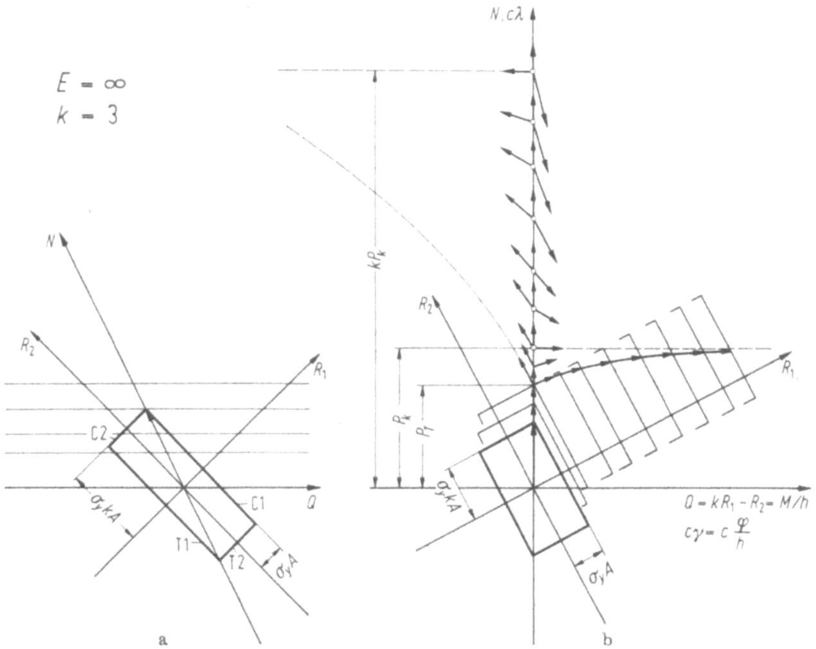

Fig. 4. Asymmetric single-cell strut. Yield profile; stress paths.

[1] When $k = 1$, this is the well-known Shanley's strut model, and the following representation coincides essentially with BOYCE's [1].

either in tension (T) or compression (C); when $k > 1$, the scales of the R_1 and R_2 axes must be adjusted in order to make the $N - Q$ axes orthogonal (Fig. 4b), while keeping the yield profile a *rectangle*, so that the rules of *kinematic hardening* can be fully applied. Elementary analysis gives the branching $N - Q$ relationships, as sketched in Figs. 3 and 4b (for $E = \infty$, and respectively $k = 1$ and $k = 3$). It is of interest to note that, when $k \neq 1$, the initial slope of the branch is different for buckling on either side of the N-axis: this type of *asymmetric buckling*, due to an asymmetry of *strength*, has no analogy in the elastic range of behavior. There are also cases of inelastic symmetric buckling due to geometry effects (analogous to the possibility first pointed out by KOITER [2] for elastic buckling): an example has been presented by the author [3].

2. Stability

To illustrate the problem of stability of a plastic structure and its relationship with the buckling load interval $[P_T, P_U]$, consider again the Shanley's strut model and assume that $L_0(N; Q_0)$ represents a possible position of equilibrium under a certain fixed load $P = N$ (Fig. 5a). Any movement from this position of equilibrium causes a variation in the bending moment

$$M - Q_0 h = PL\varphi. \tag{7}$$

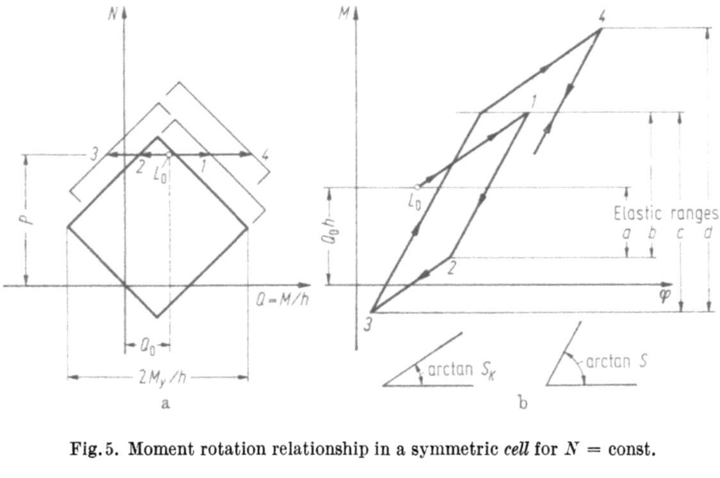

Fig. 5. Moment rotation relationship in a symmetric *cell* for N = const.

The reactive moment in the cell also varies linearly with φ (Fig. 5b): the *spring constant* $dM/d\varphi$ is equal to S_K when the movement of the stress point L causes a movement of the yield profile, and to S ($S > S_K$) when L moves within the *elastic range*, whose amplitude depends on the deformation history and cannot be larger than $2M_y/h$[1]. If $PL < S_K$, a *positive* work must be done to move the strut; also, the motion subsequent to a disturbance is bounded; the contrary is true if $PL > S_K$ [4]. Both *static* and *dynamic* criteria, thus formulated, therefore give $P_K = S_K/L > P_T$ as the limit between stability and instability: P_K (coincident with the *reduced modulus load* or Kármán load in case of symmetry) is the load at which the smaller initial slope of the branching paths of equilibrium becomes zero (Figs. 3, 4).

It is interesting to note that, from the *static* point of view, a stricter condition than simple stability (in the small) holds when $P = \text{const} < P_K$: in fact, if a small *static* disturbance M_d or φ_d is repeatedly applied and slowly removed, the residual rotation tends to a finite limit as the number of applications tends to infinity [5].

Assume now that a disturbance is applied at any stage of the *quasi-static* loading process: for instance, that a small *imperfection* is initially present[2]. The load-displacement relationship in this case is *below* the relevant lowest branch of the *perfect* strut case (Figs. 3, 4) and tends to such branch when the imperfection tends to zero. Therefore, when $P > P_T$ the *effect* (displacement) does not tend to zero with the *cause* (imperfection), and Liapunov's general definition of stability of a mechanical system is violated.

Note also that Fig. 5 holds as long as L touches only the two upper sides of the yield profile. Should L reach one of the lateral vertices (corresponding to yielding of both plates, one in compression and the other in tension), the *plastic spring constant* would reduce from S_K to $S_T = P_T L$. Therefore, if $P > P_T$ (and provided Eq. (7) remains essentially unchanged), the reactive moment in the cell becomes smaller than the overturning moment at a certain *collapse* rotation φ_c. In other words, if $P = \text{const} > P_T$, a finite amount of perturbing energy W_c is sufficient to bring the strut to *collapse*: W_c is a function of P, and tends to zero as $P \to P_K$.

In conclusion, the stability of the plastic strut is subjected to two qualifications in the range $P_T < P < P_K$: this type of stability may perhaps be denoted *partial stability*.

The combined effect of plastic constitutive relations and non-conservative external loads seems to have attracted very little attention so far. A preliminary study of this problem was carried out some years ago by the author [7], making

---

[1] If $k > 1$, S_K is different for $d\varphi > 0$ and $d\varphi < 0$, and the following arguments apply to the smaller one.

[2] Note that in elastic stability theory, at least one case is known in which the *imperfection* criterion is essential to discover an otherwise hidden instability [6]. It must also be stressed again that finite rates of loading $\dot{P}$ are excluded in this paper.

use of a combination of Shanley's and Ziegler's strut models (Fig. 6a). It was recognized that if $P > \frac{13}{3}\frac{S_K}{L}$, an infinitesimally small perturbation of the equilibrium would cause a *divergent* motion. At lower load levels, the perturbed motion (numerically determined, neglecting the possibility of tensile yielding of the cell

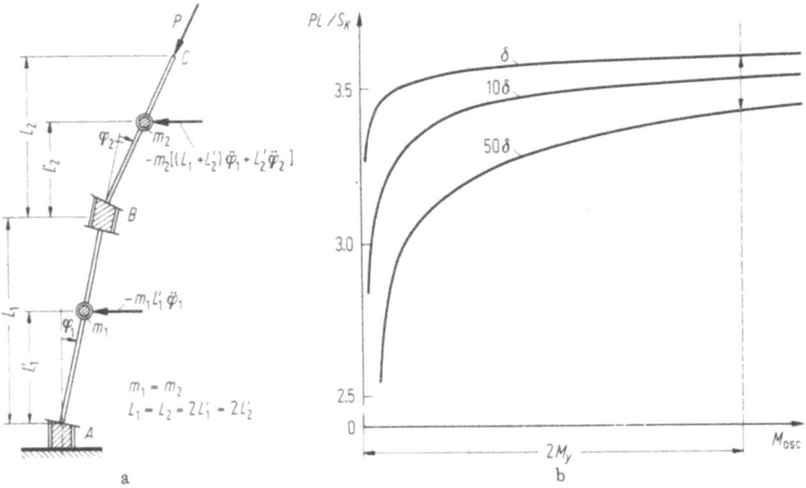

Fig. 6. a) Two-cell strut model under follower force; b) Amplitude of elastic oscillation for given perturbations in ratios 1:10:50.

plates) was found to stabilize, after few cycles, into an elastic oscillation: however, the ratio between the amplitude of this oscillation and the magnitude of the applied perturbation δ increased very rapidly above the load $P = 3\,\frac{S_K}{L}$, as qualitatively indicated in Fig. 6b.

Applying what has been said before with regard to the effect of tensile yielding, it is then easy to see that a *divergent* motion of the strut in Fig. 6a can be excited if $P > \frac{13}{3}\frac{S_T}{L}$ and a sufficiently large perturbation is applied. It so seems that also in this case a range of *partial stability* exists.

However, this research should soon be resumed, in order to get more conclusive and general results.

References

1. Boyce, W. E.: J. Aero. Sci. **24**, 332 (1957).
2. Koiter, W. T.: Doctoral Thesis, Delft, 1945 (NASA Tech. Transl. F 10833, 1967).
3. Augusti, G.: Meccanica **3**, 102 (1968).
4. Augusti, G.: Giorn. Genio Civile **103**, 396 (1965).
5. Augusti, G.: Rend. Acc. Sci. Fis. e Mat. Napoli **32**, 98 (1965).
6. Ziegler, H.: Principles of Structural Stability, Waltham, Mass.: Blaisdell 1968, p. 19.
7. Augusti, G.: Giorn. Genio Civile **104**, 7 (1966).

A Contribution to a Linearized Engineering Shell Theory

By

T. A. Barta

University College London, London, U.K.

Abstract. Earlier work has shown the possibility of formulating a consistent simple linear shell theory in terms of invariant displacement and stress functions. A similar approach is used in the proposed linearized equations which may be considered as an improvement of the Marguerre-Vlasov "shallow shell" equations. For the circular cylindrical shell and the spherical shell the results are as simple as those of the shallow-shell equations but equally accurate as those of the "rigorous" theories (Flügge, van der Neut, Vlasov).

1. The "shallow shell" equations

The linearized version of the non-linear curved-plate equations derived by Marguerre [1] and extended by Vlasov [2] to "shallow" shells can be written in terms of the normal deflection w and the stress function F:

(equilibrium) $$-DL_{11}w + L_{12}F + (N_{11}^0\varkappa_{11}^* + 2N_{12}^0\varkappa_{12}^* + N_{22}^0\varkappa_{22}^*) = 0, \tag{1a}$$

(compatibility) $$L_{21}w + \frac{1}{Et}L_{22}F = 0 \tag{1b}$$

or, reduced to a single equation

$$\frac{D}{Et}L_{11}L_{22}w + L_{12}^2 w - L_{22}(N_{11}^0\varkappa_{11}^* + 2N_{12}^0\varkappa_{12}^* + N_{22}^0\varkappa_{22}^*) = 0. \tag{1c}$$

In these equations, the flexural stiffness

$$D = \frac{Et^3}{12(1-\nu^2)} \tag{2}$$

is defined in terms of the shell thickness t, and Young modulus E and Poisson's ratio ν. The effect of finite deformations on the equilibrium equation is described by the product of the membrane forces N_{ij}^0 and the changes of curvature $\varkappa_{ij}^*$. The differential operators L_{ij} are defined as

$$L_{11} \equiv L_{22} = \nabla^4 \quad \text{and} \quad L_{12} \equiv L_{21} = \nabla_k^2, \tag{3}$$

where the Beltrami operator ∇^2 and the Pucher-Vlasov operator ∇_k^2 are given by

$$\begin{aligned} \nabla^2 &= \frac{1}{A_1 A_2}\left[\frac{\partial}{\partial\alpha_1}\left(\frac{A_2}{A_1}\frac{\partial}{\partial\alpha_1}\right) + \frac{\partial}{\partial\alpha_2}\left(\frac{A_1}{A_2}\frac{\partial}{\partial\alpha_2}\right)\right], \\ \nabla_k^2 &= \frac{1}{A_1 A_2}\left[\frac{\partial}{\partial\alpha_1}\left(\frac{A_2}{A_1}K_2\frac{\partial}{\partial\alpha_1}\right) + \frac{\partial}{\partial\alpha_2}\left(\frac{A_1}{A_2}K_1\frac{\partial}{\partial\alpha_2}\right)\right] \end{aligned} \tag{4}$$

with K_1 and K_2 the principal curvatures of the shell surface.

The changes of curvature are those known from the plate theory (in orthogonal curvilinear coordinates):

$$\begin{aligned} \varkappa_{11}^* = \varkappa_{11}^0 = \frac{\partial(A_2 w)}{A_2 A_1\, \partial\alpha_1}; \qquad \varkappa_{22}^* = \varkappa_{22}^0 = \frac{\partial(A_1 w)}{A_1 A_2\, \partial\alpha_2}; \\ \varkappa_{12}^* = \varkappa_{12}^0 = \frac{1}{2A_1A_2}\left(\frac{\partial A_1 w}{\partial\alpha_2} + \frac{\partial A_2 w}{\partial\alpha_1}\right). \end{aligned} \tag{5}$$

Due to their simplicity the "shallow-shell" equations have been widely used, although serious errors may occur when the (minimum) wave length of the deformation pattern is small in comparison with the minimum principal radius of curvature.

2. More accurate linear equations

It seems desirable to develop shell-equations which combine the simplicity of the shallow-shell equations with the accuracy of the "rigorous" theory. Using KOITER's [3] estimate of errors inherent in any engineering (two-dimensional) theory, BARTA [4, 5], derived a consistent set of linear equations, formally identical with Eq. (1), which is fully equivalent to Koiter's theory for shells with a covariantly constant second fundamental tensor, and approximately equivalent for arbitrary shells. The differential operators L_{ij} as derived in this paper are:

$$\begin{aligned} L_{11} \simeq L_{22} = \nabla_s^4 = (\nabla^2 + K_T)\,[\nabla^2 + (4K_M^2 - 2K_T)]\,^1 \\ \text{and} \quad L_{12} \equiv L_{21} = \nabla_{kT}^2 = \nabla_k^2 + 2K_M K_T, \end{aligned} \tag{6}$$

where the mean curvature $K_M = \frac{1}{2}(K_1 + K_2)$ and the total curvature $K_T = K_1 K_2 - K_{12}^2$. The equation for cylindrical shells proposed by MORLEY [6] and the equation for spherical shells derived by VLASOV [2] are special cases of this more general theory.

[1] The more rigorous expressions would be: $\left.\begin{matrix} L_{11} \\ L_{22}\end{matrix}\right\} = [\nabla^2 + (1 \pm \nu)\, K_T]$ $[\nabla^2 + (4K_M^2 - 2K_T)]$ but the omission of ν seems acceptable.

3. More accurate linearized equations

The changes of curvature $\varkappa_{ij}^*$ appearing in the supplementary terms of the equilibrium equation (1b) must be redefined. Following again KOITER's [3] argument we may add suitable terms of the form $K\varepsilon^0$ (where K is a curvature of the shell surface and ε^0 a membrane strain). Instead of ε^0 we can use $\frac{N^0}{Et}$ or $\frac{L_{ij}^0 F}{Et}$, where L_{ij} is a suitable operator of the shallow shell theory. The proposed changes of curvature in line of curvature coordinates for the additional equilibrium term are

$$\varkappa_{11}^* = \varkappa_{11}^0 + \frac{1}{Et}[2(K_2 - K_1)\nabla^2 - \nabla_k^2];$$
$$\varkappa_{22}^* = \varkappa_{22}^0 + \frac{1}{Et}[2(K_1 - K_2)\nabla^2 - \nabla_k^2]; \quad \varkappa_{12}^* = \varkappa_{12}^0, \tag{7}$$

where $\varkappa_{ij}^0$ are defined by (5).

For constant membrane forces the final equation can be written in the simple, closed form (making use of the compatibility equation (1b)).

$$\left(\frac{D}{Et}\nabla_s^8 + \nabla_{kT}^4\right)w = n_{11}^0\{\nabla_s^4\varkappa_{11}^* - \nabla_{kT}^2[2(K_2 - K_1)\nabla^2 - \nabla_k^2]\,w\}$$
$$+ n_{22}^0\{\nabla_s^4\varkappa_{22}^* - \nabla_{kT}^2[2(K_1 - K_2)\nabla^2 - \nabla_k^2]\,w\} + n_{12}^0\,2\varkappa_{12}^*, \tag{8}$$

where n_{ij}^0 are the non-dimensional membrane forces

$$n_{ij}^0 = \frac{N_{ij}^0}{Et}. \tag{9}$$

4. Applications

4.1 The circular cylindrical shell

For the circular cylindrical shell $K_1 = 0$; $K_2 = -1/R$. With $A = B = R$ the equations will result in non-dimensional form. With the shell constant

$$k = \frac{D}{EtR^2} = \frac{1}{12(1-\nu^2)}\left(\frac{t}{R}\right)^2 \tag{10}$$

and the usual notation for the derivatives: $\frac{\partial^2}{\partial\alpha_1^2} = (\,)^{\mathrm{II}}$; $\frac{\partial^2}{\partial\alpha_2^2} = (\,)^{\cdot\cdot}$, the stability equation can be written in the form

$$w^{\mathrm{IV}} + kc_0 = c_{11}n_{11}^0 + c_{22}n_{22}^0 + c_{12}n_{12}^0, \tag{11a}$$

where the differential operators are:

$$c_0 = c_M = \nabla^4(\nabla^2 + 1)^2\,w;\ c_{11} = -[\nabla^4 - (\,)^{\cdot\cdot}]\,w^{\mathrm{II}}; \tag{11b}$$
$$c_{22} = -[\nabla^2(\nabla^2 + 1) + 2(\,)^{\mathrm{II}}]\,w^{\cdot\cdot} + 3w^{\mathrm{IV}};\ c_{12} = -2\nabla^2(\nabla^2 + 1)\,w^{\cdot\mathrm{I}}.$$

(The operator c_M was first proposed by MORLEY [6].) Discussing the linear theory of cylindrical shells, BARTA [7] showed recently that the terms in w^{IV} (compared to the higher order derivatives appearing in the equations) are of the same order of magnitude as the inherent error introduced in any two-dimensional shell theory through the Kirchhoff-Love assumptions. This means that the term $3w^{IV}$ (underlined with dotted lines) can be neglected in the definition of c_{22}. A further simplification, which simplifies numerical calculations to a great extent can be achieved by putting ∇^2 as a common factor in all operators (11b). This can be easily achieved through addition of terms in w^{IV}. The new proposed operators are:

$$\begin{gathered} c_0 = c_M = \nabla^4(\nabla^2+1)^2 w; \quad c_{11} = -\nabla^2(\nabla^2-1)\, w^{II}; \\ c_{22} = -\nabla^2[(\nabla^2+1)\, w^{\cdot\cdot} + 2w^{II}]; \quad c_{12} = -2\nabla^2(\nabla^2+1)\, w^{\cdot I}. \end{gathered} \tag{12}$$

FLÜGGE [8] derived a set of three rigorous equilibrium equations which he solved after expanding the displacements into Fourier series, neglecting terms resulting from w^{IV} in the factors of the critical membrane forces. Using a more general approach, i.e. reducing the set in its differential form, and making the same simplification, we obtain

$$c_0 \equiv c_{0F} = c_{0M} + [2(1-\nu)(w^{::} + w^{:} - w^{IV})^{II} + 3(1-\nu^2)\, w^{IV}] \tag{13}$$

and the same values as in (11) for the c_{ij} operators. The neglection of the terms in square brackets in Eq. (13) has previously been shown to be legitimate [5, 6, 7] and does not affect the accuracy of results. We were interested in the actual value of the c_{ij} operators resulting from Flügge's equation, without any simplifications. After suitable arrangement of terms, we obtained

$$\begin{gathered} c_{11} = -\nabla^2(\nabla^2-1)\, w^{II} + \underline{2(1+\nu)\, w^{IV}}; \\ c_{22} = -\nabla^2[(\nabla^2+1)\, w^{\cdot\cdot} + 2w^{II}] + \underline{2(1+\nu)\, w^{IV}}; \\ c_{12} = -2\nabla^2(\nabla^2+1)\, w^{\cdot I}. \end{gathered} \tag{14}$$

The close connection to our proposed values (11) is evident.

Recently DANIELSON and SIMMONDS [9] derived accurate buckling equations for arbitrary and cylindrical elastic shells. Their equations for arbitrary shells are more complicated than those presented here, and in their equations for cylindrical shells additional simplifications have been introduced. It can be shown that their equations are obtained from Eq. (7) if the operator ∇_k^2 is multiplied by $(1+\nu)$. For practical purposes they propose the present Eqs. (10) and (11).

4.2 The spherical shell under uniform compression

For the spherical shell $K_1 = K_2$, with $A = B = R$ the non-dimensional equation for uniform critical pressure p is (with $N_{11}^0 = N_{22}^0 = -1/2\, pR$):

$$(\nabla^2 + 2)^2 \,[k(\nabla^2 + 1) + 1] - \frac{p}{2E}\left(\frac{R}{t}\right)(\nabla^2 + 2)\,(\nabla^4 + 3\nabla^2) = 0. \qquad (15)$$

Calculations will be considerably simplified if $(\nabla^2 + 2)^2$ can be taken as a common factor. Writing $\nabla^4 + 3\nabla^2 = (\nabla^2 + 2)\,(\nabla^2 + 1) - 2$ an admissible simplification is

$$(\nabla^2 + 2)^2 \left\{[k(\nabla^2 + 1) + 1] - \frac{p}{2E}\left(\frac{R}{t}\right)(\nabla^2 + 1)\right\} = 0\,[1]. \qquad (16)$$

Similar equations are due to VAN DER NEUT [10] and VLASOV [2].

References

1. MARGUERRE, K.: Zur Theorie der gekrümmten Platte großer Formänderung. In: Proc. 5th Int. Congr. Appl. Mech., Cambridge, Mass., 1938. New York: J. Wiley 1939, pp. 93–101.
2. VLASOV, V. Z.: General theory of shells and its applications in engineering. NASA Transl. N 64–19 PP 3, 1964.
3. KOITER, W. T.: A consistent first approximation in the general theory of thin elastic shells. In: Proc. IUTAM Symp. Theory of Thin Elastic Shells, Delft 1959. Amsterdam: North-Holland 1960, pp. 12–33.
4. BARTA, T. A.: An engineering theory of thin elastic shells, Part I: Linear theory. Unpublished report of Civil Engineering Dept., University College London, 1966.
5. BARTA, T. A.: A contribution to an engineering theory of thin elastic shells, Part I: Linear theory. Paper presented on 12th Int. Congr. Appl. Mech., Stanford, Calif., 1969.
6. MORLEY, L. S. D.: Quart. J. Mech. Appl. Math. **12**, 89–99 (1959).
7. BARTA, T. A.: A comparative survey of the theories for elastic circular cylindrical shells. In: IASS-Colloquium on Progress of Shell Structures, Madrid, October 1969 (to be published).
8. FLÜGGE, W.: Stresses in Shells, 2nd printing, Berlin/Göttingen/Heidelberg: Springer 1962.
9. DANIELSON, D. A., SIMMONDS, J. G.: Int. J. Engng. Sci. **7**, 459–468 (1969).
10. VAN DER NEUT, A.: The elastic stability of the thin-walled sphere (in Dutch). Dissertation, Delft, 1932.

---

[1] If the more rigorous definition of L_{22} as given in the footnote on p. 184 is used, the last round bracket in (16) would be $\nabla^2 + (1 - \nu)$. The error involved is of the order 0.5%.

Stability of the Cosserat Surface

By

M. Dikmen

Middle East Technical University, Ankara, Turkey

1. Introduction

A *Cosserat surface* is defined as a surface, to every point of which a single (deformable[1]) vector called *director*, not necessarily along the normal to the surface is assigned [1]. In this definition, within the purposes of the present paper, "surface" shall be understood as a Jordan surface portion σ embedded in a Euclidean 3-space, bounded by a curve c. The points on this surface shall be designated by means of Gaussian parameters x^α ($\alpha = 1, 2$) which shall serve as convected surface coordinates throughout *any* deformation depending upon a single parameter t. Although t is not necessarily the physical time, it will be considered so in the rest of this paper, without a loss of generality.

Hence, the surface σ can be defined by its position vectors relative to a fixed origin

$$\boldsymbol{r} = \boldsymbol{r}(x^\alpha, t). \tag{1.1}$$

The field of the directors $\boldsymbol{d}$ is similarly defined by

$$\boldsymbol{d} = \boldsymbol{d}(x^\alpha, t). \tag{1.2}$$

Thus, the pairs $\{\boldsymbol{r}, \boldsymbol{d}\}$ define the Cosserat surface. Although restrictions may be imposed to $\boldsymbol{r}$ and $\boldsymbol{d}$, this will not be necessary in the subsequent discussion.

The "*velocity*" $\boldsymbol{v}$ of a point of σ and the "*director* velocity" $\boldsymbol{w}$ (both with respect to and for a given value of t) are by definition

$$\boldsymbol{v} \overset{\text{def}}{=} \dot{\boldsymbol{r}} \tag{1.3}$$

and

$$\boldsymbol{w} \overset{\text{def}}{=} \dot{\boldsymbol{d}} \tag{1.4}$$

---

[1] In the sense that it may stretch and/or change direction.

respectively, where a superposed dot denotes the material derivative with respect to t, holding $\boldsymbol{x}$ fixed.

The mass per unit area of σ will be designated by ϱ, and conservation of mass shall be assumed.

Given a three-dimensional vector field $\boldsymbol{N}$, if the scalar $\boldsymbol{N} \cdot \boldsymbol{v}$ is a rate of work per unit length of c, for all arbitrary velocity fields $\boldsymbol{v}$, then $\boldsymbol{N}$ is called a *curve force vector* per unit length. Given a three-dimensional vector field $\boldsymbol{M}$, if the scalar $\boldsymbol{M} \cdot \boldsymbol{w}$ is a rate of work per unit length of c, for all arbitrary director velocity fields $\boldsymbol{w}$, then $\boldsymbol{M}$ is called a curve *director force vector* per unit length of c. Similarly, the three-dimensional fields $\boldsymbol{F}$ of *assigned surface force*, and $\boldsymbol{L}$ of *assigned surface director force*, per unit mass of σ (and for a value t of the parameter), are defined by considering $\varrho(\boldsymbol{F} \cdot \boldsymbol{v})$ and $\varrho(\boldsymbol{L} \cdot \boldsymbol{w})$ for all arbitrary fields $\boldsymbol{v}$ and $\boldsymbol{w}$, respectively, as rates of work per unit area.

The existence of an internal energy function U per unit mass is also assumed.

Thermal effects shall be neglected [4].

2. Statics

The equations of the statics of the Cosserat surface may, of course, be obtained by specialization of the equations given in [1]. They can, however, be derived *ab initio*, by laying down a principle of virtual rates of work for equilibrium, namely the assertion

$$\overline{\int_\sigma \varrho U \, d\sigma}^{\,\boldsymbol{\cdot}} = \int_\sigma \varrho [\boldsymbol{F} \cdot \boldsymbol{v} + \boldsymbol{L} \cdot \boldsymbol{w}] \, d\sigma + \int_c [\boldsymbol{N} \cdot \boldsymbol{v} + \boldsymbol{M} \cdot \boldsymbol{w}] \, dc \qquad (2.1)$$

as a generalization of a principle in classical mechanics [2] that a material system being in equilibrium, the sum of the virtual rates of work of all external and internal forces, as they exist for a given value of the parameter t, is equal to zero for all virtual fields of velocities.

By virtue of the conservation of mass, (2.1) reduces to

$$\int_\sigma \varrho \dot{U} \, d\sigma = \int_\sigma \varrho [\boldsymbol{F} \cdot \boldsymbol{v} + \boldsymbol{L} \cdot \boldsymbol{w}] \, d\sigma + \int_c [\boldsymbol{N} \cdot \boldsymbol{v} + \boldsymbol{M} \cdot \boldsymbol{w}] \, dc. \qquad (2.2)$$

(2.2) can, in particular, be applied to the following special kinds of velocity fields:

$$\boldsymbol{v} = \boldsymbol{v}_0 \, \delta t, \qquad \boldsymbol{w} = \boldsymbol{0}, \qquad \text{(a)}$$

where $\boldsymbol{v}_0$ is an arbitrary vector and δt is an arbitrarily small number. This corresponds to a rigid body translational velocity field.

$$\boldsymbol{v} = \boldsymbol{\omega} \times \boldsymbol{r}, \qquad \boldsymbol{w} = \boldsymbol{\omega} \times \boldsymbol{d}, \qquad \text{(b)}$$

where $\boldsymbol{\omega}$ is an axial vector standing for rigid body angular velocity.

Under consideration of (2.2) and the particular velocity fields (a) and (b), the equations of equilibrium can easily be obtained. They shall not, however, be reproduced here, for the sake of brevity.

It is to be noted however, for later reference, that the application of the field (a) yields, under the assumption that ϱ, U, $\boldsymbol{F}$, $\boldsymbol{L}$, $\boldsymbol{N}$, $\boldsymbol{M}$ remain unchanged,

$$\int_\sigma \varrho \boldsymbol{F}\, d\sigma + \int_c \boldsymbol{N}\, dc = \boldsymbol{0}. \tag{2.3}$$

3. Stability

Stability shall be verified by means of the condition

$$\int_0^t dt \Big[\int_\sigma \varrho \dot{U}\, d\sigma - \Big\{ \int_\sigma \varrho [\boldsymbol{F} \cdot \bar{\boldsymbol{v}} + \boldsymbol{L} \cdot \bar{\boldsymbol{w}}]\, d\sigma + \int_c [\boldsymbol{N} \cdot \bar{\boldsymbol{v}} + \boldsymbol{M} \cdot \bar{\boldsymbol{w}}]\, dc \Big\} \Big] \geq 0, \tag{3.1}$$

where $\bar{\boldsymbol{v}}$ and $\bar{\boldsymbol{w}}$ are any virtual velocity fields (superimposed to the equilibrium configuration) satisfying the boundary conditions.

In words, an equilibrium configuration will be said to be stable (for boundary conditions of place and traction) *if and only if* in any virtual deformation satisfying the boundary conditions of place the virtual work done by the external forces does not exceed the increase (in the same process) in the total internal energy.

This is an extension of Pearson's concept of elastic stability [3] (cf. also [4]).

In (3.1), the loads are functions of place and depend also on the parameter t:

$$\left.\begin{aligned} \boldsymbol{F} &= \boldsymbol{F}(x^\alpha, t), \\ \boldsymbol{L} &= \boldsymbol{L}(x^\alpha, t), \\ \boldsymbol{M} &= \boldsymbol{M}(x^\alpha, t), \\ \boldsymbol{N} &= \boldsymbol{N}(x^\alpha, t). \end{aligned}\right\} \tag{3.2}$$

Also U is a function of place $\boldsymbol{x}$ and depends upon the parameter t:

$$U = U(x^\alpha, t). \tag{3.3}$$

By writing

$$\int_0^t dt \Big[\int_\sigma \varrho \dot{U}\, d\sigma \Big] = \int_\sigma \varrho (U^* - U^0)\, d\sigma = \int_\sigma \varrho\, \Delta U\, d\sigma, \tag{3.4}$$

where

$$\left.\begin{aligned} U^* &= U(x^\alpha, t), \\ U^0 &= U(x^\alpha, 0), \end{aligned}\right\} \tag{3.5}$$

and

$$\Delta U \stackrel{\text{def}}{=} U^* - U^0, \tag{3.6}$$

the condition of stability (3.1) can be written as

$$\int_\sigma \varrho\, \Delta U\, d\sigma - \int_0^t dt \left\{ \int_\sigma \varrho\,[\boldsymbol{F}\cdot\bar{\boldsymbol{v}} + \boldsymbol{L}\cdot\bar{\boldsymbol{w}}]\, d\sigma + \int_c [\boldsymbol{N}\cdot\bar{\boldsymbol{v}} + \boldsymbol{M}\cdot\bar{\boldsymbol{w}}]\, dc \right\} \geq 0. \tag{3.7}$$

A condition of *infinitesimal stability* is obtained by restricting attention to small variations δt of the parameter t, and to *dead loading* (i.e. $\boldsymbol{F}$, $\boldsymbol{L}$, $\boldsymbol{N}$, and $\boldsymbol{M}$ remaining the same for any virtual deformations satisfying the boundary conditions of place). In this case, (3.1) becomes:

$$\delta t \left[\int_\sigma \varrho \dot{U}\, d\sigma - \left\{ \int_\sigma \varrho\,[\boldsymbol{F}\cdot\bar{\boldsymbol{v}} + \boldsymbol{L}\cdot\bar{\boldsymbol{w}}]\, d\sigma + \int_c [\boldsymbol{N}\cdot\bar{\boldsymbol{v}} + \boldsymbol{M}\cdot\bar{\boldsymbol{w}}]\, dc \right\} \right] \geq 0 \tag{3.8}$$

or, since (3.8) is to be valid for any small $\delta t > 0$,

$$\int_\sigma \varrho \dot{U}\, d\sigma - \left\{ \int_\sigma \varrho\,[\boldsymbol{F}\cdot\bar{\boldsymbol{v}} + \boldsymbol{L}\cdot\bar{\boldsymbol{w}}]\, d\sigma + \int_c [\boldsymbol{N}\cdot\bar{\boldsymbol{v}} + \boldsymbol{M}\cdot\bar{\boldsymbol{w}}]\, dc \right\} \geq 0. \tag{3.9}$$

It may be remarked here that both the point and director parts are responsible together, for stability.

In any solution of a boundary value problem in statics, the external forces must be equilibrated in the deformed state (*principle of solidification*). The principle of virtual rates of work, applied, as in § 2, to the Cosserat surface as a whole, yields,

$$\int_\sigma \varrho \boldsymbol{F}\, d\sigma + \int_c \boldsymbol{N}\, dc = \boldsymbol{0} \tag{2.3}$$

for a rigid translational velocity (a), and

$$\int_\sigma \varrho\,[\boldsymbol{F}\times\boldsymbol{r} + \boldsymbol{L}\times\boldsymbol{d}]\, d\sigma + \int_c [\boldsymbol{N}\times\boldsymbol{r} + \boldsymbol{M}\times\boldsymbol{d}]\, dc = \boldsymbol{0} \tag{3.10}$$

for (b).

In (3.10), $\boldsymbol{r}$ and $\boldsymbol{d}$ are the *actual* position vector and director, respectively.

The equation (3.10), which is a generalized condition of the equilibrium of torques, is to be regarded as a *condition of compatibility* [5], since it involves the position vector $\boldsymbol{r}$ of the (deformed) surface.

Now, a proximate configuration of equilibrium can be imagined, with

$$\left.\begin{aligned} \boldsymbol{r}_{\mathrm{I}} &= \boldsymbol{r} + \bar{\bar{\boldsymbol{r}}}, \\ \boldsymbol{d}_{\mathrm{I}} &= \boldsymbol{d} + \bar{\bar{\boldsymbol{d}}} \end{aligned}\right\} \tag{3.11}$$

and

$$\left.\begin{aligned} \boldsymbol{N}_{\mathrm{I}} &= \boldsymbol{N} + \Delta \boldsymbol{N}, \\ \boldsymbol{M}_{\mathrm{I}} &= \boldsymbol{M} + \Delta \boldsymbol{M}, \\ \boldsymbol{F}_{\mathrm{I}} &= \boldsymbol{F} + \Delta \boldsymbol{F}, \\ \boldsymbol{L}_{\mathrm{I}} &= \boldsymbol{L} + \Delta \boldsymbol{L}. \end{aligned}\right\} \tag{3.12}$$

Then it must be

$$\begin{aligned} &\int_\sigma \varrho [(\boldsymbol{F} + \Delta \boldsymbol{F}) \times (\boldsymbol{r} + \bar{\bar{\boldsymbol{r}}}) + (\boldsymbol{L} + \Delta \boldsymbol{L}) \times (\boldsymbol{d} + \bar{\bar{\boldsymbol{d}}})]\, d\sigma \\ &+ \int_c [(\boldsymbol{N} + \Delta \boldsymbol{N}) \times (\boldsymbol{r} + \bar{\bar{\boldsymbol{r}}}) + (\boldsymbol{M} + \Delta \boldsymbol{M}) \times (\boldsymbol{d} + \bar{\bar{\boldsymbol{d}}})]\, dc = \boldsymbol{0} \end{aligned} \tag{3.13}$$

or, subtracting (4.10) from (3.13),

$$\begin{aligned} &\int_\sigma \varrho [\Delta \boldsymbol{F} \times (\boldsymbol{r} + \bar{\bar{\boldsymbol{r}}}) + \boldsymbol{F} \times \bar{\bar{\boldsymbol{r}}} + \Delta \boldsymbol{L} \times (\boldsymbol{d} + \bar{\bar{\boldsymbol{d}}}) + \boldsymbol{L} \times \bar{\bar{\boldsymbol{d}}}]\, d\sigma \\ &+ \int_c [\Delta \boldsymbol{N} \times (\boldsymbol{r} + \bar{\bar{\boldsymbol{r}}}) + \boldsymbol{N} \times \bar{\bar{\boldsymbol{r}}} + \Delta \boldsymbol{M} \times (\boldsymbol{d} + \bar{\bar{\boldsymbol{d}}}) + \boldsymbol{M} \times \bar{\bar{\boldsymbol{d}}}]\, dc = \boldsymbol{0}. \end{aligned} \tag{3.14}$$

Now, under the restricted case of dead loading, the increments $\Delta \boldsymbol{F}$, $\Delta \boldsymbol{L}$, $\Delta \boldsymbol{N}$, $\Delta \boldsymbol{M}$, all vanish, and

$$\int_\sigma \varrho [\boldsymbol{F} \times \bar{\bar{\boldsymbol{r}}} + \boldsymbol{L} \times \bar{\bar{\boldsymbol{d}}}]\, d\sigma + \int_c [\boldsymbol{N} \times \bar{\bar{\boldsymbol{r}}} + \boldsymbol{M} \times \bar{\bar{\boldsymbol{d}}}]\, dc = \boldsymbol{0}. \tag{3.15}$$

This condition is equivalent to the condition of symmetry of

$$\int_\sigma [\varrho \boldsymbol{F} \otimes \bar{\bar{\boldsymbol{r}}} + \varrho \boldsymbol{L} \otimes \bar{\bar{\boldsymbol{d}}}]\, d\sigma + \int_c [\boldsymbol{N} \otimes \bar{\bar{\boldsymbol{r}}} + \boldsymbol{M} \otimes \bar{\bar{\boldsymbol{d}}}]\, dc. \tag{3.16}$$

Defining an *astatic load* by means of the expression in (3.16), a condition for the existence of an *axis of equilibrium* of the external loads acting on the Cosserat surface, can be found in a way similar to the one in [4].

4. Relevance to the thin shell theory

E. and F. COSSERAT were the first to recognize [6] the importance of the concept of oriented media for the construction of theories of "thin bodies", i.e. shells, plates, and rods. Later, this idea was taken up by ERICKSEN and TRUESDELL [7], GREEN, NAGHDI and WAINWRIGHT [1], and thereupon several times by GREEN and NAGHDI (cf. [8] where further references can be found). The general theory of the Cosserat surface as developed in [1] is exact from both dynamical and thermodynamical points of view. The same theory may be used for the construction of *direct* theories of thin shells, in contrast to the *derived* theories. The general theory of stability developed in this paper provides with a basic structure for stability considerations in such direct theories.

References

1. Green, A. E., Naghdi, P. M., Wainwright, W. L.: Arch. Rat. Mech. Anal. **20**, 287 (1965).
2. Cabannes, H.: Cours de Mécanique Générale, Paris: Dunod 1962.
3. Pearson, C. E.: Quart. Appl. Math. **14**, 133 (1955).
4. Beatty, M. F.: Arch. Rat. Mech. Anal. **19**, 167 (1965).
5. Truesdell, C., Noll, W.: Handbuch der Physik, III/3: The Non-Linear Field Theories of Mechanics, Berlin/Heidelberg/New York: Springer 1965.
6. Cosserat, E., Cosserat, F.: Théorie des Corps Déformables, Paris: Hermann 1909.
7. Ericksen, J. L., Truesdell, C.: Arch. Rat. Mech. Anal. **1**, 295 (1958).
8. Green, A. E., Naghdi, P. M.: Proc. IUTAM Symp. Mechanics of Generalized Continua (Freudenstadt and Stuttgart, 1967), Berlin/Heidelberg/New York: Springer 1968, pp. 36—48.

On the Stability of Periodic Solutions in Fluid Mechanics

By

W. Eckhaus

Technological University of Delft, Delft, Netherlands

1. Introduction

In various stability problems of fluid mechanics, such as the Taylor problem, the Bénard problem or the problem of plane Poiseuille flow, there exists a critical value of the governing parameter R (Taylor number, Rayleigh number, Reynolds number) above which the flow becomes unstable (according to linear theory) to spatially periodic disturbances of which the wave number k lies within an interval $k_1(R) < k < k_2(R)$. Non-linear theories have shown that spatially periodic solutions which are bounded for all time can exist in such supercritical conditions. The stability of these supercritical periodic solutions, and investigation of further possible solutions in the supercritical region, are the subject of this paper.

Instead of dealing explicitly with the fluid mechanics stability problems mentioned above, we shall define (using vector functions and matrix notation) a class of non-linear partial differential equations which contain these problems as elements. Using a method of analysis based on eigenfunction expansion and asymptotic approximation, developed previously by this author for the scalar case [1], rather general results will be obtained. We shall deal here with problems in which the unknown vector-functions are functions of two spatial variables and time.

Acknowledgement. Part of this work was supported by the Army Research Office, Durham, while the author was a visitor at Rensselaer Polytechnic Institute, Troy, N.Y., U.S.A., during the summer of 1967.

We shall first describe the stability analysis of periodic solutions subject to small perturbations, which leads to a simple and general result already obtained in [1] for the scalar case. Related results obtained by other methods will be briefly discussed and compared.

Next we shall study the problem in which the disturbance, when Fourier-analysed, has components which in the wave-number interval $k_1(R) < k < k_2(R)$ are all of the same order of magnitude, the order of magnitude being that of the periodic solutions. Using the same method of analysis (eigenfunctions expansion and asymptotic approximation) the problem will be reduced to a surprisingly tractable form and will be shown to be governed by one scalar partial non-linear differential equation in two independent variables. Analysis of this governing equation reveals the existence of solutions bounded for all time, which are different from the periodic solutions studied previously.

2. Formulation

We consider functions $\Phi_i(\boldsymbol{x}, t)$, $i = 1, \ldots, n$ defined in $\boldsymbol{x} \in D$, $t \geqq D$, where D is an open set of a m-dimensional space R_m and t is a real variable to be interpreted as time. A large class of partial differential equations for Φ_i can be written in a relatively simple form when introducing a vector $\boldsymbol{\Phi}(\boldsymbol{x}, t)$, of which the components are the functions $\Phi_i(\boldsymbol{x}, t)$.

Let $\boldsymbol{L}$, $\boldsymbol{S}$, $\boldsymbol{P}^{(l)}$ and $\boldsymbol{Q}^{(l)}$ be linear operators, represented by $n \times n$ matrices, while $\boldsymbol{K}^{(l)}$ is a linear operator represented by a $n \times 1$ matrix. The elements of these matrix operators are linear partial differential operators independent of the variable t. We shall consider partial differential equations of the following form:

$$\boldsymbol{L\Phi} - \frac{\partial}{\partial t}\boldsymbol{S\Phi} = \sum_l (\boldsymbol{P}^{(l)}\boldsymbol{\Phi}, \boldsymbol{K}^{(l)})\, \boldsymbol{Q}^{(l)}\boldsymbol{\Phi}, \tag{2.1}$$

where summation in the right-hand side is over a finite number of terms.

In accordance with the usual matrix notation, the equation should be read as follows: if, for example, the elements of the matrix $\boldsymbol{L}$ are l_{ij}, then $\boldsymbol{L\Phi} = \boldsymbol{\Psi}$ is a vector of which the components $\Psi_i = \sum_{j=1}^{n} l_{ij}\Phi_i$. Similarly, in the right-hand side if $\theta^{(l)}$ is the vector given by $\theta^{(l)} = \boldsymbol{P}^{(l)}\boldsymbol{\Phi}$, with components $\theta_i^{(l)}$, while $K_j^{(l)}$ are the components of the operator, $\boldsymbol{K}^{(l)}$, then $(\boldsymbol{P}^{(l)}\boldsymbol{\Phi}, \boldsymbol{K}^{(l)})$ is a $n \times n$ matrix with elements $\theta_i^{(l)} K_j^{(l)}$.

With suitable (homogeneous) boundary conditions on the boundary Γ of D, and with initial data $\boldsymbol{\Phi}(\boldsymbol{x}, 0)$, the equation (2.1) represents the formulation of many stability problems from fluid-dynamics. The explicit form of the operators in every particular problem can easily be deduced.

We shall, in this study, assume that $m = 2$, so that $\boldsymbol{x} = (\xi, \eta)$. The domain D will be given by

$$D = \{(\xi, \eta); 0 < \eta < \boldsymbol{x}, -\infty < \xi < \infty\}. \tag{2.2}$$

Boundary conditions will be imposed on $\eta = 0$ and $\eta = 1$. This choice of D, in particular $-\infty < \xi < \infty$, represents the often used oversimplification in the mathematical description of the stability problem and reflects the fact that one dimension of the experimental apparatus is very large as compared to the other. Thus "edge effects" are neglected. Finally we assume that all operators specified above are invariant with respect to translation along the ξ-axis. We now consider an associated linear problem, defined by

$$\boldsymbol{L}\tilde{\boldsymbol{\Phi}} - \frac{d}{dt}\boldsymbol{S}\tilde{\boldsymbol{\Phi}} = 0 \tag{2.3}$$

with the same boundary conditions on $\tilde{\boldsymbol{\Phi}}$ as the ones that are imposed on $\boldsymbol{\Phi}$. We study solutions of the form

$$\tilde{\boldsymbol{\Phi}}(\xi, \eta, t) = \exp[-ik\xi - \mu^{(k)}t]\,\varphi^{(k)}_{(\eta)}, \tag{2.4}$$

where k is a real parameter. After substitution we obtain

$$\boldsymbol{L}_k\boldsymbol{\varphi}^{(k)} + \mu^{(k)}\boldsymbol{S}_k\boldsymbol{\varphi}^{(k)} = 0; \quad 0 < \eta < 1 \tag{2.5}$$

with homogeneous boundary conditions for $\eta = 0$ and $\eta = 1$. $\boldsymbol{L}_k$ and $\boldsymbol{S}_k$ are linear ordinary differential operators with $n \times n$ matrix representation. For example, in the case of plane Poiseuille or Couette flow $n = 1$ and (2.5) is the well-known Orr-Sommerfeld equation.

Thus eigenvalues and eigenfunctions of the associated linear problem are defined. We consider here the case in which, for each value of k, the eigenvalues $\mu^{(k)}$ form a denumerable infinite sequence $\mu^{(k)}_p$, and the corresponding vector-eigenfunctions $\boldsymbol{\varphi}^{(k)}_p(\eta)$ form a complete set. If this is the case, then the eigenfunctions $\boldsymbol{\varphi}^{(k)}_p$, together with their adjoints, can be used to expand the solutions of the non-linear equation (2.1). For a class of problems of the type (2.5), including the Orr-Sommerfeld equation and the equation arising in the Taylor-Couette problem, completeness theorems have recently been proved by DiPrima and Habetler [2]. We finally formulate some hypothesis concerning the behaviour of the eigenvalues. We suppose that $\mu^{(k)}_p$ can be ordered such that for every value of k

$$\mathrm{Re}(\mu^{(k)}_{p+1}) > \mathrm{Re}(\mu^{(k)}_p); \quad p = 0, 1, \ldots \tag{2.6}$$

Furthermore we suppose that the operator $\boldsymbol{L}$ contains a parameter R and that there exists a value $R = R_{cr}$ such that:

if $R < R_{cr}$ then $\mathrm{Re}(\mu^{(k)}_p) < 0$ for all k and p;

if $R > R_{cr}$ then $\mathrm{Re}(\mu^{(k)}_0) > 0$ for $k_1(R) < k < k_2(R)$.

These assumptions are summarized in Fig. 1.

For any fixed value of k, $\mathrm{Re}(\mu_0^{(k)}) = 0$ for $R = R_0(k)$. The function $R_0(k)$ is assumed continuously differentiable in the vicinity of $k = k_{cr}$, at which point $R_0(k)$ has a local minimum.

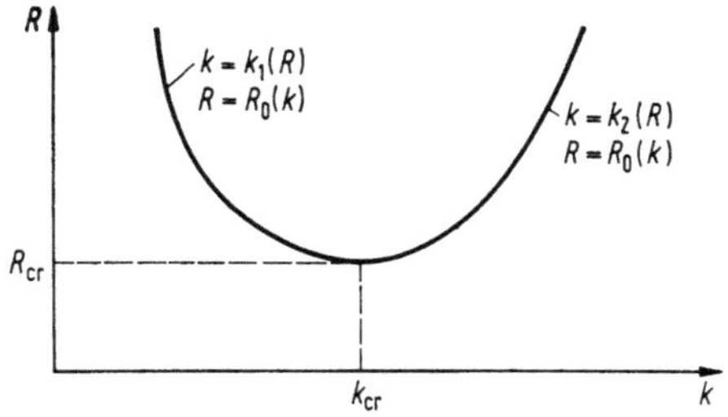

Fig. 1

3. Periodic supercritical solutions and their stability

A method of analysis of the non-linear problem (2.1) based on expansion in eigenfunctions defined by (2.5) and on asymptotic approximation with respect to a suitably defined small parameter has been developed for the scalar case $n = 1$ in Eckhaus [1]. The method has been generalized for problems involving $n > 1$ by DiPrima [3] and by Kogelman [4].

We consider $R - R_{cr} > 0$, $|R - R_{cr}|$ being very small, and define a small parameter by the relation

$$\varepsilon^2 = |\mathrm{Re}(\mu_0^{(k_{cr})})|. \tag{3.1}$$

A periodic solution of equation (2.1) can be represented by

$$\boldsymbol{\Phi}(\xi, \eta, t) = \sum_{q=-\infty}^{\infty} e^{-iqk_0\xi} \boldsymbol{\Phi}_{qk_0}(\eta, t); \quad k_1 < k_0 < k_2. \tag{3.2}$$

Using the method of [1] the following results are obtained:

$$\left.\begin{aligned} \boldsymbol{\Phi}_{k_0}(\eta, t) &= \varepsilon A_0^{(k_0)}(t)\, \boldsymbol{\varphi}_0^{(k_0)}(\eta) + O(\varepsilon^3), \\ \boldsymbol{\Phi}_0(\eta, t) &= \varepsilon^2 |A_0^{(k_0)}(t)|^2 \mathbf{g}_0(\eta) + O(\varepsilon^4), \\ \boldsymbol{\Phi}_{2k_0}(\eta, t) &= \varepsilon^2 [A_0^{(k_0)}(t)]^2 \mathbf{g}_2(\eta) + O(\varepsilon^4), \\ \boldsymbol{\Phi}_{qk_0}(\eta, t) &= O(\varepsilon^3) \text{ for } q \geqq 3. \end{aligned}\right\} \tag{3.3}$$

Here $\boldsymbol{\varphi}_0^{(k_0)}(\eta)$ is the eigenfunction defined by (2.5) with $\mu^{(k)} = \mu_0^{(k_0)}$, $\mathbf{g}_0(\eta)$ and $\mathbf{g}_2(\eta)$ are functions which can be computed explicitly, while $A_0^{(k_0)}(t)$ satisfies the equation

$$\frac{dA_0^{(k_0)}}{dt} + \mu_0^{(k_0)} A_0^{(k_0)} = -\varepsilon^2 \beta_0 A_0^{(k_0)} |A_0^{(k_0)}|^2 + O(\varepsilon^3). \tag{3.4}$$

Finally, β_0 is a constant which can be computed explicitly. The relevant formulas, for $n = 1$, have been given in Eckhaus [1]. In application to plane Poiseuille flow the results coincide with earlier results by STUART [5]. For the vector case $n > 1$ KOGELMAN [4] derived results analogous to (3.3), (3.4) while the explicit results for the Taylor-Couette problem by the method of [1] have been shown to coincide with earlier results of DAVEY [6] in DIPRIMA [3].

If now the constant β_0 is such that

$$\mathrm{Re}(\beta_0) > 0, \tag{3.5}$$

then for $R > R_{cr}$

$$\lim_{t\to\infty} |A^{(k_0)}|^2 = [\mathrm{Re}(\beta_0)]^{-1} \frac{\mathrm{Re}(\mu_0^{(k_0)})}{\varepsilon^2}. \tag{3.6}$$

Periodic supercritical solutions of (2.1), for any $k_1(R) < k_0 < k_2(R)$ are thus defined. The question that arises now can be formulated as follows: given some initial function $\boldsymbol{\Phi}(\xi, \eta, 0)$, which of the periodic solutions (if any) is the limit of $\boldsymbol{\Phi}(\xi, \eta, t)$ as $t \to \infty$. Thus one is led to study the stability of the periodic solutions. Consider now the following representation

$$\boldsymbol{\Phi}(\xi, \eta, t) = \sum_{-\infty}^{\infty} e^{-ik\xi} \boldsymbol{\Phi}_k(\eta, t), \tag{3.7}$$

where $k = \pi s/X$, X is an arbitrarily large number and the summation in (3.7) is defined with respect to all values $s = 0, \mp 1, \mp 2, \ldots$ Thus the distance Δk between two consecutive values of k in (3.7) equals π/X and can be chosen arbitrarily small. In order to study the stability of a periodic solution (3.2) we introduce:

$$\left.\begin{aligned} \boldsymbol{\Phi}_k &= \varepsilon^{|q-1|+1}\, \boldsymbol{\Psi}_q \ \text{ for } k = qk_0,\ q = 0, 1, 2, \\ \boldsymbol{\Phi}_k &= \varepsilon^3 \boldsymbol{\Psi}_k \ \text{ for } k \neq qk_0. \end{aligned}\right\} \tag{3.8}$$

The analysis, for the scalar case $n = 1$, has been performed in ECKHAUS [1], and leads to the following result:

Let the interval $k_1 \leqq k_0 \leqq k_2$ be represented by

$$|k_0 - k_{cr}| \leqq \sigma. \tag{3.9}$$

This is the interval of linear instability. If $\mathrm{Re}(\beta_0) > 0$ in that interval, then the periodic solution is stable for $|k_0 - k_{cr}| < \sigma/\sqrt{3}$ and unstable for $|k_0 - k_{cr}| > \sigma/\sqrt{3}$.

The validity of the above result for the vector case $n > 1$ has been demonstrated by KOGELMAN [4].

It is remarkable that in deriving this result no explicit information on the operators that occur in (2.1) is needed, and that the value of β_0 need not be known. However, $\mathrm{Re}(\beta_0) > 0$ in interval (3.9) is an essential condition. It is for this reason that this stability result does not apply to the case of plane Poiseuille flow. From the calculations of REYNOLDS and POTTER [7] it follows that the condition $\mathrm{Re}(\beta_0) > 0$ in $k_1 \leqq k \leqq k_2$ is not fulfilled in this case. On the other hand, in the Taylor-Couette problem, for example, the condition is satisfied, while the application of the theory outlined above can be compared with other explicit results.

KIRCHGÄSSNER and SORGER [8] have studied the stability of the branching (periodic) solution at $k_0 = k_1(R)$ and $k_0 = k_2(R)$ and found these solutions unstable, except for $k_0 = k_{cr}$. This result is in agreement with our result.

Furthermore, BUSSE [9] in the study of stability of periodic solutions in the Bénard problem has obtained by essentially numerical methods qualitatively similar results to the one formulated above.

It hence appears that in a number of problems, for $R > R_{cr}$, R being fixed and $|R - R_{cr}|$ small, when periodic solutions exist within the whole interval of linear instability $|k_0 - k_{cr}| < \sigma$, then there exists a subinterval of stable periodic solutions $|k_0 - k_{cr}| < \sigma/\sqrt{3}$. Any periodic solution from the stable subinterval is the limit of $\boldsymbol{\Phi}(\xi, \eta, t)$ as $t \to \infty$ for some (classes of) initial conditions. Experimental confirmation of this result in the case of Taylor-Couette has been given by SNYDER [10].

4. Interaction of periodic solutions

In the stability problem described above the perturbation of the periodic solution was taken $\boldsymbol{\Phi}_k = O(\varepsilon^3)$ for $k \neq q k_0$, $q = 0, 1, \ldots$ This problem is in a sense the problem of stability of a periodic solution subject to "noise". The problem is linear, as shown in [1]. Next we consider the problem in which initially within the region of possible periodic solutions, that is $|k - k_{cr}| \leqq \sigma$, all Fourier components $\boldsymbol{\Phi}_k(\eta, t)$ are of the same order of magnitude.

Introducing the Fourier decomposition (3.7) into (2.1) we get

$$\boldsymbol{L}_k \boldsymbol{\Phi}_k - \frac{\partial}{\partial t} \boldsymbol{S}_k \boldsymbol{\Phi}_k = \boldsymbol{F}_k, \tag{4.1}$$

where $\boldsymbol{F}_k$ represents non-linear interaction terms between the various Fourier components $\boldsymbol{\Phi}_k$. Explicit formulas for $\boldsymbol{F}_k$ can be found in [11]. Complete analysis of the infinite system of equations (4.1) is beyond possibilities. However, if we are particularly interested in the interaction of periodic solutions found previously, then simplification of (4.1) is possible. We now summarize the relevant results, obtained in ECKHAUS, DIPRIMA and SEGEL [11]. We observe that if $|R - R_{cr}|$ is very small, then in equation (3.9) we have

$$\sigma = O(\varepsilon). \tag{4.2}$$

We define the following regions on the k-axis:

$$J = \{k;\, |k - k_{cr}| = O(\varepsilon)\},$$
$$I = \{k;\, k = O(\varepsilon)\},$$
$$Y = \{k;\, |k - 2k_{cr}| = O(\varepsilon)\}.$$

Furthermore we introduce

$$\boldsymbol{\Phi}_k = \varepsilon\, \delta_k \boldsymbol{\Psi}_k. \tag{4.3}$$

We search to determine $\delta_k = \delta_k(\varepsilon)$ such that $\boldsymbol{\Psi}_k = 0(1)$.

Suppose that initially

$$\delta_k = 1 \quad \text{for} \quad k \in J, \tag{4.4}$$

$$\delta_k = \varepsilon \quad \text{for} \quad k \in I \cup Y, \tag{4.5}$$

$$\delta_k = \varepsilon^2 \quad \text{for} \quad k \notin I \cup Y \cup J. \tag{4.6}$$

In [11] it has been established that (4.5) and (4.6) remain true for all time for which (4.4) remains true, with the understanding that there is a very rapid "tail-off" of the orders of magnitude between the various regions. The nature of the tail-off has also been determined. These estimations of the order of magnitude of the solutions make it possible to simplify considerably the system of equation (4.1) and arrive at approximate solutions of these equations. The analysis is described in detail in [11]. The results are as follows:

If $k \in I \cup Y$ then $\boldsymbol{\Phi}_k$ can be expressed in terms of $\boldsymbol{\Phi}_{k'}$, $k' \in J$. If $k \in J$ then

$$\boldsymbol{\Phi}_k(\eta, t) = \varepsilon A^{(k)}(t)\, \boldsymbol{\varphi}_0^{(k)}(\eta) + O(\varepsilon^3), \tag{4.7}$$

where $\boldsymbol{\varphi}_0^{(k)}$ are the eigenfunctions that already have been defined, while $A^{(k)}(t)$ satisfies the following system of equations

$$\frac{dA^{(k)}}{dt} + \mu_0^{(k)} A^{(k)} = -\varepsilon^2 \beta_0 \sum_{k'} \sum_{k''} A^{(k')} \bar{A}^{(k'')} A^{(k+k''-k')}, \tag{4.8}$$

the summation in the right-hand side of (4.8) is with respect to all $(k', k'') \in J$. Furthermore, $\bar{A}^{(k)}$ denotes the complex conjugate of $A^{(k)}$, and β_0 the same constant that already appeared in the theory of periodic solutions. It is surprising that no new constants appear in the problem that we study now.

5. Analysis of the amplitude equations

We slightly change the notation and write

$$k - k_{cr} = \varepsilon q \varrho, \; q = 0, \mp 1, \mp 2, \dots \quad \text{for } k \in J. \tag{5.1}$$

Furthermore, for simplicity we use in this analysis the approximation

$$\mu_0^{(k)} = i\nu_0 + \varepsilon^2(\varrho^2 q^2 - 1) \quad \text{for } k \in J. \tag{5.2}$$

Here ϱ is a small number, so that great many components $\boldsymbol{\Phi}_k$ are located within the interval of periodic solutions $|k - k_{cr}| \leqq \sigma$.

Introducing finally

$$A^{(k)} = e^{-i\nu_0 t} a_q \text{ for } k \in J, \tag{5.3}$$

$$\tau = \varepsilon^2 t. \tag{5.4}$$

We obtain from (4.8)

$$\frac{da_q}{d\tau} + (\varrho^2 q^2 - 1)\, a_q = -\beta_0 \sum_{p=-N}^{N} \sum_{m=-N}^{N} a_p \bar{a}_m a_{q+m-p};\ q = 0, \mp 1, \ldots, \mp N. \tag{5.5}$$

Here N is a large number, so that $N\varrho = 0(1)$. Furthermore, if convenient for the analysis, we can take $a_q \equiv 0$ for $|q| > N$ or take $N \to \infty$.

The first important result is that all solutions of (5.5) are bounded if $\operatorname{Re}(\beta_0) > 0$. To establish this result we write

$$a_q = |a_q|\, e^{i\theta} q,\ \ \beta_0 = |\beta_0|\, e^{i\chi} \tag{5.6}$$

and introduce

$$\frac{1}{2} \sum_q |a_q|^2 = E. \tag{5.7}$$

It then follows by straightforward calculation that

$$\frac{dE}{dt} = \sum_q (1 - \varrho^2 q^2)\, |a_q|^2 - \cos\chi\, |\beta_0| \sum_q [C_q^2 + C_q'^2], \tag{5.8}$$

where

$$\left.\begin{aligned} C_q &= \sum_p |a_p|\, |a_{p+q}| \cos(\theta_p - \theta_{p+q}), \\ C_q' &= \sum_p |a_p|\, |a_{p+q}| \sin(\theta_p - \theta_{p+q}). \end{aligned}\right\} \tag{5.9}$$

One easily sees that $dE/dt < 0$ if $|a_q|$, for some or all values of q, is sufficiently large and $\cos\chi > 0$.

Secondly we remark that periodic solutions studied previously correspond to singular points of (5.5), given by

$$\begin{aligned} |a_q|^2 &= \frac{1 - q^2\varrho^2}{\operatorname{Re}(\beta_0)} && \text{for } q = q_0, \\ a_q &= 0 && \text{for } q \neq q_0. \end{aligned}$$

The stability analysis of the periodic solutions can be reproduced as the stability analysis of these singular points.

Finally, for further analysis the set of equation (5.5) can be transformed into a rather simple partial differential equation. For that purpose we take $N \to \infty$ and introduce

$$Z(\omega, \tau) = \sum_q e^{-iq\omega} a_q(\tau), \tag{5.10}$$

$$a_q(\tau) = \frac{1}{2\pi} \int_0^{2\pi} e^{iq\omega} Z(\omega, \tau)\, d\omega. \tag{5.11}$$

From equation (5.5) we obtain

$$\frac{\partial Z}{\partial \tau} = \varrho^2 \frac{\partial^2 Z}{\partial \omega^2} + Z[1 - \beta_0 |Z|^2], \tag{5.12}$$

while from definition (5.10), we have periodicity condition

$$Z\left(\omega + \frac{2\pi}{r}, \tau\right) = Z(\omega, \tau), \tag{5.13}$$

where r is a natural number. The equation (5.10), with condition (5.13), can be considered as the governing equation for the class of problems studied here. The periodic supercritical solutions studied in Sec. 3 can be found again as particular solutions

$$Z = \left\{\frac{1 - q_0^2 \varrho^2}{\mathrm{Re}(\beta_0)}\right\}^{1/2} e^{-i\omega q_0} \exp[-i \,\mathrm{Im}(\beta_0)\, |Z|^2 \tau]. \tag{5.14}$$

In order to demonstrate the existence of other solutions we consider the simpliest case of stationary solutions for β_0 is real. We write:

$$Z(\omega, \tau) = \frac{1}{\sqrt{\beta_0}} |f(x)|\, e^{i\gamma(x)}; \quad x = \frac{\omega}{\varrho} \tag{5.15}$$

and obtain

$$\frac{d^2 |f|}{dx^2} - \frac{C_0^2}{|f|^3} + |f|\,[1 - |f|^2] = 0; \quad \gamma = C_0 \int\limits_0^x \frac{dx}{|f|^2} \tag{5.16}$$

with conditions

$$\left|f\left(x + \frac{2\pi}{\varrho r}\right)\right| = |f(x)|; \quad C_0 \int\limits_x^{x + 2\pi/\varrho r} \frac{dx}{|f|^2} = s\pi, \tag{5.17}$$

where s is zero, or any natural number. For $s = 0$ the solution of (5.16), (5.17) follows immediately and can be expressed in terms of an elliptic function. We finally study the nature of the new solutions defined by (5.12), (5.13). Going back to the Fourier decomposition (3.7) and introducing (4.7) and (5.3) we obtain, after some simplification and making use of formula (5.10)

$$\Phi(\xi, \eta, t) = 2\varepsilon \,\mathrm{Re}\{e^{-i(k_{cr}\xi + \nu_0 t)} \varphi_0^{(k_{cr})}(\eta)\, Z(\varepsilon\varrho\xi, \varepsilon^2 t)\} + O(\varepsilon^2). \tag{5.18}$$

Thus the new solutions correspond, in the first approximation, to the supercritical periodic solution at $k = k_{cr}$ modulated by periodic functions of ξ of which the period, by (5.13), equals $2\pi/\varepsilon\varrho r$ and is therefore large. It is remarkable that the function Z is defined by equations (5.12), (5.13) for the whole class of problems studied here and does not depend on the explicit form of the operators that occur in the fundamental equation (2.1).

References

1. ECKHAUS, W.: Studies in Non-linear Stability Theory. Berlin/Heidelberg/New York: Springer 1965.
2. DIPRIMA, R. C., HABETLER, G. J.: A completeness theorem for non-self-adjoint eigenvalue problems in hydrodynamic stability. Arch. Rat. Mech. Anal. (to appear).
3. DIPRIMA, R. C.: Vector eigenfunction expansions for the growth of Taylor vortices in the flow between rotating cylinders. In: Non-Linear Partial Differential Equations, New York: Academic Press 1967.
4. KOGELMAN, S.: Stability of spatially periodic supercritical flows in hydrodynamics. Thesis, Dept. of Mathematics, Rensselaer Polytechnic Institute, Troy, N.Y., 1968. Also: Phys. Fluids (to appear).
5. STUART, J. T.: On the non-linear mechanics of wave disturbances in stable and unstable parallel flows. J. Fluid Mech. **9**, 353—370 (1960).
6. DAVEY, A.: The growth of Taylor vortices in flow between rotating cylinders. J. Fluid Mech. **14**, 336—368 (1962).
7. REYNOLDS, W. C., POTTER, M. C.: Finite-amplitude instability of parallel shear flows. J. Fluid Mech. **27**, 465—492 (1967).
8. KIRCHGÄSSNER, K., SORGER, P.: Stability analysis of branching solutions of the Navier-Stokes equations. In: Proc. 12th Int. Congr. Appl. Mech., Stanford, Calif., 1968. Berlin/Heidelberg/New York: 1969, pp. 257—268.
9. BUSSE, F. H.: On the stability of two-dimensional convection in a layer heated from below. J. Math. and Phys. **46**, 140—150 (1967).
10. SNYDER, H. A.: Wave-number selection at finite amplitude in rotating Couette flow. J. Fluid Mech. **35**, 273—298 (1969).
11. ECKHAUS, W., DIPRIMA, R. C., SEGEL, L. A.: The equations of non-linear wave-number interaction in supercritical flows (to be published).

On General Theorems for Stability

By

M. Cotsaftis

Association EURATOM—CEA, Fontenay-aux-Roses, France

I

Stability studies are very clearly of importance in applied as well as in theoretical questions, so to deal with, various approaches have been deviced, according to the chosen definition of stability. Roughly speaking, two main lines can be distinguished:

I.1 The "*normal mode*" *approach*, extending old results in Mechanics, which is well designed for linear stability when the unperturbed state $\boldsymbol{\Phi}$ of the system is time-independent; it is known that, setting

$$\varphi(\boldsymbol{x}, t) = \xi(\boldsymbol{x}) \exp \sigma t \tag{1}$$

in the variational equation giving the perturbed field $\varphi(\boldsymbol{x}, t)$:

$$\mathscr{H}(\varphi) = 0, \tag{E}$$

one gets the eigenvalue problem:

$$H(\sigma)\, \xi = 0, \tag{E'}$$

giving the growth rate $\gamma = \operatorname{Re} \sigma$, and so the instability limit when

$$\gamma > 0, \tag{S}$$

according to the definition (1). This approach, which has been extended to continuous systems, to time-periodical unperturbed state in the linear case to get the parametric or the dynamical effects, and to non-linear time-independent equations to get the non-linear saturation or coupling effects, can only, despite considerable help from high-speed computers, be termed as a *technique* in that, going from (E) to (E′) with (1), the problem is still to be solved. On the other hand, if one is only interested in the stability domain (S) in the system parameters space, the way followed here includes as a step the pattern calculation through the field $\xi(\boldsymbol{x})$ — and not a scalar as γ — and so is, though exact, unnecessarily expensive when the problem has some complexity.

I.2 The *"global" approach*, where, contrarily to the detailed previous approach I.1, solving (E) is avoided, by extracting from (E) "global" quantities containing as much information as possible on the asymptotic behavior of $\varphi(\boldsymbol{x}, t)$ when $t \to \infty$. To this class belongs the Liapunov's second *method*, initially deviced for finite dimensional systems and since very widely extended. We recall that given (E), $\varphi(\boldsymbol{x}, t)$ will be termed stable if, given $\varepsilon > 0$ and $||\varphi(\boldsymbol{x}, t_0)|| < \eta$, where $||\ ||$ represents a suitable norm, then, for any $t > t_0$,

$$||\varphi(\boldsymbol{x}, t)|| < \varepsilon. \tag{2}$$

This happens when one can construct from (E) a Liapunov function (or functional) V satisfying a set of conditions we do not reproduce here.

The wide generality of this approach contains its own weakness for, despite a huge amount of trials, it has up till now never been possible unless very precise cases to give some laws for constructing V. So in complex cases where it could have been the most useful, the results are often rather poor, especially when the system has a large number of degrees of freedom like in fluids, as it can be understood from the nature of this approach, a *topological* one, in which one tries to locate the solution of (E) in the vicinity of the origin in the configuration space, see (2), and so is quite different from I.1, which is a *functional* approach, after (1).

Related to I.2 are the "energy" method, where one has to extremalize some functional of the field related to the energy, that is having a physical meaning, and the "averaging" methods, which appear to be of interest in "gentle" problems like Bénard's. Both methods are explicit and powerful but have only been developed on specific problems.

II

On the other hand, situations are occurring, especially in complex cases, where one would like to get relatively simply some valid but not necessarily extremely precise criteria for stability, in order to orientate a choice, or to understand the effect of a parameter, before a more advanced analysis. So, to escape from the difficulties of the two previous — and extreme — ways, we intend in this survey to discuss whether it is possible, given a general equation (E), which can be time-dependent and/or non-linear, to set down directly stability conditions which, to be interesting, are furthermore required to satisfy the following three properties: (a) Explicitness, (b) Meaningfulness, (c) Tractability, which are not all satisfied by both I.1 and I.2. An alternative approach to this problem is developed here and examples are given.

II.1 From the previous discussion, one easily understands that to possibly include (a), (b), (c), a functional approach must be deviced rather than a topological one. So we shall introduce the following definition of stability: $\varphi(\boldsymbol{x}, t)$ is a stable solution of (E) if

$$\varphi(\boldsymbol{x}, t) \in \mathscr{S}, \tag{3}$$

where $\mathscr{S}$ is a space of functions correctly chosen. (3) is different from (1) and (2), but contains enough information about stability property, which is a global and a "rough" one. So now the stability problem is transformed into the following one: Given $\mathscr{S}$, to find the two sets $\mathscr{C}$ and $\mathscr{D}$, in which respectively the initial conditions $\psi(\boldsymbol{x})$ — belonging to the phase space related to (E) —, and the parameters defining (E) — we represent by $\theta(\boldsymbol{x}, t)$ —, must range such that $\varphi(\boldsymbol{x}, t) \in \mathscr{S}$. In other words, this is an embedding problem on φ:

$$\{\psi \in \mathscr{C}, \theta \in \mathscr{D} \Rightarrow \varphi \in \mathscr{S}\}, \tag{4}$$

which defines $\mathscr{C}$ and $\mathscr{D}$ — the stability domain — by a "sorting out" method on ψ and θ. For example, let $\mathscr{H} \equiv \frac{d^2}{dt^2} + a$; here $\varphi = \varphi(t)$ a scalar and $\theta = a$, a real number. Then, the equation is linear and so the corresponding embedding problem is independent of $\mathscr{C}$; furthermore, $\mathscr{S}$ is here the set $\mathscr{S}_0 \equiv \{\varphi_0 \exp i\omega t, \omega \in \mathrm{R}_1\}$. To satisfy (4), one must clearly have: $\mathscr{D} \equiv$ {real positive numbers}. Here $\mathscr{D}$ is found by inspection. In general, a procedure has to be deviced for getting $\mathscr{D}$ — and $\mathscr{C}$ when (E) is non-linear.

II.2 To simplify, suppose first that $\mathscr{H}$ is a second order in time, time-independent linear partial differential operator. So is the usual perturbed flow problem in Lagrange variables. (1) does apply, and one gets (E′), with now

$$H(\sigma) = \sigma^2 N + \sigma L + M, N, L, M \tag{5}$$

differential operators on $\boldsymbol{x}$ in a functional space $\mathscr{X}$, say $\mathscr{L}^2$. Instead of following I.1, let us introduce the equation

$$H(\sigma)\,\xi = \Lambda(\sigma)\,N\xi, \tag{E*}$$

a non-zero R.H.S. (E*) appears as an eigenvalue problem, first supposed regular, on ξ in $\mathscr{X}$ for each value of σ. Clearly σ such that $\Lambda(\sigma) = 0$ — the null set $\mathscr{N}(\Lambda)$ — is a solution of (E). The interest of (E*) over (E) is not only a formal one for, and this is the key of the approach, the set $\mathscr{N}(\Lambda)$ — or else the Characteristic Exponents (C.E.) — is controlled in some sense by the set of the eigenvalues $\{\Lambda(\sigma)\}$ of (E*), which creates an "organization" in it.

To see that, let $\mathscr{H}$ be furthermore Self-Adjoint (S.A.). Then $H(i\omega)$ is S.A. in $\mathscr{X}$, and so the $\Lambda(\omega)$ are real when ω is real. $\mathscr{N}(\Lambda)$ will be determined by plotting $\{\Lambda(\omega)\}$ vs. ω; as $\varphi(\boldsymbol{x}, t)$ is stable if all of $\mathscr{N}(\Lambda)$ is real, the reality of one C.E. will change when a Strict Extremum (S.E.) λ_{jk} on one curve $\Lambda_j(\omega)$ will cross the ω-axis, as they are continuous. Consequently, the stability of the solutions of (E) is determined by the *sign* of the S.E. λ_{jk} of $\{\Lambda(\omega)\}$, with

$$\lambda = \underset{\omega \in \mathrm{R}_1}{\mathrm{Extr}} \{\Lambda(\omega)\}. \tag{6}$$

The precise rule, which introduces the sign of the curvature of the $\Lambda_j(\omega)$ at $\Lambda_j = \lambda_{jk}$, and gives the exact Necessary and Sufficient Condition for Stability (N.S.C.S.) will not be reproduced here [1]. Instead we notice from (5) that when

$$\omega \to \pm\infty, \ (\sigma = i\omega), \ \Lambda(\omega) \to -\infty \tag{7}$$

if N positive definite. So looking at the lowest eigenvalue $\Lambda_1(\omega)$, its — only — S.E. λ_1 is the lowest of all the λ_{jk} on the $\Lambda_j(\omega)$. Furthermore, when

$$\lambda_1 > 0, \tag{8}$$

all of $\mathscr{N}(\Lambda)$ is real, and so the system is stable. This is a *Sufficient Condition* for Stability (S.C.S.). It is a N.S.C.S. for the G.U. Systems only [1], which include the equilibrium problems ($L \equiv 0$). Then (E*) writes

$$(-\omega^2 N + M)\,\xi = \Lambda N \xi. \tag{9}$$

Diagonalizing M:

$$M\xi_j = m_j N \xi_j, \tag{10}$$

one gets $\Lambda_j(\omega) = -\omega^2 + m_j$, a simple parabola. $\mathscr{N}(\Lambda_j)$ is real if $\lambda_j = \underset{\omega}{\mathrm{Extr}} \{\Lambda_j\} = m_j > 0$, which reduces to

$$\lambda_1 = m_1 > 0. \tag{11}$$

This is the Lagrange-Liapunov theorem. In fact, it is a little more, for we need only after (7) the semi-boundedness of M and *not* the regularity of the spectrum. So the S.C.S. (8) extends to singular cases as well, occurring for instance in conservative flows. According to (1) with here

$$\mathscr{S} \equiv \{\xi(\boldsymbol{x}) \cdot \exp i\omega t\colon \ \xi(\boldsymbol{x}) \in L^2, \ \omega \text{ real}\}, \tag{12}$$

this result can be alternatively said: $\lambda_1 > 0$ is a sufficient condition to embed φ into $\mathscr{S}$, with (5). Actually, one is better for generalizations to consider instead of (12) the Hilbert space $\mathscr{M}_2$, with norm $\varphi \in \mathscr{M}_2$:

$$\langle \varphi, \varphi \rangle = \lim_{T\to\infty} \frac{1}{2T} \int_{-T}^{+T} (\varphi, \varphi)\, dt < +\infty, \tag{13}$$

where $(\varphi, \varphi) = \int \varphi^* \varphi \, dx$ is a norm in $\mathscr{L}_2$[1]. Important subspaces are the almost periodical functions, represented by generalized harmonic series, with oscillating correlation functions, and the pseudo-random functions with correlation functions going to 0 when $\tau \to \infty$. $\mathscr{M}_2$ seems natural in that from (13), it physically means that the *power*, if φ is an amplitude, is bounded, and *not* the energy, as would follow from the choice $\mathscr{S} = \mathscr{L}_2$. From (1) with $\sigma = i\omega$, (6), and the possibility of deriving (E*) from a variational principle

$$\Lambda(\omega) = \operatorname*{Extr}_{\xi \in X} \frac{(\xi, H(i\omega)\,\xi)}{(\xi, \xi)}, \tag{14}$$

the previous discussion summarizes by the

Theorem I: *A S.C.S. for* (E): $\mathscr{H}\varphi = 0$ *with second order in time linear S.A. operator* $\mathscr{H}$ *is that*:

$$\lambda_1 = \operatorname*{Min}_{\varphi \in \mathscr{M}_2} \operatorname{Extr} \frac{\langle \varphi, \mathscr{H}\varphi \rangle}{\langle \varphi, \varphi \rangle} > 0. \tag{C$_1$}$$

Stability thus appears as deeply connected with Spectral Analysis.

Now when $H(\sigma)$ is not S.A., $\Lambda(\sigma)$ becomes a complex function of the complex variable $\sigma = \gamma + i\omega$; to proceed, one has to embed φ into the space $\mathscr{S} = E_{-\gamma_0}$, where $\lim_{t \to \infty} \frac{\log |\varphi(t)|}{t} = -\gamma_0 < 0$. So in the $(\omega, \gamma, \operatorname{Re} \Lambda)$-space, one has along the cylinders $\operatorname{Im} \Lambda_j = 0$ a set of space curves $\operatorname{Re} \Lambda_j$, and the points $\operatorname{Re} \Lambda_j(i\omega_0) - \omega_0 \in \mathscr{N}(\operatorname{Im} \Lambda_j)$, $\gamma = 0$ — can be shown to determine the stability property of (E). So instead of $E_{-\gamma_0}$, one reduces to $\mathscr{M}_2$, for it contains all the oscillatory type solutions, which are just the limit between growing and decaying functions. Clearly, the lowest point $\operatorname{Re} \Lambda_1(i\omega_0)$ is enough, so one gets the

Theorem II: *A S.C.S. for* (E): $\mathscr{H}\varphi = 0$ *with second order in time linear dissipative is that*

$$\operatorname{Min} \operatorname{Re} \Lambda(i\omega_0) > 0, \ \operatorname{Im} \Lambda(i\omega_0) = 0. \tag{C$_2$}$$

(C_2) generalizes (C_1), and a useful interpretation is that the C.E. of (E) are located on the left of the line $\gamma = 0$ in the σ-plane. Extension to any $\gamma = \gamma_0 \neq 0$ is tricky[2]. This is strongly connected with the problem of dichotomization we shall not touch here.

Remark that (C_1) and (C_2) satisfy properties (a) and (b), and that for (c) only a scalar is involved, which is very weak.

II.3 To extend (C_1) (C_2) to more general systems, let us analyze the process of the approach. It rests on (1), that is the representation of φ

[1] It could be different in another space, for instance W_p^l, a Sobolev space, useful in continuous systems.

[2] It must however be larger than the largest real part of the C.E. of (E).

as a function of t. So it will extend to linear periodic and even to linear periodic functional differential systems, which occur in continuous media stability with memory of the past, or else to finite difference ones, when discretizing. N.S.C.S. can be obtained: in S.A. case, it uses the sign of the eigenvalues of $\mathscr{H}$ in $\mathscr{S}$, but it does not reduce when $\mathscr{H}$ is periodic to (C_1) because of the parametric instability effects, one can study here in details. In non-S.A. case, (C_2) still applies as follows the

Theorem III: *A S.C.S. for linear functional differential periodic second order system is that it be dissipative enough and* (C_2) *be verified.*

III

In more general cases, one has to go from (3) into the corresponding singular Sturm-Liouville problem. When $\mathscr{H}$ is non-linear, (C_1) can be extended to some stationary potential operators [2], as an approach to their Spectral Theory. (C_1) (C_2) and their extensions proceed along the same way: to extract as much information as possible from the lowest eigenvalue of $\mathscr{H}$ in $\mathscr{S}$. The so-called principle of exchange of stability is an example of this.

III.1 Clearly, to progress however towards more complexity would need to leave the too much detailed approach started with. One elementary idea is, having $\mathscr{H}$ relatively close to $\mathscr{H}_0$ of previous type, to set $\mathscr{H} = \mathscr{H}_0 + \mathscr{H}_1$, and consider

$$\varphi + \mathscr{H}_0^{-1}\mathscr{H}_1(\varphi) = 0, \mathscr{H}_1 \qquad (E_1)$$

now general with $\mathscr{H}_1(0) = 0$. (E_1) in $\mathscr{S}$ — not necessarily $\mathscr{M}_2$ — is well known in Control Theory [3] (C.T.), so analyzing its loop diagram, one gets the

Theorem IV: *A S.C.S. for the solution of* (E_1) *is that*

$$||\mathscr{H}_0^{-1}\mathscr{H}_1(\varphi)||_{\mathscr{S}} < ||\varphi||_{\mathscr{S}} \qquad (C_3)$$

for any $\varphi \in \mathscr{S}$[1].

(C_3) is related to the fixed point theorem through the contraction mapping. Let $\mathscr{H}_0\, \chi_j = \Lambda_j \chi_j$, then ($C_3$) becomes

$$\sum_j \frac{1}{|\Lambda_j|^2} ||(\chi_j \mid \mathscr{H}_1(\varphi))||^2 < \sum_j |a_j|^2, \qquad (15)$$

with $\varphi = \sum_j a_j \chi_j$, $(\lambda_j \mid \chi_k) = \delta_{jk}$. (15) needs $\mathscr{H}_1$ to map $\mathscr{S}$ onto itself. Oversimplified, so rougher but more tractable conditions are

$$\operatorname{Sup}_j \frac{1}{|\Lambda_j|} ||\mathscr{H}_1(\varphi)||_{\mathscr{S}} < ||\varphi||_{\mathscr{S}} \qquad (16)$$

---

[1] (C_3) is still valid when there is a non-zero R.H.S. in (E_1), but we take it zero here.

and, with the projector $\mathscr{P}_0$ on the "vector" χ_{j_0}, corresponding to the lowest $|\Lambda_j|$,

$$\frac{1}{|\Lambda_{j_0}|} \|\mathscr{P}_0 \mathscr{H}_1(\varphi)\|_{\mathscr{S}} < \|\varphi\|_{\mathscr{S}} - \|(\mathscr{I} - \mathscr{P}_0)\mathscr{H}_0^{-1}\mathscr{H}_1(\varphi)\|_{\mathscr{S}}, \quad (17)$$

better than (16) if $\|\mathscr{P}_0\mathscr{H}_1(\varphi)\|_{\mathscr{S}}$ is large enough, $\mathscr{I}$ identity. We mention the more involved n-loops criteria like (C_3) with $(\mathscr{H}_0^{-1}\mathscr{H}_1)^{(n)}$, the n-th power of $\mathscr{H}_0^{-1}\mathscr{H}_1$. If $\mathscr{H}_1$ is linear, their sequence is monotonically upwards bounded when $n \to \infty$ by the spectral radius of $\mathscr{H}_0^{-1}\mathscr{H}_1$ in $\mathscr{S}$. If $\mathscr{H} = \mathscr{H}_0 + \mathscr{H}_1$ is of the same type as $\mathscr{H}_0$, say stationary as in II, then, from (E_1^*) — see (E^*) —, one gets $\Lambda_1 = 1 + \operatorname{Extr}_{\{\varphi\}} \frac{(\varphi, \mathscr{H}_0^{-1}\mathscr{H}_1(\varphi))}{(\varphi, \varphi)}$, and a sufficient condition to satisfy (C_2) is

$$\left|\operatorname{Extr}_{\{\varphi\}} \frac{(\varphi, \mathscr{H}_0^{-1}\mathscr{H}_1\varphi)}{(\varphi, \varphi)}\right| < 1. \quad (18)$$

(C_3) follows from Schwarz's inequality. So the "distance" between (C_2) and (C_3) lies on the difference between Re Λ and $|\Lambda|$. It is 0 if Im $\Lambda = 0$: so (C_3) is a N.S.C.S. when $\mathscr{H}_1$ is simple enough, and in general is expected to be not too bad.

III.2 The link with C.T. has been mentioned. Now to interpret (16) or (17) we plot the curve (Γ) Im Λ_{j_0} vs. Re Λ_{j_0}: then the circle at the origin with radius $\varrho = \frac{\|\mathscr{H}_1(\varphi)\|_{\mathscr{S}}}{\|\varphi\|_{\mathscr{S}}}\left(\text{or } \frac{\|\mathscr{P}_0\mathscr{H}_1(\varphi)\|_{\mathscr{S}}}{\|\varphi\|_{\mathscr{S}} - \|(\mathscr{I} - \mathscr{P}_0)\mathscr{H}_0^{-1}\mathscr{H}_1(\varphi)\|_{\mathscr{S}}}\right)$ must lie inside (Γ) for any $\varphi \in \mathscr{S}$. By classical loop diagram transformations in C.T. of the simplified systems related to (16) or (17) — as (15) is related to (E) —, one obtains other expressions.

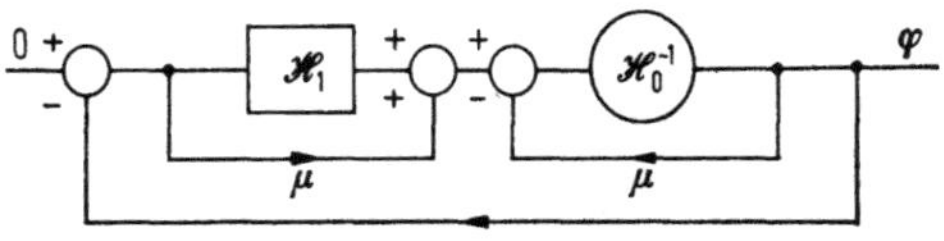

Fig. 1. μ-transform of (E).

For instance, introducing a real scalar μ (Fig. 1) and maximizing the modified criteria on μ, one gets the

Theorem V: *A S.C.S. for embedding the solution* φ *of* (E) $\mathscr{H}(\varphi) = 0$ *in* $\mathscr{S}$ *is that*

$$\operatorname{Re}\Lambda_{j_0} > \frac{\|\mathscr{H}_1(\varphi)\|_{\mathscr{S}}}{\|\varphi\|_{\mathscr{S}}}\left(\text{or } \frac{\|\mathscr{P}_0\mathscr{H}_1(\varphi)\|_{\mathscr{S}}}{\|\varphi\|_{\mathscr{S}} - \|(\mathscr{I} - \mathscr{P}_0)\mathscr{H}_0^{-1}\mathscr{H}_1(\varphi)\|_{\mathscr{S}}}\right), \operatorname{Im}\Lambda_{j_0} = 0, \quad (C_4)$$

for any $\varphi \in \mathscr{S}$.

(C_4) exactly generalizes (C_2) when $\mathscr{H}_1 \neq 0$. Similar conditions (C_4'), when exchanging Re Λ_{j_0} and Im Λ_{j_0}, are likewise obtained by introducing the pure imaginary scalar $i\mu$. The splitting of Re Λ and Im Λ

from $|\Lambda|$ can be further improved when $\mathscr{S} \equiv \mathscr{M}_2$. Let

$$\mathscr{M}\{\mathscr{H}_1(\varphi)\} = ||\varphi||_{\mathscr{M}_2} \cdot \mathscr{E}(\varphi) \tag{19}$$

define the (real) mean $\mathscr{E}(\varphi)$ of $\mathscr{H}_1(\varphi)$ in $\mathscr{M}_2$, with $\mathscr{M}(f) = \lim_{T\to\infty} \frac{1}{T}\int_0^T f(t)\,dt$.

So $\mathscr{H}_1(\varphi) = (\mathscr{E} + \mathscr{U}) \cdot ||\varphi||_{\mathscr{M}_2}$, $\mathscr{U}$ the oscillatory part of $\mathscr{H}_1$. Introducing similarity μ and maximizing, one gets the

Theorem VI: *A S.C.S. for embedding the solution φ of* (E) $\mathscr{H}(\varphi) = 0$ *in $\mathscr{M}_2$ is*

$$\operatorname{Re} \Lambda_{j_0} = -\langle \mathscr{E}(\varphi), \varphi\rangle, \operatorname{Im} \Lambda_{j_0} > ||\mathscr{U}(\varphi)||_{\mathscr{M}_2}, \tag{C$_5$}$$

for any $\varphi \in \mathscr{M}_2$.

Similar conditions (C_5'), with interchanged $\operatorname{Re}\Lambda$ and $\operatorname{Im}\Lambda$, are obtained if $\mathscr{E}$ is pure imaginary. All these criteria are easily interpreted on (Γ), as well as μ (or $i\mu$)[1]. This procedure works as well for n-loops criteria, and for (15). In this last case let $\{\alpha_j\}$ a set of real numbers such that $\sum_j \alpha_j^2 = 1$. Then one gets the set

$$\alpha_j \operatorname{Re} \Lambda_j > \frac{||(\chi, | \mathscr{H}_1(\varphi))||_{\mathscr{S}}}{||\varphi||_{\mathscr{S}}}, \operatorname{Im} \Lambda_j = 0, \tag{C$_6$}$$

extending (C_4). Even if (E) is infinite dimensional, this set can reduce to a few ones when using asymptotic theorems on j-dependence of the Λ_j in Sturm-Liouville theory and choosing an adapted set $\{\alpha_j\}$. Special cases of finite dimensional systems or typical simplifications are easily extracted from above. We only mention that when $m\varphi \leq \mathscr{H}_1(\varphi) \leq M\varphi$, m, M two numbers, one gets from (16) (or (17)) the generalization of the so-called circle criterion, and from (C_4) an extension of the well known Popov's criterion, here elementarily obtained, which sounds of very deep interest.

III.3 To understand (C_4), which we select as the more representative, were turn to (C_2). Similarly here (C_4) means that, for any $\mathscr{H}_1$ satisfying it, the C.E. of the solution of (E): $\mathscr{H}(\varphi) = 0$ do still lie on the left of the line $\gamma = 0$ in the σ-plane, $\sigma = \gamma + i\omega$, as are lying those of the solution of (E_0) $\mathscr{H}_0\varphi = 0$. Extension to any $\gamma = \gamma_0 \neq 0$ is trivial. So we got a condition of *Structural Stability*, once more related to the dichotomization of φ. Now similarly a bound on the right, to locate more strictly the C.E. in a strip of the σ-plane, can likewise be easily obtained when using the time-reserval T ($Tt \Rightarrow -t$) of (E): (E^T) $\mathscr{H}^T(\varphi^T) = 0$ and applying directly all the previous criteria. For if the C.E. of (E^T) are bounded on the left, those of (E) are bounded on the right, because $T\sigma = -\sigma$. This corresponds to an unstable situation,

[1] Once more, only the lowest eigenvalue of $\mathscr{H}_0$ is needed in these criteria.

so one gets the

Theorem VII: *A Sufficient Condition for Instability of the solution of* (E) $\mathscr{H}(\varphi) = 0$ *(or non-embedding of* φ *in* $\mathscr{S}$*) is that* (C_3), *or any transformed criterion* (C_4) (C_5) (C_6), *be satisfied for the time reversal system* (E^T) $\mathscr{H}^T(\varphi^T) = 0$.

Solving these criteria in terms of the leading parameters of (E) will give *a fork*, with surely instability above and surely stability below, the exact limit being in between. The width of the fork gives a measure of the validity of the criteria one can test.

IV

Summarizing, through the seven theorems above we have now various (a) explicit and (b) meaningful expressions from which stability or instability (more generally embedding in a given space $\mathscr{S}$ or not) of the solutions of (E) can be verified. To test (c), let us discuss some examples.

IV.1 Let $\ddot{\varphi} + 2\xi\dot{\varphi} + (1 + f(t))\,\varphi = 0$, $\varphi = \varphi(t)$ a scalar, $f \in \mathscr{M}_2$, $\mathscr{M}(f) = 0$. Following III, we set $\mathscr{H}_0 = \frac{d^2}{dt^2} + 2\xi\frac{d}{dt} + 1$. So $\Lambda_0 = \sigma^2 + 2\xi\sigma + 1$. Reading ($C_4$) ($C_4'$) gives, with $\xi > 0$, $\sigma = i\omega$, $\gamma = 0$:

$$\operatorname{Sup}(1, 2\xi) > ||f||_{\mathscr{M}_2}, \tag{20}$$

for φ not belonging to the convex-hull of f. If $\xi < 0$, $\mathscr{H}_0$ is unstable, and an instability condition for $\mathscr{H}$ is

$$+2\,|\xi| > ||f||_{\mathscr{M}_2}. \tag{21}$$

Generalizing to $\gamma_0 \neq 0$ or n-order φ is easy.

IV.2 Let the panel flutter problem: $H\varphi = 0$, $\varphi = \varphi(x, t)$, $H = H_0 + H_1$, $H_0 = \mu\frac{\partial^2}{\partial t^2} + \frac{\partial}{\partial t} + d\frac{\partial^4}{\partial x^4} - f\frac{\partial^2}{\partial x^2}$, $H_1 = M\frac{\partial}{\partial x}$, d, μ, f three parameters, $M = M(t) \in \mathscr{M}_2$ the Mach number. So

$$\Lambda_0 = \mu^2\sigma + \sigma + dk^4 + fk^2. \tag{22}$$

A S.C.S. for H_0 is

$$dk^2 + f > 0. \tag{23}$$

(16) or (C_4') gives with $\sigma = i\omega$, $\gamma_0 = 0$:

$$||M||^2_{\mathscr{M}_2} < \frac{dk^2 + f}{\mu}, \; k = \pi. \tag{24}$$

Because $H_1 = ikM$, (C_5') can apply, and, with $M(t) = M_0 + \tilde{M}(t)$, one gets

$$M_0^2 < \frac{dk^2 + f}{\mu} - \frac{||\tilde{M}||_{\mathscr{M}_2}}{k\mu}, \tag{25}$$

generalizing (24), for (Γ) is here a parabola from (22). (25) applies when the flow is oscillatory as well as turbulent from the definition of $\mathscr{M}_2$. (20) and (24) have been found [4, 5] after some calculations through Liapunov V-functions. In general, one can expect I.2, when it works, to give at most a result from one of the given criteria.

IV.3 Let the Orr-Sommerfeld (O.S.) equation with the flow $\bar{u}(y, t) = u_0(y) + \tilde{u}(y, t)$, with $\tilde{u} \in \mathscr{M}_2$. H_0 will correspond to O.S. equation with $u = u_0(y)$. Setting $\Lambda_0 = c_I - ic_R$, and using (C_2) gives the usual stability chart, when plotting the lines $c_I(\alpha, R) = \text{const.}$ in the (α, R)-plane with $c_R(\alpha, R)$. Now $H_1 = i\alpha R\left(\frac{d^2}{dy^2} - \alpha^2\right)^{-1} \cdot \left\{\tilde{u}\left(\frac{d^2}{dy^2} - \alpha^2\right) - \frac{d^2\tilde{u}}{dy^2}\right\}$. Using ($C_4$) and the Green function $G(y, y_0)$ of $\frac{d^2}{dy^2} - \alpha^2$, one obtains the condition

$$c_I(\alpha, R) + \frac{\left\|\tilde{u}\varphi - \int\limits_{y_1}^{y_2} G(y, y_0)\frac{d}{dy_0}\left(\tilde{u}\frac{d\varphi}{dy_0}\right)dy_0\right\|_{\mathscr{M}_2}}{\|\varphi\|_{\mathscr{M}_2}} > 0\,, \tag{26}$$

which simplifies into

$$c_I(\alpha, R) + \Phi(\|u\|, \alpha) > 0\,, \tag{27}$$

where $\Phi = \left\|\frac{dG}{dy_0}\right\| \cdot \left\|\frac{d\tilde{u}}{dy}\right\|_{\mathscr{M}_2} + \left(1 + \left\|\frac{d^2G}{dy_0^2}\right\|\right) \cdot \|\tilde{u}\|_{\mathscr{M}_2}$. Given $\|\tilde{u}\|$ and $\left\|\frac{d\tilde{u}}{dy}\right\|$, (27) is in the (c_I, α, R)-space the intersection of the surface

$$c_I = c_I(\alpha, R)\,, \tag{28}$$

with the cylinder $c_I = -\Phi(\alpha, \|u\|)$, giving in the (α, R)-plane a modified stability limit. Proceeding very roughly, one can take in the (c_I, R)-plane the projection (π): $c_I = c_I(R)$ parallel to α of the envelope of surface (28): let $\Phi_{\min} = \min\limits_{\alpha} \Phi(\alpha, \|u\|)$. Then

$$c_I(R) + \Phi_{\min} = 0 \tag{29}$$

gives at once the corresponding value of R for $\|\tilde{u}\|_{\mathscr{M}_2}$ and $\left\|\frac{d\tilde{u}}{dy}\right\|_{\mathscr{M}_2}$ fixed (Fig. 2). Now Theorem VI does apply, and the rough instability limit

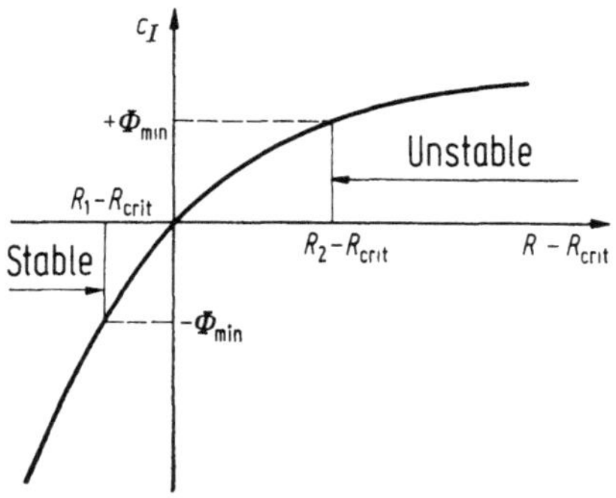

Fig. 2. Curve (π) c_I (R) projecting (28) and uncertainty interval (R_1, R_2).

follows from

$$c_I(R) - \Phi_{\min} = 0. \tag{30}$$

So having a stability chart of a flow $\bar{u} = u_0(y)$, the tops of the lines $c_I = \pm\Phi_{\min}(\|\tilde{u}\|)$ give the interval (R_1, R_2) where lies the exact stability limit with $\bar{u} = u_0(y) + \tilde{u}(y, t)$. Non-linear terms for H_1 can be discussed as well.

More detailed results will be given elsewhere.

References

1. Cotsaftis, M.: J.I.M.A. **5**, 19 (1968).
2. Cotsaftis, M.: C. R. Acad. Sci. Paris **265**, 911 (1967).
3. Hsu, J. C., Meyer, A. U.: Modern Control Principles and Applications, New York: McGraw-Hill 1968, p. 361.
4. Infante, E. F.: J. Appl. Mech., Proc. Winter Meeting of Appl. Mech. Div., Nov. 12, 1967, p. 1.
5. Parks, P. C.: Differential Equations and Dynamical Systems, New York: Academic Press 1967, p. 287.

An Invariance Principle for Dynamical Systems on Banach Space: Application to the General Problem of Thermoelastic Stability

By

M. Slemrod and **E. F. Infante**

Brown University, Providence, R.I., U.S.A.

1. Introduction

Elastic stability is usually discussed from strictly mechanical considerations. Recently, however, attempts have been made to analyze the influence of the usually neglected thermodynamics properties of elastic materials. More specifically, one may ask what effects the second law of thermodynamics has on the asymptotic stability of equilibrium of thermoelastic materials.

Several authors have investigated thermoelastic stability in the past few years. Koiter [1] has studied the general nonlinear problem with and without internal friction. Ericksen [2] has posed the question as to the asymptotic stability of the equilibrium solutions of elastic materials without imposing the assumption of internal friction. Dafermos [3] answered this question to some degree by obtaining a description of the state that the material approaches as $t \to \infty$.

This same question is studied here in a more general setting than was done by Slemrod [4], but in the same spirit: it is shown that the results of [3] can be obtained as a simple application of *an invariance principle* for abstract dynamical systems [4, 5].

Acknowledgement. This research was supported in part by the National Aeronautics and Space Administration, under Grant NGL-40-002-015, and in part by the U.S. Army Research Office, Durham, under Contract DA-31-124-ARO-D-270.

The authors would like to thank Prof. B. D. Coleman for his comments on this work.

2. Mathematical preliminaries

The principal analytical tool to be used is a generalization due to Hale [5] for abstract dynamical systems of the well known *invariance principle* of Lasalle [6] for ordinary differential equations. The following brief presentation of this tool emphasizes notation and concepts to be used in studying the problem of thermoelastic stability.

Let $R^+ = [0, \infty)$ and let $\mathscr{B}$ be a Banach space with norm $\|\varphi\|_{\mathscr{B}}$ for $\varphi \in \mathscr{B}$. Then:

Definition 2.1. *$\boldsymbol{u}$ is a* ***dynamical system on a Banach space*** *$\mathscr{B}$ if $\boldsymbol{u}$ is a function $\boldsymbol{u}$: $R^+ \times \mathscr{B} \to \mathscr{B}$ such that $\boldsymbol{u}$ is continuous, $\boldsymbol{u}(0, \varphi) = \varphi$, $\boldsymbol{u}(t + \tau, \varphi) = \boldsymbol{u}(t, \boldsymbol{u}(\tau, \varphi))$ for all t, $\tau \geqq 0$ and all φ in $\mathscr{B}$. The* ***positive orbit*** *$O^+(\varphi)$ through φ in $\mathscr{B}$ is defined as $O^+(\varphi) = \bigcup_{t \geqq 0} \boldsymbol{u}(t, \varphi)$. A point ψ in $\mathscr{B}$ is an* ***equilibrium point*** *if $O^+(\psi) = \psi$.*

This set of definitions simply generalize familiar notions from the theory of differential equations to dynamical systems.

Definition 2.2. *A set M in $\mathscr{B}$ is a* ***positively invariant set*** *of the dynamical system $\boldsymbol{u}$ if for each φ in M, $O^+(\varphi) \subset M$.*

Definition 2.3. *A set M in $\mathscr{B}$ is an* ***invariant set*** *of the dynamical system $\boldsymbol{u}$ if for each φ in M there exists a function $\boldsymbol{U}(s, \varphi)$, $\boldsymbol{U}(0, \varphi) = \varphi$ defined and in M for $s \in (-\infty, \infty)$ and such that $\boldsymbol{u}(t, \boldsymbol{U}(s, \varphi)) = \boldsymbol{U}(t + s, \varphi)$ for all $t \in R^+$.*

Definition 2.2 is well known. The second definition is used to extend backward in time those solutions of the dynamical system which lie in an invariant set. It is clear that if a set M is invariant it is positively invariant but the converse is, in general, false.

Definition 2.4. *If $\boldsymbol{u}$ is a dynamical system on $\mathscr{B}$ and V is a continuous scalar functional on $\mathscr{B}$, define the functional*

$$\dot{V}(\varphi) = \overline{\lim_{t \to 0}} \frac{1}{t} [V(\boldsymbol{u}(t, \varphi)) - V(\varphi)].$$

Then:

Definition 2.5. *$V: \mathscr{B} \to R$ is said to be a* ***Liapunov functional*** *on a set G in $\mathscr{B}$ if V is continuous on $\overline{G}$, the closure of G, and if $\dot{V}(\varphi) \leqq 0$ for φ in G. Furthermore, denote by S the set $S = \{\varphi \text{ in } \overline{G} \mid \dot{V}(\varphi) = 0\}$ and let M be the largest invariant set in S for the dynamical system $\boldsymbol{u}$. With these definitions it is then possible to prove:*

Theorem 2.1 (Hale [5]). *Let $\boldsymbol{u}$ be a dynamical system on $\mathscr{B}$. If V is a Liapunov functional on G and a positive orbit $O^+(\varphi)$ belongs to G and is in a compact set of $\mathscr{B}$ then $\boldsymbol{u}(t, \varphi) \to M$ as $t \to \infty$.*

It is self-evident that in applications to the problem of asymptotic stability of an equilibrium point ψ it is necessary to show that $M = \{\psi\}$.

Moreover, it should be emphasized that the usefulness of this theorem in applications depends on the very relaxed assumptions imposed on the Liapunov functional V and its derivative $\dot{V}$. These conditions should be compared to the much stronger conditions imposed by standard asymptotic stability theorems (see, for example, PARKS [7]).

3. Constitutive equations of linear thermoelasticity

A material point is identified by $\boldsymbol{x} = (x_1, x_2, x_3)$ in its state of equilibrium (no stresses, constant temperature $= \gamma_0$). The displacement field at some time t following an initial disturbance at time $t = 0$ is given by $\boldsymbol{u}(\boldsymbol{x}, t)$ and the temperature deviation by $T(\boldsymbol{x}, t)$; $\varrho(\boldsymbol{x})$ denotes the density at $\boldsymbol{x}$ in the equilibrium state.

Let Ω be an open, bounded, connected set in E^3 which is properly regular [8]; let $\partial\Omega$ denote the boundary of Ω. The constitutive equations of thermoelasticity are taken then in the form

$$\varrho \ddot{u}_i = (C_{ijkl} u_{k,l})_{,j} - (m_{ij} T)_{,j}, \tag{3.1}$$

$$\varrho C_D \dot{T} + m_{ij} \gamma_0 \dot{u}_{i,j} = (K_{ij} T_{,j})_{,i}; \tag{3.2}$$

where body forces and heat sources have been excluded. In these equations $C_{ijkl} = C_{jikl} = C_{klij}$, $m_{ij} = m_{ji}$, $K_{ij} = K_{ji}$ and C_D, ϱ, C_{ijkl}, m_{ij}, and K_{ij} are assumed to be smooth functions of $\boldsymbol{x}$.

Let now $t_0 > 0$. By a classical solution of the mixed initial boundary value problem in $\Omega \times (0, t_0)$ we mean a pair $(\boldsymbol{u}, T)$ satisfying equations (3.1) and (3.2) together with the boundary conditions

$$\boldsymbol{u} = 0 \text{ on } \partial\Omega \times (0, t_0) \text{ (clamped boundary)}, \tag{3.3}$$

$$T = 0 \text{ on } \partial\Omega \times (0, t_0) \text{ (constant temperature)}; \tag{3.4}$$

and with initial conditions

$$\big(\boldsymbol{u}(\boldsymbol{x}, 0), \dot{\boldsymbol{u}}(\boldsymbol{x}, 0), T(\boldsymbol{x}, 0)\big) = \big(\boldsymbol{u}_0(\boldsymbol{x}), \dot{\boldsymbol{u}}_0(\boldsymbol{x}), T_0(\boldsymbol{x})\big), \tag{3.5}$$

where $\boldsymbol{u}_0(\boldsymbol{x})$, $\dot{\boldsymbol{u}}_0(\boldsymbol{x})$ and $T_0(\boldsymbol{x})$ are given functions on Ω.

4. The thermoelastic problem as a dynamical system

In this section we show, by recalling some results of DAFERMOS [3], that the generalized solutions of the mixed initial boundary value problem described above can be viewed on an appropriate Banach space as a dynamical system (ZUBOV [9]). Once this is done, the application of Theorem 2.1 permits us to draw immediate conclusions on the asymptotic behavior of the solutions of our problem.

Consider the Sobolev space $W_2^{(k)}(\Omega)$ and $W_{20}^{(k)}(\Omega)$, $k = 1, 2, \ldots$ (see, for instance SOBOLEV [10, 11], AGMON [12]). Assume that

$$\operatorname*{ess\,inf}_{\Omega} \varrho(\boldsymbol{x}) > 0, \quad \operatorname*{ess\,inf}_{\Omega} C_D(\boldsymbol{x}) > 0, \tag{4.1}$$

$$K_{ij}\xi_i\xi_j \geqq C_1\xi_i\xi_i, \quad C_1 > 0 \text{ constant}. \tag{4.2}$$

(The second law of thermodynamics requires K_{ij} positive semidefinite at $\boldsymbol{x} \in \Omega$; we make the stronger assumption of positive definiteness.) Also for all $v_i \in W_{20}^{(1)}(\Omega)$

$$\int_\Omega C_{ijkl} v_{i,j} v_{k,l}\, d\boldsymbol{x} \geqq C_2 \int_\Omega v_{i,j} v_{i,j}\, d\boldsymbol{x}, \quad C_2 > 0 \text{ constant}. \tag{4.3}$$

(Positive semidefiniteness is expected for the tensor C_{ijkl} at each $\boldsymbol{x} \in \Omega$, TRUESDELL and NOLL [13]; we make hypothesis (4.3) which is reasonable for infinitesimal elasticity.)

Define now the spaces $H_0(\Omega) \approx W_{20}^{(1)}(\Omega) \times L_2(\Omega) \times L_2(\Omega)$ with norm $|(v_i, w_i, R)|_0^2 = \int_\Omega \left[\varrho w_i w_i + C_{ijkl} v_{i,j} v_{k,l} + \frac{\varrho C_D}{\gamma_0} R^2\right] d\boldsymbol{x}$ and $H(\Omega) = W_{20}^{(1)}(\Omega) \times W_{20}^{(1)}(\Omega) \times W_{20}^{(1)}(\Omega)$. Define the map $P\colon\ H_0(\Omega) \xrightarrow{\text{onto}} H_1(\Omega)$ sending $(v_i, w_i, R) \in H_0(\Omega)$ onto $(u_i, v_i, T) \in H_1(\Omega) \subset H(\Omega)$ where $(u_i, T) \in W_{20}^{(1)}(\Omega) \times W_{20}^{(1)}(\Omega)$ is defined by the solution of the system

$$\int_\Omega C_{ijkl} u_{k,l} \theta_{i,j}\, d\boldsymbol{x} = -\int_\Omega [\varrho w_i \theta_i - m_{ij} T \theta_{i,j}]\, d\boldsymbol{x},$$

$$\int_\Omega K_{ij} T_{,j} D_{,i}\, d\boldsymbol{x} = -\int_\Omega [\varrho C_D R + m_{ij}\gamma_0 v_{i,j}]\, D\, d\boldsymbol{x},$$

for every $D, \theta_i \in W_{20}^{(1)}(\Omega)$. The mapping P is linear, well defined on $H_0(\Omega)$ and one to one. Hence, defining $P_m = \overbrace{P \circ P \circ \cdots \circ P}^{m}$ let $H_m(\Omega)$ denote the range of the map P_m. It is clear that P_m^{-1} exists and maps $H_m(\Omega)$ onto $H_0(\Omega)$. Let $\psi \in H_m(\Omega)$ and define $|\psi|_m = |P_m^{-1}\psi|_0$. Then:

Lemma 4.1 (DAFERMOS [3]). *H_m is a Banach space with norm $|\cdot|_m$. $H_0(\Omega) \supset H(\Omega) \supset \cdots \supset H_m(\Omega)$ algebraically and topologically. Furthermore, $H_m(\Omega)$ is dense in $H_l(\Omega)$ for $m > l$ and the imbedding $I\colon H_m(\Omega) \to H_l(\Omega)$ is compact.*

Let us now define appropriately a *generalized solution* of our problem:

Definition 4.1. *$(u_i, \dot{u}_i, T)$ will be called a **generalized solution** of (3.1)–(3.5) on $\Omega \times (0, t_0)$ if for all smooth test functions (v_i, R) with* *compact support on Ω and v_i vanishing on $\Omega \times 0$,*

$$\int_0^{t_0}\int_\Omega \Big\{(t-t_0)\Big[\varrho \dot{u}_i \ddot{v}_i - C_{ijkl} u_{k,l} \dot{v}_{i,j} + m_{ij} T \dot{v}_{i,j} + \frac{\varrho C_D}{\gamma_0} T \dot{R} + m_{ij} u_{i,j} \dot{R}\Big] + \varrho \dot{u}_i \dot{v}_i + \varrho \frac{C_D}{\gamma_0} T R + m_{ij} u_{i,j} R - \frac{1}{\gamma_0}\int_0^t (K_{ij} R_{,i})_{,j}\, T\, dt\Big\}\, d\boldsymbol{x}\, dt \tag{4.4}$$

$$= -t_0 \int_\Omega \Big[\varrho \dot{u}_{0_i} \dot{v}_i\big|_{t=0} + \frac{\varrho C_D}{\gamma_0} T_0 R\big|_{t=0} + m_{ij} u_{0_{i,j}} R\big|_{t=0}\Big]\, d\boldsymbol{x}.$$

With this definition it follows that:

Theorem 4.1 (Dafermos [3]). *Under assumptions* (4.1)—(4.3) *the triple* $(u_i, \dot{u}_i, T)$ *describes a dynamical system on* $H_m(\Omega)$, $m = 0, 1, 2, \ldots$, *where* $(u_i, \dot{u}_i, T)$ *is the generalized solution to the equations of linear thermoelasticity satisfying equation* (4.4). *Furthermore, for* t *in* $(0, t_0)$,

$$|(u_i, \dot{u}_i, T)(t)|_m^2 + \frac{1}{\gamma_0}\int_0^t\int_\Omega K_{ij} T^{(m)}_{,i} T^{(m)}_{,j}\, d\boldsymbol{x}\, dt = |(u_{i_0}, \dot{u}_{i_0}, T)|_m^2, \tag{4.5}$$

where $T^{(m)}(\boldsymbol{x}, t)$ *denotes the generalized* m*-th derivative in time of* $T(\boldsymbol{x}, t)$.

5. Stability analysis

The problem of thermoelastic stability has now been put in a setting appropriate for the application of Theorem 2.1, which allows us to obtain stability results in a simple and direct manner.

For the trajectory $(u_i, \dot{u}_i, T)$ in $H_m(\Omega)$ define $P \circ (u_i, \dot{u}_i, T) \equiv (\overline{u}_i, \dot{\overline{u}}_i, \overline{T})$. It follows from the definition of the map P that $(\overline{u}_i, \dot{\overline{u}}_i, \overline{T})$ is a dynamical system on $H_{m+1}(\Omega)$ with initial data $P \circ (u_{0_i}, \dot{u}_{0_i}, T_0)$ in $H_{m+1}(\Omega)$ satisfying (4.4) and Theorem 4.1. Therefore, Theorem 4.1 and (4.5) imply that for any initial data $(u_{0_i}, \dot{u}_{0_i}, T_0)$ in $H_m(\Omega)$ the trajectory $(\overline{u}_i, \dot{\overline{u}}_i, \overline{T})(t)$ will lie in a bounded set of $H_m(\Omega)$ for all $t \geqq 0$. Hence by Lemma 4.1 the trajectory $(\overline{u}_i, \dot{\overline{u}}_i, \overline{T})$ will lie in a compact set G of $H_l(\Omega)$, $l \leqq m$. But then all the hypotheses of Theorem 2.1 are met with $\mathscr{B} = H_l(\Omega)$. For simplicity let $l = 1$ and $V = |(\overline{u}_i, \dot{\overline{u}}_i, \overline{T})|_1^2$. From (4.2) and (4.5) it immediately follows that $\dot{V} = \frac{-1}{\gamma_0}\int_\Omega K_{ij} \overset{(1)}{\overline{T}}_{,i} \overset{(1)}{\overline{T}}_{,j}\, d\boldsymbol{x} \leqq -c_3 |(0, 0, \overline{T})|_1^2$. The set S is then $S = \{(\overline{u}_i, \dot{\overline{u}}_i, \overline{T}) \in H_1(\Omega) \mid \overline{T} = 0\}$. To determine M^+, we wish to find under what conditions a motion starting on S remains in S for $t \geqq 0$. Choosing $v_i \equiv 0$ in (4.4) it follows that for $(\bar{u}_i, \dot{\bar{u}}_i, \bar{T}) \in M^+$, it is necessary that

$$\int_0^{t_0} \int_\Omega \frac{d}{dt} [(t - t_0) R]\, m_{ij} \bar{u}_{i,j}\, d\boldsymbol{x}\, dt = -t_0 \int_\Omega m_{ij} \bar{u}_{0_{i,j}} R \,|_{t=0}\, d\boldsymbol{x}$$

for every test function R. Choosing this function $R(\boldsymbol{x}, t) = \frac{\omega(t)\, \eta(\boldsymbol{x})}{(t - t_0)}$ where $\eta(\boldsymbol{x})$ is an arbitrary test function and $\omega(t)$ is the C^∞ "bump" function of SERRIN [14], it follows that $m_{ij} \dot{\bar{u}}_{i,j}(\boldsymbol{x}, t) = 0$ a.e. in Ω, $t \geqq 0$, for $(\bar{u}_i, \dot{\bar{u}}_i, \bar{T})$ in $M^+ \subset H_1(\Omega)$. Then Theorem 2.1 applied to this context yields that for any initial data $(u_{0_i}, \dot{u}_{0_i}, T_0)$ in $H_m(\Omega)$, $m \geqq 1$, and under assumptions (4.1)–(4.3), $|(\bar{u}_i, \dot{\bar{u}}_i, \bar{T})(t) - (s_i, \dot{s}_i, Q)|_1 \to 0$ as $t \to \infty$ where $(s_i, \dot{s}_i, Q) \in M^+$; $M^+ = \{(s_i, \dot{s}_i, Q) \in H_1 \mid m_{ij}, \dot{s}_{i,j} = 0, Q = 0\}$. But by the definition of $|\cdot|_1$, $P^{-1} \circ (\bar{u}_i, \dot{\bar{u}}_i, \bar{T})(t) - P^{-1} \circ (s_i, \dot{s}_i, Q)|_0 \to 0$ as $t \to \infty$; by the definitions of $(\bar{u}_i, \dot{\bar{u}}_i, \bar{T})$ and P, it then follows that $|(u_i, \dot{u}_i, T)(t) - (w_i, \dot{w}_i, Y)|_0 \to 0$ as $t \to \infty$ where $m_{ij} w_{i,j}(\boldsymbol{x}, t) = 0$, $t \geqq 0$, $Y = 0$. Doing a similar argument with $R \equiv 0$ in (4.4) yields the following theorem:

Theorem 5.1. *For any initial data* $(u_{0_i}, \dot{u}_{0_i}, T_0)$ *in* $H_m(\Omega)$, $m \geqq 1$, *and under assumptions* (4.1)–(4.3), $(u_i, \dot{u}_i, T)(t)$ *approaches the set* $M^+ = \{(w_i, \dot{w}_i, Y)$ *in* $H_0(\Omega) \mid m_{ij} w_{i,j} = 0$, $Y = 0$, $\int_0^{t_0} \int_\Omega \{(t - t_0) [\varrho \dot{w}_i \ddot{v}_i - C_{ijkl} w_{k,l} \dot{v}_{i,j}] + \varrho \dot{w}_i \dot{v}_i\}\, d\boldsymbol{x}\, dt = -t_0 \int_\Omega \varrho \dot{w}_{0_i} \dot{v}_i |_{t=0}\, d\boldsymbol{x}$ *for all* v_i *test functions with compact support on* Ω *and vanishing on* $\Omega \times 0\}$ *in the norm of the space* $H_0(\Omega)$ *as* $t \to \infty$. *The set* M^+ *represents isothermal oscillations.*

References

1. KOITER, W. T.: Report No. 360, Laboratory of Engineering Mechanics, Dept. of Mechanical Engineering, Technological University Delft, April, 1967.
2. ERICKSEN, J. L.: Int. J. Solids and Structures **2**, 573–580 (1966).
3. DAFERMOS, C. M.: Arch. Rat. Mech. Anal. **29**, 241–271 (1968).
4. SLEMROD, M.: Ph. D. Dissertation, Division of Applied Math., Brown University, June 1969.
5. HALE, J. K.: J. Math. Anal. and Appl. **26**, 39–59 (1969).
6. LASALLE, J. P.: J. Diff. Eqs. **4**, 57–65 (1968).
7. PARKS, P. C.: Int. Symp. on Diff. Eqs. and Dynamical Systems, ed. J. K. HALE and J. P. LASALLE, New York: Academic Press 1967, p. 287–298.
8. FICHERA, G.: Linear Elliptic Differential Systems and Eigenvalue Problems, Berlin/Heidelberg/New York: Springer 1965, p. 21.
9. ZUBOV, V. I.: Methods of A. M. Liapunov and their Application, Translation, Groningen: P. Noordhoff 1964.
10. SOBOLEV, S. L.: Applications of Functional Analysis in Mathematical Physics, Leningrad 1950; English translation by F. BROWDER, Amer. Math. Soc., 1963.

11. SOBOLEV, S. L.: Sur les Equations aux Dérivées Partielles Hyperboliques Non-linéaires, Rome: Cremonese 1961.
12. AGMON, S.: Lectures on Elliptic Boundary Value Problems, Princeton: Van Nostrand 1965.
13. TRUESDELL, C., NOLL, W.: The Non-linear Field Theories of Mechanics, Berlin/Heidelberg/New York: Springer 1965.
14. SERRIN, J.: Proc. Symp. on Nonlinear Prob., ed. R. E. LANGER, Madison: University of Wisconsin 1963, p. 79.

On the Possibility of Subcritical Instabilities

By

S. H. Davis

Johns Hopkins University, Baltimore, Md., U.S.A.

Let us consider a steady solution of the hydrodynamic equations governing an incompressible, Newtonian fluid. We examine the asymptotic stability of this state by superposing small disturbances and obtaining a set of partial differential equations which govern the behavior of a disturbance (also called the difference motion).

The typical disturbance equation has the form

$$-\frac{\partial \psi}{\partial t} + L\psi + \lambda M\psi = N(\psi, \psi), \tag{1}$$

where ψ is a vector function whose first three components form the divergence-free velocity and whose following coordinates may be a temperature, concentration or other field quantity. L is a real, self-adjoint, negative definite time-independent linear operator which comprises the pressure and diffusion terms. M is a real time-independent linear operator of lower order than L and λ is a positive, physically controllable constant such as the Reynolds number. $N(\psi, \psi)$ represents the convective nonlinearities which have the important property that if $\mathscr{V}$ is a representative closed volume that has zero cross-flow through its boundaries, then

$$\int_{\mathscr{V}} \psi N(\psi, \psi) = 0. \tag{2}$$

The question of stability centers about the finding of a critical value λ_c of λ separating stability from instability. The usual approach, called linear theory, is to limit consideration to only infinitesimal disturbances and neglect the nonlinear terms $N(\psi, \psi)$. One then finds a critical value λ_L of λ demarking stability from instability. Disturbances of larger

Acknowledgement. This work was supported by the Office of Naval Research under Contract Nonr 4010 (02).

size, however, may destabilize the system at some lower value of λ. Linear theory is taken to give a sufficient condition for instability. Various theories [1] of nonlinear hydrodynamic stability use the linear theory as a starting point and develop the disturbance in an asymptotic power series in amplitude.

On the other hand, SERRIN [2] and JOSEPH [3, 4] and his students have in the last few years refined the method of energy which gives a sufficient condition, $\lambda < \lambda_E$, for the stability of the basic state against arbitrary disturbances. Between these two limits, i.e. $\lambda_E < \lambda < \lambda_L$, subcritical instabilities, which are the result of disturbances of finite amplitude, can exist.

When these methods are applied to particular problems, λ_E and λ_L can sometimes differ by orders of magnitude. For example, when one considers the stability of plane Couette flow, linear theory seems to predict absolute stability, $\lambda_L = \infty$ (λ is the Reynolds number). The energy theory yields that $\lambda_E = 41.3$, [4]. The experimental observations show that plane Couette flow becomes unstable when λ is several thousand. It seems clear that at least one of the two theories fails to capture the essential physics of the onset of instability; they both seem not to describe the experiment.

The object of this paper is to consider a physically interesting class of problems in which the outcomes (i.e. λ_L and λ_E) of the linear and energy theories are predictably close. In this case the governing equations of the two theories are similar and this gives us confidence in the fact that we have captured the essential physics of the onset of instability. In fact, one can estimate λ_E knowing λ_L and obtain close upper and lower bounds to λ_c.

Let us consider Eq. (1) with given homogeneous, time-independent boundary conditions on the boundary of a closed volume $\mathscr{V}$ (which may be a representative "cell" when the domain extends to infinity).

1. *M is self-adjoint*: Bénard convection is in this class.

Linear theory:

$$-\frac{\partial\psi}{\partial t} + (L + \lambda M)\,\psi = 0. \tag{3}$$

We assume normal modes $\psi(\bar{x}, t) = e^{\sigma t}\Psi(\bar{x})$. We then have from Eq. (3) that $(L + \lambda M)\,\Psi = \sigma\Psi$. Since the operator $L + \lambda M$ is self-adjoint subject to the given boundary conditions, σ is real (the principle of exchange of stabilities is valid). Hence $\sigma = 0$ separates the regions of stability and instability. λ_L is defined as the smallest positive value of λ such that $\sigma = 0$. It can be shown [5] that the eigenvalue problem for λ_L, $L\Psi + \lambda M\Psi = 0$, is equivalent to

$$\lambda_L^{-1} = \max_{\mathbf{S}} \int_{\mathscr{V}} \Psi M\Psi \Big/ \int_{\mathscr{V}} (-\Psi L\Psi), \tag{4}$$

where $\boldsymbol{S}$ is the completed space of sufficiently smooth trial functions which satisfies the boundary conditions and whose first three components are divergence-free.

Steady-state nonlinear theory [1]:

$$L\psi + \lambda M\psi = N(\psi, \psi). \tag{5}$$

We seek a solution ψ of small amplitude ε by writing

$$\psi = \varepsilon \sum_{n=0}^{\infty} \psi_n \varepsilon^n \text{ and } \lambda = \lambda_L + \sum_{n=1}^{\infty} \lambda_n \varepsilon^n.$$

If these expansions are substituted into Eq. (5) and coefficients of like powers of ε are equated to zero, we obtain a sequence of linear boundary value problems for the ψ_n. Using now standard methods it is found that

$$\lambda_1 = \int_{\mathscr{V}} \psi_0 N(\psi_0, \psi_0) / \int_{\mathscr{V}} \psi_0 M \psi_0,$$

since $L + \lambda_L M$ is self-adjoint. $\lambda_1 = 0$ since Eq. (2) holds. Furthermore, at least for Bénard convection [6], $\lambda_2 > 0$. Thus, $\lambda = \lambda_L + \varepsilon^2 \lambda_2 + O(\varepsilon^2)$. To order ε^2, $\varepsilon^2 = \dfrac{\lambda - \lambda_L}{\lambda_2}$ which is valid only if $\lambda \geq \lambda_L$. Thus, the nonlinear theory predicts no instability of the basic state when $\lambda < \lambda_L$ (no subcritical instability).

Energy theory: Multiply Eq. (1) by ψ and integrate over $\mathscr{V}$. We obtain

$$-\frac{1}{2} \frac{d}{dt} \int_{\mathscr{V}} \psi^2 + \int_{\mathscr{V}} \psi L \psi + \lambda \int_{\mathscr{V}} \psi M \psi = \int_{\mathscr{V}} \psi N(\psi, \psi) = 0,$$

since property (2) holds.

It can be shown for the operator L indicated above that $\dfrac{d}{dt} \int_{\mathscr{V}} \psi^2 < 0$ insures asymptotic stability in the mean [4]. We have stability if $\lambda^{-1} > \int_{\mathscr{V}} \psi M \psi / \int_{\mathscr{V}} (-\psi L \psi)$. A *sufficient condition* for stability then is that

$$\lambda^{-1} > \lambda_E^{-1} \equiv \max_{\boldsymbol{S}} \int_{\mathscr{V}} \psi M \psi / \int_{\mathscr{V}} (-\psi L \psi). \tag{6}$$

If we compare Eqs. (4) and (6) we see that $\lambda_E = \lambda_L$. No subcritical instabilities are possible. This was first shown for Bénard convection by Joseph [4].

2. $M = M_1 + \gamma M_2$, where M_1 is self-adjoint and M_2 is non-self-adjoint but bounded, $||M_2|| = 1$, and $\gamma > 0$ is small. This is the case for Bénard convection subject to radiation, nonlinear equations of state, constant distributed heat sources and in some spherical geometries. Further, this is the case when one considers the instability of Couette flow between rotating cylinders [7].

Linear theory:

$$-\frac{\partial\psi}{\partial t} + L\psi + \lambda(M_1 + \gamma M_2)\,\psi = 0. \tag{7}$$

Assume that $\psi(\bar{x}, t) = e^{\sigma t}\Psi(\bar{x})$ and assume these modes are complete. Eq. (7) then becomes

$$\big(L + \lambda(M_1 + \gamma M_2)\big)\,\Psi = \sigma\Psi. \tag{8}$$

Since the operator on the left is non-self-adjoint, we must *prove* σ real. Write

$$\Psi = \sum_{n=0}^{\infty} \Psi_n\gamma^n,\quad \sigma = \sigma_0 + \sum_{n=1} \sigma_n\gamma^n \text{ and fix } \lambda.$$

If we substitute these into Eq. (8) and equate to zero coefficients of like powers of γ, we obtain a sequence of linear problems. If we assume that σ_0 is simple, it can be shown that the above expansions are convergent and hence that γ remains real within the radius of convergence which is bounded by γ_c; γ_c can be determined. (The principle of exchange of stabilities is valid, [7].) Thus, in this range of γ, $\sigma = 0$ separates unstable regions from stable ones. λ_L is defined as the smallest positive λ such that $\sigma = 0$; $L\Psi + \lambda(M_1 + \gamma M_2)\,\Psi = 0$.

Steady-state nonlinear theory:

$$L\psi + \lambda(M_1 + \gamma M_2)\,\psi = N(\psi, \psi).$$

Analogously to before, to obtain solutions for small ε and γ, we write

$$\psi = \varepsilon \sum_{n,m=0}^{\infty} \psi_{mn}\varepsilon^m\gamma^n \quad\text{and}\quad \lambda = \sum_{n,m=0}^{\infty} \lambda_{mn}\varepsilon^m\gamma^n.$$

By analogous reasoning we find that

$$\lambda = \lambda_{00} + \gamma\lambda_{01} + \varepsilon\gamma\lambda_{11} + \varepsilon^2\lambda_{20} + \cdots$$

or

$$\varepsilon = (2\lambda_{20})^{-1}\left\{-\gamma\lambda_{11} \pm [\gamma^2\lambda_{11}^2 + 4\lambda_{20}(\lambda - \lambda_{00} - \gamma\lambda_{01})]^{1/2}\right\}.$$

It can be seen that there exist solutions (which in thermal convection applications are *stable* hexagonal cells) for $\lambda < \lambda_{00} + \gamma\lambda_{01} \equiv \lambda_L$ as long as $\lambda_{11} \neq 0$ and the discriminant is non-negative. Subcritical instabilities are possible in a range of λ below λ_L whose size is $O(\gamma^2)$; γ is the measure of the non-self-adjointness of the linear, time-independent part of the governing equations.

Energy theory: Form the energy equation as before:

$$-\frac{1}{2}\frac{d}{dt}\int_{\mathscr{V}} \psi^2 + \int_{\mathscr{V}} \psi L\psi + \lambda \int_{\mathscr{V}} \psi(M_1 + \gamma M_2)\,\psi = \int_{\mathscr{V}} \psi N(\psi, \psi) = 0.$$

We have stability if

$$\lambda^{-1} > \int_{\mathscr{V}} (\psi M_1\psi + \gamma\psi M_2\psi) \Big/ \int_{\mathscr{V}} (-\psi L\psi).$$

A sufficient condition for stability follows:

$$\lambda^{-1} > \lambda_E^{-1} \equiv \max_{\mathbf{S}} \int_{\mathscr{V}} \left(\psi M_1 \psi + \frac{1}{2}\gamma\psi(M_2 + \tilde{M}_2)\psi\right) \Big/ \int_{\mathscr{V}} (-\psi L \psi),$$

where $\tilde{M}_2$ is the adjoint of M_2 and we assume that $M_2 + \tilde{M}_2$ is well-defined. The corresponding Euler-Lagrange equations are

$$L\psi + \lambda\left(M_1 + \frac{1}{2}\gamma(M_2 + \tilde{M}_2)\right)\psi = 0\,, \tag{9}$$

where we select for λ_L the smallest positive eigenvalue λ.

Conclusions

We have considered a class of nonlinear hydrodynamic stability problems where the *time-independent linear operator* of Eq. (1) differs from a self-adjoint one by a small bounded perturbation of norm and whose nonlinearities produce no energy (property (2)). Assume that we have obtained the critical value λ_L of λ using the linear theory. We find that:

(i) The nonlinear stability theories predict subcritical instabilities for small γ if $\lambda_{11} \neq 0$ within a band of width $O(\gamma^2)$ below $\lambda = \lambda_L$. If $\lambda_{11} = 0$, no subcritical instabilities are predicted. For example Bénard convection with constant heat sources [8] yields $\lambda_{11} \neq 0$ (for hexagonal cells) but in Couette flow, $\lambda_{11} = 0$.

(ii) If we develop the eigenvalue and eigenfunction of the energy theory Euler-Lagrange equation (9) in powers of γ and compare this with a similar development of the linear theory problem, we find that $\lambda_L = \lambda_E + O(\gamma)$ and a similar expression holds for the eigenfunctions. *The linear theory prediction for λ_c is correct to order γ.* The nonlinearities are relatively incapable of influencing the stability boundary. The essential physical balance that determines the onset of instability is well represented in *both* the linear and energy theories since both the corresponding eigenvalues and eigenfunctions are asymptotically close at least for small γ. It is worth emphasizing that the possibilities of subcritical instabilities predicted by both the nonlinear theories and the energy theory hinge on the *same property* of the governing system: the near self-adjointness of the time-independent linear operator of Eq. (1) which is measured by γ.

It is worth remarking that the above observation allows similar results to be obtained if we replace the Newtonian fluid by another having a constitutive equation which yields an L that is self-adjoint and nonlinearities with property (2). The flow of a porous medium subject to Darcy's law is an example.

References

1. Segel, L. A.: Non-equilibrium Thermodynamics, Variational Techniques and Stability, eds. R. J. Donnelly et al., Chicago: Univ. of Chicago Press 1966, Chapter 10.
2. Serrin, J.: Arch. Rat. Mech. Anal. **3**, 1 (1959).
3. Joseph, D. D.: Arch. Rat. Mech. Anal. **20**, 59 (1965).
4. Joseph, D. D.: Arch. Rat. Mech. Anal. **22**, 163 (1966).
5. Sani, R. L.: J. Fluid Mech. **20**, 355 (1964).
6. Fife, P. C., Joseph, D. D.: Arch. Rat. Mech. Anal. **33**, 116 (1969).
7. Davis, S. H.: Proc. Roy. Soc. **A 310**, 341 (1969).
8. Krishamurti, R.: J. Fluid Mech. **33**, 445 (1968).

Feedback Stabilization of Distributive Systems with Applications to Plasma Stabilization

By

P. K. C. Wang

University of California, Los Angeles, Cal., U.S.A.

1. Introduction

Theories for the feedback stabilization and control of dynamical systems with finite degrees of freedom have been extensively developed. The extension of some of these theories to systems with infinite degrees of freedom has been made only recently [1—2]. In the area of plasma stabilization, most studies have been associated with various thermonuclear plasma confinement schemes involving static magnetic field configurations or specified forms of high-frequency electromagnetic fields [3—6]. Recently, attempts have been made in using feedback to suppress certain forms of plasma oscillations [7—11] and to stabilize the unstable equilibria of hydromagnetic systems [12]. An important advantage of this approach is that the undesirable effects of system parameter variations may be reduced with the proper introduction of feedback and thereby relaxing the required tolerances in the system components such as the confining magnetic field structures. Further studies in this approach may lead to alternate practical means of plasma stabilization.

In this paper, simple mathematical results are obtained for the feedback stabilization of certain classes of linear distributive systems in the form of a second-order evolutional equation in a Hilbert space. The application of some of the results to the feedback stabilization of a highly conducting, fully ionized plasma is discussed. Explicit forms of feedback controls are obtained for a plasma supported against gravity by a magnetic field and also for a "linear pinch".

Acknowledgement. This work was supported by the U.S. Air Force Office of Scientific Research through Grant AFOSR-68-1547.

2. Mathematical results

Let the time interval $T^+ = [0, \infty)$ and the system's spatial domain D be a bounded open connected subset of the Euclidean space R^n with the generic point $\boldsymbol{x} = (x_1, \ldots, x_n)$. The boundary of D is denoted by ∂D and $\bar{D} = D \cup \partial D$. Let $\mathscr{H}_1(\bar{D})$ and $\mathscr{H}_2(\partial D)$ be two Hilbert spaces of real (possibly vector-valued) functions defined on $\bar{D}$ and ∂D respectively. Their inner products and induced norms are denoted by $\langle\cdot, \cdot\rangle_i$ and $||\cdot||_i$, $i = 1, 2$ respectively.

Let $\mathscr{A}$ be a linear operator with domain $\mathscr{D}(\mathscr{A}) = \{\boldsymbol{u} \in \mathscr{H}_1(\bar{D})\colon \mathscr{A}\boldsymbol{u} \in \mathscr{H}_1(\bar{D})\}$ and $\mathscr{C}$ be a linear operator with domain $\mathscr{D}(\mathscr{C}) = \{\boldsymbol{v} \in \mathscr{H}_1(\bar{D})\colon \mathscr{C}\boldsymbol{v} \in \mathscr{H}_2(\partial D)\}$. It will be assumed that $\mathscr{D}(\mathscr{A}) \subseteq \mathscr{D}(\mathscr{C})$.

Consider a time-invariant distributive system of the form:

$$\frac{d^2\boldsymbol{u}(t)}{dt^2} = \mathscr{A}\boldsymbol{u}(t) + \mathscr{B}_D\boldsymbol{f}_D(t), \tag{2.1}$$

$$\mathscr{C}\,\boldsymbol{u}(t) = \mathscr{B}_{\partial D}\boldsymbol{f}_{\partial D}(t). \tag{2.2}$$

In applications, $\mathscr{A}$ and $\mathscr{C}$ may correspond to linear differential operators with respect to the space variables. Thus, (2.2) corresponds to a boundary condition for the partial differential equation (2.1). The variables $\boldsymbol{f}_D$ and $\boldsymbol{f}_{\partial D}$ are manipulatable inputs or controls which act on the system through the interior and the boundary of D respectively. Henceforth, we shall refer to $\boldsymbol{f}_D$ as a *distributed control* and $\boldsymbol{f}_{\partial D}$ as a *boundary control*[1]. Let $\boldsymbol{f}_D(t)$ (respectively $\boldsymbol{f}_{\partial D}(t)$), for any fixed $t \in T^+$, be a member of a real Hilbert space $H_1(\bar{D})$ (respectively $H_2(\partial D)$). We shall assume that $\mathscr{B}_D$ and $\mathscr{B}_{\partial D}$ in (2.1)—(2.2) are bounded linear operators on $H_1(\bar{D})$ to $\mathscr{H}_1(\bar{D})$ and $H_2(\bar{D})$ to $\mathscr{H}_2(\partial D)$ respectively. By a *solution* of (2.1)—(2.2) *in* T^+, we mean a $\mathscr{H}_1(\bar{D})$-valued twice continuously differentiable function defined in T^+ such that $\boldsymbol{u}(t) \in \mathscr{D}(\mathscr{A})$ for all $t \in T^+$ and satisfies (2.1)—(2.2) everywhere. By a *state* of system (2.1)—(2.2), we mean an ordered pair $(\boldsymbol{u}, \dot{\boldsymbol{u}}) \in \Sigma(\bar{D})$, where $\Sigma(\bar{D})$ (the state space) is a dense subspace of $\mathscr{H}_1(\bar{D}) \times \mathscr{H}_1(\bar{D})$.

Now, the feedback stabilization problem can be stated as follows: Let the equilibrium null state $(\boldsymbol{u}, \dot{\boldsymbol{u}}) = (\boldsymbol{0}, \boldsymbol{0})$ of the uncontrolled system (2.1)—(2.2) (i.e. $\boldsymbol{f}_D(t) \equiv \boldsymbol{0}$, $\boldsymbol{f}_{\partial D}(t) \equiv \boldsymbol{0}$ on T^+) be unstable in the sense of Liapunov with respect to norm

$$||(\boldsymbol{u}, \dot{\boldsymbol{u}})||_\Sigma = \{||\boldsymbol{u}||_1^2 + ||\dot{\boldsymbol{u}}||_1^2\}^{1/2}.$$

[1] The distinction in control forms is not always necessary from the mathematical standpoint. In certain situations, it is possible to replace the boundary control by distributed controls that have the same effect on the system (see FATTORINI [13]).

Find a stabilizing feedback control of the form

$$f_D(t) = \mathscr{F}_D(u(t), \dot{u}(t)), \quad f_{\partial D}(t) = \mathscr{F}_{\partial D}(u(t), \dot{u}(t)), \tag{2.3}$$

where $\mathscr{F}_D$ and $\mathscr{F}_{\partial D}$ are bounded linear operators on $\Sigma(\bar{D})$ into $H_1(\bar{D})$ and $H_2(\partial D)$ respectively such that the null state of the resulting feedback system is stable or asymptotically stable. In the sequel, it will be shown that under certain conditions, stabilizing feedback controls can be readily constructed using either a distributed control or a boundary control.

2.1 Distributed control

(System (2.1)—(2.2) with $f_{\partial D}(t) \equiv 0$ on T^+)

Let $\hat{\mathscr{A}}$ be a linear operator with domain $\mathscr{D}(\hat{\mathscr{A}}) = \{u \in \mathscr{D}(\mathscr{A}): \mathscr{C}u = 0\}$ and $\hat{\mathscr{A}}u = \mathscr{A}u$ in $\mathscr{D}(\hat{\mathscr{A}})$. We shall assume that

(A-1) $\hat{\mathscr{A}}$ is a self-adjoint and upper-semi-bounded, i.e. there exists a real number c such that

$$\langle u, \hat{\mathscr{A}}u\rangle_1 \leq c\langle u, u\rangle_1 \quad \text{for all } u \in \mathscr{D}(\hat{\mathscr{A}}).$$

(A-2) There exists a real number $\lambda > 0$ such that

$$\langle u, \mathscr{B}_D\mathscr{B}_D^* u\rangle_1 \geq \lambda\, \|u\|_1^2 \quad \text{for all } u \in \mathscr{H}_1(\bar{D}),$$

where $\mathscr{B}_D^*$ is the adjoint of $\mathscr{B}_D$.

Let k_1 be a real number $> c/\lambda$. Then, under assumptions (A-1)—(A-2),

$$\langle u, \mathscr{S}u\rangle_1 \geq (k_1\lambda - c)\, \|u\|_1^2 \quad \text{for all } u \in \mathscr{D}(\mathscr{S}), \tag{2.4}$$

where $\mathscr{S} = (k_1\mathscr{B}_D\mathscr{B}_D^* - \hat{\mathscr{A}})$ is a positive self-adjoint operator with domain $\mathscr{D}(\mathscr{S}) = \mathscr{D}(\hat{\mathscr{A}})$.

We introduce in $\mathscr{D}(\mathscr{S})$ the new inner product and norm

$$\langle u, u'\rangle_{\mathscr{D}(\mathscr{S})} = \langle u, \mathscr{S}u'\rangle_1, \quad \|u\|_{\mathscr{D}(\mathscr{S})} = \langle u, \mathscr{S}u\rangle_1^{1/2}. \tag{2.5}$$

Let $\overline{\mathscr{D}(\mathscr{S})}$ be the completion of $\mathscr{D}(\mathscr{S})$ with respect to norm $\|\cdot\|_{\mathscr{D}(\mathscr{S})}$. Then, $\overline{\mathscr{D}(\mathscr{S})}$ is a Hilbert space. Also, $H(\bar{D}) = \overline{\mathscr{D}(\mathscr{S})} \times \overline{\mathscr{D}(\mathscr{S})}_1$ is a Hilbert space with inner product and norm defined by

$$\langle w, w'\rangle_H = \langle u, \mathscr{S}u'\rangle_1 + \langle \dot{u}, \dot{u}'\rangle_1, \quad \|w\|_H = \langle w, w\rangle_H^{1/2}, \tag{2.6}$$

where $w = (u, \dot{u})$, $w' = (u', \dot{u}')$ and $\overline{\mathscr{D}(\mathscr{S})}_1$ is the completion of $\mathscr{D}(\mathscr{S})$ with respect to norm $\|\cdot\|_1$.

Theorem 1: *Under assumption* (A-1)—(A-2), *the equilibrium null state of system* (2.1)—(2.2) *with* $f_{\partial D}(t) \equiv 0$ *on* T^+ *and distributed feedback control*

$$f_D(t) = -k_1\mathscr{B}_D^* u(t) - k_2\mathscr{B}_D^* \dot{u}(t) \tag{2.7}$$

with $k_1 > c/\lambda$ *and* $k_2 > 0$ *is asymptotically with respect to norm* $\|\cdot\|_H$ *defined in* (2.6).

Proof: Let $V = \frac{1}{2}\|(\boldsymbol{u}, \dot{\boldsymbol{u}})\|_H^2$. By direct computation and using the fact that $\hat{\mathscr{A}}$ is self-adjoint,

$$\frac{dV}{dt} = -k_2 \|\mathscr{B}_D^* \dot{\boldsymbol{u}}\|_1^2 \le 0 .$$

Since the operator $(\mathscr{A} - k_1 \mathscr{B}_D \mathscr{B}_D^*)$ is negative, $(\boldsymbol{u}(t) \not\equiv \boldsymbol{0},\ \dot{\boldsymbol{u}}(t) \equiv \boldsymbol{0})$ on T^+ is not a system motion. Thus, $dV/dt < 0$ along any system motion implying that $\|(\boldsymbol{u}(t), \dot{\boldsymbol{u}}(t)\|_H \to 0$ as $t \to \infty$ or the null state of the feedback system is asymptotically stable with respect to norm $\|\cdot\|_H$.

Remarks: (R-1) From (2.4),

$$\|(\boldsymbol{u}, \dot{\boldsymbol{u}})\|_H^2 = \langle \boldsymbol{u}, \mathscr{S}\boldsymbol{u}\rangle_1 + \|\dot{\boldsymbol{u}}\|_1^2 \ge (k_1\lambda - c)\,\|\boldsymbol{u}\|_1^2 + \|\dot{\boldsymbol{u}}\|_1^2$$
$$\ge \operatorname{Min}\{1, (k_1\lambda - c)\}\,\|(\boldsymbol{u}, \dot{\boldsymbol{u}})\|_\Sigma^2 \ge 0 .$$

Thus, feedback control (2.7) also ensures asymptotic stability of the null state with respect to norm $\|\cdot\|_\Sigma$.

(R-2) The effect of feedback control (2.7) is to shift the spectrum of $\hat{\mathscr{A}}$ into the negative real line. Clearly, stabilization by such a control is impossible if $\hat{\mathscr{A}}$ is self-adjoint but not upper-semi-bounded.

2.2 Boundary control

(System (2.1)—(2.2) with $\boldsymbol{f}_D(t) \equiv \boldsymbol{0}$ on T^+)

The following assumptions will be made:

(A-3) The operator $\hat{\mathscr{A}}$ defined in Sec. 2.1 is self-adjoint.

(A-4) For any $\boldsymbol{u} \in \mathscr{D}(\mathscr{A})$, the inner product $\langle \boldsymbol{u}, -\mathscr{A}\boldsymbol{u}\rangle_1$ can be written in the form

$$\langle \boldsymbol{u}, -\mathscr{A}\boldsymbol{u}\rangle_1 = \langle \mathscr{P}\boldsymbol{u}, \mathscr{R}\boldsymbol{u}\rangle_1 + \langle \mathscr{Q}\boldsymbol{u}, \mathscr{G}\boldsymbol{u}\rangle_2 + \langle \mathscr{Q}\boldsymbol{u}, \mathscr{C}\boldsymbol{u}\rangle_2, \qquad (2.8)$$

where $\mathscr{P}$ and $\mathscr{R}$ are linear operators on $\mathscr{D}(\mathscr{A})$ into $\mathscr{H}_1(\bar{D})$; $\mathscr{Q}$ and $\mathscr{G}$ are linear operators on $\mathscr{D}(\mathscr{A})$ into $\mathscr{H}_2(\partial D)$. Moreover, there exists a real number α such that for all $\boldsymbol{u} \in \mathscr{D}(\mathscr{A})$,

$$\langle \mathscr{Q}\boldsymbol{u}, \mathscr{G}\boldsymbol{u}\rangle_2 \ge \alpha\,\|\mathscr{Q}\boldsymbol{u}\|_2^2 \quad \text{and} \quad \langle \mathscr{P}\boldsymbol{u}, \mathscr{R}\boldsymbol{u}\rangle_1 \ge 0 . \qquad (2.9)$$

(A-5) There exists a real number $\eta > 0$ such that

$$\langle \mathscr{Q}\boldsymbol{u}, \mathscr{B}_{\partial D}\mathscr{B}_{\partial D}^* \mathscr{Q}\boldsymbol{u}\rangle_2 \ge \eta\,\|\mathscr{Q}\boldsymbol{u}\|_2^2 \quad \text{for all } \boldsymbol{u} \in \mathscr{D}(\mathscr{A}), \qquad (2.10)$$

where $\mathscr{B}_{\partial D}^*$ is the adjoint of $\mathscr{B}_{\partial D}$.

Note that assumptions (A-3)—(A-4) imply that for any $\boldsymbol{u} \in \mathscr{D}(\hat{\mathscr{A}})$,

$$\langle \boldsymbol{u}, -\mathscr{A}\boldsymbol{u}\rangle_1 = \langle \mathscr{P}\boldsymbol{u}, \mathscr{R}\boldsymbol{u}\rangle_1 + \langle \mathscr{Q}\boldsymbol{u}, \mathscr{G}\boldsymbol{u}\rangle_2 \ge \alpha\,\|\mathscr{Q}\boldsymbol{u}\|_2^2 . \qquad (2.11)$$

Let k be a real number $> (-\alpha)$ and $\tilde{\mathscr{A}}$ be a linear operator with domain $\mathscr{D}(\tilde{\mathscr{A}}) = \{\boldsymbol{u} \in \mathscr{D}(\mathscr{A}) \cap \mathscr{N}(\mathscr{Q})^\perp \colon (\mathscr{C} - k\mathscr{B}_{\partial D}\mathscr{B}_{\partial D}^*\mathscr{Q})\,\boldsymbol{u} = \boldsymbol{0}\}$

and $\tilde{\mathscr{A}}\boldsymbol{u} = \mathscr{A}\boldsymbol{u}$ in $\mathscr{D}(\tilde{\mathscr{A}})$, where $\mathscr{N}(\mathscr{Q})$ is the null space of $\mathscr{Q}$. From assumptions (A-3)—(A-5), $\tilde{\mathscr{A}}$ is a negative self-adjoint operator. Thus, we may introduce in $\mathscr{D}(\tilde{\mathscr{A}})$ the new inner product and norm:

$$\langle \boldsymbol{u}, \boldsymbol{u}' \rangle_{\mathscr{D}(\tilde{\mathscr{A}})} = \langle \boldsymbol{u}, -\tilde{\mathscr{A}}\boldsymbol{u}' \rangle_1, \quad \|\boldsymbol{u}\|_{\mathscr{D}(\tilde{\mathscr{A}})} = \langle \boldsymbol{u}, -\tilde{\mathscr{A}}\boldsymbol{u} \rangle_1^{1/2}. \qquad (2.12)$$

Also, $\tilde{H}(\overline{D}) = \overline{\mathscr{D}(\tilde{\mathscr{A}})} \times \overline{\mathscr{D}(\tilde{\mathscr{A}})}_1$ is a Hilbert space with inner product and norm defined by

$$\langle \boldsymbol{w}, \boldsymbol{w}' \rangle_{\tilde{H}} = \langle \boldsymbol{u}, -\tilde{\mathscr{A}}\boldsymbol{u}' \rangle_1 + \langle \dot{\boldsymbol{u}}, \dot{\boldsymbol{u}}' \rangle_1, \quad \|\boldsymbol{w}\|_{\tilde{H}} = \langle \boldsymbol{w}, \boldsymbol{w} \rangle_{\tilde{H}}^{1/2}, \qquad (2.13)$$

where $\overline{\mathscr{D}(\tilde{\mathscr{A}})}$, $\overline{\mathscr{D}(\tilde{\mathscr{A}})}_1$ are the completions of $\mathscr{D}(\tilde{\mathscr{A}})$ with respect to norms $\|\cdot\|_{\mathscr{D}(\tilde{\mathscr{A}})}$ and $\|\cdot\|_1$ respectively, and $\boldsymbol{w} = (\boldsymbol{u}, \dot{\boldsymbol{u}})$, $\boldsymbol{w}' = (\boldsymbol{u}', \dot{\boldsymbol{u}}')$.

Theorem 2: *Under assumptions* (A-3)—(A-5), *the equilibrium null state of system* (2.1)—(2.2) *with* $\boldsymbol{f}_D(t) \equiv \boldsymbol{0}$ *on* T^+ *and boundary feedback control*

$$\boldsymbol{f}_{\partial D}(t) = k\mathscr{B}^*_{\partial D}\mathscr{Q}\boldsymbol{u}(t) \qquad (2.14)$$

with $k\eta > -\alpha$ *is stable with respect to norm* $\|\cdot\|_{\tilde{H}}$.

By considering the functional $V = \frac{1}{2}\|(\boldsymbol{u}, \dot{\boldsymbol{u}})\|^2_{\tilde{H}}$, the above result follows trivially from the fact that $dV/dt = 0$ along any system motion (due to the self-adjointness of $\tilde{\mathscr{A}}$).

3. Application to plasma stabilization

Consider a highly conducting, fully ionized plasma immersed in a static magnetic field. Let D_{p0} be an open bounded subset of R^3 with boundary ∂D_{p0} representing the spatial domain of the plasma at static equilibrium. We shall assume that the plasma is describable by the usual hydromagnetic equations for a perfectly conducting fluid [3, 14]. Adopting the Lagrangian description of fluid motion, the displacement of a fluid element at time t from its equilibrium position $\boldsymbol{x}_0$ is denoted by $\boldsymbol{\xi}(t, \boldsymbol{x}_0)$ (a point in R^3). For sufficiently small perturbations, we may use the hydromagnetic equations for $\boldsymbol{\xi}$ linearized about a *static* equilibrium [14]. These equations can be conveniently written in the form of (2.1) by identifying $\boldsymbol{u}(t)$ with $\boldsymbol{\xi}(t, \cdot)$ and

$$\big(\mathscr{A}\boldsymbol{u}(t)\big)(\boldsymbol{x}_0) \equiv \big(\mathscr{A}\boldsymbol{\xi}(t, \cdot)\big)(\boldsymbol{x}_0) = \{\nabla_0[\gamma p_0(\boldsymbol{x}_0)\big(\nabla_0 \cdot \boldsymbol{\xi}(t, \boldsymbol{x}_0)\big)$$
$$+ \big(\boldsymbol{\xi}(t, \boldsymbol{x}_0) \cdot \nabla_0\big) p_0(\boldsymbol{x}_0)] + \boldsymbol{j}_0(\boldsymbol{x}_0) \times \boldsymbol{G}(t, \boldsymbol{x}_0) - \mu_0^{-1}\boldsymbol{B}_0(\boldsymbol{x}_0) \times \big(\nabla_0 \times \boldsymbol{G}(t, \boldsymbol{x}_0)\big)$$
$$+ [\nabla_0 \cdot \big(\varrho_{m0}(\boldsymbol{x}_0)\boldsymbol{\xi}(t, \boldsymbol{x}_0)\big)]\nabla_0\varphi_0(\boldsymbol{x}_0)\}/\varrho_{m0}(\boldsymbol{x}_0), \qquad (2.15)^1$$

where $\boldsymbol{G}(t, \boldsymbol{x}_0) \equiv \nabla_0 \times (\boldsymbol{\xi}(t, \boldsymbol{x}_0) \times \boldsymbol{B}_0(\boldsymbol{x}_0))$; γ and μ_0 are the ratio of specific heats and the permeability of free space respectively; p_0, ϱ_{m0}, $\boldsymbol{j}_0$, φ_0 and $\boldsymbol{B}_0$ are respectively the pressure, the mass and current densities,

[1] Subscript "0" will be used to denote equilibrium quantities.

the external potential energy per unit mass and the magnetic field satisfying the equations of magnetostatics [14]. The control for this system is in the form of an external time and space dependent perturbation of the magnetic field in the vacuum region about its equilibrium. Note that for a perfectly conducting fluid, the time-varying component of the external magnetic field cannot penetrate into the fluid. Also, variations of the magnetic field at any point inside the fluid are entirely due to the transport of the equilibrium magnetic field lines induced by the fluid motion. Thus, the external perturbing magnetic field corresponds to a boundary control which acts on the system throug the following linearized boundary conditions [14] in the form of (2.2) at the vacuum-fluid boundary, where

$$\big(\mathscr{C}\xi(t,\cdot)\big)(\boldsymbol{x}_0) = -\gamma p_0(\boldsymbol{x}_0)\nabla_0\cdot\xi(t,\boldsymbol{x}_0) + \mu_0^{-1}\boldsymbol{B}_0(\boldsymbol{x}_0)\cdot\big(\nabla_0\times(\xi(t,\boldsymbol{x}_0)\times\boldsymbol{B}_0(\boldsymbol{x}_0))\big) + \frac{1}{2\mu_0}\xi(t,\boldsymbol{x}_0)\cdot\big(\nabla_0|\boldsymbol{B}_0(\boldsymbol{x}_0)|^2 - \nabla_0|\hat{\boldsymbol{B}}_0(\boldsymbol{x}_0)|^2\big), \tag{2.16}$$ [1]

$$\big(\mathscr{B}_{\partial D}\delta\hat{\boldsymbol{B}}_c(t,\cdot)\big)(\boldsymbol{x}_0) = \mu_0^{-1}\hat{\boldsymbol{B}}_0(\boldsymbol{x}_0)\cdot\delta\hat{\boldsymbol{B}}_c(t,\boldsymbol{x}_0), \tag{2.17}$$

where $\boldsymbol{x}_0 \in \partial D_{p0}$, $\hat{\boldsymbol{B}}_0$ is the equilibrium magnetic field in the vacuum region and $\delta\hat{\boldsymbol{B}}_c$ is the first order perturbation of the vacuum magnetic field about $\hat{\boldsymbol{B}}_0$ due to both external current sources and the fluid motions. For simplicity, we shall regard $\delta\hat{\boldsymbol{B}}_c$ as the boundary control. Also, we let $\mathscr{H}_1(\bar{D}) = L_2^{(3)}(D_{p0})$ and $\mathscr{H}_2(\partial D) = L_2^{(3)}(\partial D_{p0})$, where $L_2^{(3)}(\cdot)$ is the three-fold cartesian product of the real Hilbert space $L_2(\cdot)$ and the inner products are defined by

$$\langle \boldsymbol{u}, \boldsymbol{u}'\rangle_1 = \int_{D_{p0}} \boldsymbol{u}(\boldsymbol{x}_0)\cdot\boldsymbol{u}'(\boldsymbol{x}_0)\,dD_{p0},\quad \langle \boldsymbol{v}, \boldsymbol{v}'\rangle_2 = \int_{\partial D_{p0}} \boldsymbol{v}(\boldsymbol{x}_0)\cdot\boldsymbol{v}'(\boldsymbol{x}_0)\,d(\partial D_{p0}).$$

From the fact that the uncontrolled system is conservative, assumption (A-3) is satisfied. A standard computation shows that the quantity $\langle \boldsymbol{u}, -\varrho_{m0}\mathscr{A}\boldsymbol{u}\rangle_1$ can be written in the form (2.8) with

$$\begin{aligned}\langle \mathscr{P}\xi(t,\cdot), \mathscr{R}\xi(t,\cdot)\rangle_1 &= \frac{1}{2}\int_{D_{p0}} dD_{p0}\{\mu_0^{-1}|\boldsymbol{G}(t,\boldsymbol{x}_0)|^2 + \gamma p_0(\boldsymbol{x}_0)\big(\nabla_0\cdot\xi(t,\boldsymbol{x}_0)\big)^2\\ &+ \big(\nabla_0\cdot\xi(t,\boldsymbol{x}_0)\big)\big(\xi(t,\boldsymbol{x}_0)\cdot\nabla_0 p_0(\boldsymbol{x}_0)\big)\\ &- \big(\xi(t,\boldsymbol{x}_0)\cdot\nabla_0\varphi_0(\boldsymbol{x}_0)\big)\big(\nabla_0\cdot(\varrho_{m0}(\boldsymbol{x}_0)\xi(t,\boldsymbol{x}_0))\big) - \xi(t,\boldsymbol{x}_0)\cdot\big(\boldsymbol{j}_0(\boldsymbol{x}_0)\times\boldsymbol{G}(t,\boldsymbol{x}_0)\big)\},\end{aligned} \tag{2.18}$$

$$\langle \mathscr{Q}\xi(t,\cdot), \mathscr{G}\xi(t,\cdot)\rangle_2 = \frac{1}{2}\int_{\partial D_{p0}} d(\partial D_{p0})\big(\boldsymbol{n}_0(\boldsymbol{x}_0)\cdot\xi(t,\boldsymbol{x}_0)\big)\,\xi(t,\boldsymbol{x}_0)\cdot\left\{\nabla_0\left(\frac{|\hat{\boldsymbol{B}}_0(\boldsymbol{x}_0)|^2}{2\mu_0}\right) - \nabla_0\left(\frac{|\boldsymbol{B}_0(\boldsymbol{x}_0)|^2}{2\mu_0} + p_0(\boldsymbol{x}_0)\right)\right\}, \tag{2.19}$$

$$\begin{aligned}\langle \mathscr{Q}\xi(t,\cdot), \mathscr{C}\xi(t,\cdot)\rangle_2 &= \langle \mathscr{Q}\xi(t,\cdot), \mathscr{B}_{\partial D} f_{\partial D}(t)\rangle_2\\ &= \frac{1}{2}\int_{\partial D_{p0}} d(\partial D_{p0})\big(\boldsymbol{n}_0(\boldsymbol{x}_0)\cdot\xi(t,\boldsymbol{x}_0)\big)\big(\mu_0^{-1}\hat{\boldsymbol{B}}_0(\boldsymbol{x}_0)\cdot\delta\hat{\boldsymbol{B}}_c(t,\boldsymbol{x}_0)\big),\end{aligned} \tag{2.20}$$

where $\boldsymbol{n}_0(\boldsymbol{x}_0)$ is the outward unit normal to ∂D_{p0} at $\boldsymbol{x}_0$.

[1] The inner product of two vectors $\boldsymbol{x}$ and $\boldsymbol{x}'$ in R^n is denoted by $\boldsymbol{x}\cdot\boldsymbol{x}'$.

From the fact that the tangential component of $\nabla_0(p_0 + |\boldsymbol{B}_0|^2/2\mu_0)$ is continuous across ∂D_{p0}, (2.19) satisfies

$$\langle \mathscr{Q}\xi(t,\cdot), \mathscr{G}\xi(t,\cdot)\rangle_2 \geq \frac{1}{2} \inf_{\boldsymbol{x}_0 \in \partial D_{p0}} \big(\boldsymbol{n}_0(\boldsymbol{x}_0)\cdot\{\cdots\}\big) \int_{\partial D_{p0}} d(\partial D_{p0}) \big(\boldsymbol{n}_0(\boldsymbol{x}_0)\cdot\xi(t,\boldsymbol{x}_0)\big)^2 \tag{2.21}$$

for all $\xi(t,\cdot) \in \mathscr{D}(\mathscr{A})$, where $(\mathscr{Q}\xi(t,\cdot))(\boldsymbol{x}_0) = \boldsymbol{n}_0(\boldsymbol{x}_0)\cdot\xi(t,\boldsymbol{x}_0)$ and the $\{\cdots\}$ term is defined in (2.19). Thus, assumption (A-4) will be satisfied if we require that (2.18) is nonnegative for all $\xi(t,\cdot)\in\mathscr{D}(\mathscr{A})$. Note that the foregoing requirement implies that instability of the null state of the uncontrolled system is attributed to the change of pressure gradient at the vacuum-fluid boundary. From (2.20), assumption (A-5) is obviously satisfied. Applying Theorem 2 to the system, the stabilizing boundary feedback control corresponding to (2.14) is given by

$$\delta\hat{\boldsymbol{B}}_c(t,\boldsymbol{x}_0) = k\big(\boldsymbol{n}_0(\boldsymbol{x}_0)\cdot\xi(t,\boldsymbol{x}_0)\big)\,\hat{\boldsymbol{B}}_0(\boldsymbol{x}_0), \tag{2.22}$$

with $k\mu_0^{-1} \inf_{\boldsymbol{x}_0\in\partial D_{p0}} |\hat{\boldsymbol{B}}_0(\boldsymbol{x}_0)|^2 + \inf_{\boldsymbol{x}_0\in\partial D_{p0}} \{\cdots\} > 0$, where the $\{\cdots\}$ term is defined in (2.19).

Remarks: (R-3) The feedback control (2.22) has the following simple physical interpretation: If $\boldsymbol{n}_0(\boldsymbol{x}_0)\cdot\xi(t,\boldsymbol{x}_0)$ at any point $\boldsymbol{x}_0\in\partial D_{p0}$ is > 0 (< 0), the control produces a corrective magnetic field in (opposite) the direction of $\hat{\boldsymbol{B}}_0(\boldsymbol{x}_0)$, thereby increasing (decreasing) the local magnetic pressure to restore the fluid surface to its equilibrium position. Obviously, the control has no effect on those fluid motions with $\boldsymbol{n}_0(\boldsymbol{x}_0)\cdot\xi(t,\boldsymbol{x}_0) = 0$ for all t and $\boldsymbol{x}_0\in\partial D_{p0}$.

(R-4) Physical realization of feedback control (2.22) requires spatially discrete approximations of the corrective magnetic field. This can be regarded as a constraint on the control form. Theoretical treatment of the feedback stabilization problem taking such constraints into account will be discussed elsewhere.

Now, explicit results will be given for the following simple specific systems:

(1) *Plasma supported against gravity by a magnetic field*: Consider a perfectly conducting, incompressible plasma lying above the plane $x_1 = 0$ as shown in Fig. 1. A uniform static magnetic field $\hat{\boldsymbol{B}}_0$ is applied in the x_3 direction and the gravity is in the $-x_1$ direction. Moreover, there is no magnetic field inside the plasma. In this case,

$$\langle \mathscr{P}\xi(t,\cdot),\ \mathscr{R}\xi(t,\cdot)\rangle_1 = 0,$$

$$\langle \mathscr{Q}\xi(t,\cdot), \mathscr{G}\xi(t,\cdot)\rangle_2 = -\frac{1}{2}\varrho_{m0}g\int\big(\xi_1(t,x_3)\big)^2\,dx_3 \leq 0,$$

$$\langle \mathscr{Q}\xi(t,\cdot), \mathscr{B}_{\partial D} f_{\partial D}(t)\rangle_2 = -\frac{1}{2\mu_0}\int \xi_1(t,x_3)\,\delta\hat{\boldsymbol{B}}_c(t,x_3)\cdot\hat{\boldsymbol{B}}_0\,dx_3,$$

where $\xi = (\xi_1, \xi_2, \xi_3)$. The stabilizing boundary feedback control corresponding to (2.22) is given by

$$\delta\hat{\boldsymbol{B}}_c(t, x_3) = -k\xi_1(t, x_3)\,\hat{\boldsymbol{B}}_0, \quad k > \mu_0 \varrho_{m0} g\,|\hat{\boldsymbol{B}}_0|^{-2}. \tag{2.23}$$

A crude physical realization of (2.23) is shown in Fig. 1.

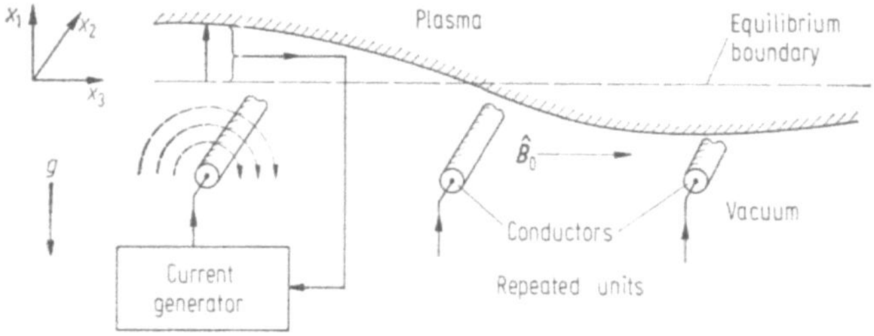

Fig. 1

(2) *"Linear pinch"*: Consider an infinitely long perfectly conducting plasma cylinder of radius a confined by its own magnetic field produced by a current sheet at its surface (Fig. 2). We shall assume that there is also a uniform axial magnetic field both inside and outside the plasma.

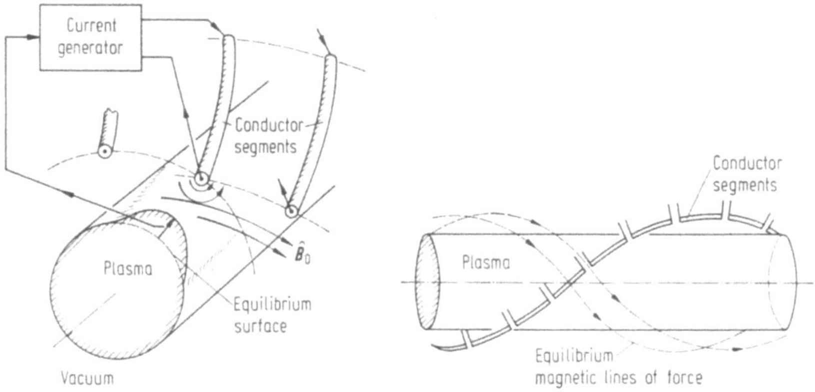

Fig. 2

Let $\boldsymbol{x}_0 = (r, \theta, z)$. The magnetic field at equilibrium is given by

$$\boldsymbol{B}_0 = (0, 0, B_{z0}) \text{ for } 0 < r < a \text{ (plasma region)},$$
$$\hat{\boldsymbol{B}}_0(r) = (0, \mu_0 a j^*_{0z}/r, \hat{B}_{z0}) \text{ for } r > a \text{ (vacuum region)}$$

satisfying the following pressure balance condition at equilibrium:

$$p_0 = \frac{1}{2\mu_0}\left(\mu_0^2 (j^*_{0z})^2 + \hat{B}^2_{z0} - B^2_{z0}\right) > 0,$$

where j^*_{0z} is a constant surface current density in the z-direction; B_{z0} and $\hat{B}_{z0}$ are positive constants. For this case, the quantities corresponding to (2.18)—(2.20) are given by

$$\langle \mathscr{P}\xi(t,\cdot), \mathscr{R}\xi(t,\cdot)\rangle_1 = \frac{1}{2}\int dD_{p0}\{\mu_0^{-1}|\boldsymbol{G}(t,\boldsymbol{x}_0)|^2 + \gamma p_0(\nabla_0\cdot\xi(t,\boldsymbol{x}_0))^2\} \geq 0,$$

$$\langle \mathscr{Q}\xi(t,\cdot), \mathscr{G}\xi(t,\cdot)\rangle_2 = -\frac{1}{2}\mu_0(j_{0z}^*)^2\iint(\xi_r(t,\theta,z))^2\,d\theta\,dz \leq 0,$$

$$\langle \mathscr{Q}\xi(t,\cdot), \mathscr{B}_{\partial D}f_{\partial D}(t)\rangle_2 = \frac{a}{2\mu_0}\iint \xi_r(t,\theta,z)\,\delta\hat{\boldsymbol{B}}_c(t,\theta,z)\cdot\hat{\boldsymbol{B}}_0(a)\,d\theta\,dz,$$

where $\xi = (\xi_r, \xi_\theta, \xi_z)$.

The boundary feedback control corresponding to (2.22) is given by

$$\delta\hat{\boldsymbol{B}}_c(t,\theta,z) = k\xi_r(t,\theta,z)\,\hat{\boldsymbol{B}}_0(a), \tag{2.24}$$

with $k > a^{-1}(\mu_0 j_{0z}^*)^2\,|\hat{\boldsymbol{B}}_0(a)|^{-2}$. Note that the equilibrium magnetic lines of force in the vacuum region describe helices about the z-axis with pitches that depend on r. A crude physical realization of (2.24) is shown in Fig. 2 in which the control current-carrying conductors lie in helices orthogonal to those corresponding to the equilibrium magnetic lines of forces at $r = a$.

References

1. Wang, P. K. C.: Theory of stability and control of distributed parameter systems (a bibliography). Int. J. Control **7**, 101—116 (1968).
2. Athans, M. (ed.): Control of Distributed Parameter Systems, New York: ASME Publ., August 1969.
3. Jeffrey, A., Taniuti, T. (eds.): Magnetohydrodynamic Stability and Thermonuclear Containment, New York: Academic Press 1966.
4. Furth, H. P.: Minimum-average-B stabilization for toruses. In: Advances in Plasma Physics, Vol. 1, eds. A. Simon and W. B. Thompson, New York: Interscience 1968, pp. 67—100.
5. Demirkhanov, R. A.: Stabilization of a plasma by high-frequency electromagnetic fields. Preprint, Second International Colloquium on the Interaction of High-Frequency Electromagnetic Fields with a Plasma, Paris-Saclay, Jan. 1968.
6. Plasma Physics and Controlled Nuclear Fusion Research (Proc. Conf. on Plasma Physics and Controlled Nuclear Fusion, Culham, England, Sept. 6—10, 1965), U.S. AEC Publication AEC-tr-6760, 1966.
7. Morozov, A. I., Solov'ev, L. S.: Cybernetic stabilization of plasma instabilities. Soviet Physics-Tech. Physics **9**, 1214—1220 (March 1965).
8. Arsenin, V. V., Chuyanov, V. A.: A possible stabilization of the trough-shaped plasma instability through a feedback system, Soviet Physics Doklady **13**, 570—571 (Dec. 1968).
9. Arsenin, V. V., Zhilkov, V. A., Chuyanov, V. A.: The suppression of flute instability with a feedback circuit. Proc. Third IAEA Conf. on Plasma Physics and Controlled Fusion Research, Novosibirsk, U.S.S.R., August 1—7, 1968.
10. Paslavskii, E. S., Samoilenko, Yu. I.: Distributed control system for plasma stabilization. Soviet Physics-Tech. Physics **12**, 707—708 (Nov. 1967).

11. CHURCH, M. J., et al.: Stabilization of ion cyclotron instabilities by electron heating and flute instability by feedback in magnetic well/mirror geometries. Proc. Third European Conf. on Controlled Fusion and Plasma Physics, Utrecht, Netherlands, June 23—27, 1969, Groningen: Wolters-Noordhoff 1970, p. 12.
12. MELCHER, J. R.: Continuum feedback control of instabilities on an infinite fluid interface. Phys. Fluids **9**, 1973—1982 (1966).
13. FATTORINI, H. O.: Boundary control systems. SIAM J. Control **6**, 349—385 (1968).
14. BERNSTEIN, I. B., FRIEMAN, E. A., KRUSKAL, M. D., KULSRUD, R. M.: An energy principle for hydromagnetic stability problems. Proc. Roy. Soc. **244**, 17—40 (1958).

Determinism and Uncertainty in Stability

By

G. Herrmann

Stanford University, Stanford, Cal., U.S.A.

Abstract. The destabilizing effects produced in mechanical systems with follower forces by a variety of physical phenomena are shown to result in a lack of determinism in the analysis of stability. Several possibilities of removing the uncertainties are suggested.

1. Introduction

We are concerned with the analysis of stability of equilibrium of mechanical systems subjected to follower (circulatory) forces, i.e., to forces which depend on position of the system. Attention is confined to loss of stability by flutter (oscillations with increasing amplitude). With the aid of an example, the destabilizing effect of small and vanishing viscous damping is discussed, as well as the recently uncovered similar effects produced by a variety of other physical phenomena. In all these instances the inclusion of a vanishing quantity leads to considerably lower critical loads than complete omission of the quantity.

The determinism of circulatory stability problems becomes thus questionable, not only on the practical level of measuring experimentally the values of various parameters associated with very small and vanishing effects, but, perhaps more strikingly, on the conceptual level, since one can never be certain that the most severe destabilizing effect is included in the analysis. Several possibilities of rendering the stability investigations of mechanical systems under consideration more meaningful are discussed. One of these consists in modifying physically the system in such a manner as to make it strongly asymptotically stable in the

---

Acknowledgement. This work was supported by the National Aeronautics and Space Administration under Grant NGR 14-007-067.

absence of circulatory loads. Another replaces Liapunov's definition of stability by a consideration of "patterns of behavior" based on practical engineering requirements. It is shown finally how by introducing an additional parameter the results of the mathematical limiting processes may be modified.

Solely linear dynamic systems are treated here, whose stability (or instability) is studied in terms of the nature of the characteristic roots applying Routh-Hurwitz criteria.

2. Destabilizing effects of linear viscous damping

It has been noted by ZIEGLER[1] that the presence of linear viscous damping in an elastic system may be accompanied by the so-called destabilizing effect. ZIEGLER considered a double pendulum with visco-elastic hinges as a model of a continuous elastic bar with internal damping and let a compressive follower force act at the free end. It was found that the critical force is considerably higher in the absence of viscosity than in the case of small or even vanishing (arbitrarily small) viscosity, see Fig. 1.

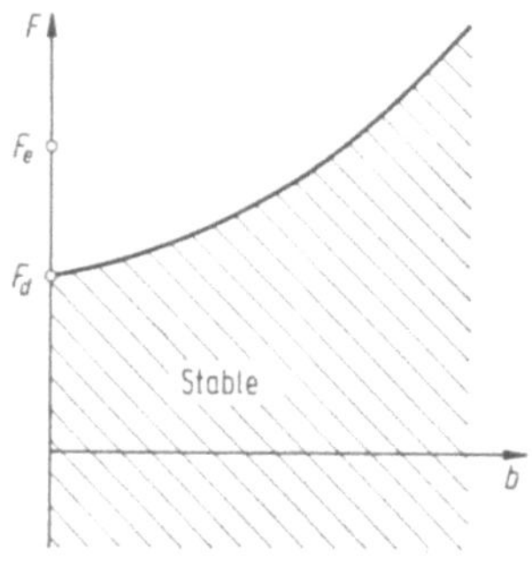

Fig. 1. Region of stability in the presence of damping b. F_d is the critical force for vanishing damping and F_e for no damping.

In numerous later studies, various features of this destabilizing effect were explored further. BOLOTIN [2] showed that the magnitude of the destabilizing effect depends (for vanishing damping) on the ratio of the damping coefficients in the two degrees of freedom. A generic relationship between critical loadings for no damping and for slight damping as well as vanishing viscous damping was established by HERRMANN and JONG [3] for the case of a system with two degrees of freedom shown in Fig. 2, which is almost the same as that of Ref. [1]. The equations of small motions about the position of equilibrium $\varphi_1 = \varphi_2 = 0$ of this system are

$$3ml^2\ddot{\varphi}_1 + (b_1 + b_2)\,\dot{\varphi}_1 - (Pl - 2c)\,\varphi_1 + ml^2\,\ddot{\varphi}_2 - b_2\dot{\varphi}_2 + (Pl - c)\,\varphi_2 = 0\,,$$
$$ml^2\ddot{\varphi}_1 - b_2\dot{\varphi}_1 - c\varphi_1 + ml^2\ddot{\varphi}_2 + b_2\dot{\varphi}_2 + c\varphi_2 = 0\,. \tag{1}$$

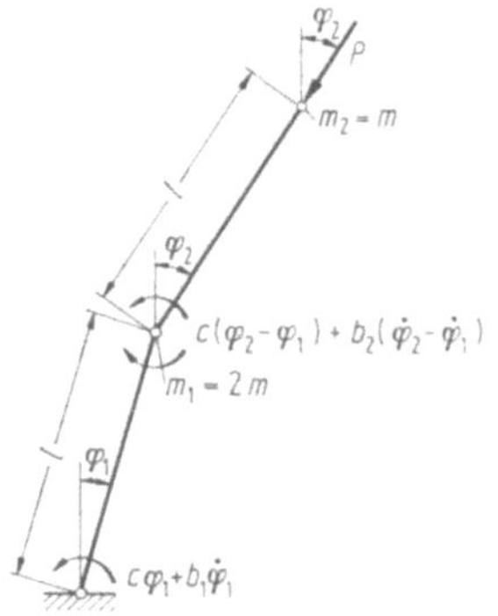

Fig. 2. Two-degree-of-freedom model.

The system parameters are indicated in Fig. 2 and a dot denotes differentiation with respect to time t. Stipulating solutions of the form

$$\varphi_i = A_i e^{\omega t} \qquad (i = 1, 2) \tag{2}$$

one is lead to the characteristic equation

$$p_0\Omega^4 + p_1\Omega^3 + p_2\Omega^2 + p_3\Omega + p_4 = 0 \tag{3}$$

with the coefficients

$$\begin{aligned} &p_0 = 2\,; \qquad p_1 = B_1 + 6B_2\,, \\ &p_2 = 7 - 2F + B_1B_2\,; \quad p_3 = B_1 + B_2\,; \quad p_4 = 1\,, \end{aligned} \tag{4}$$

where

$$\begin{aligned} &\Omega^2 = l^2\omega^2 m/c\,; \qquad F = Pl/c\,, \\ &B_i = b_i/(l^2cm)^{1/2} \quad (i = 1, 2)\,. \end{aligned} \tag{5}$$

Application of Routh-Hurwitz criteria supplies the critical load F_d (for small damping)

$$F_d = \frac{4\beta^2 + 33\beta + 4}{2(\beta^2 + 7\beta + 6)} \tag{6}$$

(where $\beta = B_1/B_2$) which separates unstable (in the sense of Liapunov) from asymptotically stable behavior of the system. For $\beta = 1$ ZIEGLER's value of $F = 41/28$ is recorded. In the absence of damping ($B_1 = B_2 = 0$), it is recalled that the characteristic equation is a biquadratic

$$2\Omega^4 + (7 - 2F)\,\Omega^2 + 1 = 0\,, \tag{7}$$

which results in the critical loads

$$F_{e1,2} = 7/2 \mp \sqrt{2} = 2.086;\ 4.914. \tag{8}$$

If subjected to a force which is larger than $F_{e1} = 7/2 - \sqrt{2}$ the system is again unstable. If $F < 7/2 - \sqrt{2}$ the system is "marginally stable" (corresponding to the "doubtful" or "critical" case of Liapunov), since the roots of the characteristic equation, even though distinct, are pure imaginary, and no statement regarding stability or instability can be made. Several aspects of the relationship between F_e and F_d were discussed by HERRMANN and JONG in Refs. [3] and [4], where it was shown, for example, that for certain systems instability either by divergence (static buckling) or by flutter may occur first, as the compressive load increases, depending upon the ratio of the damping coefficients in the two degrees of freedom.

Among the other studies concerned with the destabilizing effects of linear viscous damping mention should be made here of the papers by LEIPHOLZ [5], NEMAT-NASSER, PRASAD and HERRMANN [6], NEMAT-NASSER and HERRMANN [7], and BOLOTIN and ZHINZHER [8].

3. Other destabilizing effects

It is of interest to note that not only linear viscous damping, but other types of dissipation mechanisms are associated with destabilizing effects of the type discussed in the previous section. In Ref. [9] a general formulation of the stability analysis of elastic continuous systems subjected to nonconservative forces in the presence of thermomechanical coupling was presented and applied to the problem of a cantilever under a tangential follower force at the free end. A pronounced destabilizing effect of thermoelastic dissipation was found to exist. Bilinear hysteretic damping was studied in Ref. [10] where it was shown that it may have a destabilizing effect similar to linear viscous damping, but that this effect disappears for a large class of hysteretic systems.

Similarly, so-called magnetic damping may be associated with a destabilizing effect, as shown in Ref. [11]. If the bars in Fig. 2 constitute portions of electrical circuits having resistances R_1 and R_2, respectively, and the system is placed in a uniform magnetic field of strength B_0 which acts normal to the plane of motion, the equations of small motion about the position of equilibrium $\varphi_1 = \varphi_2 = 0$ will be

$$\begin{aligned} 3ml^2\ddot{\varphi}_1 + (r_1 + 4r_2)\,\dot{\varphi}_1/2 - (Pl - 2c)\,\varphi_1 + ml^2\ddot{\varphi}_2 + r_2l^2\dot{\varphi}_2 \\ + (Pl - c)\,\varphi_2 = 0, \end{aligned} \tag{9}$$

$$ml^2\ddot{\varphi}_1 + r_2l^2\dot{\varphi}_1 - c\varphi_1 + ml^2\ddot{\varphi}_2 + r_2l^2\dot{\varphi}_2/2 + c\varphi_2 = 0,$$

where

$$r_j = B_0^2 l^2 / 2R_j \ (j = 1, 2). \tag{10}$$

The critical load F_{md} which separates unstable from asymptotically stable behavior of the system in this case, for $r_1 = r_2 \ll 1$, is calculated to be $F_{md} = 5/3$ which again is considerably smaller than $F_e = 7/2 - \sqrt{2}$.

The stability of the system with two degrees of freedom mentioned before under the action of a retarded follower force was analyzed by Kiusalaas and Davis [12]. The direction of the follower force with constant time lag τ was specified as

$$\theta(t) = \varphi_2(t - \tau). \tag{11}$$

Solutions to the linearized equations for small motions are sought again in the form (2) and lead to a frequency equation with exponential terms requiring the application of Pontryagin's stability criteria rather than those of Routh-Hurwitz. After a comprehensive and rather elaborate analysis the authors arrive to the important result that a very small (vanishing) time lag renders the system unstable for all positive (compressive) values of the applied force P. Even under the most favorable time lag the critical load was found to be $F = Pl/c = 0.177$, as compared to $F = 2.086$ for the same system without any retardation τ.

In supplementing the analysis of Ref. [12], it may be remarked here that for small time lag the stability investigation can be readily carried out employing the simpler Routh-Hurwitz criteria if $\varphi_2(t - \tau)$ is expanded into a Taylor series about $\varphi_2(t)$ (see Ref. [13]) and if only the first two terms are retained. Details of this analysis will be included in a forthcoming paper by the author.

The destabilizing effect is in general introduced by any sufficiently small, velocity-dependent forces, such as, for example, Coriolis forces. Some types of such forces are realized by fluid jets; they have received considerable attention and have been referred to as "jet damping" [14]. Others are produced e.g. by flow through pipes [15].

4. Determinism and uncertainty

The foregoing examples of various destabilizing effects amply illustrate the necessity of a firmer grasp of certain aspects in the analysis of stability problems as applied to systems subjected to follower forces. What is required is a better understanding of the physical relevance of the critical forces, calculated with and without destabilization. Some observations in this regard were made in Ref. [3] where the dependence of F on the positive real part of Ω was investigated for various values of the (small) damping coefficients. It was shown that for relatively large values of damping, F_d represents the physically significant critical load, because a small increase of the load above F_d will be associated with a relatively large Re Ω. For very small damping, however, the physical significance of F_d will be considerably lessened because a small increase of the load above F_d will result in a very small Re Ω. A large increase of Re Ω will now be associated with a small increase of a load which is slightly lower than F_e. It was thus concluded that in a limiting process of vanishing damping the physical significance of F_d as a critical load is gradually transferred to F_e. No suggestions were made in Ref. [3] however, as to how assess quantitatively this shift in significance.

Another aspect which requires additional insight is that of experimental determination of system parameters. If very small, even vanishing quantities which induce a destabilizing effect have such a decisive influence on the critical loads calculated analytically, how should these quantities be measured with required accuracy? Further, how can one be sure that the "correct" or "right" parameters have been included? It is even conceivable that (vanishing) destabilizing effects exist which have never been thought of as yet, and which have perhaps an even stronger influence on the stability boundaries of a given system than any of those mentioned in the preceding section. The following section will describe some attempts to remedy this obviously unsatisfactory state of affairs.

5. Removal of uncertainties

If we are dealing with a man-made system, it would probably be desirable, if possible, to make it well-behaved by means of a suitable choice of system parameters and, in particular, by making it strongly asymptotically stable to begin with. This is done sometimes in control systems where the "doubtful", "critical" or "marginal" case of Liapunov (pure imaginary roots of the characteristic equation) is interpreted as describing an inherently unstable system. It is well known that if a system is asymptotically stable, small "destabilizing" quantities will have but a small effect on the critical loads; this effect will vanish with the vanishing of the "destabilizing" quantity.

In many man-made systems and in given natural systems the uncertainty cannot be circumvented in this manner. It may then be suggested that the analysis of stability be replaced (or supplemented) by an analysis of "patterns of behavior" of the disturbed system for various ranges of the controlling parameter (force). Since we are interested here only in oscillatory response to a disturbance, three types of behavior are qualitatively sketched in Fig. 3. Let it be our aim to classify the response in just two categories. Depending upon the specific performance requirements of the system at hand, it may be meaningful to place the response in Figs. 3a and b into one category and the response in Fig. 3c into the other. In the first category the disturbance remains small during a certain interval of time, while growing fairly large in the second category during the same interval.

As a measure of the rate of growth of the oscillations it is convenient to introduce the largest real part α of the relevant root of the characteristic equation. This is analogous to the introduction of the smallest negative part as the "absolute stability margin" [13].

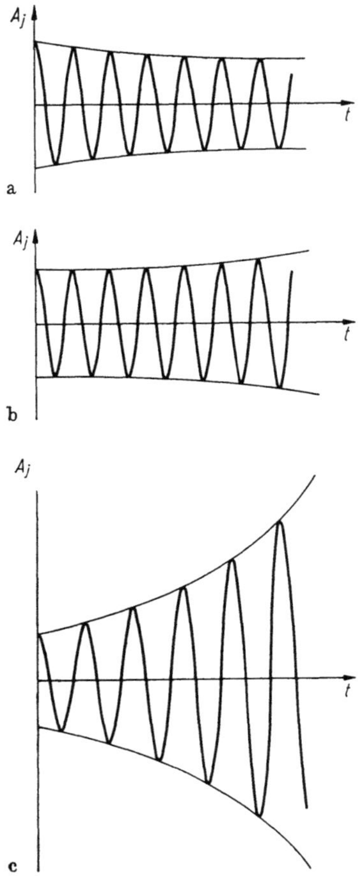

Fig. 3. Three types of oscillatory response to a disturbance.

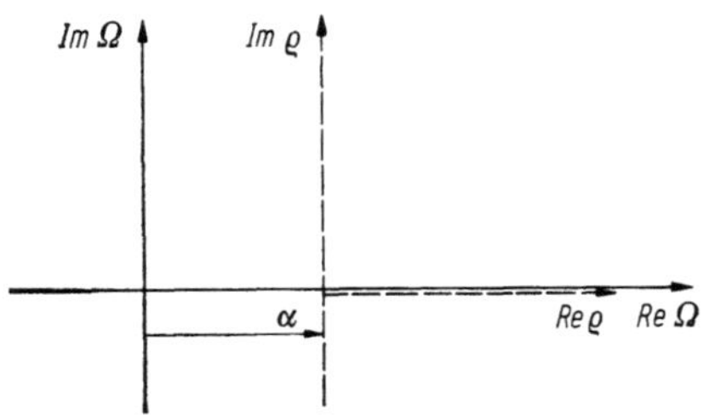

Fig. 4. Transformation $\Omega = \varrho + \alpha$.

The stability analysis corresponding to a given $\alpha > 0$ can be carried out by introducing the transformation (Fig. 4)

$$\Omega = \varrho + \alpha \tag{12}$$

into the characteristic equation, e.g., (3) which yields the modified characteristic equation for ϱ,

$$a_0\varrho^4 + a_1\varrho^3 + a_2\varrho^2 + a_3\varrho + a_4 = 0, \tag{13}$$

where

$$\begin{gathered} a_0 = p_0; \quad a_1 = p_1 + 4\alpha p_0; \qquad a_2 = p_2 + 3\alpha p_1 + 6\alpha^2 p_0, \\ a_3 = p_3 + 2\alpha p_2 + 3\alpha^2 p_1 + 4\alpha^3 p_0, \\ a_4 = p_4 + \alpha p_3 + \alpha^2 p_2 + \alpha^3 p_1 + \alpha^4 p_0. \end{gathered} \tag{14}$$

Applying the usual Routh-Hurwitz criteria to this modified equation, the critical force can be calculated. For the system of Fig. 2 the critical force F_c is to be calculated from

$$X \equiv a_1 a_2 a_3 - a_0 a_3^2 - a_1^2 a_4 \tag{15}$$

and is found to be

$$F_c = \left(B^{(+)}\sqrt{B^2 - AC}\right)/A\,, \tag{16}$$

where

$$\begin{aligned} A &= 8\alpha(d_1 - 2\alpha d_0)\,, \\ B &= d_1 d_3 + 2\alpha d_1 d_2 - 4\alpha d_0 d_3 - \alpha^2 d_1^2\,, \\ C &= d_1 d_2 d_3 - d_0 d_3^2 - d_1^2 d_4 \end{aligned} \tag{17}$$

and

$$\begin{aligned} &d_0 = p_0; \quad d_1 = p_1 + 4\alpha p_0; \quad d_2 = 7 + B_1 B_2 + 3\alpha p_1 + 6\alpha^2 p_0\,, \\ &d_3 = p_3 + 2\alpha(7 + B_1 B_2) + 3\alpha^2 p_1 + 4\alpha^3 p_0\,, \\ &d_4 = p_4 + \alpha p_3 + \alpha^2(7 + B_1 B_2) + \alpha^3 p_1 + \alpha^4 p_0\,. \end{aligned} \tag{18}$$

The results of the numerical calculations are displayed in Fig. 5. Thin solid lines represent the critical force F_c as function of the growth para-

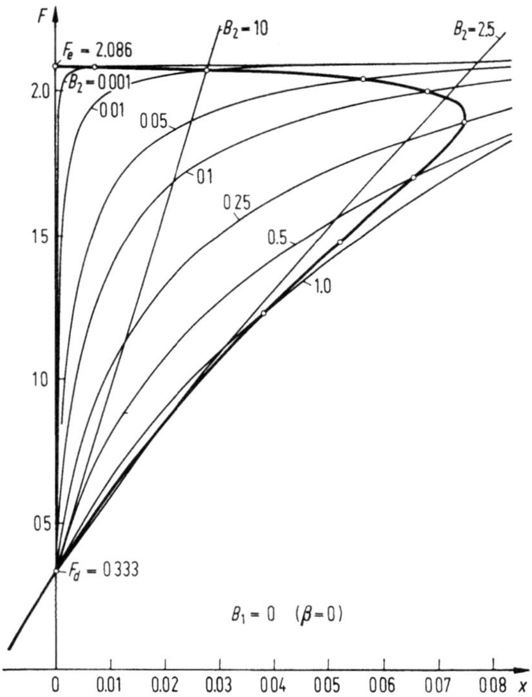

Fig. 5. Results of numerical calculations. Thin lines: critical force F_c as a function of degree of instability α for $B_1 = 0$ and various values of B_2. Thick line: transition force F_t for $B_1 = 0$ and various values of B_2.

meter α for given damping coefficient B_i. The thin curves in Fig. 5 are the same as those in Fig. 5 of Ref. [3], but they have been calculated in a different manner and their interpretation is also entirely different.

The critical force as defined with the aid of the growth parameter α is not entirely satisfactory because it does not separate the different types of behavior illustrated in Fig. 3. We seek now to define what we may call a "transition" force F_t below which α (whether positive or negative) would be relatively small and above which it would be relatively large. It appears to be reasonable to define the transition force F_t as the force for which the absolute value of the curvature of a given curve $F(\alpha) = 0$ attains a maximum. Corresponding calculations have been carried out and the values of the transition force F_t for various values of the damping coefficient have been joined by a thick solid line in Fig. 5. It is noted that for given damping coefficients B_i there exists an associated transition force F_t which in turn corresponds to a certain particular value of growth parameter α_t. The experimental determination of system parameters associated with F_t appears to be feasible. It should be also observed that as damping decreases F_t approaches F_e, while F_t approaches F_d as damping increases. The relationship between critical and transition forces is thus clarified.

It is rather evident that the two types of system behavior which are separated by F_t can be less or more different and thus it may be appropriate to introduce the notion of degree of separation σ associated with any particular value of F_t. This additional characteristic separating "patterns of behavior" could be made to depend on the magnitudes of curvature and slope of the function $F(\alpha)$ at α_t. It is expected that details of these considerations will be presented in a forthcoming paper by W. HAUGER and the author.

The parameter α can be employed yet for another, purely mathematical purpose. For small α, B_1 and B_2 one may write

$$\begin{aligned}&[\alpha^2 + \alpha(B_1 + 6B_2)/4]\,(F - F_{e1})\,(F - F_{e2})\\ &\quad - (B_1 + B_2)\,(B_1 + 6B_2)\,(F - F_d) = 0.\end{aligned} \tag{19}$$

Here F_{e1} and F_{e2} are the critical loads obtained in the absence of damping, Eq. (8), while F_d is the critical load for vanishing damping, Eq. (6). It is observed that either critical load may be obtained by a limiting process in Eq. (19), which gives either α, F, B_1 or B_2 in terms of the remaining three quantities. If the growth parameter $\alpha \to 0$, it is seen that $F \to F_d$, which depends only on $\beta = B_1/B_2$ but not on B_1 and B_2 itself. By contrast, if the damping coefficients are made to vanish first, then $F \to F_{e1}$ or $F \to F_{e2}$, regardless of the (small) value of α. Thus the introduction of the growth parameter α permits to approach the critical load for no damping F_{e1} even in the presence of vanishing damping, removing mathematically any destabilizing effects.

In closing it is deemed appropriate to quote the sentence: "The difficulty of framing a definition of kinetic stability which shall be comprehensive and at the same time conform to natural prepossessions has long been recognized" from a paper by LAMB [16]. The difficulty which LAMB spoke of over sixty years ago appears to be still with us.

References

1. ZIEGLER, H.: Ing.-Arch. **20**, 49 (1952).
2. BOLOTIN, V. V.: Nonconservative Problems of the Theory of Elastic Stability, New York: Pergamon Press 1963.
3. HERRMANN, G., JONG, I. C.: J. Appl. Mech. **32**, 592 (1965).
4. HERRMANN, G., JONG, I. C.: J. Appl. Mech. **33**, 125 (1966).
5. LEIPHOLZ, H.: Ing.-Arch. **33**, 308 (1964).
6. NEMAT-NASSER, S., PRASAD, S. N., HERRMANN, G.: AIAA J. **4**, 1276 (1966).
7. NEMAT-NASSER, S., HERRMANN, G.: ZAMP **17**, 305 (1966).
8. BOLOTIN, V. V., ZHINZHER, N. I.: Int. J. Solids and Structures **5**, 965 (1969).
9. HUANG, N. C., SHIEH, R. C.: Thermomechanical Coupling Effect on the Stability of Nonconservative Elastic Continuous Systems. A report (undated) of the University of California at San Diego, La Jolla.
10. JONG, I. C.: J. Appl. Mech. **36**, 76 (1969).
11. SMITH, T. E., HERRMANN, G.: Stability of Circulatory Elastic Systems in the Presence of Magnetic Damping (to be published).
12. KIUSALAAS, J., DAVIS, H. E.: On the Stability of Elastic Systems under Retarded Follower Forces. Int. J. Solids and Structures (in press).
13. PORTER, B.: Stability Criteria for Linear Dynamical Systems, New York: Academic Press 1968.
14. THOMSON, W. T., REITER, G. S.: AIAA J. **3**, 413 (1965).
15. HERRMANN, G., NEMAT-NASSER, S.: Int. J. Solids and Structures **3**, 39 (1967).
16. LAMB, H.: Proc. Roy. Soc. **A 80**, 168 (1908).

Hölder Stability and Logarithmic Convexity

By

R. J. Knops

University of Newcastle upon Tyne, Newcastle upon Tyne, U.K.

and

L. E. Payne

Cornell University, Ithaca, N.Y., U.S.A.

1. Introduction

The usual concept employed in most stability investigations is that due to Liapunov and may be interpreted as the ordinary continuity in suitable norms of the solution upon its initial data (see, e.g., Gilbert and Knops [1]). However, in this paper we study a device that yields the special case of Liapunov stability corresponding to Hölder continuity of the solution upon its data on compact subintervals of the finite-closed time interval. To be precise, let $\varphi(\cdot, t)$ and $\psi(\cdot, t)$ be solutions of the given problem in a class C, and let $\varphi(\cdot, 0)$ and $\psi(\cdot, 0)$ denote their initial values, with t denoting the time variable. Let $\|\cdot\|_0$ be the norm defined on the initial data and $\|\cdot\|_t$ the norm defined on the solutions in C. (More generally, positive definite measures may replace the norms.) For our purposes, Hölder stability is then expressed by

Definition 1.1. *The solution $\varphi(\cdot, t)$ is Hölder stable on the interval $0 \leq t \leq T$ if and only if for every ε (> 0) the following implication is valid:*

$$\|\varphi(\cdot, 0) - \psi(\cdot, 0)\|_0 < \varepsilon \Rightarrow \sup_{0 \leq t \leq t_1 < T} \|\varphi(\cdot, t) - \psi(\cdot, t)\|_t < A\varepsilon^{\alpha} \quad (1.1)$$

for arbitrary admissible $\psi(\cdot, 0)$, where $0 < \alpha \leq 1$ and A is a positive constant independent of ε.

Definition 1.1, in a more general context, was proposed by F. John [2] in connection with certain non-well posed problems, i.e., problems in which the solution does not depend continuously on its data. He inquired what *a priori* restriction must be imposed upon the solutions in order to recover continuous dependence and in what sense is continuity thereby obtained. For most of the problems he examined, it is sufficient to impose a uniform bound on the solution, the appropriate continuity then being in the sense of Hölder. This is the precise situation facing us in the problems we discuss here. Although in general the solution of a given problem may be unstable, we show that in a certain class of bounded perturbations the solution is in fact Hölder stable.

To provide a clear intuitive picture of Hölder stability and the conditions producing it, let us imagine a linear system whose null solution is unstable due to the presence of increasing exponential eigenfunctions of type e^{nt}. The eigenfunctions bounded at $t = T$ but arbitrarily small at $t = 0$ are then of the form $e^{-n(T-t)}$, and thus remain arbitrarily near the null solution for $0 \leq t < T$. However, the eigenfunctions cannot remain arbitrarily close to the null solution for $0 \leq t \leq T$, since in neighborhoods of $t = T$, no matter how large n is, the eigenfunctions must increase to their (bounded) value of 1. Thus there is stability on all compact sub-intervals of $[0, T]$, but not on $[0, T]$ itself. These remarks illustrate the necessity of admitting only bounded solutions and also why compact sub-intervals are introduced in Definition 1.1.

Hölder stability is of course implied by the inequality

$$\sup_{0 \leq t \leq t_1 < T} \|\varphi(\cdot, t) - \psi(\cdot, t)\|_t \leq A(\|\varphi(\cdot, 0) - \psi(\cdot, 0)\|_0)^\alpha \times \Big(\sup_{0 \leq t \leq T} \|\varphi(\cdot, t) - \psi(\cdot, t)\|_t\Big)^{1-\alpha}, \tag{1.2}$$

provided

$$\sup_{0 \leq t \leq T} \|\varphi(\cdot, t) - \psi(\cdot, t)\|_t < M \tag{1.3}$$

for some positive bounded constant M. We shall be interested not only in the derivation of an inequality of type (1.2) but also in some of the interesting consequences which follow from such a result.

The argument we employ is based upon proving the convexity with respect to time of the logarithm of a certain positive definite function, whose exact choice is determined by the problem. This approach is not new, but has previously been used for instance in establishing the Hadamard's three circle theorem and various interpolation formulae. Recent applications have been made by e.g., Laurentiev [3], Pucci [4], Payne [5], Agmon [6], Agmon and Nirenberg [7], and Levine

[8, 9]. We first describe the steps necessary to establish the logarithmic convexity arguments and then consider some implications of this property. Apart from Hölder stability, these are concerned with uniqueness and more significantly, with possible growth patterns of the solution, which in turn yield information about instability and non-existence of solutions of various types. Finally, we consider briefly the theory of linear elasticity without body force (including the corresponding theory of small elastic deformations superposed upon large) and show how to choose an appropriate positive definite function and prove that its logarithm is a convex function of time. Other applications studied elsewhere are to linear elasticity with body force (basic for the study of continuous dependence on boundary data, elasticities, initial data and geometry); linear coupled thermoelasticity; the Navier-Stokes equation backward in time; and the mixture of two linear elastic solids. Accounts are given respectively by KNOPS and PAYNE [10, 11, 12] and KNOPS and STEEL [13, 14].

2. Logarithmic convexity and its consequences

In using the technique of logarithmic convexity, the basic idea is to construct a real-valued continuous function $F(t)$ defined on the solutions $\varphi(\cdot, t)$[1] and possessing the following three properties:

(i) $F(t) \geq 0$ for $0 < t < T$,

(ii) $F(t) = 0 \Leftrightarrow \varphi(\cdot, t) = 0$ for $0 < t < T$,

(iii) there exist constants a_1, a_2 such that

$$FF'' - (F')^2 \geq -a_1 FF' - a_2 F^2, \quad 0 < t < T, \tag{2.1}$$

where primes denote differentiation with respect to time. Properties (i) and (ii) follow from the actual construction of $F(t)$, whereas property (iii) is established by means of the governing equation and data. The choice of $F(t)$ clearly depends upon the particular problem under study, but apart from additive data terms, usually consists of L_2 — integrals of the solution over either space or space — time.

To see that inequality (2.1) is equivalent to the convexity of the logarithm of $F(t)$, let us suppose that $F(t) > 0$ in the open interval $0 < t_1 < t < t_2 < T$. Then (2.1) may be rewritten

$$\{\ln F(t)\}'' + a_1\{\ln F(t)\}' + a_2 \geq 0, \quad 0 < t_1 < t < t_2 < T, \tag{2.2}$$

which, if $a_1 = a_2 = 0$, is the desired result. Otherwise, we put

$$\sigma = e^{-a_1 t} \tag{2.3}$$

[1] In the case of linear systems, we may consider perturbations about the null solutions. Otherwise, $\varphi(\cdot, t)$ here represents the difference of two solutions.

and then (2.2) reduces to

$$\frac{d^2}{d\sigma^2}\{\ln\,[F(\sigma)\,\sigma^{-a_2/a_1^2}]\} \geq 0, \quad \sigma_2 < \sigma < \sigma_1, \tag{2.4}$$

$$\sigma_\alpha = e^{-a_1 t_\alpha}, \quad \alpha = 1, 2, \tag{2.5}$$

demonstrating the convexity with respect to σ of the logarithm of $F(\sigma)\,\sigma^{-a_2/a_1^2}$. Clearly, in view of the similar structure between (2.4) and the special case of (2.2) with $a_1 = a_2 = 0$, we may always take without loss the values $a_1 = a_2 = 0$ when discussing general consequences of logarithmic covexity. We now consider what are these consequences and accordingly confine our attention to the *convexity inequality*

$$\{\ln F(t)\}'' \geq 0, \quad 0 < t < T. \tag{2.6}$$

2.1 Uniqueness

Let us first examine the question of uniqueness under the assumption that $F(0) = 0$. We wish to show that (2.6) implies that $F(t)$ vanishes identically for $0 \leq t \leq T$ and hence, by property (ii) and the assumed continuity of $\varphi(\cdot, t)$, that $\varphi(\cdot, t)$ likewise vanishes identically for $0 \leq t \leq T$. Let us assume the contrary, namely that $F(t) > 0$ on some open interval $0 \leq t_1 < t < t_2 \leq T$. By means of finite Taylor series expansions, or alternatively, by Jensen's inequality, we obtain from (2.6) the inequality

$$F(t) \leq F(t_1)^{\frac{t_2-t}{t_2-t_1}} F(t_2)^{\frac{t-t_1}{t_2-t_1}}, \quad 0 \leq t_1 < t < t_2 \leq T. \tag{2.7}$$

Now, either $t_1 = 0$, or by continuity $F(t_1) = 0$. If $F(t_1) = 0$ then it may be shown (see Agmon and Nirenberg [7], Levine [8]) that

$$F(t) = 0, \quad 0 \leq t_1 < t < t_2 \leq T,$$

and so by continuity $F(t_2) = 0$. Thus, $F(t) \equiv 0$ for $0 \leq t \leq T$. In the case of $t_1 = 0$, we have $F(0) = 0$, and a repetition of the argument again leads to the same result. Uniqueness is therefore established.

2.2 Hölder stability

Secondly, let us investigate the consequences of the convexity inequality (2.6) for the Hölder stability of the solution. We assume that $F(t) > 0$ for $0 < t < T$ (otherwise $F(t) = 0$, $0 < t < T$, by the uniqueness just proved) and then inequality (2.7) becomes

$$F(t) \leq [F(0)]^{1-t/T}\,[F(T)]^{t/T}, \quad 0 < t < T. \tag{2.8}$$

Let us restrict ourselves to solutions satisfying the condition $F(T) < M$, where M is some positive bounded constant, and let us put $\alpha = 1 - t/T$. Then (2.8) may be rewritten

$$F(t) < M^{1-\alpha}\,[F(0)]^\alpha, \tag{2.9}$$

which is in the desired form of inequality (1.2), and therefore establishes the Hölder stability on $0 \leq t \leq t_1 < T$ of the solution in the norm $\|\cdot\|_t^2 = F(t)$. Notice that when $t = T$, we have $\alpha = 0$, and (2.9) merely confirms the boundedness hypothesis on $F(t)$.

Unequality (2.8) may be alternatively written in the form

$$F(t) \leq F(0) \exp\left\{\frac{t}{T} \ln \frac{F(T)}{F(0)}\right\}, \qquad 0 < t < T, \tag{2.10}$$

clearly demonstrating that a function whose logarithm is convex on a (finite) time interval cannot be greater than some time-increasing exponential function. In fact, as $F(0)$ decreases to zero, this upper bound for $F(t)$ behaves in a manner similar to the exponential function described in the introduction, even though the corresponding operator may not admit exponential growth. We may thus derive information about the growth behaviour of $F(t)$ on finite time intervals. This result may be considerably extended when semi-infinite time intervals are examined.

2.3 Growth of $F(t)$ and non-existence

To gather from the convexity inequality (2.6) how $F(t)$ behaves for large (possibly semi-infinite) time intervals, we need first to obtain from (2.6) an inequality complementing (2.7), or (2.8). This is achieved by a finite Taylor series expansion of $\ln F(t)$ about $t = 0$ and leads to

$$F(t) \geq F(0) \exp\left\{t \frac{F'(0)}{F(0)}\right\}, \qquad t > 0. \tag{2.11}$$

Several deductions may now be made from (2.11) for the three cases: $F'(0) > 0$, $F'(0) = 0$, $F'(0) < 0$.

2.3 (i) $F'(0) > 0$. Obviously, from (2.11), $F(t)$ is bounded below for $t > 0$ by an increasing exponential function of time and therefore becomes increasingly large with time.

This result may be used in the following computation, based upon inequality (2.8). Let us assume that $F'(0)$ is arbitrary and[1]

$$\lim_{T\to\infty} \left(\frac{\ln F(T)}{T}\right) = 0. \tag{2.12}$$

On substituting into (2.8) and letting $T \to \infty$, we get

$$F(t) \leq F(0), \qquad t > 0. \tag{2.13}$$

On the other hand, if (2.13) is not true then by the assumed continuity of $F(t)$, there is some instant t, at which $F'(t_1) > 0$ and hence by an

[1] An alternative condition useful when the limit in (2.12) does not exist, is $F(T) < \beta e^{\gamma T^{1-\varepsilon}} T^N$, where β, γ, ε are positive constants and N is any positive integer. T is assumed large.

obvious extension of the argument used in 2.3 (i), we see that for $t > t_1$, $F(t)$ is bounded below by an exponentially increasing function. Thus we conclude that either

(a) $F(t)$ grows at least exponentially for sufficiently large time,

or

(b) $F(t)$ remains less than or equal to its initial value for all time.

Observe that the last result is independent of the initial value of $F'(t)$.

2.3 (ii) $F'(0) = 0$. For this condition (2.11) yields

$$F(t) \geq F(0), \qquad t > 0, \tag{2.14}$$

which from (2.13) and the arguments above leads to either

(c) $F(t)$ grows at least exponentially for sufficiently large time,

or

(d) $F(t)$ remains equal to its initial value for all time.

In general, the ambiguity represented by alternatives (c) and (d) remains, although in certain linear problems a complete resolution is possible. See Knops and Payne [15].

2.3 (iii) $F'(0) < 0$. Particular examples may be easily constructed showing that in this case either alternative (a) or (b) is possible, and hence without further restrictions nothing more definite can be said. However, where $F(t)$ is known to decay, inequality (2.11) in fact states that the decay cannot be faster than a decreasing exponential function.

The above remarks made in 2.3 (i)—(iii) demonstrate the non-existence of certain smooth solutions for which it is possible to define a corresponding $F(t)$. In particular, we see that when the convexity inequality is satisfied, there are no solutions with a polynomial growth behaviour for large values of time.

Naturally, for particular operators and definite choices of $F(t)$ it is possible to make more detailed statements about the growth properties of $F(t)$. These results are reported elsewhere in the literature (especially in Knops and Payne [15]). Where, however, an inequality of the form (2.2) with non-zero a_1, a_2 is established, growth statements are modified in the light of the time transformation (2.3).

3. Linear elasticity without body force

We give an example of one possible choice of the function $F(t)$ which satisfies the convexity inequality (2.6), in the special case of linear elasticity without body force. Because of linearity we need consider only perturbations about the null solution. Hence, in the standard notation (see, e.g., Knops and Payne [16]), we examine classical solutions of

$$(c_{ijkl} u_{k,l})_{,j} = \varrho \frac{\partial^2 u_i}{\partial t^2} \quad \text{on } B \times (0, T), \tag{3.1}$$

where the density ϱ and elasticities c_{ijkl} are non-uniform and satisfy

$$\varrho > 0, \qquad c_{ijkl} = c_{klij}. \tag{3.2}$$

The boundary conditions are homogeneous, typical ones being

$$\left.\begin{aligned} u_i &= 0 \text{ on } \overline{\partial B_1} \times (0, T) \\ u_i c_{ijkl} u_{k,l} &= 0 \text{ on } \partial B_2 \times (0, T) \end{aligned}\right\} \partial B = \overline{\partial B_1} \cup \partial B_2, \tag{3.3}$$

while the initial conditions are

$$u_i(x, 0) = f_i(x), \qquad \frac{\partial u_i}{\partial t}(x, 0) = g_i(x). \tag{3.4}$$

A consequence of (3.1)—(3.4) is the conservation of energy, expressed by

$$E(t) = \frac{1}{2} \int_{B(t)} \left(\varrho \frac{\partial u_i}{\partial t} \frac{\partial u_i}{\partial t} + c_{ijkl} u_{i,j} u_{k,l} \right) dx = E(0). \tag{3.5}$$

In what follows, we assume for convenience that $E(0) \leq 0$, treatments with $E(0) > 0$ being given in KNOPS and PAYNE [15, 16]. We now take

$$F(t) = \int_{B(t)} \varrho u_i u_i \, dx + \gamma (t + t_0)^2, \tag{3.6}$$

in which γ, t_0 are non-negative constants to be determined later. Immediately from (3.6) we obtain

$$F'(t) = 2 \int_{B(t)} \varrho u_i \frac{\partial u_i}{\partial t} dx + 2\gamma (t + t_0) \tag{3.7}$$

and

$$F''(t) = 2 \int_{B(t)} \varrho \left(\frac{\partial u_i}{\partial t} \frac{\partial u_i}{\partial t} + u_i \frac{\partial^2 u_i}{\partial t^2} \right) dx + 2\gamma. \tag{3.8}$$

Upon substituting for the acceleration from (3.1) and integrating by parts, we get after some rearrangement and use of (3.6),

$$F''(t) = 4 \left[\int_{B(t)} \varrho \frac{\partial u_i}{\partial t} \frac{\partial u_i}{\partial t} dx + \gamma \right] - 2 [E(0) + \gamma].$$

Hence,

$$FF'' - (F')^2 \equiv S^2(t) - 2F(t) [E(0) + \gamma],$$

where

$$S^2(t) = 4 \left[\int_{B(t)} \varrho u_i u_i \, dx + \gamma (t + t_0)^2 \right] \left[\int_{B(t)} \varrho \frac{\partial u_i}{\partial t} \frac{\partial u_i}{\partial t} + \gamma \right] - 4 \left[\int_{B(t)} \varrho u_i \frac{\partial u_i}{\partial t} dx + \gamma (t + t_0) \right]^2$$

is non-negative by virtue of Schwarz's inequality. Hence, on choosing $\gamma \leq -E(0)$, we recover (2.1) with $a_1 = a_2 = 0$, and therefore are led to (2.6)[1].

Uniqueness and Hölder stability may now be established, as described above, on taking $\gamma = 0$ in (3.6) (see also [16, 17]). For growth estimates, however, non-zero values of γ and t_0 must be retained. Details are described in [15] which also deals with the case $E(0) > 0$ and removal of the ambiguities encountered in 2.3 (i), (ii).

References

1. GILBERT, J. E., KNOPS, R. J.: Arch. Rat. Mech. Anal. **25**, 271 (1967).
2. JOHN, F.: Comm. Pure and Appl. Math. **13**, 551 (1960).
3. LAURENTIEV, M. M.: Dokl. Akad. Nauk S.S.S.R. **112**, 195 (1957).
4. PUCCI, C.: Rend. Acad. Naz. Lincei **18**, 473 (1955).
5. PAYNE, L. E.: In: Numerical Solutions of Non-linear Differential Equations, New York: Wiley 1966, pp. 239–263.
6. AGMON, L.: Unicité et Convexité dans les Problèmes Différentiels, Sem. Math. Sup., Univ. of Montreal Press 1966.
7. AGMON, L., NIRENBERG, L.: Comm. Pure and Appl. Math. **20**, 207 (1967).
8. LEVINE, H. A.: J. Diff. Eqs. (in print).
9. LEVINE, H. A.: (in preparation).
10. KNOPS, R. J., PAYNE, L. E.: Proc. Camb. Phil. Soc. (in print).
11. KNOPS, R. J., PAYNE, L. E.: (in preparation).
12. KNOPS, R. J., PAYNE, L. E.: Arch. Rat. Mech. Anal. **29**, 331 (1968).
13. KNOPS, R. J., STEEL, T. R.: Int. J. Eng. Sci. **7**, 571 (1969).
14. KNOPS, R. J., STEEL, T. R.: J. Composite Mat. (in print).
15. KNOPS, R. J., PAYNE, L. E.: (in preparation).
16. KNOPS, R. J., PAYNE, L. E.: Int. J. Solids and Structures **4**, 1233 (1968).
17. KNOPS, R. J., PAYNE, L. E.: Arch. Rat. Mech. Anal. **27**, 349 (1969).

[1] In other applications listed in the references, the inequality (2.2) with nonzero a_1, a_2 is established.

Thermoelastic Stability of a Finitely Deformed Solid under Nonconservative Loads

By

S. Nemat-Nasser

University of California, San Diego, La Jolla, Cal., U.S.A.

1. Introduction

Local stability of a finitely deformed solid with internal dissipation, subjected to followed type loads, has been studied by Nemat-Nasser [1], neglecting all thermal effects. On the other hand, thermoelastic stability under *conservative* loads was considered by Ericksen [2] and Koiter [3]. In the present paper, we combine the results in [1—3] to develop conditions for thermoelastic stability of solids subjected to nonconservative surface tractions. We ignore body forces, although they can easily be included.

2. Formulation

We consider a body B with a regular boundary ∂B deformed from its unstrained (virgin) configuration C_0 to an equilibrium (rest) state C with a constant absolute temperature $\theta > 0$ by the application of surface tractions $\boldsymbol{T}$ on part of its boundary ∂B^T, on the remaining part of ∂B the displacements being prescribed so as to preclude all possible rigid-body motions. We employ a rectangular Cartesian coordinate system, and denote particle positions in C_0 and C by X_α and x_i, respectively, where α, $i = 1, 2, 3$; capital letters bearing Greek subscripts refer to C_0, and lower case ones having italic subscripts refer to C. Since C is an equilibrium state, we have

$$T_{ij,i} = T^R_{\alpha j,\alpha} = 0 \text{ in } B, \quad T_{ij}\nu_i = T_j \text{ on } \partial B^T, \tag{2.1}$$

Acknowledgement. This research was supported by the National Science Foundation under Grant GK-1831, and by the Office of Naval Research under Contract N 00014-67-A-0109-0003, Task NR 064-496.

where T_{ij} and $T^R_{\alpha j}$ are Cauchy's and the first Piola-Kirchhoff (or Lagrangian) stress tensors, respectively, which are assumed to be smooth functions; a comma followed by a subscript denotes partial differentiation with respect to the corresponding coordinate, ν_i is the exterior unit normal on ∂B^T in C, and the summation convention on repeated subscripts is used. We assume that the mapping from C_0 to C is one-to-one and invertable; $J = \det |x_{i,\alpha}|$ is finite. Green's deformation tensor is

$$C_{\alpha\beta} = x_{i,\alpha} x_{i,\beta}. \tag{2.2}$$

If $\varphi = \varphi(C_{\alpha\beta}, \theta)$ denotes the Helmholtz free energy, we have

$$T_{ij} = \frac{\varrho}{\varrho_0} x_{i,\alpha} T^R_{\alpha j} = 2\varrho x_{i,\alpha} x_{j,\beta} \frac{\partial \varphi}{\partial C_{\alpha\beta}}, \quad \eta = -\frac{\partial \varphi}{\partial \theta}, \tag{2.3}$$

where η is entropy per unit mass, and ϱ and ϱ_0 are mass-densities in C and C_0, respectively.

As is stressed in [1], to study the stability of C when applied loads do not admit a potential, we must consider the actual motion of B from C to the non-equilibrium, time-dependent configuration C_t^*. Let starred letters denote quantities in C_t^*. The equation of energy conservation referred to C with surface S and volume V, then is

$$\frac{d}{dt} \int_B \left[e^*(C^*_{\alpha\beta}, \eta^*) + \frac{1}{2} \dot{x}_i^* \dot{x}_i^* \right] dm - \int_{S^T} T_i^* \dot{x}_i^* \, dS = \int_S -q_i^* \nu_i \, dS, \tag{2.4}$$

where e^* is the internal energy per unit mass, dm is the element of mass, q_i^* is the heat flux, and T_i^* is the external surface traction on ∂B^T measured per unit surface area in C; a superposed dot denotes the material-time-derivative. We also have an entropy equation:

$$\frac{d}{dt} \int_B \eta^* \, dm = \int_B (\gamma_1^* + \gamma_2^*) \, dm - \int_S \frac{q_i^* \nu_i}{\theta^*} dS = \int_B \left[\gamma_2^* - \frac{q^*_{i,i}}{\varrho \theta^*} \right] dm, \tag{2.5}$$

where

$$\gamma_1^* = -\frac{1}{\varrho \theta^{*2}} \varrho_i^* \theta^*_{,i} \geq 0 \text{ and } \gamma_2^* = \dot{\eta}^* + \frac{q^*_{i,i}}{\varrho \theta^*} \geq 0 \tag{2.6}$$

are entropy production per unit mass due to heat conduction and energy dissipation, respectively.

Following Ericksen [2] and Koiter [3], we write

$$e^*(C^*_{\alpha\beta}, \eta^*) - \theta \eta^* = \varphi^*(C^*_{\alpha\beta}, \theta) + \frac{1}{2} K \tau^2, \; \tau = \theta^* - \theta, \tag{2.7}$$

where K is $-\partial^2 \varphi^* / \partial \theta^{*2}$ evaluated at some temperature between θ and θ^* (see [3]), and $e^* - \theta^* \eta^* = \varphi^*$. With the aid of (2.7), (2.4) and (2.5) yield

$$\frac{d}{dt} \int_B \left[\varphi^*(C^*_{\alpha\beta}, \theta) + \frac{1}{2} K \tau^2 + \frac{1}{2} \dot{x}_i^* \dot{x}_i^* \right] dm - \int_{S_T} T_i^* \dot{x}_i^* \, dS = \dot{\mathscr{D}}, \tag{2.8}$$

$$\dot{\mathscr{D}} = \int_S \left(\frac{\theta}{\theta^*} - 1 \right) q_i^* \nu_i \, dS + \int_B \left[\frac{\theta}{\theta^{*2}} q_i^* \theta^*_{,i} - \theta \gamma_2^* \right] dm \leq 0, \tag{2.9}$$

where, as is usual, it is assumed that $K > 0$, and that a local increase in the surface temperature leads to an outward flow of heat. In all subsequent analyses we assume that K has a positive greatest lower bound $K^\circ > 0$.

For a *local* stability analysis, we may consider slight deviations of B from C, and write

$$\begin{aligned}&\frac{d}{dt}\left\{\int_B (\varphi^* - \varphi)\, dm - \int_{S^T} T_i u_i\, dS\right\}\\&= \frac{1}{2}\frac{d}{dt}\int_V [\gamma_{ijkl} u_{i,j} u_{k,l} + T_{ij} u_{k,i} u_{k,j} + f]\, dV,\end{aligned} \tag{2.10}$$

where

$$\gamma_{ijkl} = 4\varrho \frac{\partial^2 \varphi}{\partial C_{\alpha\beta}\, \partial C_{\gamma\delta}} x_{i,\alpha} x_{j,\beta} x_{k,\gamma} x_{l,\delta} \tag{2.11}$$

is a 9×9 positive-definite matrix whose elements are functions of x_i, denoting the isothermal elastic moduli for small displacements u_i superimposed on the finitely deformed equilibrium state C. f in (2.10) stands for terms depending on second order displacement gradients and higher order terms in $u_{i,j}$. We shall ignore these terms here, although for certain cases they may be used advantageously as is done by KOITER [4]. With the aid of (2.10), (2.8) becomes

$$\dot{\mathscr{K}} + \frac{1}{2}\frac{d}{dt}\int_V T_{ij} u_{k,i} u_{k,j}\, dV - \int_{S^T} t_i \dot{u}_i\, dS = \dot{\mathscr{D}}, \tag{2.12}$$

$$\mathscr{K} = \frac{1}{2}\int_V [\gamma_{ijkl} u_{i,j} u_{k,l} + \varrho \dot{u}_i \dot{u}_i + \varrho K \tau^2]\, dV, \tag{2.13}$$

where $t_i = T_i^* - T_i$ is the change in the surface tractions per unit area in C due to deformation from C to C_t^*. When *dead loads* are considered, t_i is identically zero [1]. But in general it is not. The explicit form of t_i depends on the manner in which surface loads follow the deformation; we do not exclude the possibility that t_i may depend on the velocity or the velocity gradients (velocity-dependent loads). In [1], a class of deformation-dependent loads is examined in the context of small deformations superimposed on C (see SEWELL [5] for a comprehensive account). As is shown in [1, 5], *deformation-dependent loads are not necessarily nonconservative*. We assume that these loads can be separated into a conservative and a nonconservative part; this is not, however, always uniquely possible, although a class of such loads is studied in detail in [1]. Due to the lack of space, we shall demonstrate our results assuming that

$$t_i = \mu T_{jk} u_{i,j} \nu_k, \tag{2.14}$$

where $0 \le \mu \le 1$; $\mu = 0$ corresponds to dead loading. It should, however, be stressed that our results are valid for more complicated deformation- and velocity-dependent, or any general loading conditions. Substitution from (2.14) into (2.12) yields

$$\dot{\mathscr{K}} + \frac{1-\mu}{2}\frac{d}{dt}\int_V T_{ij}u_{k,i}u_{k,j}\,dV = \dot{\mathscr{D}} + \mu\dot{\mathscr{W}}^N = \mathscr{G}(t), \qquad (2.15)$$

$$\dot{\mathscr{W}}^N = \int_V T_{ki}u_{j,ik}\dot{u}_j\,dV. \qquad (2.16)$$

3. Thermoelastic stability

We regard C locally stable if $\forall\varepsilon > 0$, $\exists\delta(\varepsilon) > 0 \ni \mathscr{K}(t) < \varepsilon$ for $t \geq 0$ whenever $\mathscr{K}(0) < \delta$. C is called *locally asymptotically* stable if it is stable, and also $\mathscr{K}(t) \to 0$ as $t \to \infty$. $\mathscr{K}$ is an appropriate *squared energy norm*[1]. Norms of this kind are commonly used in Applied Mathematics [6, 7]. Although it may sometimes be useful to employ other norms, such as those discussed by KNOPS and WILKES [8] and others[2] [9], I see no compelling reason for abandoning the energy norm which appears to be an adequate measure for all practical purposes, and which fits quite nicely with energy type Liapunov functionals. Note that, since γ_{ijkl} is a continuous tensor-valued function with finite positive values everywhere inside and on the closed, bounded region V, using Korn's inequality we may write $\int_V \gamma_{ijkl}u_{i,j}u_{k,l}\,dV \geq K_1 \int_V u_{i,j}u_{i,j}\,dV$, where K_1 is a constant, independent of u_i, with the dimension of force divided by squared length. Hence, stability with respect to $\mathscr{K}$ implies also stability with respect to $\mathscr{R} = \int_V [K_1 u_{i,j}u_{i,j} + \varrho\dot{u}_i\dot{u}_i + \varrho K\tau^2]\,dV$ which is *equivalent* to $\mathscr{K}$.

We denote by (Φ, Θ) the set of all sufficiently smooth functions (u_i, τ) defined in V for $t \geq 0$, and satisfying all the geometrical and thermal conditions on ∂B. We let $S(C, r)$ denote the r-neighborhood of C, consisting of all $(u_i, \tau) \in (\Phi, \Theta) \ni \mathscr{K} < r$. Then by theorem 12, p. 41 of ZUBOV's book [10], we conclude that C is locally stable if $\exists$ a functional $\mathscr{V} \ni$: (a) $\forall c_1 > 0$, $\exists c_2 > 0 \ni \mathscr{V}(u_i, \tau) > c_2$ for $(u_i, \tau) \in S(C, r)$ and[3] $\mathscr{K}(u_i, \tau) > c_1$; (b) $\forall\gamma_2 > 0$, $\exists\gamma_1 > 0 \ni$ if $\mathscr{K}(u_i, \tau) < \gamma_1$, then $\mathscr{V}(u_i, \tau) \leq \gamma_2$; and (c) $\dot{\mathscr{V}}(u_i, \tau) \leq 0$ for all $t \geq 0$.

Consider now a purely nonconservative loading, set $\mu = 1$ in (2.15), and obtain $\dot{\mathscr{K}} = \dot{\mathscr{D}} + \dot{\mathscr{W}}^N = \mathscr{G}(t)$. With $\mathscr{V} = \mathscr{K}$, conditions (a)

[1] One may use $\hat{\varrho} = [\mathscr{K}/V]^{1/2}$ or its equivalents instead of $\mathscr{K}$, but this would not affect the analysis, since V is finite.

[2] A discussion of the usefulness of various norms can be found in [9]. However, I regretfully note that there are several errors in [9]; for example, the word "sup" is missing in the definition of ϱ_1 and ϱ_2, and the word "positive-definite" is omitted in the stability theorem stated at the end of p. 19; also, most of $t > 0$ should be changed to $t \geq 0$.

[3] Condition (a) states that $\mathscr{V}$ is positive-definite with respect to $\mathscr{K}$.

and (b) are trivially satisfied, and, hence, for stability it is sufficient that

$$\mathscr{G}(t) = \dot{\mathscr{D}} + \dot{\mathscr{W}}^N \leq 0, \tag{3.1}$$

which states that *the rate at which energy is being lost because of heat conduction and internal dissipation should exceed the rate at which nonconservative loads supply energy to the solid. Since the rate of energy loss* $\dot{\mathscr{D}}$ *relates directly to the rate of entropy production* (Eqs. (2.5), (2.6) and (2.9)), *we observe that the behavior of a solid subjected to nonconservative external loads must be highly affected by the manner in which entropy is produced in the body.* To my knowledge, this fact has been first stressed in [11]. We note that (3.1) is too strong a condition for stability in most practical cases. But I know of no better results, heuristic and engineering approaches [12—14] being excepted. In this regard, it may be useful to point out that a necessary condition for $\mathscr{K}$ to remain bounded and possibly approach zero is that $\int_{t_0}^{t_0+T} \mathscr{G}(t)\, dt \leq 0$ for all t_0 finite and T sufficiently large. A condition of this kind has been used in [1, 13, 14] to develop certain approximate stability analyses.

To generalize the above results to cases where both conservative and nonconservative loads act on the solid, let $0 < \mu < 1$ in (2.15), and consider the functional

$$\mathscr{V} = \frac{1}{2} \int_V [c_{ijkl} u_{i,j} u_{k,l} + \varrho \dot{u}_i \dot{u}_i + \varrho K \tau^2]\, dV; \; c_{ijkl} = \gamma_{ijkl} + (1-\mu)\, T_{jl} \delta_{ik}. \tag{3.2}$$

We require that $\mathscr{V}$ be positive-definite[1] with respect to $\mathscr{K}$ in the sense defined by MIKHLIN [7] so that $\forall (u_i, \tau) \in S(C, r) \,\exists$ a constant $\Gamma(V) \ni \mathscr{V} \geq \Gamma \mathscr{K}$ for all $t \geq 0$. Moreover, we have $|\mathscr{V}| \leq \mathscr{K} + \frac{1-\mu}{2} \int_V |T_{ij} u_{k,i} u_{k,j}| \leq \mathscr{K} + (1-\mu)\, T^\circ \int_V \left[\frac{1}{2} u_{i,j} u_{i,j} + |u_{1,1} u_{1,2}| + |u_{1,1} u_{1,3}| + \cdots + |u_{3,2} u_{3,3}|\right] dV \leq \mathscr{K} + \frac{73}{2} (1-\mu)\, T^\circ \int_V u_{i,j} u_{i,j}\, dV \leq [1 + 73(1-\mu)\, T^\circ/2K_1]\, \mathscr{K}$, where $T^\circ = \max |T_{ij}|$, K_1 is a constant entering in Korn's inequality, and the Schwartz inequality is also used. Therefore, if $\mathscr{V}$ is positive-definite with respect to $\mathscr{K}$ (or its equivalent $\mathscr{R}$), conditions (a) and (b) of Zubov's theorem would be satisfied. Note that, since the last two terms in the integrands of $\mathscr{K}$ and $\mathscr{V}$ are identical, $\mathscr{V}$ is posi-

[1] For $\mu = 0$, c_{ijkl} reduces to that used by HILL [15], and, unless $T_{jl} \equiv 0$, it may not be taken as the modulus of anisotropic linear elasticity as is stated in [8].

tive-definite with respect to $\mathcal{K}$ if $\mathcal{U} = \frac{1}{2} \int_V c_{ijkl} u_{i,j} u_{k,l} \, dV$ is so (namely $\mathcal{U} \geq \Gamma_1 \mathcal{S}$) with respect to $\mathcal{S} = \frac{1}{2} \int_V \gamma_{ijkl} u_{i,j} u_{k,l} \, dV$ for all virtual or actual admissible displacement fields for which $\mathcal{S} < r$. It now remains to satisfy condition (c).

When the applied load is conservative, condition (c) reduces to $\dot{\mathcal{V}} = \dot{\mathcal{D}} \leq 0$ which is fulfilled by virtue of the entropy inequality (2.9). Thus, for conservative loading, we arrive at the usual Pearson-Hill energy criterion [15, 16] *which guarantees stability with respect to an energy norm*[1]. In a general loading case, on the other hand, we must require more. Hence, for the present case with $0 < \mu < 1$, we may require that

$$\dot{\mathcal{D}} + \mu \dot{\mathcal{W}}^N = \mathcal{G}(t) \leq 0 , \tag{3.3}$$

which together with

$$\mathcal{U} \geq \Gamma_1 \mathcal{S} \tag{3.4}$$

is sufficient to ensure stability of C with respect to $\mathcal{K}$. Note that when $\mu = 1$ (purely nonconservative loading), (3.4) is trivially satisfied and (3.3) becomes a sufficient condition for stability, whereas when $\mu = 0$ (conservative loading), (3.3) is trivially satisfied and (3.4) must be required to ensure stability. The latter condition is of course static in character. In this case no additional condition is required if stability with respect to an energy norm is being sought. In particular, we do not include the so-called *moment condition* even if $\partial B^T = \partial B$ (see TRUESDELL and NOLL [17] for references), since it (incorrectly) regards stable an unrestrained bar under small compressive end loads that remain parallel to their initial direction. The fact that such a system is indeed (rotationally) unstable for any non-zero loads sheds doubts on the merits (if any) of this condition, see HILL [18], footnote p. 376.

References

1. NEMAT-NASSER, S.: Quart. Appl. Math. **26**, 119 (1968).
2. ERICKSEN, J. L.: Int. J. Solids Structures **2**, 573 (1966).
3. KOITER, W. T.: Rept. 360, Lab. Engng. Mech., Tech. Univ. Delft, April 1967.
4. KOITER, W. T.: Proc. Kon. Ned. Ak. Wet. **B 68**, 178, 190 (1965).
5. SEWELL, M. J.: Arch. Rat. Mech. Anal. **23**, 327 (1967).
6. MIKHLIN, S. G.: Variational Methods in Mathematical Physics, Translation, New York: Pergamon Press 1964.
7. MIKHLIN, S. G.: The Problem of the Minimum of a Quadratic Functional, Translation, London: Holden-Day 1965.

---

[1] In this regard the assertion made in [8] that Pearson-Hill criterion is not sufficient because it does not yield stability with respect to a pointwise metric, $\sup |u_{i,j}| + \sup |\dot{u}_i|$, appears misleading; the criterion is certainly sufficient with respect to $\mathcal{K}$ (for $\tau = 0$) or its equivalents.

8. KNOPS, R. J., WILKES, E. W.: Int. J. Engng. Sci. **4**, 303 (1966).
9. NEMAT-NASSER, S., HERRMANN, G.: Ing.-Arch. **35**, 17 (1966).
10. ZUBOV, V. I.: Methods of A. M. Lyapunov and Their Application, Translation, Groningen: P. Noordhoff 1964.
11. NEMAT-NASSER, S., ROORDA, J.: Acta Mechanica **15**, 296 (1967).
12. BOLOTIN, V. V.: Nonconservative Problems of the Theory of Elastic Stability, Translation, New York: Pergamon Press 1963.
13. NEMAT-NASSER, S.: J. Appl. Mech. **34**, 344 (1967).
14. SHIEH, R. C., MASUR, E. F.: J. Appl. Math. Phys. (ZAMP) **19**, 927 (1968).
15. HILL, R.: J. Mech. Phys. Solids **14**, 229 (1957).
16. PEARSON, C. E.: Quart. Appl. Math. **14**, 133 (1956).
17. TRUESDELL, C., NOLL, W.: The Non-linear Field Theories of Mechanics, Berlin/Heidelberg/New York: Springer 1965.
18. HILL, R.: J. Mech. Phys. Solids **15**, 371 (1967).

Examples on the Stabilizing and Destabilizing Effects

By

J. Barta

Budapest, Hungary

Two examples will be presented here concerning the equilibrium stability of elastic structures. The loading force is non-conservative in the first example, and conservative in the second one. Both examples deal with plane problems.

First example

The structure (Figs. 1 and 2) consists of two mass elements m_1 and m_2, two hinges Γ_1 and Γ_2, two massless bars and a fixed base. The spring constants of the hinges will be denoted also by Γ_1 and Γ_2. The length of each bar is l. The loading force P is acting on the free end of the upper bar, always in the direction of the upper bar. Fig. 1 shows the equilibrium position. In order to determine the critical value P_k, we consider the small oscillations of the system. Let the motion be characterized by the deflections $u_1(t)$ and $u_2(t)$, (Fig. 2). Then the rotating moments of the hinges Γ_1 and Γ_2 are $\Gamma_1 u_1/l$ and $\Gamma_2(u_2 - 2u_1)/l$. Consequently, the differential equations of motion are

$$-(u_2 - 2u_1)\,P - m_1\ddot{u}_1 l - m_2\ddot{u}_2\,2l = \Gamma_1 u_1/l,$$
$$-m_2\ddot{u}_2 l = \Gamma_2(u_2 - 2u_1)/l.$$

Substituting here $u_1 = A_1 e^{\lambda t}$ and $u_2 = A_2 e^{\lambda t}$, where A_1 and A_2 are constants, we obtain the equations

$$-(A_2 - 2A_1)\,P - m_1 A_1 l\lambda^2 - m_2 A_2\,2l\lambda^2 = \Gamma_1 A_1/l,$$
$$-m_2 A_2 l\lambda^2 = \Gamma_2(A_2 - 2A_1)/l.$$

Hence, the frequency equation is

$$\begin{vmatrix} 2P - m_1 l\lambda^2 - \Gamma_1/l & -P - 2m_2 l\lambda^2 \\ 2\Gamma_2/l & -m_2 l\lambda^2 - \Gamma_2/l \end{vmatrix} = 0. \qquad (1)$$

This is an algebraical equation of the fourth degree. From $u_1 = A_1 e^{\lambda t}$ and $u_2 = A_2 e^{\lambda t}$ follows that the equilibrium position shown in Fig. 1 is stable if and only if none of the roots $\lambda_1, \lambda_2, \lambda_3, \lambda_4$ of Eq. (1) has a positive real part.

Fig. 1. Equilibrium position.

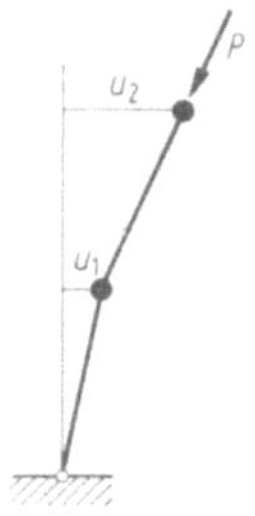

Fig. 2. Deflection during the oscillation.

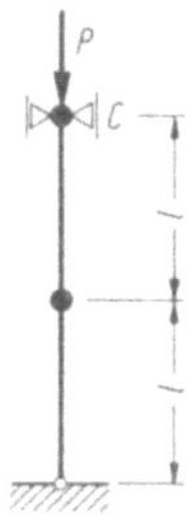

Fig. 3. Equilibrium position.

Fig. 4. Deflection during the oscillation.

In the numerical case

$$\Gamma_1 = 1,\ \Gamma_2 = 1,\ m_1 = 1,\ m_2 = 1,\ l = 1, \tag{2}$$

Eq. (1) becomes $\lambda^4 + (6 - 2P)\,\lambda^2 + 1 = 0$. Hence,

$$\lambda^2 = P - 3 \pm \sqrt{(P-3)^2 - 1}.$$

This leads to the condition

$$P - 3 \leqq 0, \quad (P-3)^2 - 1 \geqq 0,$$

that is $P \leqq 2$. Thus we have

$$P_k = 2. \tag{3}$$

In the numerical case

$$\Gamma_1 = 2, \quad \Gamma_2 = 4, \quad m_1 = 8, \quad m_2 = 1, \quad l = 1, \tag{4}$$

we proceed similarly. Eq. (1) becomes $4\lambda^4 + (25 - P)\,\lambda^2 + 4 = 0$. Hence

$$\lambda^2 = \frac{1}{8}\left\{P - 25 \pm \sqrt{(P-25)^2 - 64}\right\}.$$

This leads to the condition

$$P - 25 \leqq 0, \quad (P-25)^2 - 64 \geqq 0,$$

that is $P \leqq 17$. Thus we have

$$P_k = 17. \tag{5}$$

Let us now apply the constraint C at the upper end of the structure (Figs. 3 and 4). Then we obtain in the very same manner as before, in the numerical case (2),

$$P_k = 2.5, \tag{6}$$

and in the numerical case (4),

$$P_k = 9. \tag{7}$$

By comparing (3) with (6), and (5) with (7), we see that the effect of the constraint C is stabilizing in the numerical case (2), and destabilizing in the numerical case (4). Thus we can draw the conclusion that the effect of constraint applied against displacement is not always stabilizing but sometimes destabilizing.

Second example

By means of ideal hinges, a vertical slender bar is joined to the ends of two cantilever beams (Fig. 5). P_k is the slightest value of the loading force P which may produce a buckling indicated by dotted line in Fig. 5.

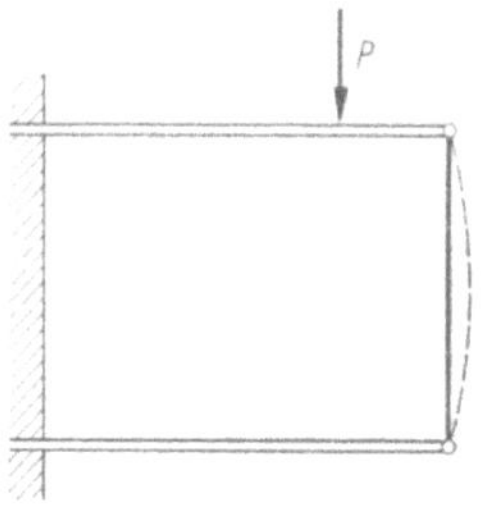

Fig. 5. The loading force P produces a buckling of the vertical bar.

If the upper cantilever beam will be stiffened, then P_k becomes greater. If the lower cantilever beam will be stiffened, then P_k becomes smaller. Thus we can draw the conclusion that the effect of stiffening is not always stabilizing but sometimes destabilizing.

Stability Conditions of Rigid-plastic Solids with Extension to Visco-plasticity

By

B. Storåkers

Royal Institute of Technology, Stockholm, Sweden

Summary. Drucker's postulate for stability in the small of a rigid-plastic solid is extended to take geometry effects into consideration. The result is compared with Hill's condition for stability and the two approaches are found compatible. Application of the postulate to a certain class of visco-plastic material is suggested.

Introduction

In two recent papers [1, 2] a condition for stability of equilibrium has been established for a rigid-plastic circular tube under internal pressure, torsion and axial tension. The final results for a von Mises solid are identical although the approaches to the problem are different. This event calls for a comparative study of the two methods of approach.

The first treatment is based on papers by Hill [3, 4] who established a sufficient condition for the stability of a rigid-plastic solid by requiring that if a structure under external dead loading and rigid constraints is displaced from its static equilibrium position by any kind of virtual disturbance, the internal dissipated energy should exceed the work done by the external loads during the virtual motion. Geometrical changes of the structure caused by the disturbance are considered and it is shown that the requirement of a certain functional to be positive for any virtual mode of deformation yields a sufficient condition for stability.

The second treatment is based on Drucker's postulate for stability in the small. Drucker [5, 6] considers a structure in stable equilibrium under external loading and postulates that for any external perturbing agency which is added to the system and then removed, the work done by the agency on the displacements it produces should be non-negative for an elastic-plastic and positive for a rigid-plastic structure. If any conceivable infinitesimal agency fulfils this requirement, the structure is then said to be stable in the small. DRUCKER has, however, in his treatises excluded the event of geometric changes during deformation and his postulate is mostly concerned with material stability and has found applications in establishing restrictions for the shapes of yield surfaces and flow rules of materials.

The postulate may however be applied also when geometric effects play an essential role in a stability investigation which will be shown below. It must only be remembered that the kind of perturbing agency must be more clearly specified.

Analysis

In mathematical form the postulate for stability in the small of a rigid-plastic structure reads

$$dP_k\, du_k + \int_A dT_k\, du_k\, dA + \int_V dF_k\, du_k\, dV > 0\,, \tag{1}$$

where dP_k, dT_k, dF_k is the perturbing agency, consisting of discrete forces, surface tractions and body forces respectively, causing displacements du_k. A is the surface area and V the volume of the structure in the undisturbed configuration.

Permissible agencies are according to DRUCKER such, which do not violate equilibrium, compatibility and stress-strain equations. Boundary conditions for displacements must also be fulfilled.

The changes in surface tractions and body forces must now be specified whether they are nominal or true changes. A nominal definition seems reasonable since the changes of true tractions and body forces are not a priori known when geometry changes are considered. With this definition, points of attack of discrete forces may be regarded as points of infinite surface traction. This contribution to the surface traction may however for physical reasons be considered finite and thus the first term in inequality (1) may be formally included in the second one.

With the postulate formulated in this way, and proceeding in the spirit of HILL, it may be shown compatible with Hill's resulting condition for stability.

For a body disturbed at time t by nominal surface tractions dT_k and body forces dF_k, the boundary conditions in a Cartesian reference frame are

$$\frac{dT_j}{dt} = l_i \frac{ds_{ij}}{dt} = l_i \dot{s}_{ij}\,, \tag{2}$$

where $\boldsymbol{l} = (l_1, l_2, l_3)$ is the normal (with $\boldsymbol{l} \cdot \boldsymbol{l} = 1$) to the body surface and $\dot{s}_{ij}$ denotes material rate of change of the Lagrangian stress tensor s_{ij} defined in the configuration at time t. At this instant s_{ij} equals the Cauchy (true) stress tensor σ_{ij}.

A general relation between the material rate of change of the Lagrangian stress tensor $\dot{s}_{ij}$ and that of the Cauchy stress tensor $\dot{\sigma}_{ij}$ has been given by PRAGER [7] in the form

$$\dot{s}_{ij} = \frac{\varrho_0}{\varrho} \frac{\partial X_i}{\partial x_p} (\dot{\sigma}_{pj} + v_{k,k}\sigma_{pj} - v_{p,k}\sigma_{jk}), \tag{3}$$

where ϱ_0 and ϱ are the specific masses in the initial and the instantaneous configurations respectively, X_i and x_i are Lagrangian and Eulerian coordinates respectively, $v_i(x_j)$ particle velocities and index i after a comma denotes partial differentiation with respect to x_i.

As the Lagrangian coordinate system was introduced in the configuration at time t, Eq. (3) in this case reduces to

$$\dot{s}_{ij} = \delta_{ip}(\dot{\sigma}_{pj} + v_{k,k}\sigma_{pj} - v_{p,k}\sigma_{jk}). \tag{4}$$

Introduction of the incompressibility condition $v_{k,k} = 0$ then yields

$$\dot{s}_{ij} = \dot{\sigma}_{ij} - v_{i,k}\sigma_{jk}. \tag{5}$$

The stability condition (1) may now be rewritten as

$$\int_A l_i \dot{s}_{ij} v_j \, dA + \int_V \dot{F}_j v_j \, dV > 0, \tag{6}$$

where the first term in inequality (1) has been included in the surface integral and v_j denotes du_j/dt.

Application of Gauss' theorem then yields

$$\int_V [(\dot{s}_{ij} v_j)_{,i} + \dot{F}_j v_j] \, dV > 0. \tag{7}$$

Carrying out differentiation and rearranging yields

$$\int_V [(\dot{s}_{ij,i} + \dot{F}_j) \, v_j + \dot{s}_{ij} v_{j,i}] \, dV > 0. \tag{8}$$

According to prescribed equilibrium however

$$\dot{s}_{ij,i} + \dot{F}_j = 0. \tag{9}$$

Introduction of Eqs. (5) and (9) into inequality (8) now yields

$$\int_V (\dot{\sigma}_{ij} v_{j,i} - \sigma_{jk} v_{i,k} v_{j,i}) \, dV > 0. \tag{10}$$

As $\dot{\sigma}_{ij}$ is a symmetric tensor

$$\dot{\sigma}_{ij} v_{j,i} = \frac{1}{2} \dot{\sigma}_{ij} (v_{j,i} + v_{i,j}) \tag{11}$$

or

$$\dot{\sigma}_{ij} v_{j,i} = \dot{\sigma}_{ij} \eta_{ij}, \tag{12}$$

where η_{ij} is the rate of deformation tensor.

It is now of interest to express the right member of Eq. (12) in terms of velocity gradients and material properties.

If, for the material in question, plastic and viscous strain increments are simply additive and the yield function $f(\sigma_{ij})$ and the plastic potential coincide, the constitutive equation may be written in a general form with time as an essential variable as

$$\eta_{ij} = \begin{cases} \left[g \cdot \frac{\partial f}{\partial \sigma_{kl}} \tilde{\sigma}_{kl} \Big/ \left(\frac{\partial f}{\partial \sigma_{mn}} \frac{\partial f}{\partial \sigma_{mn}} \right) + h \Big/ \left(\frac{\partial f}{\partial \sigma_{mn}} \frac{\partial f}{\partial \sigma_{mn}} \right)^{1/2} \right] \frac{\partial f}{\partial \sigma_{ij}}, \\ \qquad f(\sigma_{ij}) = 0 \text{ and } \frac{\partial f}{\partial \sigma_{kl}} \tilde{\sigma}_{kl} > 0, \\ h \cdot \frac{\partial f}{\partial \sigma_{ij}} \Big/ \left(\frac{\partial f}{\partial \sigma_{mn}} \frac{\partial f}{\partial \sigma_{mn}} \right)^{1/2}, \quad f(\sigma_{ij}) < 0 \text{ or } f(\sigma_{ij}) = 0 \text{ and } \frac{\partial f}{\partial \sigma_{kl}} \tilde{\sigma}_{kl} \leq 0. \end{cases} \tag{13}$$

This kind of constitutive equation is frequently employed in the solution of engineering creep problems.

Here g and h are scalar functions of stress and strain history describing the current rate of strain-hardening and viscous strain-rate respectively. For a von Mises solid, g equals 3/2 of the inverse of the rate of strain-hardening and h equals $\sqrt{3/2}$ of the effective viscous strain-rate. $\tilde{\sigma}_{ij}$ is the Jaumann derivative of stress which is an objective stress-rate related to $\dot{\sigma}_{ij}$ by

$$\tilde{\sigma}_{ij} = \dot{\sigma}_{ij} - \sigma_{ik}\omega_{jk} - \sigma_{jk}\omega_{ik}, \tag{14}$$

where ω_{pq} are the components of the skew-symmetric part of the tensor $v_{p,q}$.

The consequence of loss of stability involves in the treated case a spontaneous change of the structural configuration. This means that with the employed material model no viscous strain contribution could be expected during this motion and consequently the term $h(\partial f/\partial \sigma_{ij})/(\partial f/\partial \sigma_{mn} \cdot \partial f/\partial \sigma_{mn})^{1/2}$ should then be neglected in Eq. (13).

This proposed treatment of stability in the small for a structure of the considered special class of visco-plastic material means that, in the case of finite straining prior to time t, the influence upon the true stress state σ_{ij} at t due to the viscous part of the deformation prior to t should of course be considered although viscous contribution is discarded when the structure is perturbed.

Multiplying both members of Eq. (13) with $\tilde{\sigma}_{ij}$ with the mentioned restriction in mind yields

$$\tilde{\sigma}_{ij} \eta_{ij} = \left[g \cdot \frac{\partial f}{\partial \sigma_{kl}} \tilde{\sigma}_{kl} \Big/ \left(\frac{\partial f}{\partial \sigma_{mn}} \frac{\partial f}{\partial \sigma_{mn}} \right) \right] \frac{\partial f}{\partial \sigma_{ij}} \sigma_{ij}. \tag{15}$$

Making use of Eq. (13) once again then yields

$$\tilde{\sigma}_{ij}\eta_{ij} = \frac{1}{g}\,\eta_{ij}\eta_{ij}\,. \tag{16}$$

Introduction of Eqs. (12), (14) and (16) into inequality (10) now yields

$$\int\limits_V \left(\frac{1}{g}\,\eta_{ij}\eta_{ij} + \sigma_{ik}\omega_{jk}\eta_{ij} + \sigma_{jk}\omega_{ik}\eta_{ij} - \sigma_{jk}v_{i,k}v_{j,i}\right) dV > 0\,. \tag{17}$$

Interchange of dummy subscripts noting the symmetry of η_{ij} yields

$$\int\limits_V \left[\frac{1}{g}\,\eta_{ij}\eta_{ij} + \sigma_{jk}(2\eta_{ij}\omega_{ik} - v_{i,k}v_{j,i})\right] dV > 0\,. \tag{18}$$

The scalar $\sigma_{jk}\eta_{ij}\omega_{ik}$ is an invariant and it has been shown by Hill [8] to vanish for an isotropic solid (in which the principal directions of strain-rate and stress tensors coincide).

Thus for an isotropic solid the condition for stability under Drucker's assumptions is

$$\int\limits_V \left[\frac{1}{4g}\,(v_{j,i} + v_{i,j})\,(v_{j,i} + v_{i,j}) - \sigma_{jk}v_{i,k}v_{j,i}\right] dV > 0\,. \tag{19}$$

The functional appearing in the stability condition (19) is identically the same as the one in Hill's condition in which all virtual velocity fields v_k are admissible in the functional. The same interpretation may be given in the present derivation.

As the choice of perturbing body forces is arbitrary, it should in principle always be possible to construct a body force distribution so that every kinematically admissible motion may be executed in a quasi-static manner with Eq. (9) fulfilled.

Conclusions

It has been shown that extension of Drucker's postulate to cover also geometric effects when investigating the stability of a rigid-plastic structure leads to a final requirement of a certain functional to be positive for all virtual velocity fields and that this result is compatible with Hill's earlier condition for stability.

For simple problems (e.g. when homogeneous instability modes may be expected) the original form of Drucker's postulate is easy to apply but for complicated problems the derived functional condition is a more convenient basis of a stability analysis.

Inclusion of a viscous term in the constitutive equation offers a possibility to predict ductile creep rupture as this kind of failure may often be interpreted as creep-induced instability. An exemplifying treatment of this phenomenon for a von Mises visco-plastic solid under complex stress has earlier been given by Storåkers [9].

References

1. YAMADA, Y., AOKI, I.: Int. J. Mech. Sci. **8**, 665 (1966).
2. STORÅKERS, B.: Int. J. Mech. Sci. **10**, 519 (1968).
3. HILL, R.: J. Mech. Phys. Solids **5**, 153 (1957).
4. HILL, R.: J. Mech. Phys. Solids **6**, 1 (1957).
5. DRUCKER, D. C.: Proc. 1st Nat. Congr. Appl. Mech. ASME, 1951, p. 487.
6. MAIER, G., DRUCKER, D. C.: Schweizerische Bauzeitung **84**, 23 (1966).
7. PRAGER, W.: Introduction to Mechanics of Continua, New York: Ginn and Comp. 1961.
8. HILL, R.: J. Mech. Phys. Solids **5**, 302 (1957).
9. STORÅKERS, B.: J. Mécanique **6**, 449 (1967).

On the Dynamical Stability of Fluid Phases

By

B. D. Coleman

Carnegie-Mellon University, Pittsburgh, Pa., U.S.A.

1. Introduction

In his essay, "A Method of Graphical Representation of the Thermodynamic Properties of Substances by Means of Surfaces," published in 1873, GIBBS [1] proposed and studied a criterion for the stability of a fluid surrounded by a medium held at fixed temperature and pressure. Assuming that the specific internal energy ε of a fluid in thermodynamic equilibrium is given by a function $\bar{\varepsilon}$ of the specific entropy η and specific volume v of the fluid, GIBBS gave a heuristic argument to the effect that a uniform phase with specific entropy η° and specific volume v° is stable in an environment at temperature θ° and pressure p° if the inequality

$$\bar{\varepsilon}(\eta, v) - \theta^\circ \eta + p^\circ v > \bar{\varepsilon}(\eta^\circ, v^\circ) - \theta^\circ \eta^\circ + p^\circ v^\circ \tag{1.1}$$

holds for all pairs (η, v) not equal to (η°, v°).[1] Criteria of this type occur also in GIBBS' memoir "On the Equilibrium of Heterogeneous Substances" [2] and are there related to his concept of the stability of an isolated system.

[1] If (1.1) holds with $>$ replaced by $\geq$ and reduces to an equality for some pair $(\eta, v) \neq (\eta^\circ, v^\circ)$, then there are two or more stable uniform phases possible at the temperature θ° and pressure p°. A beautiful statical theory of coexistent phases was developed at length by GIBBS; in the present work, however, I have tried to avoid the formidable mathematical problems which arise in any attempt to formulate a dynamical theory of coexistent phases.

Acknowledgement. This research was supported in part by the U.S. Air Force Office of Scientific Research through Grant AFOSR-68-1538.

The present note is concerned with the dynamical significance of the criterion (1.1) for stability. The research I describe was done in an attempt to find an answer to the question: *If a uniform equilibrium state of a fluid body is stable according to the definitions of classical thermostatics, and if the body is, in some precise sense, in a "fixed environment", is it then true that every thermodynamic process in the body which passes near to the equilibrium state at one time must remain near to that state at all future times?*

I have attempted to answer this question for a broad class of materials called "regular fluids".[1] Each regular fluid has an equilibrium response function $\bar{\varepsilon}$ which for states of equilibrium gives ε as a function of η and v,

$$\varepsilon = \bar{\varepsilon}(\eta, v), \tag{1.2}$$

but which outside of equilibrium yields only a lower bound for ε:

$$\varepsilon \geq \bar{\varepsilon}(\eta, v). \tag{1.3}$$

For many materials, including simple fluids with fading memory, (1.3) is a consequence of the second law of thermodynamics.[2] The class of regular fluids includes not only fluids with memory, but also the perfect fluids and linearly viscous fluids of classical hydrodynamics.

The main results of this study are the theorems of Sec. 4. These theorems give a *dynamical* significance to the stability criterion (1.1) from thermostatics and, I believe, supply an answer to the question raised above. Assuming that the equilibrium response function $\bar{\varepsilon}$ for a regular fluid body $\mathscr{B}$ is convex outside a compact set, i.e. that the points in the domain D of $\bar{\varepsilon}$ which are not points of convexity for $\bar{\varepsilon}$ are contained in a compact subset of D, I show that if (η°, v°) is such that (1.1) holds for all $(\eta, v) \neq (\eta^\circ, v^\circ)$, then the uniform equilibrium state at (η°, v°) is dynamically stable in the following sense: If a process $\mathscr{C}$ of $\mathscr{B}$ is compatible with immersion of $\mathscr{B}$ in an environment at temperature θ° and pressure p° and is such that the fields over $\mathscr{B}$ describing the spatial distribution of internal energy, kinetic energy, specific volume, and entropy are close, in $\mathscr{L}_1$, to the corresponding uniform equilibrium fields at some time t, then in the process $\mathscr{C}$ these fields remain close, in $\mathscr{L}_1$, to the equilibrium fields at *all* times after t. This is the content of Theorem 4.2. This theorem and the closely related Theorem 4.1 follow, by easy arguments, from Remark 3.2, which, in turn, rests on a lemma from the theory of functions with points of convexity.

---

[1] Coleman and Greenberg [3].

[2] Cf. Coleman [4], Remark 22, p. 35.

The present Theorems 4.1 and 4.2 generalize and extend in two important respects the stability theorems which GREENBERG and I gave in Ref. [3]. First, in the earlier work we proved stability only against the class of processes compatible with isolation of the fluid body, and that class is smaller than the class of processes compatible with immersion in an environment at fixed temperature and pressure.[1] Second, to prove theorems about dynamical stability, GREENBERG and I assumed in Ref. [3] that $\bar{\varepsilon}$ is strictly convex throughout its domain; here I assume only that (1.1) holds at the point (η°, v°) under consideration and that $\bar{\varepsilon}$ is convex outside a compact set. In other words, instead of assuming that every uniform equilibrium state is thermostatically stable, the present theory takes into account the physical observation that for most substances there exists in D a relatively compact set S of pairs (η, v) for which the corresponding uniform phases are unstable or only metastable.

In an article to be published soon in the Archive for Rational Mechanics and Analysis, I discuss in detail the role played by the inequality (1.1) in GIBBS' classical theory of thermostatics and treat more fully the dynamical theorems presented here.

2. Thermodynamic processes

Here a *body* $\mathscr{B}$ is a smooth manifold, of material points X, imbedded in a fixed Euclidean space $\mathscr{E}$ and possessing an intrinsic mass-measure m. A *thermodynamic process* of $\mathscr{B}$ is a collection $\mathscr{C}$ of functions, of X and the time t, compatible with the laws of balance of momentum and energy. At the level of generality which I seek, each thermodynamic process consists of six functions: (1) the *motion* $\boldsymbol{\chi}$, with $\boldsymbol{x} = \boldsymbol{\chi}(X, t)$ called the *position* at time t of the material point X; (2) the *specific internal energy* ε; (3) the *specific entropy* η; (4) the *temperature* θ; (5) the *heat flux vector* $\boldsymbol{q}$; and (6) the symmetric *Cauchy stress tensor* $\boldsymbol{T}$.[2] The laws of *balance of momentum* and *balance of energy* take the forms

$$\frac{d}{dt}\int_{\mathscr{P}} \dot{\boldsymbol{x}}\, dm = \int_{\partial\mathscr{P}} \boldsymbol{T}\boldsymbol{n}\, da, \tag{2.1}$$

$$\frac{d}{dt}\int_{\mathscr{P}} \left(\varepsilon + \frac{1}{2}\dot{x}^2\right) dm = \int_{\partial\mathscr{P}} (\dot{\boldsymbol{x}} \cdot \boldsymbol{T}\boldsymbol{n} - \boldsymbol{q} \cdot \boldsymbol{n})\, da, \tag{2.2}$$

[1] This may not be obvious until the concept of "immersion" is rendered mathematical. See the discussion following Definition 2.1.

[2] It is here assumed that gravitational body forces are absent and that there is no heat transfer by radiation.

and must hold at all times t and in all parts $\mathscr{P}$ of $\mathscr{B}$;[1] $\boldsymbol{n}$ is the exterior unit normal vector to the surface $\partial\mathscr{P}$ of $\mathscr{P}$ in the configuration at time t, da is the element of surface area in this configuration, $\dot{x} = (\dot{\boldsymbol{x}} \cdot \dot{\boldsymbol{x}})^{1/2}$, and the superposed dot denotes the material time-derivative.

Many types of processes are imaginable. The theorems of Sec. 4 concern processes possible in fluid bodies which from some time, say $t = 0$, onward are either isolated or immersed in an environment held at fixed temperature and pressure.

Definition 2.1. *If a process $\mathscr{C}$ of a body $\mathscr{B}$ is such that for some positive pair (θ°, p°),*

$$(\theta - \theta^\circ)\, \boldsymbol{q} \cdot \boldsymbol{n} \geq 0 \qquad \text{for all } X \in \partial\mathscr{B} \text{ and all } t \geq 0 \tag{2.3}$$

and

$$\int_{\partial\mathscr{B}} \dot{\boldsymbol{x}} \cdot \boldsymbol{T}\boldsymbol{n}\, da = -Vp^\circ \left(\text{with } V = \int_{\mathscr{B}} v\, dm\right) \qquad \text{for all } t \geq 0, \tag{2.4}$$

then $\mathscr{C}$ is said to be ***compatible with immersion of $\mathscr{B}$ in an environment at temperature*** θ° ***and pressure*** p°, *from time zero onward.*

The condition (2.3) asserts that if a material point on the surface of $\mathscr{B}$ has a temperature $\left\{\begin{matrix}\text{greater}\\ \text{less}\end{matrix}\right\}$ than θ°, then heat does not flow $\left\{\begin{matrix}\text{into}\\ \text{out of}\end{matrix}\right\}$ $\mathscr{B}$ at that point. This condition is met, trivially, in the extreme cases in which (1) $\mathscr{B}$ is thermally isolated, ($\boldsymbol{q} \cdot \boldsymbol{n} \equiv 0$), and (2) the surface of $\mathscr{B}$ is held at the constant temperature θ°.[2] If $\mathscr{B}$ is in an environment which is held at the constant and uniform temperature θ°, then (2.3) is expected to be a consequence of the second law of thermodynamics, under reasonable constitutive assumptions for the transfer of heat across the surface of $\mathscr{B}$. For example, if one assumes that for each point on the surface of $\mathscr{B}$,

$$\boldsymbol{q} = (\theta - \theta^\circ)\, k\boldsymbol{n}, \quad k = k(\theta, v),$$

then the second law requires that k be not negative, and (2.3) is obeyed.

The condition (2.4) asserts that Vp° is a potential for the total work done by the contact forces acting on $\mathscr{B}$. This condition is met, for example, when $\mathscr{B}$ is mechanically isolated, i.e. when $\dot{V} = 0$ and $\dot{\boldsymbol{x}} \cdot \boldsymbol{T}\boldsymbol{n} = 0$ on the surface of $\mathscr{B}$; a special case of a process compatible with mechanical isolation is one in which $\mathscr{B}$ fills a rigid container to which it adheres, so that the velocity $\dot{\boldsymbol{x}}$ is zero at the bounding surface of $\mathscr{B}$. Processes which are not compatible with isolation, but yet obey (2.4), occur in a body confined in a piston chamber, provided that the force F exerted by the piston on $\mathscr{B}$ is positive and constant in time; then $p^\circ = F/A$, with A the cross-sectional area of the piston.

The *Clausius-Duhem inequality* asserts that, at each instant of time and for each part $\mathscr{P}$ of $\mathscr{B}$,

$$\frac{d}{dt} \int_{\mathscr{P}} \eta\, dm \geq -\int_{\partial\mathscr{P}} \frac{1}{\theta}\, \boldsymbol{q} \cdot \boldsymbol{n}\, da. \tag{2.5}$$

---

[1] A part $\mathscr{P}$ of $\mathscr{B}$ is a subset of the closure of $\mathscr{B}$ with certain properties of regularity which I need not list here. Of course, $\mathscr{B}$ belongs to the set of its parts.

[2] Cf. KOITER [5].

Given two positive numbers θ°, p°, and a process $\mathscr{C}$ of $\mathscr{B}$, one can define, at each time t, a number $\Phi(t)$ by

$$\Phi(t) = \int_{\mathscr{B}} \left[\varepsilon(X,t) - \theta^\circ \eta(X,t) + p^\circ v(X,t) + \frac{1}{2}\dot{x}^2(X,t)\right] dm. \quad (2.6)$$

I call $\Phi(t)$ the *canonical free-energy of* $\mathscr{B}$*, at time* t*, under an environment at temperature* θ° *and pressure* p°.

Proposition 2.1.[1] *It follows from the balance law* (2.2) *and the Clausius-Duhem inequality* (2.5) *that if* $\mathscr{C}$ *is a process of* $\mathscr{B}$ *compatible with immersion of* $\mathscr{B}$ *in an environment at temperature* θ° *and pressure* p° *(from time zero onward), and if* Φ *is the canonical free-energy of* $\mathscr{B}$ *under such an environment, then in the process* $\mathscr{C}$

$$\frac{d}{dt}\Phi(t) \leq 0,$$

for each $t \geq 0$.

Proof. When (2.4) holds, the law of balance of energy, (2.2), reduces to

$$\frac{d}{dt}\int_{\mathscr{B}} \left(\varepsilon + \frac{1}{2}\dot{x}^2 + p^\circ v\right) dm = -\int_{\partial\mathscr{B}} \boldsymbol{q}\cdot\boldsymbol{n}\, da. \quad (2.7)$$

Since θ and θ° are positive, (2.3) yields

$$-\int_{\partial\mathscr{B}} \boldsymbol{q}\cdot\boldsymbol{n}\, da \leq -\theta^\circ \int_{\partial\mathscr{B}} \frac{1}{\theta}\boldsymbol{q}\cdot\boldsymbol{n}\, da, \quad (2.8)$$

and the Clausius-Duhem inequality, (2.5), can be written

$$-\theta^\circ \frac{d}{dt}\int_{\mathscr{B}} \eta\, dm \leq \theta^\circ \int_{\partial\mathscr{B}} \frac{1}{\theta}\boldsymbol{q}\cdot\boldsymbol{n}\, da. \quad (2.9)$$

Adding (2.7), (2.8), and (2.9), we obtain

$$\frac{d}{dt}\int_{\mathscr{B}} \left(\varepsilon + \frac{1}{2}\dot{x}^2 + p^\circ v - \theta^\circ\eta\right) dm \leq 0;$$

q.e.d.

In continuum physics each *material* is characterized by a list of constitutive assumptions which are interpreted as restrictions on the class of processes that can occur in a body $\mathscr{B}$ comprised of the material. The thermodynamic processes which obey the constitutive assumptions are called *admissible processes* of $\mathscr{B}$. The concept of a "uniform static state", which plays a central role in classical thermostatics, can be given a meaning in thermodynamics through the following definition.

---

[1] Propositions related to 2.1 are given by Ericksen [6], Koiter [5], and Coleman and Dill [7].

Definition 2.2. *A* ***uniform static state*** *of a body* $\mathscr{B}$ *is an admissible process of* $\mathscr{B}$ *for which*

$$\eta(X,t)=\eta^\circ=\text{const.},\quad v(X,t)=v^\circ=\text{const.},\quad\text{and}\quad \dot{\boldsymbol{x}}(X,t)=\mathbf{0}, \tag{2.10}$$

for all X *in* $\mathscr{B}$ *and all* t *in* $(-\infty,\infty)$.

The apparatus required to specify with precision the class of materials called "regular fluids" has now been assembled.

Definition 2.3.[1] *A body* $\mathscr{B}$ *is comprised of a* ***regular fluid*** *if there exists a differentiable function* $\bar{\varepsilon}$ *mapping* $(0,\infty)\times(0,\infty)$ *into* $(0,\infty)$ *such that*

(i) *for each pair* (η°, v°) *of positive numbers, there is a unique uniform static state of* $\mathscr{B}$ *with* $\eta=\eta^\circ$ *and* $v=v^\circ$;[2] *in this state (regarded as a process), for all* X *and* t,

$$\left.\begin{aligned}
&\varepsilon(X,t)=\varepsilon^\circ \overset{\text{def}}{=} \bar{\varepsilon}(\eta^\circ,v^\circ),\\
&\theta(X,t)=\theta^\circ \overset{\text{def}}{=} \bar{\theta}(\eta^\circ,v^\circ),\quad \textit{with}\quad \bar{\theta}=\partial_\eta\bar{\varepsilon},\\
&\boldsymbol{T}(X,t)=-p^\circ\mathbf{1},\quad \textit{with}\quad p^\circ \overset{\text{def}}{=} \bar{p}(\eta^\circ,v^\circ),\quad \bar{p}=-\partial_v\bar{\varepsilon},\\
&\boldsymbol{q}(X,t)=\mathbf{0};
\end{aligned}\right\} \tag{2.11}$$

(ii) *in* ***every*** *admissible thermodynamic process of* $\mathscr{B}$

$$\varepsilon(X,t)\geq\bar{\varepsilon}\big(\eta(X,t),v(X,t)\big), \tag{2.12}$$

for all X *and* t.

The function $\bar{\varepsilon}$ *in* (2.11) *and* (2.12) *is called the* ***equilibrium energy function*** *for the regular fluid.*

3. Preliminary observations

The ideas introduced in the previous section are brought together in the following remark, which shows that for a regular fluid body $\mathscr{B}$ the inequality (1.1) implies that the canonical free energy under an environment at temperature θ° and pressure p° has a strict minimum at the uniform static state which gives $\mathscr{B}$ the temperature θ° and the pressure p°.

---

[1] This definition is essentially the same as that employed by COLEMAN and GREENBERG [3].

[2] *Unique* here means the following: If $\boldsymbol{\chi}$, ε, η, θ, $\boldsymbol{q}$, and $\boldsymbol{T}$ are the six functions of X and t describing an admissible process of $\mathscr{B}$ with $\eta(X,t)\equiv\eta^\circ$, $v(X,t)\equiv v^\circ$, and $\partial_t\boldsymbol{\chi}(X,t)=\mathbf{0}$, then ε, η, $\boldsymbol{q}$, and $\boldsymbol{T}$ are completely determined once η° and v° are given, and $\boldsymbol{\chi}$ is determined to within a constant unimodular transformation, i.e. to within a transformation of the form $\boldsymbol{\chi}(X,t)\to\boldsymbol{U}\boldsymbol{\chi}(X,t)+\boldsymbol{c}$, with $\boldsymbol{c}$ a vector and $\boldsymbol{U}$ a tensor obeying $|\det\boldsymbol{U}|=1$.

Remark 3.1. *Let $\mathscr{B}$ be a regular fluid body, let (η°, v°) be a pair of positive numbers, put*

$$\left.\begin{aligned} \varepsilon^\circ = \bar{\varepsilon}(\eta^\circ, v^\circ), \quad \theta^\circ = \bar{\theta}(\eta^\circ, v^\circ), \quad p^\circ = \bar{p}(\eta^\circ, v^\circ), \\ \varphi^\circ = \varepsilon^\circ - \theta^\circ\eta^\circ + p^\circ v^\circ, \\ \Phi^\circ = \int_{\mathscr{B}} \varphi^\circ \, dm = M\varphi^\circ, \quad M = \int_{\mathscr{B}} dm, \end{aligned}\right\} \tag{3.1}$$

and suppose that the inequality (1.1) *holds for every pair (η, v) not equal to (η°, v°). If $\varepsilon(X,t)$, $\eta(X,t)$, $v(X,t)$, and $\dot{\boldsymbol{x}}(X,t)$ are the values assumed by ε, η, v, and $\dot{\boldsymbol{x}}$ in an arbitrary admissible process $\mathscr{C}$ of $\mathscr{B}$, and if*

$$\zeta(X,t) = \varepsilon(X,t) - \theta^\circ\eta(X,t) + p^\circ v(X,t), \tag{3.2}$$

then

$$\zeta(X,t) \geq \varphi^\circ \tag{3.3}$$

for all X and all t, and consequently, at each time t, for the process $\mathscr{C}$,

$$\Phi(t) \geq \Phi^\circ, \tag{3.4}$$

where

$$\begin{aligned} \Phi(t) &\stackrel{\text{def}}{=} \int_{\mathscr{B}} \left[\varepsilon(X,t) - \theta^\circ\eta(X,t) + p^\circ v(X,t) + \frac{1}{2}\dot{x}^2(X,t)\right] dm \\ &= \int_{\mathscr{B}} \left[\zeta(X,t) + \frac{1}{2}\dot{x}^2(X,t)\right] dm. \end{aligned} \tag{3.5}$$

Moreover, equality holds in (3.3) *only when*

$$\eta(X,t) = \eta^\circ, \quad v(X,t) = v^\circ, \quad \text{and} \quad \varepsilon(X,t) = \varepsilon^\circ,$$

and hence equality holds in (3.4) *only when this condition and the condition $\dot{\boldsymbol{x}}(X,t) = \boldsymbol{0}$ hold almost everywhere in $\mathscr{B}$.*

Of course, in Remark 3.1 $\Phi(t)$ is the canonical free energy of $\mathscr{B}$ in an environment at temperature θ° and pressure p°. The number Φ° is the canonical free energy of $\mathscr{B}$, under the same environment, in the uniform static state (2.10).

Proof of Remark 3.1. The only thing that requires proof here is the relation (3.3). Let $\bar{\varphi}$ be the function on $(0,\infty)\times(0,\infty)$ defined by

$$\bar{\varphi}(\eta, v) = \bar{\varepsilon}(\eta, v) - \theta^\circ\eta + p^\circ v \tag{3.6}$$

with $\bar{\varepsilon}$ the equilibrium energy function for $\mathscr{B}$. Then

$$\varphi^\circ = \bar{\varphi}(\eta^\circ, v^\circ), \tag{3.7}$$

and since, by (2.12),

$$\bar{\varepsilon}(\eta(X,t), v(X,t)) \leq \varepsilon(X,t),$$

we have

$$\bar{\varphi}(\eta(X,t), v(X,t)) \leq \zeta(X,t) \tag{3.8}$$

with equality holding only when $\varepsilon(X,t) = \bar{\varepsilon}(\eta(X,t), v(X,t))$. Clearly, (1.1) implies

$$\bar{\varphi}(\eta^\circ, v^\circ) \leq \bar{\varphi}(\eta(X,t), v(X,t)) \tag{3.9}$$

with equality holding only when $\eta(X,t)=\eta^\circ$ and $v(X,t)=v^\circ$. In view of (3.7) and (3.9), the relation (3.8) yields

$$\varphi^\circ \leq \zeta(X,t)$$

with equality holding only when $\eta(X,t)=\eta^\circ$, $v(X,t)=v^\circ$, and $\varepsilon(X,t)=\bar{\varepsilon}(\eta^\circ, v^\circ)=\varepsilon^\circ$; q.e.d.

The conclusion (3.4) above is strengthened in Remark 3.2 below. To formulate and prove that remark, however, I must first state some definitions and results from the general theory of functions with points of convexity.

Let R be the set of real numbers, let R^n, $n \geq 1$, be the space of n-tuples of real numbers, let D be a convex open subset of R^n, and let f be a function mapping D into R. A point $\boldsymbol{z}$ in D is called a *point of convexity for* f if

$$f(\boldsymbol{z}) \leq \alpha f(\boldsymbol{x}) + (1-\alpha) f(\boldsymbol{y}),$$

whenever $\boldsymbol{x}$, $\boldsymbol{y}$ in D and α in $[0, 1]$ are such that

$$\alpha \boldsymbol{x} + (1-\alpha)\boldsymbol{y} = \boldsymbol{z}.$$

It is easy to show that, when f is differentiable, $\boldsymbol{z}$ is a point of convexity for f if and only if, for each $\boldsymbol{x}$ in D,

$$f(\boldsymbol{x}) \geq f(\boldsymbol{z}) + (\boldsymbol{x}-\boldsymbol{z}) \cdot \nabla f(\boldsymbol{z})$$

with $\nabla f(\boldsymbol{z})$ the gradient of f at $\boldsymbol{z}$. If D has a compact subset S such that every point in $\mathsf{D}-\mathsf{S}$ is a point of convexity for f, then f is said to be *convex outside a compact set*. When $\boldsymbol{0}=(0,\ldots,0)$ is in D, f is said to be *positive definite* on D if $f(\boldsymbol{x})>0$ for all $\boldsymbol{x}$ in D with $\boldsymbol{x} \neq \boldsymbol{0}$.

In the lemma which I now state, a norm $\|\cdot\|$ on R^n is assumed given. Although the lemma is valid regardless of the choice of this norm, in applications $\|\cdot\|$ is defined by

$$\|\boldsymbol{y}\| = |y_1| + |y_2| + \cdots + |y_n|,$$

where $\boldsymbol{y}=(y_1,\ldots,y_n)$ and $|y_i|$ is the absolute value of y_i.

Lemma. *Suppose* D *is a convex open subset of* R^n *containing the point* $\boldsymbol{0}$. *Let* f *be a continuous function mapping* D *into* R *with* $f(\boldsymbol{0})=0$, *and let* $\mathscr{B}$ *be a set with a finite, positive measure* m. *If* f *is convex outside a compact set and positive definite on* D, *then for each* $\varepsilon>0$ *there exists a* $\delta>0$ *such that every* m*-measurable function* g *mapping* $\mathscr{B}$ *into* D *with*

$$\int_{\mathscr{B}} f(g(X))\, dm < \delta,$$

obeys

$$\int_{\mathscr{B}} \|g(X)\|\, dm < \varepsilon.$$

A proof of this lemma will be published elsewhere.[1]

[1] Vid. Coleman [8], Lemmata A.1 and A.2.

Remark 3.2. *For each regular fluid body $\mathscr{B}$ whose equilibrium energy function $\bar{\varepsilon}$ is convex outside a compact set, there exists a positive function $\delta(\omega, \eta^\circ, v^\circ)$ on $(0,\infty)\times(0,\infty)\times(0,\infty)$ such that* ***if*** *ε°, θ°, p°, and Φ° are given by* (3.1), ***if*** *η° and v° are such that* (1.1) *holds for every pair (η, v) not equal to (η°, v°), and* ***if*** *$\mathscr{C}$ is an admissible process of $\mathscr{B}$ with*

$$\Phi(t) - \Phi^\circ < \delta(\omega, \eta^\circ, v^\circ), \tag{3.10}$$

where $\Phi(t)$ is defined by (3.5), ***then*** *in the process $\mathscr{C}$, at time t,*

$$\left.\begin{aligned} \theta^\circ \int_{\mathscr{B}} |\eta(X,t) - \eta^\circ|\, dm < \omega, \quad p^\circ \int_{\mathscr{B}} |v(X,t) - v^\circ|\, dm < \omega, \\ \frac{1}{2}\int_{\mathscr{B}} \dot{x}^2(X,t)\, dm < \omega, \quad \text{and} \int_{\mathscr{B}} |\varepsilon(X,t) - \varepsilon^\circ|\, dm < \omega. \end{aligned}\right\} \tag{3.11}$$

Proof. Let $\omega > 0$, $\eta^\circ > 0$, and $v^\circ > 0$ be given, and let ε°, θ°, p°, Φ°, $\Phi(t)$, φ°, and $\zeta(X,t)$ be as in Remark 3.1. Then

$$\Phi(t) - \Phi^\circ = \int_{\mathscr{B}} [\zeta(X,t) - \varphi^\circ]\, dm + \frac{1}{2}\int_{\mathscr{B}} \dot{x}^2(X,t)\, dm. \tag{3.12}$$

Suppose now that η° and v° are such that (1.1) holds for all $(\eta, v) \neq (\eta^\circ, v^\circ)$. By the conclusion (3.3) of Remark 3.1,

$$\zeta(X,t) - \varphi^\circ \geq 0,$$

and therefore (3.12) yields

$$\Phi(t) - \Phi^\circ \geq \frac{1}{2}\int_{\mathscr{B}} \dot{x}^2(X,t)\, dm. \tag{3.13}$$

Furthermore, since $\dot{x}^2$ is never negative, (3.12) yields also

$$\Phi(t) - \Phi^\circ \geq \int_{\mathscr{B}} [\zeta(X,t) - \varphi^\circ]\, dm,$$

and, by (3.8), this implies

$$\Phi(t) - \Phi^\circ \geq \int_{\mathscr{B}} [\bar{\varphi}(\eta, v) - \varphi^\circ]\, dm, \tag{3.14}$$

where I have written η for $\eta(X,t)$ and v for $v(X,t)$, and $\bar{\varphi}$ is as in (3.6). Let

$$y_1 = (\eta - \eta^\circ)\,\theta^\circ, \quad y_2 = (v - v^\circ)\,p^\circ, \quad \boldsymbol{y} = (y_1, y_2), \tag{3.15}$$

$$f(\boldsymbol{y}) \stackrel{\text{def}}{=} \bar{\varphi}\left(\frac{y_1}{\theta^\circ} + \eta^\circ, \frac{y_2}{p^\circ} + v^\circ\right) - \varphi^\circ = \bar{\varphi}(\eta, v) - \varphi^\circ. \tag{3.16}$$

In terms of the function f, (3.14) becomes

$$\Phi(t) - \Phi^\circ \geq \int_{\mathscr{B}} f(\boldsymbol{y})\, dm. \tag{3.17}$$

By (3.16), (3.6), and (3.1),

$$f(\boldsymbol{y}) = \bar{\varepsilon}\left(\frac{y_1}{\theta^\circ} + \eta^\circ, \frac{y_2}{p^\circ} + v^\circ\right) - \varepsilon^\circ + y_1 + y_2, \tag{3.18}$$

and, because $\bar{\varepsilon}$ is defined and differentiable on $(0,\infty)\times(0,\infty)$, f is defined and differentiable for all $\boldsymbol{y}$ in the set

$$\mathsf{D} = (-\eta^\circ\theta^\circ, \infty)\times(-v^\circ p^\circ, \infty).$$

Of course, since $\bar{\varphi}(\eta^\circ, v^\circ) = \varphi^\circ$, (3.16) yields $f(\mathbf{0}) = 0$. It follows from (1.1) that f is positive definite on D. [See (3.9).] As it is here assumed that $\bar{\varepsilon}$ is convex outside a compact set, it is clear from (3.18) that f is too. Thus, f meets the hypothesis of the lemma stated above, and there exists a $\lambda > 0$ such that

$$\int_{\mathscr{B}} f(\mathbf{y})\, dm < \lambda \Rightarrow \int_{\mathscr{B}} \|y\|\, dm < \frac{\omega}{2}.$$

Hence, if we use the norm $\|\cdot\|$ defined before the statement of the lemma, and note (3.17) and (3.15), we have

$$\Phi(t) - \Phi^\circ < \lambda \Rightarrow \theta^\circ \int_{\mathscr{B}} |\eta - \eta^\circ|\, dm + p^\circ \int_{\mathscr{B}} |v - v^\circ|\, dm < \frac{\omega}{2}. \qquad (3.19)$$

It follows from (3.1), (3.5), and (3.3) that

$$\Phi(t) - \Phi^\circ = \int_{\mathscr{B}} [\varphi - \varphi^\circ]\, dm = \int_{\mathscr{B}} |\varphi - \varphi^\circ|\, dm, \qquad (3.20)$$

where

$$\varphi = \varepsilon - \theta^\circ \eta + p^\circ v + \frac{1}{2} \dot{x}^2. \qquad (3.21)$$

Since

$$\varepsilon - \varepsilon^\circ = \varphi - \varphi^\circ + (\eta - \eta^\circ)\, \theta^\circ - (v - v^\circ)\, p^\circ - \frac{1}{2} \dot{x}^2,$$

the triangle inequality, (3.20), and (3.13) yield

$$\begin{aligned} \int_{\mathscr{B}} |\varepsilon - \varepsilon^\circ|\, dm &\leq \int_{\mathscr{B}} |\varphi - \varphi^\circ|\, dm + \theta^\circ \int_{\mathscr{B}} |\eta - \eta^\circ|\, dm + p^\circ \int_{\mathscr{B}} |v - v^\circ|\, dm \\ &\quad + \frac{1}{2} \int_{\mathscr{B}} \dot{x}^2\, dm \qquad (3.22) \\ &\leq 2[\Phi(t) - \Phi^\circ] + \theta^\circ \int_{\mathscr{B}} |\eta - \eta^\circ|\, dm + p^\circ \int_{\mathscr{B}} |v - v^\circ|\, dm. \end{aligned}$$

It is a consequence of (3.13), (3.19), and (3.22) that if one puts

$$\delta = \min(\lambda, \omega/4),$$

then δ is positive, and all four of the inequalities in (3.11) hold whenever $\Phi(t) - \Phi^\circ$ is less than δ. Of course the δ so obtained depends on not only ω but also η° and v°, i.e. $\delta = \delta(\omega, \eta^\circ, v^\circ)$. The argument just given rests on the assumption that (1.1) holds; if η° and v° are not such that (1.1) holds for all $(\eta, v) \neq (\eta^\circ, v^\circ)$, then let $\delta(\omega, \eta^\circ, v^\circ)$ equal ω. Thus one obtains a function $\delta(\omega, \eta^\circ, v^\circ)$ with the desired properties; q.e.d.

4. Dynamical stability

In the two theorems of this section, a regular fluid body $\mathscr{B}$ is supposed assigned in advance, and, for each pair (θ°, p°) of positive numbers, $\mathfrak{C}(\theta^\circ, p^\circ)$ denotes the class of admissible thermodynamic processes of $\mathscr{B}$ that are compatible with immersion of $\mathscr{B}$ in an environment at temperature θ° and pressure p° from time $t = 0$ onward.

Theorem 4.1. *Let θ°, p°, η°, v°, and ε° be the values of the temperature, pressure, specific entropy, specific volume, and specific internal energy in a uniform static state of a regular fluid body $\mathscr{B}$ for which the equilibrium energy function is convex outside a compact set. If* (1.1) *holds for every pair (η, v) not equal to (η°, v°), then given any $\omega > 0$, there exists a $\delta = \delta(\omega, \eta^\circ, v^\circ) > 0$ such that each process in $\mathfrak{C}(\theta^\circ, p^\circ)$ which, at any* *one time* $t \geq 0$, *satisfies*

$$\Phi(t) - \Phi^\circ < \delta, \tag{4.1}$$

with $\Phi(t)$ *and* Φ° *defined as in Remark* 3.1, *must also satisfy*

$$\left.\begin{aligned} &\frac{1}{2}\int_{\mathscr{B}} \dot{x}^2(X,\tau)\,dm < \omega, \qquad \int_{\mathscr{B}} |\varepsilon(X,\tau) - \varepsilon^\circ|\,dm < \omega, \\ &\theta^\circ \int_{\mathscr{B}} |\eta(X,\tau) - \eta^\circ|\,dm < \omega, \quad \text{and} \quad p^\circ \int_{\mathscr{B}} |v(X,\tau) - v^\circ|\,dm < \omega, \end{aligned}\right\} \tag{4.2}$$

for all $\tau \geq t$.

Proof. Let $\delta = \delta(\omega, \eta^\circ, v^\circ)$ be as in Remark 3.2. Of course, δ is then always positive. Now, according to Proposition 2.1, for $\tau \geq 0$, $\Phi(\tau)$ does not increase with τ in any process $\mathscr{C}$ belonging to $\mathfrak{C}(\theta^\circ, p^\circ)$. Hence if $\mathscr{C}$ in $\mathfrak{C}(\theta^\circ, p^\circ)$ obeys (4.1) at some time $t \geq 0$, then for the process $\mathscr{C}$,

$$\Phi(t) - \Phi^\circ < \delta, \tag{4.3}$$

at each time $\tau \geq t$. But, by Remark 3.2, if (4.3) holds at time τ, then (4.2) also holds at time τ; q.e.d.

Theorem 4.2. *Let* θ°, p°, η°, v°, *and* ε° *be the values of the temperature, pressure, specific entropy, specific volume, and specific internal energy in a uniform static state of a regular fluid body* $\mathscr{B}$ *for which the equilibrium energy function* $\bar{\varepsilon}$ *is convex outside a compact set. Suppose* (1.1) *holds for every pair* (η, v) *not equal to* (η°, v°). *Then, given any* $\omega > 0$, *there exists a* $\lambda = \lambda(\omega, \theta^\circ, p^\circ) > 0$ *such that if a process* $\mathscr{C}$ *in* $\mathfrak{C}(\theta^\circ, p^\circ)$ *has*

$$\left.\begin{aligned} &\frac{1}{2}\int_{\mathscr{B}} \dot{x}^2(X,t)\,dm < \lambda, \qquad \int_{\mathscr{B}} |\varepsilon(X,t) - \varepsilon^\circ|\,dm < \lambda, \\ &\theta^\circ \int_{\mathscr{B}} |\eta(X,t) - \eta^\circ|\,dm < \lambda, \quad \text{and} \quad p^\circ \int_{\mathscr{B}} |v(X,t) - v^\circ|\,dm < \lambda, \end{aligned}\right\} \tag{4.4}$$

at any one time $t \geq 0$, *then* $\mathscr{C}$ *must obey*

$$\left.\begin{aligned} &\frac{1}{2}\int_{\mathscr{B}} \dot{x}^2(X,\tau)\,dm < \omega, \qquad \int_{\mathscr{B}} |\varepsilon(X,\tau) - \varepsilon^\circ|\,dm < \omega, \\ &\theta^\circ \int_{\mathscr{B}} |\eta(X,\tau) - \eta^\circ|\,dm < \omega, \quad \text{and} \quad p^\circ \int_{\mathscr{B}} |v(X,\tau) - v^\circ|\,dm < \omega, \end{aligned}\right\} \tag{4.5}$$

for all $\tau \geq t$.

Proof. It is clear from Remark 3.1 [see also (3.20)] that in the process $\mathscr{C}$,

$$\Phi(t) - \Phi^\circ = \int_{\mathscr{B}} |\varphi(X,t) - \varphi^\circ|\,dm,$$

where

$$\begin{aligned} \varphi(X,t) - \varphi^\circ = \frac{1}{2}\dot{x}^2(X,t) + \varepsilon(X,t) - \varepsilon^\circ - [\eta(X,t) - \eta^\circ]\,\theta^\circ \\ + [v(X,t) - v^\circ]\,p^\circ. \end{aligned}$$

Hence, the triangle inequality yields

$$\Phi(t) - \Phi^\circ \leq \int_{\mathscr{B}} \frac{1}{2} \dot{x}^2(X,t)\, dm + \int_{\mathscr{B}} |\varepsilon(X,t) - \varepsilon^\circ|\, dm + \theta^\circ \int |\eta(X,t) - \eta^\circ|\, dm$$
$$+ p^\circ \int |v(X,t) - v^\circ|\, dm,$$

and, if we put

$$\lambda = \lambda(\omega, \theta^\circ, p^\circ) = \frac{1}{4} \delta(\omega, \eta^\circ, v^\circ)$$

with $\delta(\omega, \eta^\circ, v^\circ)$ as in Theorem 4.1, then λ is not only positive but is also such that (4.4) implies (4.1). But, Theorem 4.1 asserts that if $\mathscr{C}$ satisfies (4.1) at some time $t \geq 0$, then $\mathscr{C}$ must obey (4.5) for all $\tau \geq t$; q.e.d.

References

1. Gibbs, J. W.: Trans. Conn. Acad. **2**, 382–404 (1873). The Scientific Papers of J. Willard Gibbs, Vol. I, London/New York: Longmans, Green 1906, pp. 33–54.
2. Gibbs, J. W.: Trans. Conn. Acad. **3**, 108–248, 343–524 (1875–1878). The Scientific Papers, Vol. I, pp. 55–353.
3. Coleman, B. D., Greenberg, J. M.: Arch. Rat. Mech. Anal. **25**, 321–341 (1967).
4. Coleman, B. D.: Arch. Rat. Mech. Anal. **17**, 1–46, 230–254 (1964).
5. Koiter, W. T.: Report 360, Laboratory of Engineering Mechanics, Technological University, Delft, 1967.
6. Ericksen, J. L.: Int. J. Solids Structures **2**, 573–580 (1966).
7. Coleman, B. D., Dill, E. H.: Arch. Rat. Mech. Anal. **30**, 197–224 (1968).
8. Coleman, B. D.: Arch. Rat. Mech. Anal. **36**, 1–32 (1970).

Review of the Finite Bandwidth Concept

By

A. C. Newell and **J. A. Whitehead**

University of California, Los Angeles, Cal., U.S.A.

I. Introduction

The neutral stability curves of many problems in continuum physics have a remarkable similarity. The curves usually relate the value of some parameter to some length scale, and if the instability sets in as growing oscillations there is an additional dispersion relation. Previous studies have attempted to describe the time behavior of a number of discrete scales of motion when the value of the parameter is slightly supercritical. However, it is intuitively clear that in general a continuous range of scales are involved and the following analysis describes how this concept may be readily incorporated. As a prototype we choose the Bénard-Rayleigh problem whose stability diagram appears below (Fig. 1). In the final section we sketch what happens when growing oscillations are the initial instability.

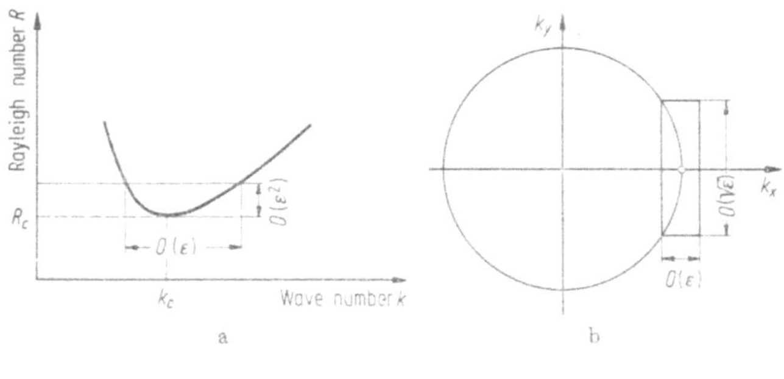

Fig. 1. a) Rayleigh number vs. wave number diagram; b) Critical circle $k_x^2 + k_y^2 = \pi^2/2$.

II. Stability analysis

The governing equation for the motion in the Boussinesq approximation in non-dimensional form is

$$\left(\frac{1}{\mathscr{P}}\frac{\partial}{\partial t} - \nabla^2\right)\left(\frac{\partial}{\partial t} - \nabla^2\right)\nabla^2 w - R\,\nabla_1^2 w = -\nabla_1^2(u\cdot\nabla\theta) + \frac{1}{\mathscr{P}}\left(\frac{\partial}{\partial t} - \nabla^2\right)\{(\nabla\times\nabla\times(\nabla\times\boldsymbol{u})\times\boldsymbol{u})\cdot\hat{z}\}, \tag{2.1}$$

with boundary conditions

$$w = \frac{\partial^2 w}{\partial z^2} = \frac{\partial^4 w}{\partial z^4} = 0, \qquad z = 0, 1, \tag{2.2}$$

where (2.1) is the scalar product of the unit vector in the direction of the gravity field and the curl of the vorticity equation. The velocity field is $\boldsymbol{u}(u, v, w)$ and θ denotes the difference between the true temperature and that corresponding to the purely conductive solution. R is the Rayleigh number which is a non-dimensional measure of the temperature difference and $\mathscr{P}$, the Prandtl number, is the ratio of the coefficients of viscous to thermal diffusion. The linear stability analysis shows that solutions of the form

$$e^{\sigma t}\, e^{i\boldsymbol{k}\cdot\boldsymbol{x}} \sin \pi z, \qquad \boldsymbol{k} = (k_x, k_y), \qquad \boldsymbol{x} = (x, y)$$

first become unstable when $R \geq R_c\,(= 27\pi^4/4)$ and for $|\boldsymbol{k}| = \pi/\sqrt{2}$. The principle of exchange of stabilities holds and the first entry into the unstable region occurs when $\operatorname{Im}\sigma = 0$. A glance at the operator on the left-hand side shows that for $R > R_c$ the infinitesimal solution grows like $e^{\frac{R-R_c}{R_c}t}$. Accordingly, when describing the finite amplitude time behavior of a single mode one is led to try the description

$$w_0(x, y, z, t) = \varepsilon\left(W(\varepsilon^2 t)\, e^{ikx} + (*)\right) \sin \pi z \tag{2.3}$$

as a first approximation where (*) denotes the complex conjugate and $\frac{R - R_c}{R_c} = \varepsilon^2\chi$. The procedure for incorporating a multimode description up till now has been to add solutions of the type (2.3) together,

$$w_0 = \varepsilon \sum_m \left(Wm(\varepsilon t)\, e^{i\boldsymbol{k}_m\cdot\boldsymbol{x}} + (*)\right) \sin \pi z, \tag{2.4}$$

even though the $\boldsymbol{k}_m$ may be close together. Let us for a moment concentrate on the two-dimensional case where all quantities only depend on x. Then the modes which are likely to appear are of the type $\{k_c + \varepsilon L_M, 0\}$ as all modes outside this range (see Fig. 1) will decay on the thermal diffusion time scale t_0 and will not play a role in the much longer t_0/ε^2 time scale. Therefore, instead of writing the basic solution as a sum of such modes (the number of which rapidly increases due to nonlinear coupling) we simply write the basic approximation as

$$w_0 = \varepsilon\big(W(\varepsilon x, \varepsilon^2 t)\, e^{ik_c x} + (*)\big) \sin \pi z. \tag{2.5}$$

We obtain an equation describing the behavior of W by inserting (2.5) in (2.1) and solving iteratively,

$$w = w_0 + \varepsilon^2 w_1 + \varepsilon^3 w_2. \tag{2.6}$$

In order that solutions exist for the successive iterates a certain solvability criterion must be met, which is

$$2\frac{\mathscr{P}+1}{\mathscr{P}}\frac{\partial W}{\partial T} - 8\frac{\partial^2 W}{\partial X^2} = (3\pi^2\chi - WW^*)\,W, \quad X = \varepsilon x, \quad T = \varepsilon^2 t. \tag{2.7}$$

Before we note some properties of this equation we emphasise the following: the packet idea is rather essential for it takes into account all possible modal interactions from the physical space point of view. Dealing with the problem using a discrete multimode approach as in (2.4), leads to the difficulty of deciding which interactions to take into account. The interaction giving the nonlinear response is of the type $k_1 + k_2 + k_3 = k_4$ where k_1, k_2, k_3 are close to k_c. However, it is not just the couplings between k_1, k_2 and k_3 which yield exactly k_c that matter; for if

$$k_1 + k_2 + k_3 = k_4 = k_c + O(\delta),$$

then one would find

$$w_1 = O\left(\frac{1}{\delta^2}\right),$$

and the asymptotic expansion (2.6) loses its uniformity. The packet concept removes the possibility of ill-conditioning in the successive iterates.

The difference between (2.7) and previous approaches is the presence of the diffusion term. In its absence the single mode solution is driven to its finite amplitude state $WW^* = 3\pi^2\chi$. However, its presence shows that nonuniformities in space are propagated by diffusion according to the law

$$x \sim 4\sqrt{\frac{\mathscr{P}}{\mathscr{P}+1}\varkappa t}.$$

However, we note that once a roll has been triggered its subsequent evolution is determined forcing terms on the right-hand side of (2.7). We note in addition that single mode solutions

$$W \propto e^{iKX}$$

are possible for $K^2 < \frac{3\pi^2}{8}\chi$. It is readily verified that $K^2 = \frac{3\pi^2}{8}\chi$ is the equation of the parabola near the minimum of the Rayleigh number vs.

wave number diagram in Fig. 1. Certain other steady solutions of the equation are also readily found [1, 2] and in some cases their stability has been tested [1].

In order to take account of the possible three-dimensional nature of packet we write

$$W = W(\varepsilon x, \sqrt{\varepsilon} y, \varepsilon^2 t)$$

as an $O(\varepsilon)$ spread in the direction of the vector $(\pi/\sqrt{2}, 0)$ leads to an $O(\sqrt{\varepsilon})$ spread in the k_y direction (see Fig. 1b). The solvability condition in nondimensional and normalized form now becomes

$$\frac{\partial W}{\partial T} - \left(\frac{\partial}{\partial X} - \frac{i}{\sqrt{2\pi}} \frac{\partial^2}{\partial Y^2}\right)^2 W = W - W^2 W^*, \qquad Y = \sqrt{\varepsilon} y, \quad (2.8)$$

and if we have N discrete packets centred on the mode vectors $\left\{\boldsymbol{k}_m; |\boldsymbol{k}_m| = \frac{\pi}{\sqrt{2}}, |\boldsymbol{k}_l - \boldsymbol{k}_n| = O(1)\right\}_{m=1}^{N}$ the solvability equations are

$$\frac{\partial W_m}{\partial T} - \left(\frac{\partial}{\partial X_m} - \frac{i}{\sqrt{2\pi}} \frac{\partial^2}{\partial Y_m^2}\right)^2 W_m = W_m \left(1 - \sum_{n=1}^{N} \beta_{mn} W_n W_n^*\right), \quad (2.9)$$

where $X_m = \varepsilon(x \cos\theta_m + y \sin\theta_m)$, $Y_m = \sqrt{\varepsilon}(-x \cos\theta_m + y \sin\theta_m)$, $\tan\theta_m = \frac{k_{my}}{k_{mx}}$. The nonlinear couplings in (2.9) arise from the type

$$\boldsymbol{k}_1 + \boldsymbol{k}_2 - \boldsymbol{k}_2 = \boldsymbol{k}_1$$

as there is no other way in which four distinct wave vectors of the same modulus can satisfy

$$\boldsymbol{k}_1 + \boldsymbol{k}_2 + \boldsymbol{k}_3 = \boldsymbol{k}_4,$$

and by distinct we mean $|\boldsymbol{k}_1 - \boldsymbol{k}_2| = O(1)$.

Multimodal steady solutions can be found by setting the right-hand side of (2.9) zero and looking for X, Y and T independent solutions. In the particular model we are considering here the β_{mn} have the property which causes multimodal solutions to be unstable; this is not the case when we have a slight vertical asymmetry in the problem (introduced by mean temperature being slightly dependent on time, viscosity being temperature dependent) for then the nonlinear terms can play a crucial role by reproducing the vertical eigenfunction structure of the neutral mode by a quadratic interaction. The horizontal scales yield a $\boldsymbol{k}_1 + \boldsymbol{k}_2 + \boldsymbol{k}_3 = 0$ interaction which interactions can lead to the predominance of the hexagonal structure in the final solution.

The stability of the sideband mode solution $W_1 = (1 - L^2)^{1/2} e^{iLX}$, $W_m = 0$, $m \neq 1$ can be readily analysed. There are three types of instability. The first is a full three-dimensional instability which arises

[1] These solutions are also important in describing the effect of boundaries on shell buckling [6].

from a $\boldsymbol{k}_1 + \boldsymbol{k}_2 - \boldsymbol{k}_2 = \boldsymbol{k}_1$ interaction and puts a stability band $|L| < \beta(\mathscr{P})$ on the range of L. The other two are local instabilities and can be found by considering the stability of the above solution in (2.8). Both of these arise from a $2\boldsymbol{k}_1 = \boldsymbol{k}_2 + \boldsymbol{k}_3$ interaction. The first exhibits a slight three-dimensionality and may be called an oblique mode instability. It occurs whenever $L < 0$ as $(k_c + \varepsilon L, 0)$ can interact with the two modes $\left(k_c + \varepsilon L, \pm \sqrt{\varepsilon(-2\sqrt{\pi L})}\right)$ which lie on the critical circle. The second of the local instabilities is purely two-dimensional and puts the stability bounds $|L| < 1/\sqrt{3}$ on the range of L. Thus the absolute stability bound on L is $0 < L < \mathrm{Min}\left(\beta(\mathscr{P}), 1/\sqrt{3}\right)$ and the right-hand bound depends on the Prandtl number. The three-dimensional and the oblique mode instabilities were first discussed by SCHLÜTER, LORTZ, and BUSSE [3]; the two-dimensional one was first discussed by ECKHAUS [4]. However, we note that the solvability equations using the finite bandwidth concept facilitate the stability analysis considerably; in fact, *the solvability equations are a uniformly valid approximation to the Boussinesq equations for* $\frac{R - R_c}{R_c} \ll 1$ and so the stability analysis can be performed directly on them without returning to the full Boussinesq equations. This enable sone to automatically choose the correct modal interaction yielding the instability.

Recent work of NEWELL and WHITEHEAD [1] and from the viewpoint of the statistical initial value problem by NEWELL, LANGE and AUCOIN [5] suggests that in most cases the most critical mode k_c is the one which is *realised*. The conclusion is that if the initial disturbance has a sufficiently smooth spectrum, then a natural statistical selection process chooses the roll with horizontal wave number k_c. If, on the other hand, the initial spectrum is sufficiently weighted in favour of a discrete side-band mode, lying within the stability bounds, then the sideband mode solution can be achieved.

III. The case of overstability

The technique for obtaining the solvability condition in the overstable case follows closely the analysis of Sec. II. However, considerable algebraic manipulation can be saved in the construction of the linear operator, analogous to $\partial/\partial T - \partial^2/\partial X^2 - 1$ in the overstable case. The governing equation can be written schematically

$$\mathscr{L}\left(\frac{\partial}{\partial t}, -\frac{\partial^2}{\partial x^2}, R\right) \cdot w = N w^2,$$

where for simplicity we assume the vertical eigenfunction structure has been removed. Assuming a modified neutral solution

$$w_0 = W(\varepsilon x, \varepsilon t, \varepsilon^2 t, \ldots)\, e^{i(k_c x + \sigma_i(k_c) t)} + (*), \tag{3.1}$$

we again expand

$$w = \varepsilon (w_0 + \varepsilon w_1 + \varepsilon^2 w_2 + - \cdots).$$

The operator $\mathscr{L}$ acting on w_0

$$\mathscr{L}\left(\frac{\partial}{\partial t}, -\frac{\partial^2}{\partial x^2}, R\right) w_0 = \mathscr{L}\left(\sigma_i(k_c) - i\varepsilon \frac{\partial}{\partial T_1}\right.$$
$$\left. - i\varepsilon^2 \frac{\partial}{\partial T_2}, k_c^2 - 2i\varepsilon k_c \frac{\partial}{\partial X} - \varepsilon^2 \frac{\partial^2}{\partial X^2}, R_c(1 + \varepsilon^2\chi)\right) W e^{i(k_c x + \sigma_i t)}, \tag{3.2}$$

where

$$T_1 = \varepsilon t,\ T_2 = \varepsilon^2 t,\ X = \varepsilon x.$$

From the linear eigenvalue problem, it is true that

$$\mathscr{L}\left(\sigma_i(k^2), R(k^2), k^2\right) = 0. \tag{3.3}$$

Eq. (3.3) has real and complex parts which determine $R(k^2)$ and $\sigma_i(k^2)$ and is true for all k^2. Differentiating, we obtain

$$\frac{\partial \mathscr{L}}{\partial \sigma_i} \frac{d\sigma_i}{dk^2} + \frac{\partial \mathscr{L}}{\partial R} \frac{dR}{dk^2} + \frac{\partial \mathscr{L}}{\partial k^2} = 0 \tag{3.4}$$

and the choice of k_c^2 by $dR/dk^2 = 0$ ensures the choice of the minimum Rayleigh number. Differentiating once again we have the relations

$$\frac{\partial^2 \mathscr{L}}{\partial \sigma_i^2}\left(\frac{d\sigma_i}{dk^2}\right)^2 + 2\frac{\partial^2 \mathscr{L}}{\partial \sigma_i\, \partial k^2}\frac{d\sigma_i}{dk^2} + \frac{\partial \mathscr{L}}{\partial R}\frac{d^2 R}{d(k^2)^2} + \frac{\partial^2 \mathscr{L}}{\partial (k^2)^2} + \frac{\partial \mathscr{L}}{\partial \sigma_i}\frac{d^2\sigma_i}{d(k^2)^2}\bigg|_{k=k_c} = 0, \tag{3.5}$$

and from (3.4)

$$\frac{\partial \mathscr{L}}{\partial k^2} + \frac{\partial \mathscr{L}}{\partial \sigma_i}\frac{d\sigma_i}{dk^2}\bigg|_{k=k_c} = 0. \tag{3.6}$$

Expanding the operator $\mathscr{L}$ in (3.2) we find in order to eliminate secular terms, $W = W(\varepsilon(x + C_g t), \varepsilon^2 t)$ and

$$-i\frac{\partial \mathscr{L}}{\partial \sigma_i}\left(\frac{\partial}{\partial T_2} + \frac{i\delta}{2}\frac{\partial^2}{\partial X^2}\right) W + 2k_c^2 \frac{\partial \mathscr{L}}{\partial R}\frac{d^2 R}{d(k^2)^2}\frac{\partial^2 W}{\partial X^2} + R_c\chi \frac{\partial \mathscr{L}}{\partial R} W = \beta W^2 W^*,$$

"dispersion" "diffusion" "excitation" "nonlinear response"

where C_g is the group velocity and δ the dispersion. Eq. (2.7) can be deduced from (3.7) by a careful evaluation of the limit.

References

1. Newell, A. C., Whitehead, A. J.: J. Fluid Mech. **38**, 279 (1969).
2. Segel, L. A.: J. Fluid Mech. **38**, 203 (1969).
3. Schlüter, A., Lortz, D., Busse, F.: J. Fluid Mech. **23**, 129 (1965).
4. Eckhaus, W.: Studies in Non-Linear Stability Theory, Berlin/Heidelberg/New York: Springer 1965.
5. Newell, A. C., Lange, C. G., Aucoin, P. A.: J. Fluid Mech. **40**, 513 (1969).
6. Lange, C. G., Newell, A. C.: (to be published).

A Class of Unsteady Nonlinear Waves in Parallel Flows

By

R. F. Bergeron, Jr.

Bell Telephone Laboratories, Whippany, N.J., U.S.A.

Introduction

Studies of instability waves in parallel shear flows generally begin with linearized governing equations, justified by the small amplitude (ε) of the disturbances considered. The fundamental problem of what we will call the viscous theory is the eigenvalue problem for the fourth order Orr-Sommerfeld equation [1]. At high Reynolds number (R), a case of particular interest, the coefficient on the fourth order term is small and the second order Rayleigh equation (inviscid approximation) is valid except near the boundaries and, most important, at the critical layer, where the phase speed c is equal to the flow velocity. The inviscid solution is singular there; viscous effects must be considered in a layer of thickness $R^{-1/3}$ to resolve the singularity.

There are small amplitude waves at high Reynolds number that are not described correctly by the viscous theory, for it is also possible that the critical layer singularity of the inviscid, linearized equation is resolved by the nonlinear terms neglected at the outset. The parameter $\lambda = (R\varepsilon^{3/2})^{-1}$ measures the relative importance of viscous and nonlinear effects near the critical layer. A necessary condition for validity of the viscous theory is $\lambda \gg 1$; we will consider the "nonlinear theory", treating the case $\lambda \ll 1$.

Neutral waves of the nonlinear type have been found for a number of velocity profiles, and described in detail by Benney and Bergeron [2, 3]. The nonlinear critical layer, connecting outer regions governed by the Rayleigh equation, is of thickness $\varepsilon^{1/2}$: the basic flow within the layer is the Kelvin "cat's eye" flow (Fig. 1); the first vorticity correction is odd and changes sharply across viscous layers on the cell boundaries;

the connection established between the outer solutions allows no phase change, hence no Reynolds stress.

By contrast, the phase change across a viscous critical layer is finite, and this difference has two significant consequences: the viscous critical layer is constrained to lie near a boundary, but the nonlinear layer, with no phase change, is not. Accordingly, the nonlinear neutral curves (Fig. 2) show solutions far into the uniform flow region ($c \to 1$).

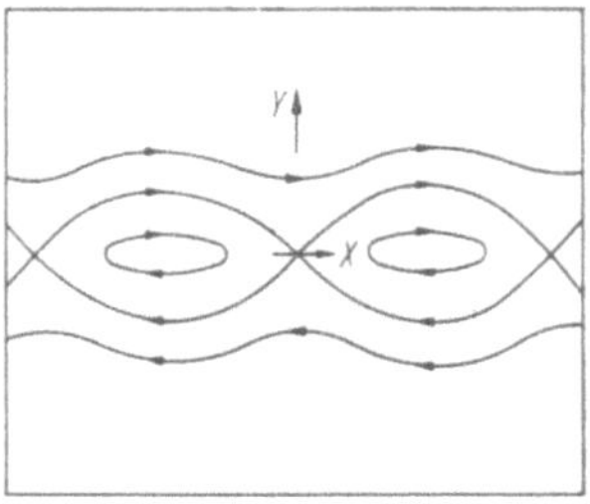

Fig. 1. The Kelvin "cat's eye" flow.

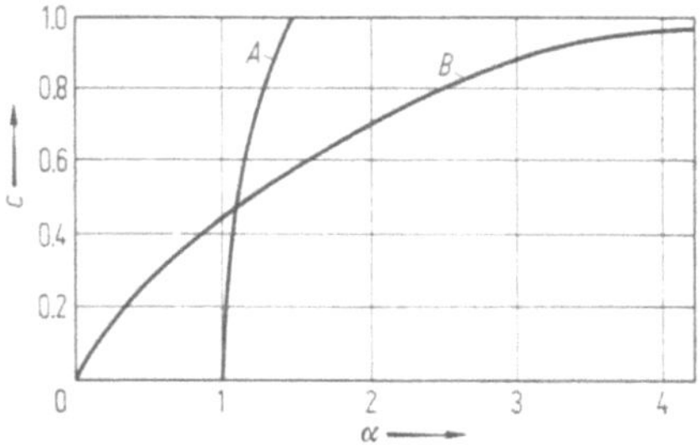

Fig. 2. Nonlinear neutral curves for hyperbolic tangent (A) ($\bar{u}(y) = \tanh y$) and Blasius model (B) $\left(\bar{u}(y) = 3\left(-\frac{2}{3}\,!\right)^{-1} \int_0^y e^{-z^3}\,dz\right)$ velocity profiles.

Second, the Reynolds stress provides a mechanism of energy transfer to the viscous waves that does not exist for the nonlinear waves. Therefore, unsteady effects associated with a nonlinear critical layer will not include the exponential time growth or decay characteristic of the viscous theory.

In the following, we treat one aspect of unsteady behavior in the nonlinear theory: the steady solutions represented by points on the neutral curves may evolve on a diffusion time scale from more general initial conditions. This may have physical significance.

Nonlinear critical layer

The stream-function for an incompressible two-dimensional flow consisting of a parallel stream with a small wavelike disturbance can be written

$$\Psi = \int_{y_c}^{y} \left(\bar{u}(y') - c\right) dy' + \varepsilon\psi(x, y, t), \tag{1}$$

where the frame of axes is translating downstream with real wave-speed c. With the usual parallel flow hypothesis, the vorticity equation is

$$\Delta\Psi_t + \Psi_y \Delta\Psi_x - \Psi_x \Delta\Psi_y = R^{-1} \Delta\Delta\Psi \tag{2}$$

or

$$\Delta\psi_t + \left(\bar{u}(y) - c\right) \Delta\psi_x - \bar{u}''\psi_x + \varepsilon(\psi_y \Delta\psi_x - \psi_x \Delta\psi_y) = R^{-1} \Delta\Delta\psi. \tag{3}$$

We assume $\varepsilon \ll 1$, $R \gg 1$, and time dependence through the slow variable $T = \lambda \varepsilon^{1/2} t$. The first term in a perturbation expansion of ψ is assumed sinusoidal in x ($\psi^{(0)} = \cos(\alpha x)\, \varphi(y, T)$); φ is governed by the Rayleigh equation,

$$\left(\bar{u}(y) - c\right)\left(\frac{d^2}{dy^2} - \alpha^2\right)\varphi - \frac{d^2\bar{u}}{dy^2}\varphi = 0. \tag{4}$$

Two independent solutions are

$$\begin{aligned} \varphi_a(y) &= y - y_c + a_1(y - y_c)^2 + \cdots, \\ \varphi_b(y) &= 1 + b_2(y - y_c)^2 + \cdots + \frac{\bar{u}_c''}{\bar{u}_c'}\varphi_a(y)\log|y - y_c|. \\ \bar{u}(y_c) &= c;\ \bar{u}_c' = \bar{u}'(y_c), \text{ etc.} \end{aligned} \tag{5}$$

Then, unless the critical layer is near a boundary, the stream function is

$$\psi_s^{(0)}(x, y, T) = A_s(T)\cos\left(\alpha x + \eta_s(T)\right)\left[\varphi_b(y) + C_s\varphi_a(y)\right], \tag{6}$$

where $s = \pm$, for $y - y_c \gtrless 0$. The C_s are constants, determined by boundary conditions.

The logarithmic singularity in φ_b shows the need for new scaling in the neighborhood of $y = y_c$. Because $\lambda \ll 1$, there is a nonlinear critical layer of thickness $\varepsilon^{1/2}$. The proper inner variable is $Y = \varepsilon^{-1/2}(y - y_c)$; substitution into Eq. (2) yields

$$\begin{aligned} &\Psi_Y\Psi_{YYx} - \Psi_x\Psi_{YYY} + \varepsilon(\Psi_Y\Psi_{xxx} - \Psi_x\Psi_{Yxx}) \\ &= \lambda\varepsilon(\Psi_{YYYY} - \Psi_{YYT} + 2\varepsilon\Psi_{YYxx} + \varepsilon^2\Psi_{xxxx}). \end{aligned} \tag{7}$$

Standard matching conditions indicate an inner expansion of the form

$$\begin{aligned} \Psi &= \varepsilon(\Psi^{(0)} + \varepsilon^{1/2}\ln\varepsilon\Psi^{(1)} + \varepsilon^{1/2}\Psi^{(2)}) \\ &+ \cdots + \lambda(\Psi^{(3)} + \varepsilon^{1/2}\ln\varepsilon\Psi^{(4)} + \varepsilon^{1/2}\Psi^{(5)} + \cdots) + \cdots. \end{aligned} \tag{8}$$

The zero order problem is

$$\begin{aligned} &\Psi_Y^{(0)}\Psi_{YYx}^{(0)} - \Psi_x^{(0)}\Psi_{YYY}^{(0)} = 0, \\ &\Psi^{(0)} \sim \bar{u}_c'\frac{Y^2}{2} + A_s(T)\cos\left(\alpha x + \eta_s(T)\right), \quad sY \to \infty. \end{aligned} \tag{9}$$

Immediately, $\Psi_{YY}^{(0)} = \dfrac{\partial G}{\partial \Psi^{(0)}}(\Psi^{(0)}, T)$, so upon integration

$$\left(\frac{\partial \Psi^{(0)}}{\partial Y}\right)^2 = 2G(\Psi^{(0)}, T) - 2\bar{u}_c' A(T)\cos\left((\alpha x + \eta(T)\right), \tag{10}$$

which operation has imposed the conditions $A_s(T) = A(T)$, $\eta_s(T) = \eta(T)$, $s = \pm$. As there is no energy transfer, we may take $A(T) \equiv 1$; also, we have specified the phase velocity (c), so $\eta(T) \equiv 0$. Beyond that, however, time dependence in G is arbitrary within the constraints of asymptotic conditions (9).

Development of steady solutions

The equation for time dependence follows from the condition of periodicity in x, applied to $\Psi^{(3)}$. It is parabolic and nonlinear. The steady solution with proper asymptotic behavior is

$$G(\Psi^{(0)}) = \bar{u}_c' \Psi^{(0)}, \tag{11}$$

corresponding to

$$\Psi^{(0)} = \bar{u}_c' \frac{Y^2}{2} + \cos \alpha x, \tag{12}$$

the Kelvin "cat's eye" flow (Fig. 1).

To show that there is evolution from an arbitrary initial distribution of G (consistent with asymptotic conditions) to the steady solution (12), we proceed indirectly, considering the equation

$$\Psi_Y \Psi_{YYx} - \Psi_x \Psi_{YYY} = \lambda(\Psi_{YYYY} - \Psi_{YYT}), \tag{13}$$

which links the $\Psi^{(0)}$ and $\Psi^{(3)}$ terms. Let

$$\Psi = \bar{u}_c' \frac{Y^2}{2} + \cos \alpha x + Q(x, Y, T). \tag{14}$$

Asymptotic conditions are

$$Q(x, Y, T) \to 0, \quad Y \to \pm \infty. \tag{15}$$

We substitute form (14) into Eq. (13), multiply by Q_{YY}, then integrate over x and Y (applying asymptotic and periodicity conditions) to obtain

$$\frac{\partial}{\partial T} \int_0^{2\pi/\alpha} \int_{-\infty}^{\infty} Q_{YY}^2 \, dY \, dx = -2 \int_0^{2\pi/\alpha} \int_{-\infty}^{\infty} (Q_{YXY})^2 \, dY \, dx. \tag{16}$$

In view of Eq. (15), this implies $Q \to 0$ as $T \to \infty$. The solution approaches the steady state, Eq. (12).

Similar results hold for $\Psi^{(1)}$ and (taking the steady solution for $\Psi^{(0)}$), for $\Psi^{(2)}$. In the latter case, however, an additional constraint is imposed on the initial values. (It is the $\Psi^{(2)}$ term that completes the steady eigenvalue problem with the relation $C_+ = C_-$, see Eq. (6).) In general, the matching condition is

$$\int_{-\infty}^{+\infty} \Psi_{YY}^{(2)} \, dY = (C_+ - C_-) \cos \alpha x. \tag{17}$$

But the governing equation at this order implies that

$$\Psi_{YY}^{(2)} = \Psi_{YY}^{(2)}(\Psi^{(0)}, T). \tag{18}$$

By changing the integration variable to $\Psi^{(0)}$, Eq. (17) can be converted to an Abel integral equation, which is easily shown to have only the trivial solution. Thus $C_+ = C_-$ is necessary initially if time dependence is to be slow: the initial conditions correspond to a point on the neutral curve. That given, there will be evolution in time to the steady solution.

This result can be extended to the problem of an oblique wave. There is an exact mathematical correspondence between the cross-wave velocity in such a problem and the first vorticity correction ($\Psi_{YY}^{(2)}$) in the eigenvalue problem. The steady cross-wave velocity has layers of intense shear across the cat's eye boundaries, analogous to the sharp changes there in $\Psi_{YY}^{(2)}$. An initially odd, but smooth, distribution of cross-wave velocity would, on the time scale T, evolve to the steady solution, generating the intense shears in the process.

A matter still not satisfactorily understood is the meaning of "off-critical" points and the dynamics of associated initial value problems.

References

1. Lin, C. C.: The Theory of Hydrodynamic Stability, Cambridge: Cambridge University Press 1966.
2. Bergeron, R. F., Jr.: A nonlinear critical layer theory for shear flow stability. Thesis, Massachusetts Institute of Technology, 1968.
3. Benney, D. J., Bergeron, R. F., Jr.: A new class of nonlinear waves in parallel flows. Studies in Applied Mathematics, 3, Sept. 1969.

Optimal Structural Design in Non-conservative Problems of Elastic Stability

By

M. Życzkowski and **A. Gajewski**

Politechnika Krakowska (Technical University of Cracow), Kraków, Poland

1. Introductory remarks

The problem of optimal shape of an elastic column subject to buckling under a compressive force with constant direction (Eulerian force) was formulated by LAGRANGE; basic solutions were given by CLAUSEN, BLASIUS, NICOLAI and TCHENTSOV. Such a behaviour of force corresponds to the existence of a potential and the system is conservative. In many cases, however, the behaviour of applied loading leads to non-conservative problems of elastic stability. One of the simplest examples — prismatic bar compressed by a tangential (follower) force — was solved by M. BECK [1] who used the kinetic criterion of stability. Z. KORDAS and M. ŻYCZKOWSKI [2] generalized this solution for any value of the tangency coefficient η, determined by Fig. 1.

Fig. 1

The present paper constitutes a first step towards optimization of such problems. We seek the shape of non-prismatic bars with minimal volume (minimal weight) corresponding to a given critical force with prescribed behaviour during buckling. This problem belongs to the class A 1 according to the classification of problems of optimal structural design proposed by M. ŻYCZKOWSKI [3]. Related problems of optimization in aeroelasticity were formulated by H. ASHLEY [4]. Inasmuch as the kinetic criterion of stability is used, the problem is also related to optimal design of a vibrating beam (F. Niordson [5]). Let us also mention the problem of optimization of pseudo-nonconservative systems, e.g. the force directed towards a constant pole (A. Gajewski and M. Życzkowski [6]). A certain general method of solution of non-conservative problems of stability (but without optimization) was suggested by H. Leipholz [7]. Some solutions for non-conservative problems of non-prismatic bars were given by A. Gajewski [8] and A. Kowalski [9].

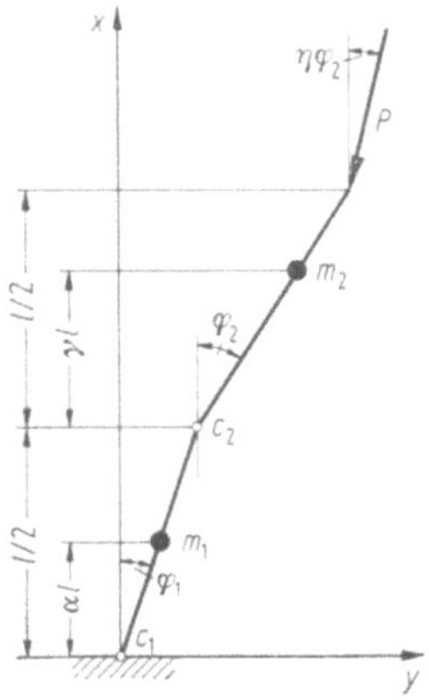

Fig. 2

2. Optimization of Ziegler's model with two degrees of freedom

To obtain a relatively simple analysis of the problem we start with the optimization of the system shown in Fig. 2. This system, consisting of two rigid bars of the length $l/2$ and two elastic hinges and approximating the real deformable bar, was introduced by H. Ziegler [10] and used later by numerous investigators for various purposes. Assume the force P to follow the deflections with the tangency coefficient η; according to the nomenclature proposed in [2] we call the force in the case $\eta = 1$ "tangential", $0 < \eta < 1$ "subtangential", $\eta > 1$ "supertangential", $\eta < 0$ "antitangential". The motion of the system is determined by the equations of Lagrange of the second kind:

$$\left.\begin{aligned} &\left(m_1\alpha^2 + \frac{1}{4}m_2\right)l^2\ddot{\varphi}_1 + \left(c_1 + c_2 - \frac{1}{2}Pl\right)\varphi_1 + \frac{1}{2}m_2\gamma l^2\ddot{\varphi}_2 \\ &\qquad\qquad + \left(\frac{1}{2}Pl\eta - c_2\right)\varphi_2 = 0, \\ &\frac{1}{2}m_2 l^2\gamma\ddot{\varphi}_1 - c_2\varphi_1 + m_2\gamma^2 l^2\ddot{\varphi}_2 + \left[c_2 - \frac{1}{2}Pl(1-\eta)\right]\varphi_2 = 0, \end{aligned}\right\} \qquad (2.1)$$

where φ_i are the deflection angles of the bars (assumed to be small), c_i the elastic constants of the hinges, and m_i the masses of the bars fixed at the distances αl and γl (Fig. 2).

Assume the solution of (2.1) in the form $\varphi_i = C_i e^{rt}$ and introduce dimensionless quantities as follows:

$$\beta = \frac{Pl}{c_2}, \quad \lambda = \frac{m_2 r^2 l^2}{c_2}, \quad \psi = \frac{c_1}{c_2}, \quad \mu = \frac{m_1}{m_2}. \qquad (2.2)$$

We arrive at a system of two linear algebraic equations with respect to the constants C_i. Equating the determinant of that system to zero we obtain a quadratic equation with respect to λ:

$$a\lambda^2 + (b - c\beta)\lambda + (d - e\beta + f\beta^2) = 0, \qquad (2.3)$$

where

$$\left.\begin{aligned} a &= 8\mu\alpha^2\gamma^2, & d &= 8\psi, \\ b &= 2(1+4\mu\alpha^2)+8\gamma^2(1+\psi)+8\gamma, & e &= 4(1-\eta)(2+\psi), \\ c &= (1-\eta)(1+4\mu\alpha^2)+4\gamma^2+2\eta\gamma, & f &= 2(1-\eta). \end{aligned}\right\} \quad (2.4)$$

As far as all roots of (2.3) yield various and imaginary values of r, the motion is stable. It may become unstable if $r = 0$ (static criterion), or if the discriminant $\Delta = 0$ (kinetic criterion of stability). The general equation (2.3) contains, as particular cases, those derived by I. A. Birger, B. F. Shorr and R. M. Shneyderovitch [11] ($\mu = 0$, $\gamma = 1/2$, $\psi = 1$), and by G. Herrmann and J. C. Jong [12] ($\mu = 2, \alpha = \gamma = 1/2$, $\psi = 1$).

Putting subsequently $\lambda = 0$ and $\Delta = 0$ we determine the dimensionless critical force β corresponding to the static criterion of stability

$$\beta_{1,2} = \frac{e \mp \sqrt{e^2 - df}}{2f} = \frac{(1-\eta)(2+\psi) \mp \sqrt{(1-\eta)[(1-\eta)(2+\psi)^2 - 4\psi]}}{1-\eta} \quad (2.5)$$

and to the kinetic criterion of stability

$$\beta_{3,4} = \frac{(bc - 2ae) \mp 2\sqrt{a[c(cd - be) + a(e^2 - 4df) + b^2 f]}}{c^2 - 4af}. \quad (2.6)$$

The static criterion is valid inside the intervals $-\infty < \eta \leq \eta^*$ and $\eta > 1$, where

$$\eta^* = \frac{4+\psi^2}{(2+\psi)^2}. \quad (2.7)$$

Inside the interval $\eta^* < \eta \leq 1$ the kinetic criterion must be used. If both criteria lead to the result, the smaller value of β should be taken.

To reduce the number of parameters we assume further $\alpha = \gamma = 1/2$ (masses concentrated at mid-points of rigid bars).

The optimal design of the system is here of parametrical type (this term was introduced in the paper by W. Krzyś and M. Życzkowski [13]). We look for an optimal value of the parameter $\psi = c_1/c_2$ for which the volume of the system will be minimal at a given value of β. It will be more convenient here to use the dual formulation (maximal force at a constant volume). At first we have to define the volume of the system. This should be done so as to approximate the real deformable bar. Assume the rigid bars to be non-prismatic with the cross-sections F_1 and F_2 in the hinges and $F_3 = 0$ at the free end (this feature is characteristic for optimal bars). Consider two cases: (1) the lower bar with linearly changing cross-sectional area, and the upper bar accordingly to the equation $F(x) = F_2(1 - x^n)$, then

$$V = V_1 = \frac{1}{4} l \left(F_1 + \frac{3n+1}{n+1} F_2\right); \quad (2.8)$$

(2) the cross-sectional area changes according to the quadratic parabola

$$F(x) = F_1 + (4F_2 - 3F_1)\,x + (2F_1 - 4F_2)\,x^2, \tag{2.9}$$

then

$$V = V_2 = \frac{1}{6}\,l(F_1 + 4F_2). \tag{2.10}$$

Further, we have to connect the areas F_i with the elastic stiffnesses c_i of the hinges. The following assumptions are here the most justified: (a) local relation $F_1 = kc_1^\varkappa$, $F_2 = kc_2^\varkappa$, where k is a certain constant, and the exponent $\varkappa$ depends on the type of the change of the cross-section, namely $\varkappa = 1$ for plane-tapered bar, buckling perpendicular to the plane of tapering, $\varkappa = 1/3$ for the same bar but buckling in the plane of tapering, finally $\varkappa = 1/2$ for spatially tapered bar (cone or pyramide); for example, considering the case (1) we arrive at

$$V = V_{1a} = \frac{1}{4}\,lkc_2^\varkappa\left(\frac{3n+1}{n+1} + \psi^\varkappa\right),\ \mu = \mu_{1a} = \frac{n+1}{2n}(1 + \psi^\varkappa); \tag{2.11}$$

(b) integral relation

$$kc_1^\varkappa = 4\int_0^{1/4} F(x)\,dx,\ \ kc_2^\varkappa = 2\int_{1/4}^{3/4} F(x)\,dx, \tag{2.12}$$

with the same sense of k and $\varkappa$, x being dimensionless variable. For example, combining (2.12) with (2.9) we obtain

$$V = V_{2b} = \frac{4}{19}\,lkc_2^\varkappa(3 + \psi^\varkappa),\ \ \mu = \mu_{2b} = \frac{13 + 17\psi^\varkappa}{23 - 5\psi^\varkappa}. \tag{2.13}$$

In the range of static criterion of stability, (2.5), we express the stiffness c_2 by means of the volume V and substitute it into the formula for the critical force; in the case (1) combined with (a) we obtain

$$P_{\text{cr}} = \left(\frac{4V}{lk}\right)^{1/\varkappa} \frac{(1-\eta)(2+\psi) - \sqrt{(1-\eta)\left[(1-\eta)(2+\psi)^2 - 4\psi\right]}}{l(1-\eta)\left(\dfrac{3n+1}{n+1} + \psi^\varkappa\right)^{1/\varkappa}}. \tag{2.14}$$

The analytical condition of maximum, $dP_{\text{cr}}/d\psi = 0$, leads here to the equation

$$\begin{aligned}(1-\eta)\,\psi^2 - \frac{n+1}{3n+1}(1-2\eta)\,\psi^{1+\varkappa} - 2\psi \\ - (1-2\eta)\frac{3n+1}{n+1}\psi^{1-\varkappa} + 4(1-\eta) = 0.\end{aligned} \tag{2.15}$$

Considering, for example, plane-tapered bar with $\varkappa = 1$ we obtain a quadratic equation with respect to ψ, whence

$$\psi = \psi_{\text{opt}} = \frac{\dfrac{(n+1)(3n+1)}{n-1} + \sqrt{(n+1)(3n+1)(1-\eta)(1-2\eta)}}{(n+1)\left(\dfrac{2n}{n-1} - \eta\right)}. \tag{2.16}$$

Similar procedure may be used in the range of kinetic criterion of stability, but the results are more complicated. In some cases the analytical condition of maximum must be replaced by the upper bound reached for $\psi = 0$.

In the most interesting case $\eta = 1$ (tangential force) for $\varkappa = 1$ the formula (2.6) yields

$$\beta = \frac{2}{3}(9 + \mu + \psi) - \frac{4}{3}\sqrt{\mu\psi}\,. \qquad (2.17)$$

Taking, for example, $n = 2$ we obtain $\psi_{\text{opt}} = 0$ and the profit in material $Z = 42.1\%$. Further numerical examples will be published separately (A. Gajewski [14]).

3. Optimization of an elastic bar compressed by antitangential force

Consider a bar shown in Fig. 1. We confine ourselves to the range of static stability criterion. The differential equation of the deflection line will be written in the form

$$(gy'')'' + \beta y'' = 0, \qquad (3.1)$$

where x and y are dimensionless variables $x = \xi/l$, $y = w/l$, g dimensionless rigidity $g(x) = EJ(x)/EJ_0$ (J_0 being the moment of inertia at a certain point x_0), β dimensionless force $\beta = Pl^2/EJ_0$. Four boundary conditions corresponding to the considered problem are of the form

$$\begin{aligned} y(0) &= 0, \qquad g(1)\,y''(1) = 0,\\ y'(0) &= 0, \qquad [(gy'')' + \beta(1-\eta)\,y']_{x=1} = 0. \end{aligned} \qquad (3.2)$$

Integrating (3.1) twice and introducing the new dependent variable

$$v(x) = y(x) - \frac{C_1}{\beta} - \frac{C_2}{\beta}x \qquad (3.3)$$

(C_i being the constants of integration) we obtain

$$gv'' + \beta v = 0 \qquad (3.4)$$

with two following boundary conditions:

$$v(1) = 0, \quad \eta v'(1) + (1-\eta)\,v'(0) = 0. \qquad (3.5)$$

To optimize the bar we seek the minimum of the functional (volume)

$$V = \int_0^l F(\xi)\,d\xi = F_0 l \int_0^1 [g(x)]^{\varkappa} dx \qquad (3.6)$$

with the auxiliary condition in the form of the differential equation (3.4). The exponent $\varkappa$ is equal to 1, 1/3 or 1/2 according to the remarks given in p. 298. Using the method of N. G. Tchentsov [15] we eliminate

g from (3.4) as to obtain

$$V = F_0 l \beta \int_0^1 \left(-\frac{v}{v''}\right)^{\varkappa} dx. \tag{3.7}$$

Following Tchentsov we write down the equation of Euler-Lagrange for this functional and integrate it once; this procedure leads to

$$v^{\varkappa}[(1-\varkappa)\,v'v'' + (1+\varkappa)\,vv'''] = Cv''^{\varkappa+2}. \tag{3.8}$$

We should fulfil two boundary conditions (3.5); further, Eq. (3.8) is homogeneous of the degree $\varkappa + 2$ and the functional homogeneous of the degree zero, thus the choice of one non-homogeneous boundary condition is meaningless and free — take it in the form $v(0) = 1$. We arrive therefore at the variational problem with one degree of freedom at the end of interval. The corresponding well-known condition of optimum in such a case

$$\begin{aligned} &-\left[\left(F_{v'} - \frac{d}{dx}F_{v''}\right)_0 \delta v_0 + (F_{v''})_0\, \delta v_0'\right] \\ &+\left[\left(F_{v'} - \frac{d}{dx}F_{v''}\right)_1 \delta v_1 + (F_{v''})_1\, \delta v_1'\right] = 0 \end{aligned} \tag{3.9}$$

leads in the case of Eulerian force and in the conservative problem of the force directed to a constant pole to the result $C = 0$, but in the considered problem the conditions (3.5) and the condition (3.9) with

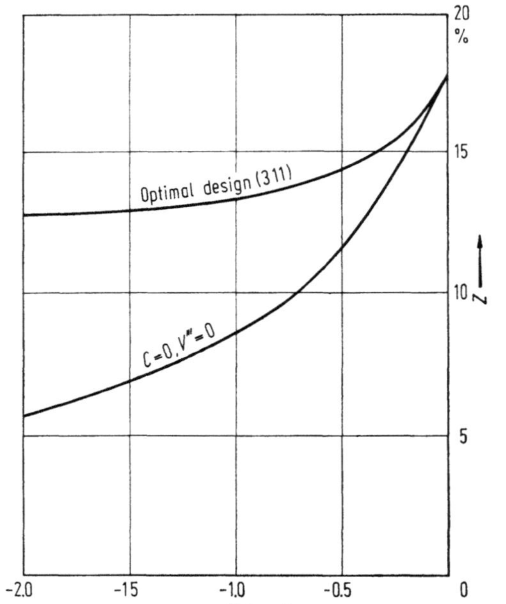

Fig. 3

substituted function $F = (-v/v'')^{\varkappa}$ yield (in the simplest case $\varkappa = 1$)

$$\left(\frac{v}{v''^2}\right)_0 + \frac{1-\eta}{\eta}\left(\frac{v}{v''^2}\right)_1 = 0. \tag{3.10}$$

The functions v and v'' are equal to zero at the point $x = 1$. Using (3.8) and de l'Hospital's rule we obtain finally

$$v_0' = C v_0''^2. \tag{3.11}$$

Thus the integration constant C in Eq. (3.8) is, in general, different from zero. Eq. (3.10) may be fulfilled only for negative values of η, thus we restricted our considerations to antitangential forces. For positive values of η we obtain no analytical solutions realizing the optimum (basing on the static criterion of stability).

In the case $C \neq 0$ Eq. (3.8) cannot be integrated exactly in a closed form. The results of numerical integration (using the computer "Odra 1013") are presented in Fig. 3, which shows the profit in material Z as a function of the tangency coefficient η. The solution of (3.8) for $C = 0$ is also shown for comparison; it is much worse than the solution making use of the condition (3.11).

References

1. Beck, M.: Z. angew. Math. Phys. **3**, 225–228 (1952).
2. Kordas, Z., Życzkowski, M.: Arch. Mech. Stos. **15**, 7–31 (1963).
3. Życzkowski, M.: J. Appl. Mech. (in print).
4. Ashley, H., McIntosh, S. C., Jr.: Proc. 12th Int. Congr. Appl. Mech., Stanford, August 1968. Berlin/Heidelberg/New York: Springer 1969, pp. 100–113.
5. Niordson, F.: Quart. Appl. Math. **23**, 47–53 (1965).
6. Gajewski, A., Życzkowski, M.: Rozpr. Inżyn. **17**, 479–488 (1969).
7. Leipholz, H.: Ing.-Arch. **34**, 56–68 (1965).
8. Gajewski, A.: Rozpr. Inżyn. **14**, 591–608 (1966); English summary: Bull. Acad. Pol., Sci. Tech. **15**, 245–253 (1967).
9. Kowalski, A.: Rozpr. Inżyn. **15**, 197–209 (1967).
10. Ziegler, H.: Z. angew. Math. Mech. **31**, 265 (1951); Advances in Appl. Mech. **4**, 351–403.
11. Birger, I. A., Shorr, B. F., Shneyderovitch, R. M.: Raschet na protchnost detaley mashin, Moskva 1966.
12. Herrmann, G., Jong, J. C.: J. Appl. Mech. **33**, 125–133 (1966).
13. Krzyś, W., Życzkowski, M.: Rozpr. Inżyn. **11**, 643–666 (1963); English summary: Bull. Acad. Pol., Sci. Tech. **11**, 335–345 (1963).
14. Gajewski, A.: Prace Komisji Mechaniki Oddz. Krak. PAN (in print).
15. Tchentsov, N. G.: Trudy CAGI **265** (1936).

Zur Stabilität des schwingenden Tragflügels im Unterschallbereich

Von

F. Wille

Universität Freiburg und Deutsche Forschungs- und Versuchsanstalt für Luft- und Raumfahrt (DFVLR), Freiburg i. Br., Deutschland

Summary. The oscillating wing with control surfaces in two-dimensional, compressible, subsonic flow is investigated. Computed numerical results are given for the forces and moments, useful for flutter calculations.

1. Einleitung

Bei Stabilitätsuntersuchungen an Flugzeugen (Flatterrechnungen) ist die Kenntnis der Trägheitskräfte, Elastizitätskräfte, strukturellen Dämpfungskräfte und Luftkräfte erforderlich, s. [3]. Am schwierigsten zu bestimmen sind die Luftkräfte. In diesem Artikel werden die Ergebnisse von Luftkraftberechnungen bei schwingenden dünnen Tragflügeln in ebener kompressibler Unterschallströmung angegeben. Dabei liegt die von H. G. Küssner in [5] angegebene Kernfunktionsmethode zugrunde, sowie das in [6—9] niedergelegte numerische Verfahren.

2. Vereinfachende Annahmen

Es soll der Luftdruck auf einen harmonisch schwingenden dünnen Rechteckflügel bestimmt werden, der mit konstanter Geschwindigkeit V angeströmt wird, wobei bei Durchgang durch die Ruhelage die Richtung von V mit der Flügelprofilachse zusammenfällt. Dabei werden folgende vereinfachende Annahmen gemacht.

Die Arbeit wurde zum überwiegenden Teil in der Aerodynamischen Versuchsanstalt Göttingen der DFVLR durchgeführt und von der Deutschen Forschungsgemeinschaft unterstützt.

1. Die Strömung ist reibungsfrei und zweidimensional.

2. Man ersetzt das Flügelprofil durch ein *Streckenprofil*, s. Bild 1, wobei Flügelflosse, Ruder und Hilfsruder einzelne starre Strecken sind, die Drehungen um die in Bild 1 skizzierten Drehpunkte ausführen und außerdem gemeinsam in Richtung der z-Achse verschoben werden können (Translation des Flügels).

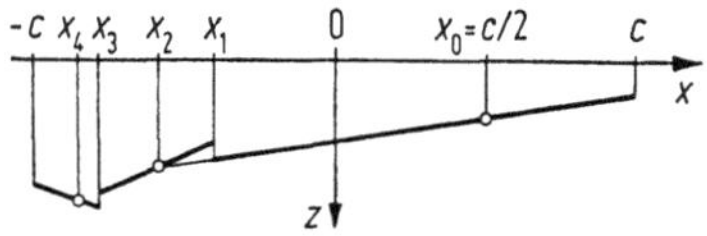

Bild 1. Auslenkung des Streckenprofils mit Ruder und Hilfsruder, überhöht dargestellt.

3. Der Luftdruck am Flügel hängt nur vom Abwind ab, das ist die z-Komponente der Strömung am Profil. Der Abwind $w(x, t)$ (t = Zeit) ergibt sich aus der Bewegung

$$z = \zeta(x)\, e^{\omega t}, \quad \omega \text{ imaginär}, \tag{1}$$

des Profils durch

$$w(x, t) = \frac{\partial z}{\partial t} - V \frac{\partial z}{\partial x} = [\omega \zeta(x) - V \zeta'(x)]\, e^{\omega t}, \tag{2}$$

s. [1, 5]. Darin liegt die weitere Vereinfachung, daß die x-Komponente der Strömungsgeschwindigkeit am Profil konstant gleich V ist, unabhängig von der Bewegung des Profils.

4. Die Schwingungsamplituden seien so klein, daß mit den linearisierten Eulerschen Gleichungen gearbeitet werden darf. Dies hat zur Folge, daß der Druck am Flügel linear von der Auslenkung $\zeta(x)$, s. (1), abhängt.

H. G. Küssner hat in [5] durch die von ihm geschaffene Kernfunktionsmethode angegeben, wie die Druckdifferenz $\Pi(x, t)$ zwischen Unter- und Oberseite des Profils bei gegebenem Abwind, also bei gegebener Bewegung (1), gewonnen wird. In [6, 7, 9] wird diese Methode in ein numerisches Verfahren verwandelt.

3. Freiheitsgrade

Bei den Flügelbewegungen $z = \zeta(x)\, e^{\omega t}$ werden die Auslenkungen $\zeta(x)$ linear aus acht Grundauslenkungen, den *Freiheitsgraden* a, b, c, d^0, $\bar{d}$, e, f^0, $\bar{f}$, zusammengesetzt, die aus Bild 2 zu ersehen sind. Bei Ruderschlag wird zwischen offener Stufe d^0 und geschlossener Stufe $\bar{d}$ unterschieden. Bei der offenen Stufe strömt die Luft ungehindert durch den Spalt zwischen Flosse und Ruder hindurch. Die geschlossene Stufe ist als Grenzfall der „schrägen" Stufe aufzufassen, bei der Flossenhinterkante und Rudervorderkante durch eine schräge Platte verbunden sind, vgl. [1, 7, 8]. Entsprechendes gilt für das Hilfsruder.

Alle Auslenkungen $\zeta(x)$, die nach Bild 1 möglich sind, lassen sich aus den 8 Freiheitsgraden linear zusammensetzen. Es genügt in der linearisierten Theorie also, die Drücke für die 8 Freiheitsgrade zu berechnen.

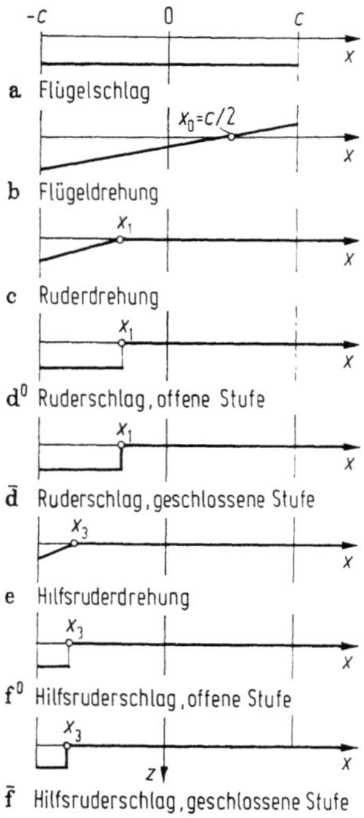

Bild 2. Freiheitsgrade des schwingenden Streckenprofiles.

Die zu den Freiheitsgraden $g = a, b, \ldots, \bar{f}$ gehörenden Druckdifferenzen bezeichnen wir mit

$$\Pi_g(x,t), \quad g = a, b, \ldots, \bar{f}. \tag{3}$$

Die Kraft auf einen Flügelteil von der Hinterkante $-c$ bis zu einer Stelle x, sowie der Spannweite b, ist für den Freiheitsgrad g:

$$K_g(x,t) = b \int_{-c}^{x} \Pi_g(\xi,t)\, d\xi. \tag{4}$$

Entsprechend ist das Moment dieses Flügelteiles um eine zur Profilebene senkrechte Achse durch x_g gleich

$$M_g(x,t) = b \int_{-c}^{x} (x_g - \xi)\, \Pi_g(\xi,t)\, d\xi, \tag{5}$$

wobei wir

$$x_g = \begin{cases} \frac{c}{2} & \text{für } g = a, b, \\ x_1 & \text{für } g = c, d^0, \bar{d}, \\ x_3 & \text{für } g = e, f^0, \bar{f} \end{cases}$$

wählen, s. Bild 1.

4. Derivativa

Die *Kraft-* und *Momentenderivativa* $k_a, k_b, \ldots, q_{\bar{f}}$ des schwingenden Streckenprofils werden durch folgende Gleichungen definiert [1, 7]:

Kräfte: Momente:

$$K_g(c,t) = \pi\varrho V^2 c b e^{\omega t} G \cdot k_g, \qquad M_g(c,t) = \pi\varrho V^2 c^2 b e^{\omega t} G \cdot m_g,$$
$$K_g(x_1,t) = \pi\varrho V^2 c b e^{\omega t} G \cdot r_g, \qquad M_g(x_1,t) = \pi\varrho V^2 c^2 b e^{\omega t} G \cdot n_g,$$
$$K_g(x_3,t) = \pi\varrho V^2 c b e^{\omega t} G \cdot p_g, \qquad M_g(x_3,t) = \pi\varrho V^2 c^2 b e^{\omega t} G \cdot q_g,$$
$$g = a, b, c, d^0, \bar{d}, e, f^0, \bar{f}.$$

Dabei ist ϱ die Luftdichte und G eine dimensionslose Amplitude in folgendem Sinne:

Für die Freiheitsgrade der Translation a, d^0, $\bar{d}$, f^0, $\bar{f}$ ist $c \cdot G$ der Maximalausschlag, für Drehungen b, c, e ist arctan (G) der maximale Winkelausschlag.

Die Derivativa k_g, m_g gehören also zu Kräften und Momenten des ganzen Flügels, r_g und n_g gehören zu Kräften und Momenten auf das Ruder und entsprechend p_g, q_g zum Hilfsruder. Die Derivativa hängen nur noch von der Machzahl M und *reduzierten Frequenz*

$$\omega^* = \frac{\omega \cdot c}{\sqrt{(1 - M^2)}}$$

ab, s. [7]. Die genaue Kenntnis der Derivativa setzt uns also instand, Kräfte und Momente an einem beliebig schwingenden Streckenprofil auf einfache Weise zu gewinnen.

Die numerische Berechnung der Derivativa erfordert einen hohen Rechenaufwand, da unendliche Reihen Mathieuscher Funktionen vorkommen, bei denen Konvergenzschwierigkeiten auftreten. Wie diese Schwierigkeiten überwunden werden können, ist in [6, 7, 9] dargestellt.

5. Numerische Ergebnisse

Ausführliche Tabellen numerisch berechneter Derivativa, die für Flatterrechnungen im Machzahlbereich $0 \leq M < 1$ gebraucht werden,

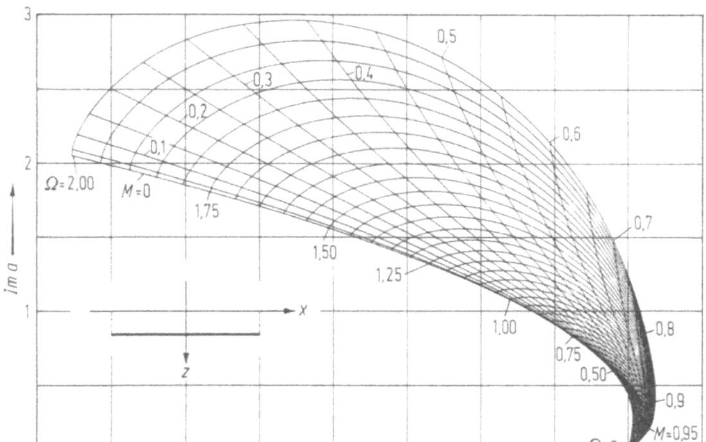

Bild 3. Derivativ k_a der Gesamtkraft bei Flügelschlagschwingung, dargestellt in der komplexen Ebene in Abhängigkeit von M und Ω.

sind in [8] wiedergegeben. Diese Tabellen sind umfangreicher und genauer als die früher von van de Vooren [2] herausgegebenen.

Einige Bilder, entnommen aus [7], sollen die Resultate veranschaulichen.

Die Bilder 3—11 zeigen graphische Darstellungen der für Flatterrechnungen wichtigsten Derivativa

$$k_a, m_a, k_b, m_b, k_c, m_c, n_a, n_b, n_c,$$

wobei eine relative Rudertiefe von $(x_1 + c)/(2c) = 0{,}3$ gewählt wurde.

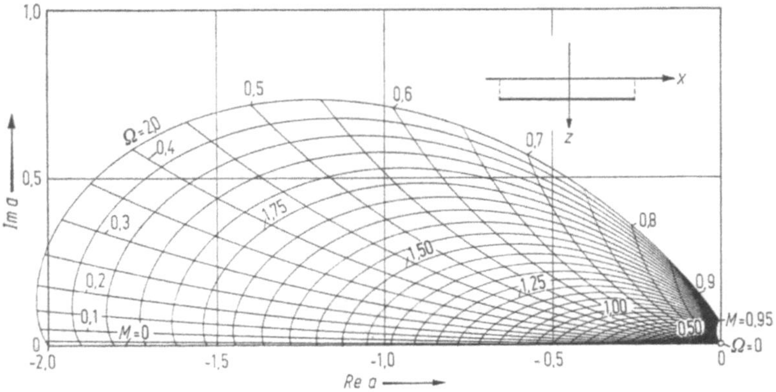

Bild 4. Derivativ m_a des Gesamtmomentes um den vorderen Neutralpunkt $c/2$ bei Flügelschlagschwingung.

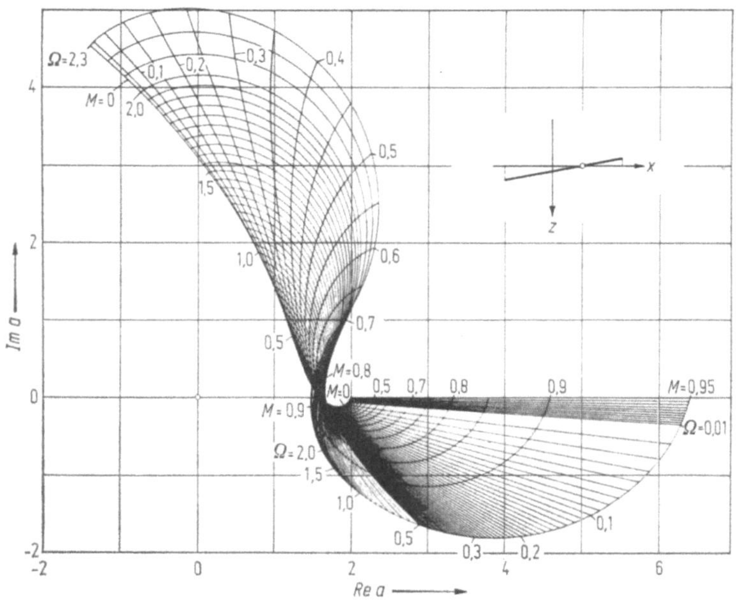

Bild 5. Derivativ k_b der Gesamtkraft bei Flügeldrehschwingung um den vorderen Neutralpunkt.

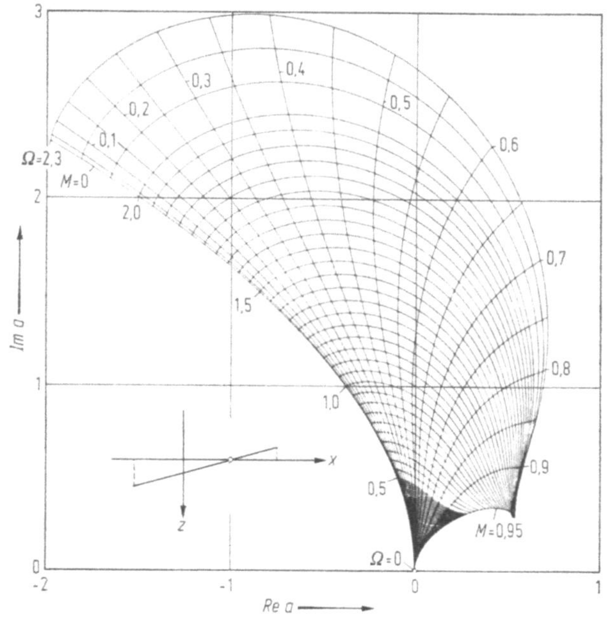

Bild 6. Derivativ m_b des Gesamtmomentes um den vorderen Neutralpunkt $c/2$ bei Flügeldrehschwingung um den vorderen Neutralpunkt.

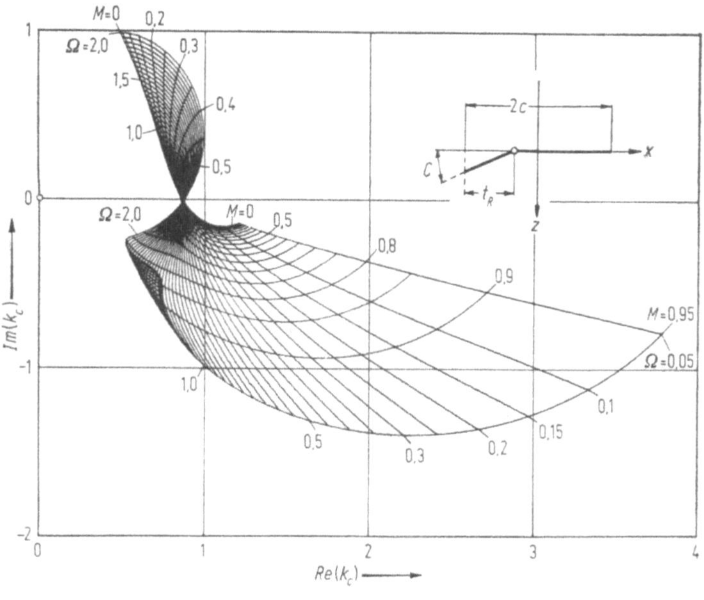

Bild 7. Derivativ k_c der Gesamtkraft bei Ruderdrehschwingung um die Rudervorderkante. Relative Rudertiefe $t_R/(2c) = 0{,}3$.

Die Werte der Derivativa sind in der komplexen Zahlenebene durch die Kurven konstanter Machzahl M und konstanter reduzierter Frequenz ω^*, $\Omega = |\omega^*|$, aufgetragen (vgl. [7]).

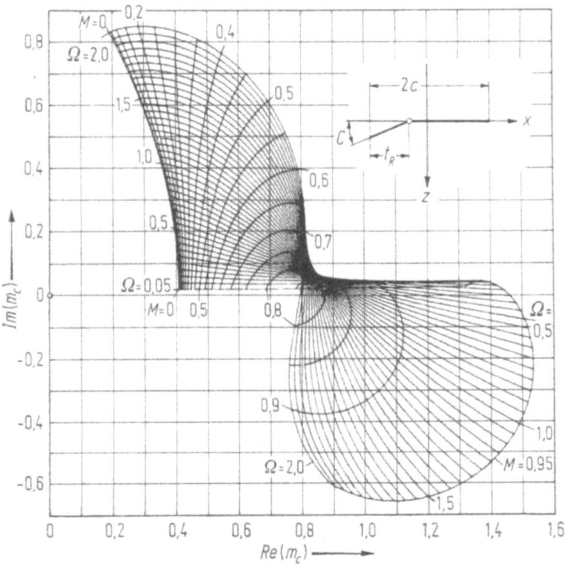

Bild 8. Derivativ m_c des Gesamtmomentes um den vorderen Neutralpunkt $c/2$ bei Ruderdrehschwingung um die Rudervorderkante. Relative Rudertiefe 0,3.

Stellt man die Derivativa in der Form $re^{i\varphi}$ $(r \geq 0)$ dar, so bedeutet φ die Phasenverschiebung, um die die zugehörige Kraft bzw. das Moment der Bewegung des Flügels *nachläuft*. Negative Imaginärteile bei den Derivativa bewirken also Anfachung, positive dagegen Dämpfung.

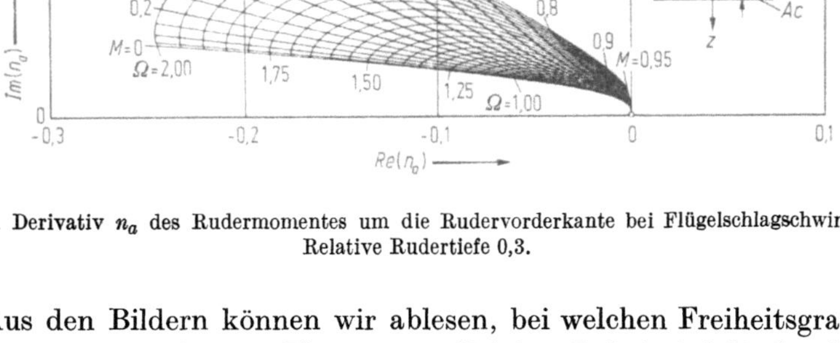

Bild 9. Derivativ n_a des Rudermomentes um die Rudervorderkante bei Flügelschlagschwingung. Relative Rudertiefe 0,3.

Aus den Bildern können wir ablesen, bei welchen Freiheitsgraden, wenn sie allein auftreten, Flattern möglich ist. Dabei wird die Strukturdämpfung vernachlässigt. Bei Flügelschlag oder Flügeldrehung um $c/2$

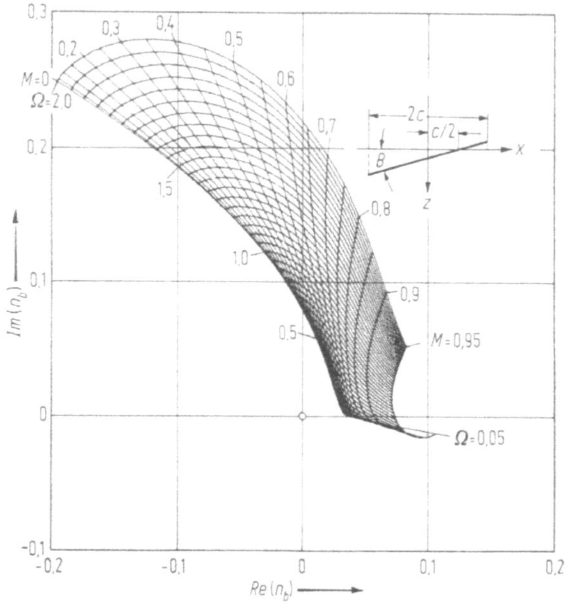

Bild 10. Derivativ n_b des Rudermomentes um die Rudervorderkante bei Flügeldrehschwingung um den vorderen Neutralpunkt $c/2$. Relative Rudertiefe 0,3.

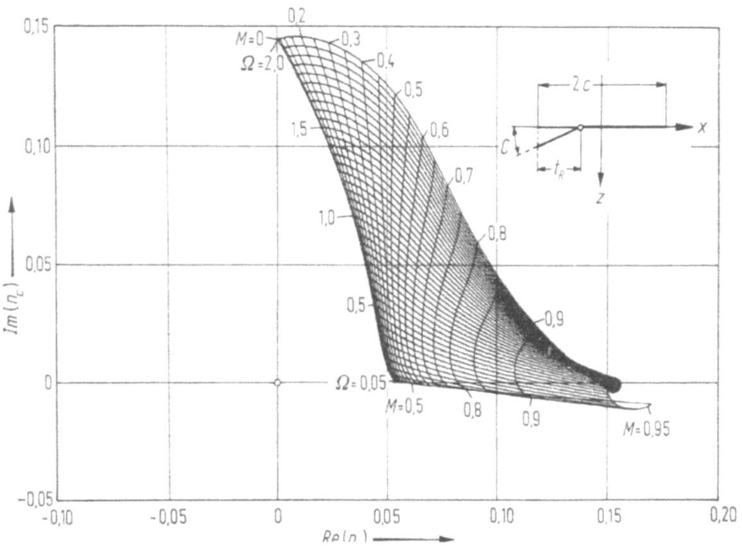

Bild 11. Derivativ n_c des Rudermomentes um die Rudervorderkante bei Ruderdrehschwingung um die Rudervorderkante. Relative Rudertiefe 0,3.

tritt kein Flattern auf, da bei k_a als auch bei m_b keine negativen Imaginärteile vorkommen (Bilder 3 und 6). Dagegen ist die Drehschwingung des Ruders um die Rudervorderkante bei hohen Machzahlen und niedrigen reduzierten Frequenzen instabil, wie aus Bild 11 hervorgeht, da dort Im $(n_c) < 0$ ist. Dies ist in der Praxis auch beobachtet worden.

Literatur

1. Küssner, H. G., Schwarz, L.: Der schwingende Flügel mit aerodynamisch ausgeglichenem Ruder. Luftfahrtforschung **17**, 337—354 (1940).
2. van de Vooren, A. I.: Collected Tables and Graphs. AGARD Manual on Aeroelasticity, Vol. VI/1 (1961).
3. Küssner, H. G., Göllnitz, H.: Theorie und Methode der Flatterrechnung von Flugzeugen unter Benutzung des Standschwingungsversuchs. Aerodynamische Versuchsanstalt Göttingen, Forsch.-Ber. 64-01 (1964).
4. Küssner, H. G., Göllnitz, H.: Tabellen der aerodynamischen Derivativa des schwingenden Streckenprofils mit Knicken und Stufen. Deutsche Luft- und Raumfahrt, Forsch.-Ber. 64-05 (1964).
5. Küssner, H. G.: Theorie des schwingenden Streckenprofils bei kompressibler Unterschallströmung. Teil I: Herleitung der geschlossenen Lösung für den Stördruck. Aerodynamische Versuchsanstalt Göttingen, Ber. 67 J 03 (1967) und Z. Flugwiss. **15**, 315—323 (1967).
6. Wille, F.: Theorie des schwingenden Streckenprofils bei kompressibler Unterschallströmung. Teil II: Berechnung der Küssnerschen Wirbelschleppenfunktion (T-Funktion). Deutsche Luft- und Raumfahrt, Forsch.-Ber. 68-04 (1968).
7. Wille, F.: Theorie des schwingenden Streckenprofils bei kompressibler Unterschallströmung. Teil III: Berechnung der instationären aerodynamischen Derivativa. Deutsche Luft- und Raumfahrt, Forsch.-Ber. 68-05 (1968).
8. Küssner, H. G., Wille, F.: Theorie des schwingenden Streckenprofils bei kompressibler Unterschallströmung. Teil IV: Tabellen der aerodynamischen Derivativa des schwingenden Streckenprofils mit Knick und Stufe. Deutsche Luft- und Raumfahrt, Forsch.-Ber. 68-76 (1968).
9. Wille, F.: Programm zur Berechnung instationärer aerodynamischer Derivativa eines schwingenden Streckenprofils in kompressibler Unterschallströmung. Aerodynamische Versuchsanstalt Göttingen, Ber. 68 R 01 (1968).

Finite Amplitude Response of Circular Plates Subject to Dynamic Loading

By

W. A. Nash and **H. Kanematsu**

University of Massachusetts, Amherst, Mass., U.S.A.

Abstract. The nonlinear equations describing axisymmetric finite amplitude response of a thin elastic clamped edge circular plate to pulse loading are solved by finite difference techniques for the case of a spatially uniform load which time-wise is represented as a step function. Damping of the plate is considered.

Introduction

The problem of finite amplitude dynamic response of thin elastic plates has been attacked only in recent years, although the corresponding static problem has been the subject of many investigations during the past two decades. In the present paper the equations governing dynamic response are considered to be the dynamic analog of the von Karman large-deflection plate equations which are valid for lateral deflections of moderate amplitude, i.e. of the order of magnitude of the plate thickness.

To date, the use of the finite deflection equations for dynamic considerations has been extremely limited because of the severe mathematical difficulties encountered. Chu and Herrmann [1] have investigated the large deflections during free vibration of a rectangular plate with hinged immovable edges. Yamaki [2] has studied one-term solutions for both clamped as well as simply supported plates for free and forced vibrations. The first author of the present paper has presented an approximate analysis of finite amplitude undamped free vibrations of circular plates based upon the technique of neglect of the second invariant of the strain energy [3]. Very recently, Bauer [4] investigated

Acknowledgement. This work was sponsored by the Air Force Office of Scientific Research under Grant AFOSR 68-1527.

the flexural vibrations of both simply supported as well as clamped edge rectangular plates subject to two pulse loads: (a) a step function, and (b) an exponentially decaying pulse. The solution was assumed to be space-time separable and numerical results were presented only for the case of a square plate.

The objective of the present paper is to present a solution that does not pre-suppose a separable nature for the case of a step-function type loading on a thin, elastic clamped edge circular plate.

Analysis

The basic equations governing the axisymmetric nonlinear finite amplitude motions of a thin elastic circular plate subject to a normal pressure $\bar{q}(r, t)$ are

$$D\nabla^4 \bar{w} = h\frac{\partial^2 \bar{w}}{\partial \bar{r}^2}\left(\frac{1}{\bar{r}}\frac{\partial \bar{F}}{\partial \bar{r}}\right) + h\frac{\partial^2 \bar{F}}{\partial \bar{r}^2}\frac{1}{\bar{r}}\frac{\partial \bar{w}}{\partial \bar{r}} + \bar{q}, \tag{1}$$

$$\nabla^4 \bar{F} = -E\frac{1}{\bar{r}}\frac{\partial^2 \bar{w}}{\partial \bar{r}^2}\frac{\partial \bar{w}}{\partial \bar{r}}, \tag{2}$$

where h represents the plate thickness, $\bar{w}$ is the transverse displacement of a point in the middle surface of the plate, $\bar{F}$ is the Airy function of the middle surface forces per unit length of the middle surface, $\bar{q}$ is the lateral load per unit area, $D = Eh^3/12(1-\nu^2)$ is the flexural rigidity of the plate where E is Young's modulus and ν represents Poisson's rate and

$$\nabla^4 = \frac{\partial^4}{\partial \bar{r}^4} + \frac{2}{\bar{r}}\frac{\partial^3}{\partial \bar{r}^3} - \frac{1}{\bar{r}^2}\frac{\partial^2}{\partial \bar{r}^2} + \frac{1}{\bar{r}^3}\frac{\partial}{\partial \bar{r}}.$$

Eqs. (1) and (2) may be employed to treat dynamic problems if the lateral load $\bar{q}$ is taken to represent the transverse inertia loading $\varrho\, \partial^2\bar{w}/\partial\bar{t}^2$ where ϱ denotes plate density and $\bar{t}$ is time together with the driving force $\bar{p}(r, t)$ which acts normal to the middle surface of the plate. Further, viscous damping is readily introduced by considering another factor $\bar{c}\;\partial\bar{w}/\partial\bar{t}$ in the lateral load $\bar{q}$. The final equation of motion corresponding to Eq. (1) is thus

$$\varrho\frac{\partial^2 \bar{w}}{\partial \bar{t}^2} + \bar{c}\frac{\partial \bar{w}}{\partial \bar{t}} = -D\nabla^4 \bar{w} + h\frac{\partial^2 \bar{w}}{\partial \bar{r}^2}\left(\frac{1}{\bar{r}}\frac{\partial \bar{F}}{\partial \bar{r}}\right) + h\frac{\partial^2 \bar{F}}{\partial \bar{r}^2}\frac{1}{\bar{r}}\frac{\partial \bar{w}}{\partial \bar{r}} + \bar{p}(r, t). \tag{3}$$

To render Eqs. (2) and (3) dimensionless we introduce the following dimensionless variables:

$$\begin{aligned} w &= \bar{w}/h; & t &= \bar{t}\left(\frac{D}{\varrho h a^4}\right)^{1/2}, \\ r &= \bar{r}/a; & c &= \bar{c}\left(\frac{\varrho h a^4}{D}\right)^{1/2}, \\ F &= \bar{F}/Eh^2; & p &= \bar{p}a^4/Eh^4, \end{aligned} \tag{4}$$

where a denotes the radius of the plate.

The dimensionless forms of Eqs. (3) and (2) thus become

$$\frac{\partial^2 w}{\partial t^2} + c\,\frac{\partial w}{\partial t} = -\nabla^4 w + 12(1-\nu^2)\left\{N_\theta\,\frac{\partial^2 w}{\partial r^2} + \frac{N_r}{r}\frac{\partial w}{\partial r}\right\} + 12(1-\nu^2)\,p, \quad (5)$$

$$\nabla^4 F = -\frac{1}{r}\,\frac{\partial^2 w}{\partial r^2}\,\frac{\partial w}{\partial r}, \quad (6)$$

where N_θ and N_r represent membrane forces per unit length of the middle surface in the tangential and radial directions respectively as found from the relations

$$N_\theta = \frac{\partial^2 F}{\partial r^2}; \qquad N_r = \frac{1}{r}\,\frac{\partial F}{\partial r}. \quad (7)$$

The boundary conditions at the clamped edge may be expressed in dimensionless form as

$$w = \frac{\partial w}{\partial r} = 0 \quad \text{at} \quad r = 1, \quad (8)$$

$$\frac{\partial^2 F}{\partial r^2} - \frac{\nu}{r}\,\frac{\partial F}{\partial r} = 0 \quad \text{at} \quad r = 1. \quad (9)$$

In the formulation of the finite-difference representation of the problem it will become necessary to consider a fictitious point outside the boundary of the plate. For this reason it is necessary to develop an additional relation involving the variables w and F. One convenient way to do this is to consider the two representations of the radial strain, viz.:

$$\frac{1}{r}\,\frac{\partial F}{\partial r} - \nu\,\frac{\partial^2 F}{\partial r^2} = \frac{\partial u}{\partial r} + \frac{1}{2}\left(\frac{\partial w}{\partial r}\right)^2, \quad (10)$$

where u denotes the radial displacement of a point in the middle surface. Integrating over the radius and taking advantage of boundary conditions this furnishes the desired relation:

$$\int_0^1 \left(\frac{1}{r}\,\frac{\partial F}{\partial r} - \nu\,\frac{\partial^2 F}{\partial r^2}\right) dr = \frac{1}{2}\int_0^1 \left(\frac{\partial w}{\partial r}\right)^2 dr. \quad (11)$$

Eqs. (5) and (6) may now be formulated in finite difference form but it is desirable to first transform Eq. (4) to two lower order equations through introduction of new variables U and V defined as follows

$$U = \frac{\partial^2 u}{\partial r^2}, \quad (12)$$

$$V = \frac{\partial w}{\partial t}. \quad (13)$$

If Eq. (12) is differentiated with respect to t we get

$$\frac{\partial U}{\partial t} = \frac{\partial^2 V}{\partial r^2}, \quad (14)$$

and it is possible to write Eq. (5) in the form

$$V_{,t} + cV = -\left(U_{,rr} + \frac{2}{r} U_{,r} - \frac{1}{r^2} U + \frac{1}{r^3} w_{,r}\right) + 12(1-\nu^2)\, p + 12(1-\nu^2)\left(N_\theta U + \frac{N_r}{r} w_{,r}\right) \tag{15}$$

where a comma denotes partial differentiation with respect to the variable following the comma. This equation may be put in matrix form as follows:

$$\begin{pmatrix}1 & 0\\ 0 & 1\end{pmatrix}\begin{pmatrix}U\\ V\end{pmatrix}_{,t} + \begin{pmatrix}0 & 0\\ 0 & c\end{pmatrix}\begin{pmatrix}U\\ V\end{pmatrix} = -\begin{pmatrix}0 & -1\\ 1 & 0\end{pmatrix}\begin{pmatrix}U\\ V\end{pmatrix}_{,rr} - \begin{pmatrix}0 & 0\\ \frac{2}{r} & 0\end{pmatrix}\begin{pmatrix}U\\ V\end{pmatrix}_{,r} + \begin{pmatrix}0 & 0\\ \frac{1}{r^2} & 0\end{pmatrix}\begin{pmatrix}U\\ V\end{pmatrix} + 12(1-\nu^2)\, N_\theta \begin{pmatrix}0 & 0\\ 1 & 0\end{pmatrix}\begin{pmatrix}U\\ V\end{pmatrix} + \begin{pmatrix}0\\ 12(1-\nu^2)\, p + \frac{N_r}{r} w_{,r}\, 12(1-\nu^2) - \frac{1}{r^3} w_{,r}\end{pmatrix}. \tag{16}$$

If we define $\boldsymbol{U} = \begin{pmatrix}U\\ V\end{pmatrix}$; Eq. (16) becomes

$$\begin{pmatrix}1 & 0\\ 0 & 1\end{pmatrix}\boldsymbol{U}_{,t} + \begin{pmatrix}0 & 0\\ 0 & c\end{pmatrix}\boldsymbol{U} = -\begin{pmatrix}0 & -1\\ 1 & 0\end{pmatrix}\boldsymbol{U}_{,rr} - \begin{pmatrix}0 & 0\\ \frac{2}{r} & 0\end{pmatrix}\boldsymbol{U}_{,r} + \begin{pmatrix}0 & 0\\ 12(1-\nu^2)\, N_\theta + \frac{1}{r^2} & 0\end{pmatrix}\boldsymbol{U} + \begin{pmatrix}0\\ 12(1-\nu^2)\, p + \left\{12(1-\nu^2)\frac{N_r}{r} - \frac{1}{r^3}\right\} w_{,r}\end{pmatrix}. \tag{17}$$

In central finite difference form Eq. (17) can be written as

$$\begin{aligned}&\begin{pmatrix}1 & 0\\ 0 & 1\end{pmatrix}\left(\frac{\boldsymbol{U}_m^{n+1} - \boldsymbol{U}_m^n}{\Delta t}\right) + \begin{pmatrix}0 & 0\\ -12(1-\nu^2)\, N_\theta - \frac{1}{r^2} & 0\end{pmatrix}\left(\frac{\boldsymbol{U}_m^{n+1} + \boldsymbol{U}_m^n}{2}\right)\\ &= \begin{pmatrix}0 & -1\\ 1 & 0\end{pmatrix}\left(\frac{\boldsymbol{U}_{m+1}^{n+1} - 2\boldsymbol{U}_m^{n+1} + \boldsymbol{U}_{m-1}^{n+1} + \boldsymbol{U}_{m+1}^n - 2\boldsymbol{U}_m^n + \boldsymbol{U}_{m-1}^n}{2(\Delta r)^2}\right)\\ &\quad - \begin{pmatrix}0 & 0\\ \frac{2}{r} & 0\end{pmatrix}\left(\frac{\boldsymbol{U}_{m+1}^{n+1} - \boldsymbol{U}_{m-1}^{n+1} + \boldsymbol{U}_{m+1}^n - \boldsymbol{U}_{m-1}^n}{4(\Delta r)}\right)\\ &\quad + \begin{pmatrix}0\\ 12(1-\nu^2)\, p + \left[\frac{N_r}{r} 12(1-\nu^2) - \frac{1}{r^3}\right]\left[\frac{w_{m+1}^n - w_{m-1}^n}{2(\Delta r)} + \frac{\Delta t\left(V_{m+1}^{n+1} - V_{m-1}^{n+1}\right)}{4(\Delta r)}\right]\end{pmatrix},\end{aligned} \tag{18}$$

where the superscripts represent the time level and the subscripts represent location of the point along the radius as indicated below:

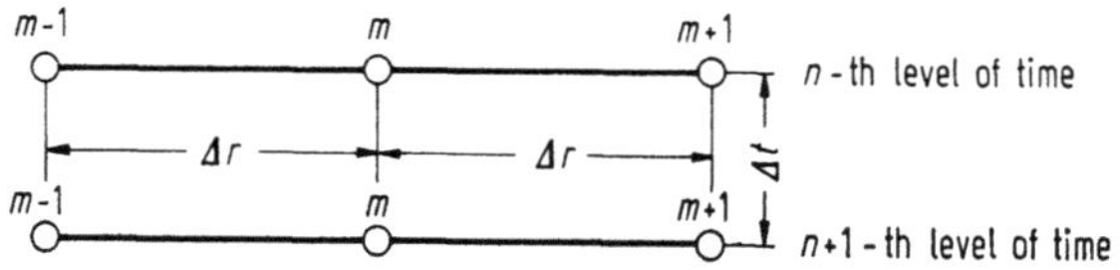

Eq. (15) cannot be used at $r = 0$ but an expression analogous to Eq. (18) may be derived for that point and its neighbors through use of l'Hospital's rule.

Similarly, the compatibility equation (6) may be expressed in finite difference form as:

$$\begin{aligned} &[F_{m+2} - 4F_{m+1} + 6F_m - 4F_{m-1} + F_{m-2}] \frac{1}{(\Delta r)^4} \\ &+ [F_{m+2} - 2F_{m+1} + 2F_{m-1} - F_{m-2}] \frac{1}{(r_m)(\Delta r)^3} \\ &- [F_{m+1} - 2F_m + F_{m-1}] \frac{1}{(r_m)^2 (\Delta r)^2} \\ &+ [F_{m+1} - F_{m-1}] \frac{1}{2(r_m)^3 \Delta r} \\ &= \frac{1}{(r_m)} [w_{m+1} - 2w_m + w_{m-1}] [w_{m+1} - w_{m-1}] \frac{1}{2(\Delta r)^3}, \end{aligned} \tag{19}$$

where the subscripts refer to points along the radius as indicated below:

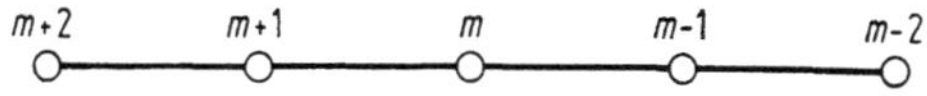

Lastly, the boundary conditions (8), (9) and the relation (11) may readily be expressed in finite difference forms.

The procedure for solving Eqs. (18) and (19) by digital computer techniques is to first specify the initial conditions of the plate so as to determine the initial membrane stresses. Next, the equation of motion is integrated along the time axis from t to $(t + \Delta t)$. The result of this step is then substituted into the right side of Eq. (19) and this equation solved for F by a matrix inverse method. The membrane forces corresponding to this value of F are then substituted into the right side of Eq. (18) and this equation solved for $\boldsymbol{U}$. The value of w is then determined at every point in the grid. The computation is then repeated for the next time increment.

Results

The finite difference analysis was carried out on a CDC 3600 digital computer for the case of a step function type loading applied uniformly over the entire surface of the plate. For the case of a dimensionless load $\bar{p}a^4/Eh^4 = 10$ and a dimensionless damping coefficient $c = 16$ (corresponding to moderate damping) it was found satisfactory to employ a dimensionless time increment $\Delta t = 0.002$ together with a dimensionless radial coordinate $\Delta r = 0.2$ to obtain the central deflection as a function of time as indicated in Fig. 1. The value of this deflection approaches 0.97 as time increases and the initial transients have

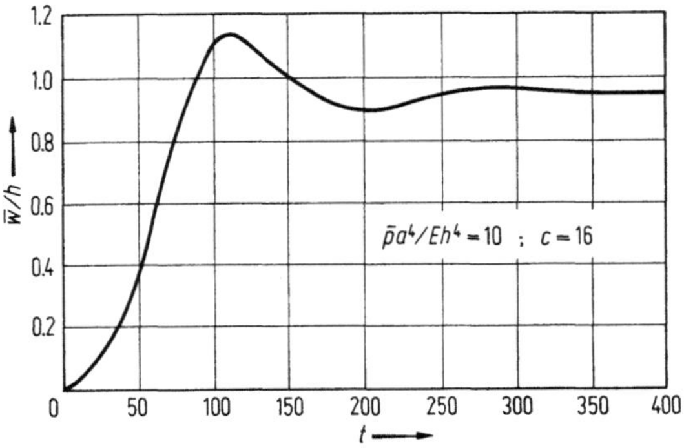

Fig. 1

damped out. This compares with the value of 1.05 reported by Kornishin and Isanvaeva [5] in their digital solution of the case of static loading of like magnitude. For the dynamic loading treated in the present paper, the magnitude of the initial transient response as well as suitable value of Δr and Δt to obtain numerical convergence depend strongly upon the damping factor involved.

References

1. Chu, H. N., Herrmann, G.: Influence of large amplitudes on free flexural vibrations of rectangular elastic plates. J. Appl. Mech., Trans. ASME, **23**, 532–540 (1956).
2. Yamaki, N.: Influence of large amplitudes on flexural vibrations of elastic plates. Z. angew. Math. Mech. **41**, 501–510 (1961).
3. Nash, W. A., Modeer, J. R.: Certain approximate analyses of the nonlinear behavior of plates and shallow shells. Proc. of the IUTAM Symposium on the Theory of Thin Elastic Shells, Amsterdam: North-Holland Publ. Co. 1960, pp. 331–354.
4. Bauer, H.: Nonlinear response of elastic plates to pulse excitations. J. Appl. Mech., Trans. ASME, **35**, 47–52 (1968).
5. Kornishin, M. S., Isanvaeva, F. S.: Flexible Plates and Shells, Moscow: Izd. "Nauka" 1968, 258 pages.

Equilibrium and Stability of Elastic-plastic Bodies

By

J. Christoffersen

Technical University of Denmark, Copenhagen, Denmark

1. Introduction

The existence of a yield-surface in stress-space, is generally taken to be the most fundamental feature of elastic-plastic materials. This means that stresses σ_{ij} satisfy a yield-condition

$$\psi(\boldsymbol{\sigma}) \leq 0 \tag{1.01}$$

(VON MISES [9]). Stresses are determined by the elastic strains

$$\sigma_{ij} = A_{ijkl}\gamma_{kl}. \tag{1.02}$$

The total strain ε_{ij} is the sum of elastic strains and plastic strains α_{ij},

$$\varepsilon_{ij} = \gamma_{ij} + \alpha_{ij}. \tag{1.03}$$

The plastic strain is assumed to be constant when the inequality sign applies in (1.01), whereas for $\psi(\boldsymbol{\sigma}) = 0$ the increment of α_{ij} is determined so as to insure that an inadmissible state of stress will not occur.

The condition of normality states that the change of plastic strain $\delta\alpha_{ij}$ is proportional with a positive factor to the outward pointing normal ν_{ij} of the yield-surface

$$\nu_{ij} = \frac{\partial\psi}{\partial\sigma_{ij}} \tag{1.04}$$

(VON MISES [9]).

These assumptions form the basis on which the constitutive equations for elastic-plastic materials are constructed relating stress-increments to strain-increments,

$$\delta\varepsilon_{ij} = B_{ijkl}\,\delta\sigma_{kl} + \nu_{ij}\,\delta\alpha, \tag{1.05}$$

where B_{ijkl} is the inverse tensor of elastic moduli. In the case of non-hardening materials

$$\begin{aligned} &\delta\alpha = 0 \text{ for either } \psi(\boldsymbol{\sigma}) < 0 \text{ or } (\psi(\boldsymbol{\sigma}) = 0 \text{ and } \nu_{ij}\,\delta\sigma_{ij} < 0); \\ &\delta\alpha \geq 0 \text{ for } \psi(\boldsymbol{\sigma}) = 0 \text{ and } \nu_{ij}\,\delta\sigma_{ij} = 0 \end{aligned} \tag{1.06}$$

(PRANDTL [11], REUSS [12]), whereas in the case of work-hardening

$$\begin{aligned} &\delta\alpha = 0 \text{ for either } \psi(\boldsymbol{\sigma}) < 0 \text{ or } (\psi(\boldsymbol{\sigma}) = 0 \text{ and } \nu_{ij}\,\delta\sigma_{ij} \leq 0); \\ &\delta\alpha = \frac{1}{\chi}\nu_{ij}\,\delta\sigma_{ij} \text{ for } \psi(\boldsymbol{\sigma}) = 0 \text{ and } \nu_{ij}\,\delta\sigma_{ij} > 0. \end{aligned} \tag{1.07}$$

Here χ is the work-hardening function satisfying

$$\chi > 0 \tag{1.08}$$

(MELAN [8], PRAGER [10]).

The inversion of Eq. (1.05) reads

$$\delta\sigma_{ij} = \begin{cases} A_{ijkl}\,\delta\varepsilon_{kl} \text{ for either } \psi(\boldsymbol{\sigma}) < 0 \text{ or } (\psi(\boldsymbol{\sigma}) = 0 \text{ and } A_{ijkl}\nu_{kl}\,\delta\varepsilon_{ij} \leq 0) \\ \left(A_{ijkl} - \dfrac{A_{ijmn}\nu_{mn}A_{klpq}\nu_{pq}}{\chi + A_{rstu}\nu_{tu}\nu_{rs}}\right)\delta\varepsilon_{kl} \text{ for } \psi(\boldsymbol{\sigma}) = 0 \text{ and } A_{ijkl}\nu_{kl}\,\delta\varepsilon_{ij} > 0. \end{cases} \tag{1.09}$$

In the case of non-hardening materials

$$\chi = 0 \tag{1.10}$$

(HILL [6]).

It is worth while noting that the constitutive equations (1.09) may violate the yield-condition (1.01). A set of equations consistent with this condition is

$$\delta\sigma_{ij} = \begin{cases} A_{ijkl}\,\delta\varepsilon_{kl} \text{ for } A_{ijkl}\nu_{kl}\,\delta\varepsilon_{ij} + \psi \leq 0 \\ \left(A_{ijkl} - \dfrac{A_{ijmn}\nu_{mn}A_{klpq}\nu_{pq}}{\chi + A_{rstu}\nu_{tu}\nu_{rs}}\right)\delta\varepsilon_{kl} - \dfrac{A_{ijkl}\nu_{kl}\psi}{A_{mnpq}\nu_{pq}\nu_{mn} + \chi} \\ \qquad \text{for } A_{ijkl}\nu_{kl}\,\delta\varepsilon_{ij} + \psi > 0. \end{cases} \tag{1.11}$$

The present paper contains a closer investigation of the background of the yield-condition (1.01) and the constitutive equations (1.11).

The results are obtained by an application of classical thermodynamics. The idea of approaching problems of elastic-plastic deformation from the thermodynamical side is not new. Investigations along these lines are performed by BRIDGMAN [1], DRUCKER [2], GREEN and NAGHDI [4], and ZIEGLER [13]. In the present paper the classical method of GIBBS [3] is employed to provide conditions of equilibrium and stability. The effects of large deformations are incorporated in the theory. This aspect has been treated earlier by HILL [7] and GREEN and NAGHDI [4].

Considering thermodynamics a discipline which applies to real bodies the author could hardly claim to have contributed substantially to a deeper understanding of the physical background of elastic-plastic deformation. The body to which thermodynamics is applied is defined as a continuum. As such it could hardly be characterized as a real body, so the method is in fact purely formal.

On the other hand classical thermodynamics may be formulated in entirely macroscopical terms, and as such it lends itself well to an application within the context of a continuous description. This is at the expence of a bit of clarity as regards the real nature of the inherent thermodynamical concepts entropy, temperature, and heat. The mystery may, however, be resolved when proper attention is paid to the atomistic structure of matter.

2. Thermodynamical concepts

To a body in equilibrium we may ascribe values of the internal energy U, the entropy S, the absolute temperature T, and the free energy F. These are all state functions, and they are interrelated through the formula

$$F = U - TS. \tag{2.01}$$

The values of the state functions are generally changed when work dW is performed on the body and heat dQ is supplied to it. The change is subject to certain restrictions known as the 1st and 2nd laws of thermodynamics. The 1st law is the Mayer-Joule conservation principle

$$dU = dW + dQ, \tag{2.02}$$

and the 2nd law may be stated as the Clausius-Kelvin inequality

$$dS \geq \frac{dQ}{T}. \tag{2.03}$$

When the equality sign applies the process is reversible, otherwise it is irreversible or spontaneous.

When the body is brought from one state of equilibrium into another with the same temperature we speak of an isothermal process. In this case the two expressions (2.02) and (2.03) may be combined into one which states that

$$dW \geq dF, \tag{2.04}$$

i.e. the increase of free energy of the body cannot exceed the amount of work performed on it. The excess of work appearing in the irreversible case is dissipated.

Following Gibbs we may state the following criterion for stable equilibrium in the diathermal case,

$$\delta F - \delta W > 0. \tag{2.05}$$

Here δF is the change of free energy associated with virtual changes within the body, and δW is the work performed on the body by external forces through virtual displacements. It is both a necessary and sufficient condition for stable equilibrium that the inequality (2.05) holds for all sufficiently small admissible virtual changes.

3. Displacements and loading

The body to which we intend to apply the criterion (2.05) is shown in Fig. 1 in two positions, a reference position and a displaced position.

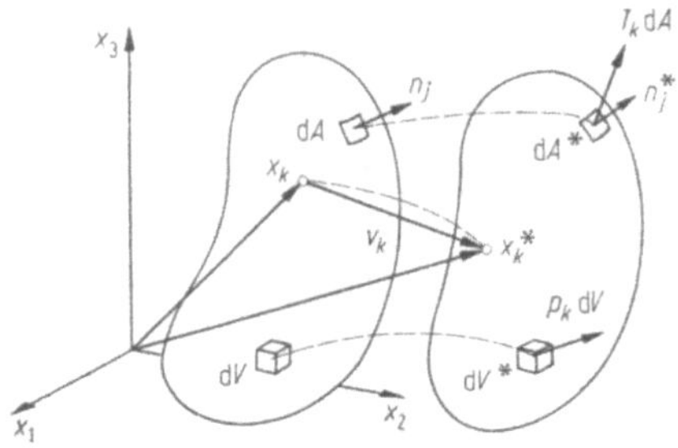

Fig. 1. Body B with surface C.

The relation between the two positions is determined by the displacing functions $f_k(\boldsymbol{x})$. In a fixed Cartesian system the coordinates x_k^* of a particle of the body in the displaced position are obtained from the coordinates x_k of the particle in the reference position by

$$x_k^* = f_k(\boldsymbol{x}). \tag{3.01}$$

The coordinates x_k of the particle in the reference position serve as labels which identify the particle uniquely. The displacement is the vector field

$$v_k = f_k(\boldsymbol{x}) - x_k. \tag{3.02}$$

The external forces acting on the body in the displaced position are surface tractions T_k and body forces p_k. These are defined as nominal quantities so that $T_k\,dA$ is the force acting on a surface element the area of which is dA in the reference position, and $p_k\,dV$ is the force on a body element the volume of which is dV in the reference position.

We consider a case of dead loading with displacements v_k prescribed on a part of the surface C_u, T_k prescribed on the remaining surface $C_T = C - C_u$, and p_k prescribed everywhere within the body B. The case of dead loading is chosen as an illustration for the sake of simplicity. The applicability of the theory is, however, not restricted to this case.

4. Constitutive equations for elastic-plastic materials

The constitutive equations for elastic-plastic materials appear as lower bounds of the increase of free energy in terms of intrinsic parameters. These parameters are the displacements v_k and the plastic strains α_{ij} which are the components of a symmetric tensor,

$$\alpha_{ij} = \alpha_{ji}. \tag{4.01}$$

The Lagrangean strain ε_{ij} has the components

$$\varepsilon_{ij} = \frac{1}{2}(f_{k,i}f_{k,j} - \delta_{ij}) \tag{4.02}$$

where f_k are the displacing functions and δ_{ij} is Kronecker's delta. The subscript "comma i" denotes partial differentiation with respect to the labels x_i. Alternatively the Lagrangean strain may be expressed in terms of displacement derivatives as

$$\varepsilon_{ij} = \frac{1}{2}(v_{i,j} + v_{j,i} + v_{k,i}v_{k,j}). \tag{4.03}$$

The elastic strain γ_{ij} is defined as the excess of Lagrangean strain over plastic strain

$$\gamma_{ij} = \varepsilon_{ij} - \alpha_{ij}. \tag{4.04}$$

We assume that the free energy of the body F may be written as the integral of the sum of two densities, the elastic potential $\Phi(\boldsymbol{\gamma})$ and the density of plastic free energy R

$$F = \int_B \{\Phi(\boldsymbol{\gamma}) + R\}\, dV. \tag{4.05}$$

The elastic potential (density of elastic free energy) $\Phi(\boldsymbol{\gamma})$, which is a function of elastic strains γ_{ij}, and the density of plastic free energy R are subject to the non-coupling conditions

$$\frac{\partial \Phi}{\partial \alpha_{ij}} = \frac{\partial R}{\partial \gamma_{ij}} = 0. \tag{4.06}$$

The density of plastic free energy R is related to plastic strains α_{ij} as follows. The increase of plastic strains $\delta\alpha_{ij}$ is written

$$\delta\alpha_{ij} = \mu_{ij}\,\delta\alpha \tag{4.07}$$

where the plastic directional tensor μ_{ij} is normalized according to

$$\mu_{ij}\mu_{ij} = 1, \tag{4.08}$$

and the magnitude of plastic deformation $\delta\alpha$ is non-negative,

$$\delta\alpha \geq 0. \tag{4.09}$$

It is assumed that the increase of density of plastic free energy δR has the following lower bound,

$$\delta R \geq \varphi(\boldsymbol{\mu})\,\delta\alpha + \frac{1}{2}\xi(\boldsymbol{\mu})\,\delta\alpha^2. \tag{4.10}$$

This expression may be considered a Taylor-series approximation to a more general relation.

5. The change of free energy

With the assumptions contained in the expressions (4.05), (4.06), and (4.10) we may calculate a lower bound of the increase of free energy F associated with the plastic direction μ_{ij} and arbitrary changes of displacements δv_k and magnitudes of plastic deformation $\delta\alpha$. Expanding the increase of elastic potential $\delta\Phi$ in a Taylor series, retaining only the 1st and 2nd order terms, we obtain

$$\delta\Phi = \sigma_{ij}\,\delta\gamma_{ij} + \frac{1}{2}\,A_{ijkl}\,\delta\gamma_{kl}\,\delta\gamma_{ij} \tag{5.01}$$

where

$$\sigma_{ij} = \frac{\partial\Phi}{\partial\gamma_{ij}} \tag{5.02}$$

is the stress-tensor and

$$A_{ijkl} = \frac{\partial^2\Phi}{\partial\gamma_{ij}\,\partial\gamma_{kl}} \tag{5.03}$$

is the tensor of elastic moduli. The change of elastic strains $\delta\gamma_{ij}$ is

$$\delta\gamma_{ij} = \delta\varepsilon_{ij} - \mu_{ij}\,\delta\alpha \tag{5.04}$$

where $\delta\varepsilon_{ij}$ is the change of Lagrangean strain associated with the change of displacements δv_k,

$$\delta\varepsilon_{ij} = \frac{1}{2}\,(f_{k,i}\,\delta v_{k,j} + f_{k,j}\,\delta v_{k,i} + \delta v_{k,i}\,\delta v_{k,j})\,. \tag{5.05}$$

The change of free energy

$$\delta F = \int_B (\delta\Phi + \delta R)\,dV \tag{5.06}$$

has the following lower bound

$$\delta F \geq \int_B \left\{\sigma_{ij}\,\delta\varepsilon_{ij} - \omega\,\delta\alpha + \frac{1}{2}\,[A_{ijkl}(\delta\varepsilon_{kl} - \mu_{kl}\,\delta\alpha)(\delta\varepsilon_{ij} - \mu_{ij}\,\delta\alpha) + \xi\,\delta\alpha^2]\right\} dV \tag{5.07}$$

where

$$\omega(\boldsymbol{\sigma}, \boldsymbol{\mu}) = \sigma_{ij}\mu_{ij} - \varphi(\boldsymbol{\mu})\,. \tag{5.08}$$

The lower bound of δF depends both on the parameters δv_k, $\delta\alpha$ and on the stress σ_{ij} and plastic directional tensor μ_{ij}. For σ_{ij} fixed and δv_k and $\delta\alpha$ arbitrarily small the lowest of the lower bounds is obtained for that value ν_{ij} of μ_{ij} which maximizes ω, the coefficient of $\delta\alpha$. The function

$$\psi(\boldsymbol{\sigma}) = \max\,\{\omega(\boldsymbol{\sigma}, \boldsymbol{\mu})\}_{\boldsymbol{\sigma}} = \omega(\boldsymbol{\sigma}, \boldsymbol{\nu}) \tag{5.09}$$

is the yield-function, and

$$\chi(\boldsymbol{\sigma}) = \xi(\boldsymbol{\nu}) \tag{5.10}$$

is the work-hardening function. According to the expression (5.07) the lowest bound of δF is

$$\delta F \geq \int_B \left\{\sigma_{ij}\,\delta\varepsilon_{ij} - \psi\,\delta\alpha + \frac{1}{2}\,[A_{ijkl}(\delta\varepsilon_{kl} - \nu_{kl}\,\delta\alpha)(\delta\varepsilon_{ij} - \nu_{ij}\,\delta\alpha) + \chi\,\delta\alpha^2]\right\} dV. \tag{5.11}$$

6. Properties of the yield-function

The yield-function $\psi(\boldsymbol{\sigma})$ is the maximum value of

$$\omega(\boldsymbol{\sigma}, \boldsymbol{\mu}) = \sigma_{ij}\mu_{ij} - \varphi(\boldsymbol{\mu}) \qquad [5.08]$$

for σ_{ij} fixed. The arguments μ_{ij} are bounded by

$$\mu_{ij}\mu_{ij} = 1. \qquad [4.08]$$

To obtain stationary values of $\omega(\boldsymbol{\sigma}, \boldsymbol{\mu})$ for σ_{ij} fixed we apply the Lagrangean multiplier method introducing

$$\Omega(\boldsymbol{\sigma}, \boldsymbol{\mu}, \lambda) = \sigma_{ij}\mu_{ij} - \varphi(\boldsymbol{\mu}) - \frac{1}{2}\lambda(\mu_{ij}\mu_{ij} - 1). \qquad (6.01)$$

The value of the function $\varphi(\boldsymbol{\mu})$ is defined for arguments μ_{ij} which do not satisfy the condition (4.08) by the requirement of 1st order homogeneity,

$$\varphi(c\boldsymbol{\mu}) = c\varphi(\boldsymbol{\mu}), \qquad c \geq 0. \qquad (6.02)$$

The variation of Ω for σ_{ij} fixed is

$$\delta\Omega = \left(\sigma_{ij} - \frac{\partial\varphi}{\partial\mu_{ij}} - \lambda\mu_{ij}\right)\delta\mu_{ij} - \frac{1}{2}(\mu_{ij}\mu_{ij} - 1)\,\delta\lambda.$$

Stationary values of Ω and hence of ω are obtained for values ν_{ij} of μ_{ij} which are solutions of the system of equations

$$\sigma_{ij} - \frac{\partial\varphi}{\partial\mu_{ij}} - \lambda\mu_{ij} = 0 \qquad (6.03)$$

together with (4.08). To be able to interprete the multiplier λ we multiply the expression (6.03) by μ_{ij} to obtain

$$\sigma_{ij}\mu_{ij} - \varphi(\boldsymbol{\mu}) - \lambda = 0$$

where the condition (4.08) and Euler's formula for homogeneous functions

$$\frac{\partial\varphi}{\partial\mu_{ij}}\mu_{ij} = \varphi(\boldsymbol{\mu}) \qquad (6.04)$$

is applied. Hence λ is the stationary value of ω. In particular the yield-function is

$$\psi = \omega. \qquad (6.05)$$

This value of ω is obtained for $\mu_{ij} = \nu_{ij}$. From

$$\psi(\boldsymbol{\sigma}) = \sigma_{ij}\nu_{ij} - \varphi(\boldsymbol{\nu}) \qquad (6.06)$$

we obtain by differentiation

$$\left(\sigma_{ij} - \left(\frac{\partial\varphi}{\partial\mu_{ij}}\right)_{\mu=\nu}\right)d\nu_{ij} + \left(\nu_{ij} - \frac{\partial\psi}{\partial\sigma_{ij}}\right)d\sigma_{ij}.$$

According to Eq. (6.03)

$$\left(\sigma_{ij} - \left(\frac{\partial\varphi}{\partial\mu_{ij}}\right)_{\mu=\nu}\right)d\nu_{ij} = \psi\nu_{ij}\,d\nu_{ij},$$

and so

$$\left(\nu_{ij} - \frac{\partial\psi}{\partial\sigma_{ij}}\right) d\sigma_{ij} = 0$$

on account of

$$\nu_{ij}\nu_{ij} = 1. \tag{6.07}$$

Hence we obtain

$$\nu_{ij} = \frac{\partial\psi}{\partial\sigma_{ij}} \tag{6.08}$$

which is the condition of normality.

The yield-function is normalized according to Eq. (6.07) and from (6.06) we obtain the condition of convexity

$$(\sigma_{ij} - \sigma_{ij}^0)\, \nu_{ij}^0 \leq 0 \quad \text{for} \quad \psi(\boldsymbol{\sigma}) \leq \psi(\boldsymbol{\sigma}^0). \tag{6.09}$$

A geometrical interpretation in a stress-space of the conditions of normality, convexity, and normalization is shown in Fig. 2.

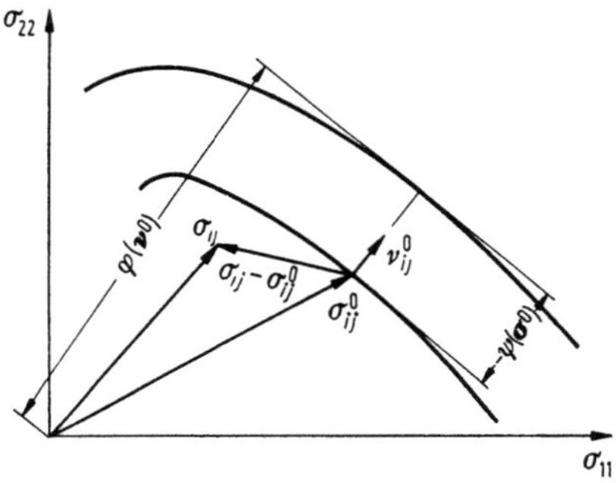

Fig. 2. Geometrical interpretation.

The ideas according to which the yield-surface is constructed in this section is due to Gvozdev [5].

7. Equilibrium and stability

Let δv_k be virtual displacements of the body satisfying

$$\delta v_k = 0 \quad \text{on} \quad C_u. \tag{7.01}$$

The work performed on the body through virtual displacements is

$$\delta W = \int_C T_k\, \delta v_k\, dA + \int_B p_k\, \delta v_k\, dV. \tag{7.02}$$

The change of free energy of the body associated with virtual displacements and virtual changes of plastic strains has a lower bound

$$\delta F \geq \int_B \left\{\sigma_{ij}\,\delta\varepsilon_{ij} - \psi\,\delta\alpha + \frac{1}{2}\left[A_{ijkl}(\delta\varepsilon_{kl} - \nu_{kl}\,\delta\alpha)(\delta\varepsilon_{ij} - \nu_{ij}\,\delta\alpha) + \chi\,\delta\alpha^2\right]\right\} dV \tag{7.03}$$

where

$$\delta\varepsilon_{ij} = \frac{1}{2}\,(f_{k,i}\,\delta v_{k,j} + f_{k,j}\,\delta v_{k,i} + \delta v_{k,i}\,\delta v_{k,j})\,. \tag{7.04}$$

According to the divergence theorem

$$\int\limits_B \sigma_{ij}\,\delta\varepsilon_{ij}\,dV = \int\limits_C f_{k,i}\sigma_{ij}n_j\,\delta v_k\,dA - \int\limits_B (f_{k,i}\sigma_{ij})_{,j}\,\delta v_k\,dV + \frac{1}{2}\int\limits_B \sigma_{ij}\,\delta v_{k,i}\delta v_{k,j}\,dV. \tag{7.05}$$

Applying the criterion (2.05) for equilibrium and stability with the expression (7.05) substituted for $\int \sigma_{ij}\,\delta\varepsilon_{ij}\,dV$ in (7.03) we obtain

$$\int\limits_C \{f_{k,i}\sigma_{ij}n_j - T_k\}\,\delta v_k\,dA - \int\limits_B \{(f_{k,i}\sigma_{ij})_{,j} + p_k\}\,\delta v_k\,dV - \int\limits_B \psi\,\delta\alpha\,dV$$

$$+ \frac{1}{2}\int\limits_B \{A_{ijkl}(\delta\varepsilon_{kl} - \nu_{kl}\,\delta\alpha)(\delta\varepsilon_{ij} - \nu_{ij}\,\delta\alpha) + \chi\,\delta\alpha^2 + \sigma_{ij}\,\delta v_{k,i}\,\delta v_{k.j}\}\,dV > 0\,. \tag{7.06}$$

From this we immediately obtain the conditions for elastic equilibrium

$$\left.\begin{aligned} f_{k,i}\sigma_{ij}n_j - T_k &= 0 \text{ on } C_T \\ (f_{k,i}\sigma_{ij})_{,j} + p_k &= 0 \text{ in } B, \end{aligned}\right\} \tag{7.07}$$

and the condition for plastic equilibrium

$$\psi(\boldsymbol{\sigma}) \leq 0 \text{ in } B. \tag{7.08}$$

Assuming that the conditions for equilibrium are satisfied there still remains the question of stability. This may be stated as the requirement

$$\int\limits_B \{A_{ijkl}(\delta\varepsilon_{kl} - \nu_{kl}\,\delta\alpha)(\delta\varepsilon_{ij} - \nu_{ij}\,\delta\alpha) + \chi\,\delta\alpha^2 + \sigma_{ij}\,\delta v_{k,i}\,\delta v_{k,j}\}\,dV > 0 \tag{7.09}$$

for all admissible δv_k and all admissible $\delta\alpha$ vanishing at points in which $\psi(\boldsymbol{\sigma}) < 0$. Here

$$\delta\varepsilon_{ij} = \frac{1}{2}\,(f_{k,i}\,\delta v_{k,j} + f_{k,j}\,\delta v_{k,i})\,. \tag{7.10}$$

The quadratic terms are suppressed.

Putting $\delta v_k = 0$ we obtain as a necessary condition for stability

$$A_{ijkl}\nu_{kl}\nu_{ij} + \chi(\boldsymbol{\sigma}) > 0 \quad \text{for} \quad \psi(\boldsymbol{\sigma}) = 0\,. \tag{7.11}$$

It is generally assumed that the first term of this expression is positive and the second is non-negative. It is, however, only necessary for stability that their sum is positive.

Introducing the tensor of tangential moduli

$$\bar{A}_{ijkl} = \begin{cases} A_{ijkl} & \text{for } \psi(\boldsymbol{\sigma}) < 0 \\ A_{ijkl} - \dfrac{A_{ijmn}\nu_{mn}A_{klpq}\nu_{pq}}{A_{rstu}\nu_{tu}\nu_{rs} + \chi} & \text{for } \psi(\boldsymbol{\sigma}) = 0 \end{cases} \tag{7.12}$$

we may write the condition (7.09) for stability

$$\int_B \left\{ \bar{A}_{ijkl}\, \delta\varepsilon_{kl}\, \delta\varepsilon_{ij} + \sigma_{ij}\, \delta v_{k,i}\, \delta v_{k,j} + \Lambda \left(\frac{\delta \bar{E}}{\Lambda} - \delta\alpha \right)^2 \right\} dV > 0\,. \quad (7.13)$$

Here

$$\Lambda = A_{ijkl} \nu_{kl} \nu_{ij} + \chi \quad (7.14)$$

and

$$\delta \bar{E} = \begin{cases} 0 & \text{for } \psi(\boldsymbol{\sigma}) < 0 \\ A_{ijkl} \nu_{kl}\, \delta\varepsilon_{ij} & \text{for } \psi(\boldsymbol{\sigma}) = 0\,. \end{cases} \quad (7.15)$$

So we arrive eventually at the following condition for elastic-plastic stability,

$$\int_B (\bar{A}_{ijkl}\, \delta\varepsilon_{kl}\, \delta\varepsilon_{ij} + \sigma_{ij}\, \delta v_{k,i}\, \delta v_{k,j})\, dV > 0 \quad (7.16)$$

for all admissible δv_k.

8. Stress-strain relations

The change of the yield-function $\psi(\boldsymbol{\sigma})$ is

$$\delta\psi = \frac{\partial \psi}{\partial \sigma_{ij}}\, \delta\sigma_{ij} + (\delta\psi)_{\boldsymbol{\sigma}} \quad (8.01)$$

where $(\delta\psi)_{\boldsymbol{\sigma}}$ is the change of ψ for constant σ_{ij} due to plastic deformation. From Eq. (6.06)

$$(\delta\psi)_{\boldsymbol{\sigma}} = -(\delta\varphi)_{\boldsymbol{\nu}} \quad (8.02)$$

where $(\delta\varphi)_{\boldsymbol{\nu}}$ is the change of the function $\varphi(\boldsymbol{\nu})$ for constant ν_{ij} due to plastic deformation. From the expression (4.10) of the lower bound of the change of plastic free energy we obtain

$$(\delta\psi)_{\boldsymbol{\sigma}} = -\chi\, \delta\alpha\,. \quad (8.03)$$

Hence

$$\delta\psi = \nu_{ij}\, \delta\sigma_{ij} - \chi\, \delta\alpha\,. \quad (8.04)$$

We consider a situation in which the body is brought from one state of equilibrium into another with the values of σ_{ij} and α_{ij} increased by $\delta\sigma_{ij}$ and $\nu_{ij}\, \delta\alpha$. It is assumed that no plastic deformation occurs at points in which the value of ψ is negative after the change of state has taken place. Hence

$$\delta\alpha = 0 \text{ if otherwise } \nu_{ij}\, \delta\sigma_{ij} - \chi\, \delta\alpha + \psi < 0\,. \quad (8.05)$$

A 1st order approximation to the increase of stress is

$$\delta\sigma_{ij} = A_{ijkl}(\delta\varepsilon_{kl} - \nu_{kl}\, \delta\alpha) \quad (8.06)$$

where

$$\delta\alpha = 0 \text{ if otherwise } \delta\alpha > \frac{A_{ijkl} \nu_{kl} \delta\varepsilon_{ij} + \psi}{A_{mnpq} \nu_{pq} \nu_{mn} + \chi}\,. \quad (8.07)$$

The problem stated by the expressions (8.06) and (8.07) has the solution

$$\left.\begin{aligned}
&\delta\sigma_{ij} = A_{ijkl}\,\delta\varepsilon_{kl};\ \ \delta\alpha = 0\\
&\text{for}\ \ A_{ijkl}\nu_{kl}\,\delta\varepsilon_{ij} + \psi \le 0;\\
&\delta\sigma_{ij} = \left(A_{ijkl} - \frac{A_{ijmn}\nu_{mn}A_{klpq}\nu_{pq}}{A_{rstu}\nu_{tu}\nu_{rs} + \chi}\right)\delta\varepsilon_{ij} - \frac{A_{ijkl}\nu_{kl}\psi}{A_{mnpq}\nu_{pq}\nu_{mn} + \chi};\\
&\delta\alpha = \frac{A_{ijkl}\nu_{kl}\delta\varepsilon_{ij} + \psi}{A_{mnpq}\nu_{pq}\nu_{mn} + \chi}\\
&\text{for}\ \ A_{ijkl}\nu_{kl}\,\delta\varepsilon_{ij} + \psi > 0.
\end{aligned}\right\}\qquad (8.08)$$

This formulation of stress-strain relations may be compared to the well known formulation due to von Mises [9], Melan [8], Prager [10], and Hill [6],

$$\left.\begin{aligned}
&\delta\sigma_{ij} = A_{ijkl}\,\delta\varepsilon_{kl};\ \delta\alpha = 0\\
&\text{for either}\ \psi < 0\ \text{or}\ (\psi = 0\ \text{and}\ A_{ijkl}\nu_{kl}\,\delta\varepsilon_{ij} \le 0);\\
&\delta\sigma_{ij} = \left(A_{ijkl} - \frac{A_{ijmn}\nu_{mn}A_{klpq}\nu_{pq}}{A_{rstu}\nu_{tu}\nu_{rs} + \chi}\right)\delta\varepsilon_{ij};\\
&\delta\alpha = \frac{A_{ijkl}\nu_{kl}\delta\varepsilon_{ij}}{A_{mnpq}\nu_{pq}\nu_{mn} + \chi}\\
&\text{for}\ \psi = 0\ \text{and}\ A_{ijkl}\nu_{kl}\,\delta\varepsilon_{ij} > 0.
\end{aligned}\right\}\qquad (8.09)$$

It should be noted that this expression may violate the condition (7.08) for plastic equilibrium.

The difference between the two sets of Eqs. (8.08) and (8.09) is perhaps not quite as sophisticated as it might seem at a first glance. In fact the two sets are entirely different of nature, one being homogeneous, the other inhomogeneous in $\delta\sigma_{ij}$ and $\delta\varepsilon_{ij}$. The homogeneity of Eqs. (8.08) may be preserved together with the equilibrium conditions when the value of ψ appearing in Eqs. (8.08) tends towards zero along with $\delta\varepsilon_{ij}$. The question remains, however, how this requirement may be met with. A typical stress-strain diagram is shown in Fig. 3.

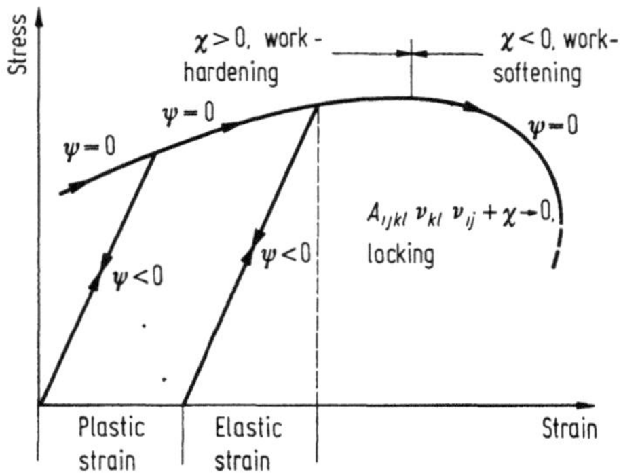

Fig. 3. Stress-strain diagram.

9. Conclusion

It is hoped that the derivations of the previous sections may serve as a convincing demonstration of the usefulness of thermodynamical techniques in the context of elasto-plasticity. The particular assumptions of Sec. 4 are chosen so as to imply the well known features of theories of elastic-plastic materials such as the existence of a yield-surface and the conditions of normality and convexity. It is emphasized that a proof of the validity of these conditions is not provided. Other assumptions regarding the constitution of the material will of course imply other results. However, the implications of such assumptions as regards equilibrium and stability may be tested according to the scheme outlined in the present paper.

Only the aspects of equilibrium thermodynamics are incorporated in the forwarded theory. Hence non-equilibrium phenomena such as dynamical effects and viscosity are excluded from consideration. It is, however, possible that the scope of the theory may be extended in these directions by the introduction of non-equilibrium- or irreversible thermodynamics. The concept of the yield-condition as a condition of equilibrium may serve as a clue in this respect.

References

1. Bridgman, P. W: The thermodynamics of plastic deformation and generalized entropy. Rev. Mod. Phys. **22**, 56 (1950).
2. Drucker, D. C.: A more fundamental approach to plastic stress-strain relations. Proc. 1st U.S. Nat. Congr. Appl. Mech., 1951, p. 487.
3. Gibbs, J. W.: Collected Works, Vol. I, Yale University Press 1948.
4. Green, A. E., Naghdi, P. M.: A general theory of an elastic-plastic continuum. Arch. Rat. Mech. Anal. **18**, 251 (1965).
5. Gvozdev, A. A.: The determination of the value of the collapse load for statically indeterminate systems undergoing plastic deformation. Int. J. Mech. Sci. **1**, 322 (1960).
6. Hill, R.: Mathematical Theory of Plasticity, Oxford University Press 1950, Ch. III.
7. Hill, R.: A general theory of uniqueness and stability in elastic-plastic solids. J. Mech. Phys. Solids **6**, 236 (1958).
8. Melan, E.: Zur Plastizität des räumlichen Kontinuums. Ing.-Arch. **9**, 116 (1938).
9. Mises, R. von: Mechanik der plastischen Formänderung von Kristallen. Z. A. M. M. **8**, 161 (1928).
10. Prager, W.: Recent developments in the mathematical theory of plasticity. J. Appl. Phys. **20**, 235 (1949).
11. Prandtl, L.: Spannungsverteilung in plastischen Körpern. Proc. 1st Int. Congr. Appl. Mech., 1924, p. 43.
12. Reuss, A.: Berücksichtigung der elastischen Formänderungen in der Plastizitätstheorie. Z. A. M. M. **10**, 266 (1930).
13. Ziegler, H.: An attempt to generalize Onsager's principle, and its significance for rheological problems. Z. A. M. P. **9 b**, 748 (1958).

Instability under Cycles of Plastic Deformation

By

K. Neale and **J. Schroeder**

University of Waterloo, Waterloo, Ont., Canada

1. Introduction

Instability of metals due to cyclic strain accumulation in the plastic range is relatively unknown and has not been thoroughly analysed. One of the simplest forms of cyclic strain accumulation, referred to as

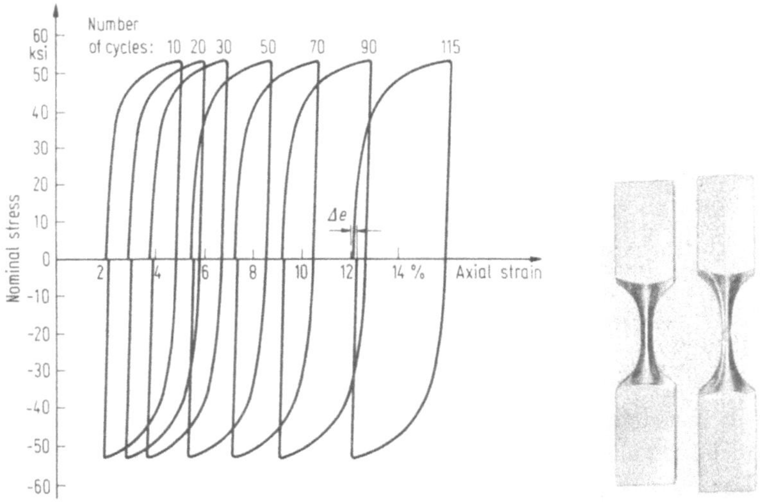

Fig. 1. Results from load-controlled test on uniaxial specimen.

cyclic creep by metallurgists, is indicated in Fig. 1 by results of a load controlled test on a uniaxial mild steel specimen. Fig. 1 shows that irreversible increments Δe of permanent axial strain accumulate cycle by cycle until the gradual elongation of the specimen causes necking and fracture as observed in monotonic tension tests. It is worthwhile noting that even in this case, fracture is caused by necking, i.e., tensile instability, and not by low cycle fatigue cracks. Additional test results indicate that cyclic strain accumulation takes precedence over fatigue in causing failure for the higher stress ranges and that Δe increases with increasing stress range.

A form of structural instability due to cyclic strain accumulation is illustrated by the following experimental observation. A column was subjected to cycles of controlled axial displacements as shown in Fig. 2.

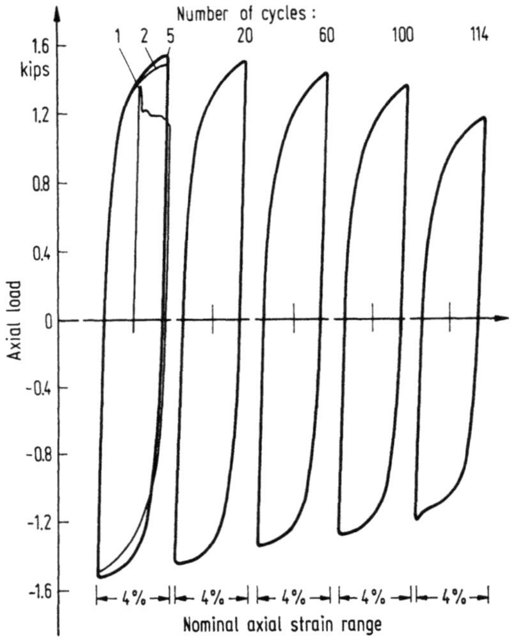

Fig. 2. Hys' resis loops of steel column (0.2 in. diameter, 1.1 in. height) in displacement-controlled test.

Tensile instability described previously was thus prevented and cyclic strain accumulation caused a gradual accumulation of transverse displacement of the column as indicated in Fig. 3. In Fig. 2 the origins for cycles 20 onwards have been displaced to the right for the sake of clarity. Although the first five cycles show the column to be initially stable for this particular range of axial deformation, the load limits decreased from cycle to cycle indicating a gradual reduction in the critical compressive load for instability. This reduction in critical load was also confirmed by monotonic compression tests on specimens after they were exposed to cycles of controlled axial displacements.

Conclusions drawn from these results are twofold:

(i) plastic stability analysis based on monotonic loading alone may be unsafe, and

(ii) a structure designed against low cycle fatigue failure may become unstable due to a gradual reduction in stability after a number of load applications in the plastic range long before the fatigue life is reached.

In the following, a method is proposed to analyse both cyclic strain accumulation and its effect on stability in general.

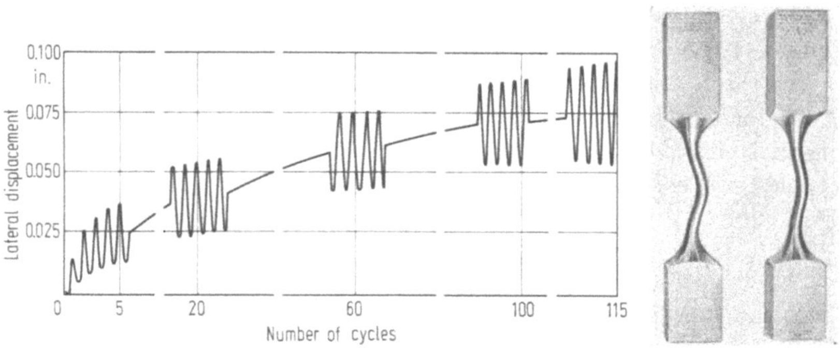

Fig. 3. Lateral displacement of steel column in displacement-controlled test.

2. Constitutive relation

A number of models has been proposed to describe cyclic strain accumulation from both the continuum mechanics and metallurgical points of view. In accordance with the theory of FREUDENTHAL and RONAY [1] accumulation of second-order plastic strain increments, irreversible under reversals of stress, is predicted by the proper form of the plastic constitutive relation proposed by HILL [2]

$$d\varepsilon_{ij}^{P} = h\left[\frac{\partial g}{\partial J_2}\sigma'_{ij} + \frac{\partial g}{\partial J_3}\left(\sigma'_{ik}\sigma'_{kj} - \frac{2}{3}J_2\,\delta_{ij}\right)\right]df, \tag{1}$$

where $d\varepsilon_{ij}^{P}$ represents the tensor of plastic strain increments, σ'_{ij} is the deviatoric stress tensor, δ_{ij} designates the Kronecker delta, the material parameters h and g denote a hardening function and a dissipative potential respectively, and both are considered to be scalar functions of the second and third invariants of the deviatoric stress tensor denoted respectively by J_2 and J_3, and $f = f(J_2, J_3)$ is a yield criterion.

The following analysis of cyclic strain accumulation is restricted to steady-state conditions under quasi-static loading where temperature and time effects are negligible. A steady-state response of this nature is usually attained after a few cycles as for example is shown in Fig. 1. For these conditions, the following basic assumptions are proposed:

(i) the dissipative potential

$$g = f = J_2 + bJ_3, \tag{2}$$

where in view of (1) b appears in the capacity of a parameter of cyclic, steady-state strain accumulation,

(ii) both h and b can be determined from a number of stress controlled uniaxial tests on cyclically stable specimens exposed to full reversals of axial loads as for example shown in Fig. 1,

(iii) elastic strain increments are fully reversible and vanish if accumulated over one cycle, and

(iv) the Bauschinger effect is negligible.

By including elastic effects and assuming both isotropy and incompressibility, the Jaumann rate [3] of the deviatoric stress tensor becomes in accordance with (1) and (2)

$$\frac{D\sigma'_{ij}}{Dt} = \frac{2E}{3}\left\{\dot{\varepsilon}_{ij} - \frac{[\sigma'_{ij} + b(\sigma'_{ik}\sigma'_{kj} - 2/3\delta_{ij}J_2)]\,(\sigma'_{mn} + b\sigma'_{mv}\,\sigma'_{vn})\,\dot{\varepsilon}_{mn}}{3/2Eh + (\sigma'_{pq} + b\sigma'_{ps}\sigma'_{sq})\,[\sigma'_{pq} + b(\sigma'_{pr}\sigma'_{rq} - 2/3\delta_{pq}J_2)]}\right\}, \tag{3}$$

where the superposition principle of elastic and plastic strain rates is adopted, E is the modulus of elasticity and $\dot{\varepsilon}_{ij}$ denotes the rate of deformation tensor. Eq. (3) may be used to find h and b as suggested in (ii) and may be employed as a constitutive equation to predict cyclic strain accumulation under multiaxial states of stress.

3. Stability analysis

Instability may be determined using the stability criterion for elastic-plastic solids due to HILL [4]

$$\int_V \left(\dot{\sigma}_{ij} + \sigma_{ij}\frac{\partial v_k}{\partial x_k} - \sigma_{jk}\frac{\partial v_i}{\partial x_k}\right)\frac{\partial v_j}{\partial x_i}\,dV > 0 \tag{4}$$

in connection with the constitutive equation (3). In expression (4), $\dot{\sigma}_{ij}$ represents the material rate of change of the true (Eulerian) stress tensor σ_{ij}, v_i are the components of an admissible velocity field, x_i are Cartesian coordinates in the Eulerian description, and V denotes the volume of the body. Although influence of terms containing b is important in the *accumulation* of second-order effects, these terms may be neglected in the analysis of stability in a particular cycle indicated for example in Fig. 2. Using this approximation, the relationship between Jaumann rates and material derivatives [3], and the constitutive equation (3), Hill's stability criterion becomes:

$$\int\left\{\frac{2E}{3}\left[\dot{\varepsilon}_{ij}\dot{\varepsilon}_{ij} - \frac{(\sigma_{mn}\dot{\varepsilon}_{mn})^2}{3/2Eh + \sigma'_{pq}\sigma'_{pq}}\right] - 2\sigma_{ij}\dot{\varepsilon}_{ki}\dot{\varepsilon}_{kj} + \sigma_{ij}\frac{\partial v_k}{\partial x_i}\frac{\partial v_k}{\partial x_j}\right\}dV > 0. \tag{5}$$

For a given distribution of stress an admissible velocity field satisfying incompressibility and rendering the above integral a minimum, corresponds to the exact solution. However, any admissible velocity field will yield an upper bound to the critical stress for instability.

References

1. Freudenthal, A. M., Ronay, M.: Proc. Roy. Soc. London **A 292**, 14 (1966).
2. Hill, R.: The Mathematical Theory of Plasticity, Oxford: Clarendon Press 1964.
3. Prager, W.: Introduction to Mechanics of Continua, New York: Ginn and Comp. 1961.
4. Hill, R.: J. Mech. Phys. Solids **6**, 236 (1958).

Coupled Modes of Buckling in Some Continuous Systems

By

A. H. Chilver and **K. C. Johns**

University College London, London, U.K.

Coupled modes of elastic buckling, or simultaneous buckling in structures, may occur where two or more buckling loads (or eigenvalues) of a structure are identical or nearly so. Post-buckling in mixed modes may be the result, as well as an increased number of post-buckling paths all passing through the same branching point. This is the conclusion of earlier theoretical studies of general discrete coordinate systems, and of experimental studies involving simple models.

These studies indicated that the maximum number of paths generated by a general n-degree-of-freedom system with n simultaneous branching points is $2^n - 1$. Complete symmetry of the total potential energy of the system with respect to all coordinates is found to raise this number to $\frac{1}{2}(3^n - 1)$. The existence of all these paths in real coordinates in all cases of simultaneous buckling was not necessarily established by the general study, so it was thought useful to examine some individual continuous systems to determine whether actual behaviour could conform to the complex patterns indicated by the general theory.

A strut, pin-ended each end, and supported laterally along its length by an elastic medium, will have coincident buckling loads if the ratio of stiffness of the strut itself to that of the medium is one of a number of critical values. Such a strut, supported by a medium which has a force response per unit length of column equal to $k_1 w + k_2(x)\, w^2$, can be analysed, where w is lateral deflection of the column, k_1 is a constant and $k_2(x)$ varies with position (x) along the column. k_2 is taken as $(A/l^4 + (B/l^5)\, x)$ and mode shapes for w are taken as $U_1 l \sin(\pi x/l) + U_2 l \sin(2\pi x/l)$, where U_1 and U_2 are non-dimensional amplitudes and l is the column length. If $k_1 = 4\pi^2 EI/l^4$ then the loads corresponding to buckling into modes U_1 and U_2 are both equal to $5\pi^2 EI/l^2$. E is Young's modulus and l is moment of inertia of the strut. The equilibrium equations governing post-buckling in the near vicinity of this simultaneous branching point are

$$\begin{aligned}(-15\pi^5 EI/2l)\,\lambda u_1 + 20(2A + B)\,u_1^2 + (45B/4)\,u_1 u_2 \\ + 16(2A + B)\,u_2^2 = 0, \\ (-30\pi^5 EI/l)\,\lambda u_2 + (45B/8)\,u_1^2 + 32(2A + B)\,u_1 u_2 \\ - 10Bu_2^2 = 0.\end{aligned} \tag{1}$$

λ is the variation of the load from the critical value, u_1 and u_2 are incremental variations of the amplitudes U_1 and U_2. One or three equilibrium paths will exist, as predicted by the theory, depending on the value of A/B. To take two examples: (i) if $A/B = 1/2$ then one path exists in the direction $u_2/u_1 = 17.7$; (ii) if $A/B = -5/8$ then three paths exist in directions $u_2/u_1 = -4.465$, $+0.319$, and $+2.016$.

The classical buckling pressure of a spherical shell under external pressure is well known to be the same for many buckling modes. In a paper by HUTCHINSON [1], a shell representation of a portion of the complete sphere is used. Modes of radial displacement w are of the form $\cos(k_x x/R)\cos(k_y y/R)$, where x and y are cartesian coordinates in the base plane of the portion of the sphere, R is the radius of curvature and k_x and k_y are wave numbers. The classical critical pressure is associated with all modes of buckling whose wave numbers satisfy $k_x^2 + k_y^2 = q_0^2$, where q_0^2 is $2[3(1-\nu)^2]^{1/2} R/h$, and h is shell thickness. Coupling between modes is found to depend on relationships between wave numbers such that the equilibrium equations always decouple into separate sets of equations corresponding to interaction between either two or three of these modes. Two modes which couple are found to be

$$w_1 = \xi_1 h \cos(q_0 x/R)$$

and (2)

$$w_2 = \xi_2 h \sin(q_0 x/2R)\sin(\sqrt{3}\,q_0 y/2R),$$

where ξ_1 and ξ_2 are amplitudes. The equilibrium equations are

$$\begin{aligned}(1 - p/p_c)\,\xi_1 - \frac{9C}{32}\xi_2^2 = 0, \\ (1 - p/p_c)\,\xi_2 - \frac{9C}{32}\xi_1\xi_2 = 0\end{aligned} \tag{3}$$

for the perfect case. HUTCHINSON gives two solutions to these equations:

$$\xi_1 = \frac{8}{9C}(1 - p/p_c), \quad \xi_2 = \pm\frac{16}{9C}(1 - p/p_c), \tag{4}$$

but, clearly, a third solution also exists, namely

$$\xi_2 = 0, \quad (1 - p/p_c) = 0, \quad \xi_1 \neq 0; \tag{5}$$

a neutral equilibrium path through the same branching point. Thus the two mode case yields the full $2^n - 1$ paths possible. Two modes only are operative, however, only when one wave number is zero as for w_1, above, where $k_y = 0$. Otherwise three modes operate and the three equations are

$$\begin{aligned} (1 - p/p_c)\,\xi_1 - \frac{9C}{16}\xi_2\xi_3 &= 0, \\ (1 - p/p_c)\,\xi_2 - \frac{9C}{16}\xi_1\xi_3 &= 0, \\ (1 - p/p_c)\,\xi_3 - \frac{9C}{16}\xi_2\xi_1 &= 0 \end{aligned} \tag{6}$$

for the perfect case. Hutchinson gives the simple solution:

$$\xi_1 = \xi_2 = \xi_3 = \frac{16}{9C}(1 - p/p_c) \tag{7}$$

but six others also exist, namely:

$$\begin{aligned} &\xi_1 = -\xi_2 = \xi_3 = -\frac{16}{9C}(1 - p/p_c); \\ &\xi_1 = -\xi_2 = -\xi_3 = \frac{16}{9C}(1 - p/p_c); \\ &\xi_1 = \xi_2 = -\xi_3 = -\frac{16}{9C}(1 - p/p_c); \\ &\xi_2 = \xi_3 = 0, \quad (1 - p/p_c) = 0, \quad \xi_1 \neq 0; \\ &\xi_3 = \xi_1 = 0, \quad (1 - p/p_c) = 0, \quad \xi_2 \neq 0; \\ &\xi_2 = \xi_1 = 0, \quad (1 - p/p_c) = 0, \quad \xi_3 \neq 0. \end{aligned} \tag{8}$$

This is therefore an example of a three-degree-of-freedom system generating its 7 maximum possible simultaneous buckling paths.

Long flat plates, compressed at the ends, are known to generate simultaneous buckling phenomena. Supple [2] has treated the buckling behaviour of rectangular plates in edge compression, with length to breadth ratio 2 : 1, as two-degree-of-freedom systems, where the degrees of freedom correspond to buckling into one and two half sine waves axially. Both theoretically and experimentally, coupled modes of buckling were shown to exist, taking the form of two hyperbolic shaped, upward-turned, symmetric paths, passing through the higher uncoupled or pure post-buckling path at secondary points of bifurcation. At these points of bifurcation the higher, unstable, uncoupled path becomes stable. Investigations of plates with imperfections in both modes were carried out. It was found that the simultaneous buckling phenomena accounted for (i) imperfect behaviour involving buckling into either mode of deflection, (ii) post-buckling paths losing stability at snapping points, and (iii) changes in buckle pattern after initial post-buckling as well as bifurcations on imperfect paths.

The buckling behaviour of a model thin-walled compression member subject to both column (bending) buckling and local buckling of the flanges is discussed in a paper by VAN DER NEUT [3]. Interest has developed in this case due to the optimization criterion which requires that these two buckling loads be equal or nearly so. The findings of this paper indicate that this may be considered to be a coincidence of column and local buckling modes which gives rise to a mixed form. The two uncoupled modes are symmetric and stable, but on coincidence they are found to give rise to a symmetric unstable mixed form. When the loads are not quite coincident, but nearly so, SUPPLE's findings indicate this unstable coupled path will branch off the lower of the two uncoupled paths. This appears also to be the case in the example discussed by VAN DER NEUT, where the unstable mixed form dominates the imperfect behaviour both with Euler load just above and just below local buckling load. Consequently, imperfection sensitivity is found to be very high in this region and the behaviour is aptly described as "shell-like".

We conclude, at this stage, that multiple post-buckling path generation can be important in continuous systems. In some cases, it can have an important effect on the type of instability which occurs.

References

1. HUTCHINSON, J. W.: J. Appl. Mech. **34**, 49 (1967).
2. SUPPLE, W. J.: Coupled Buckling Modes of Structures. Thesis, University of London, 1966.
3. VAN DER NEUT, A.: The Interaction of Local Buckling and Column Buckling of Thin-walled Compression Members. Technological University Delft, Report VTH 149.

Perturbation Patterns in Nonlinear Branching Theory

By

J. M. T. Thompson and **G. W. Hunt**

University College London, London, U.K.

1. Introduction

The work of KOITER in the field of structural stability has provoked a great deal of interest in the nonlinear analysis of bifurcating systems and within the discrete representation of such systems a particular form of perturbation scheme has been developed by SEWELL [1] and THOMPSON [2, 3]. The benefits from such a scheme are considerable since the process reduces the original nonlinear equation to an ordered series of linear equations. Past interest in *initial* post-buckling and imperfection sensitivity has led to some emphasis being placed on the lower order equations rather than the patterns developed in a continuing scheme but it is these patterns which are investigated here.

2. Branching phenomena

Let us first consider two types of bifurcation problem which are of particular interest. These are shown in Figs. 1 and 2, the former illustrating the asymmetric point of bifurcation and the latter the unstable symmetric point of bifurcation. In each case the left-hand diagram shows a loading parameter Λ plotted against a generalised coordinate Q_i. The perfect system exhibits a fundamental path which loses its stability at a discrete critical point and a post-buckling path which emerges from this point. A family of imperfect systems is generated by the variation of an imperfection parameter ε. The two arrowheads indicate the paths sought by two perturbation schemes, the "perfect" scheme finding the post-buckling path of the perfect system and the "imperfect" scheme the locus of critical points. The right-hand diagram shows the variation of critical load with imperfection parameter which would be derivable from the latter. In the case of the asymmetric point of bifurcation the curve approaches a parabolic variation at $\varepsilon = 0$ but is contaminated by higher derivatives as ε increases. In contrast

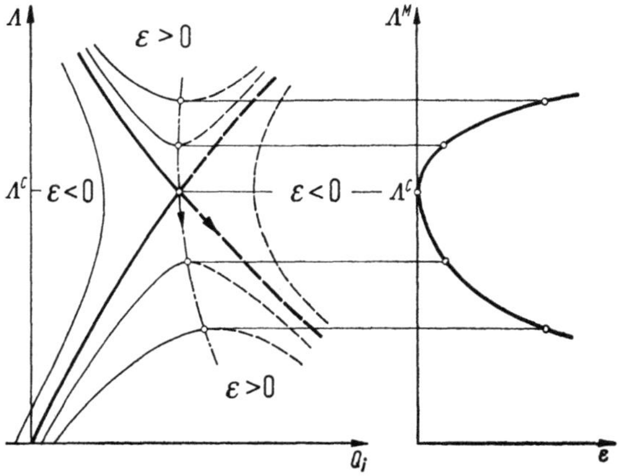

Fig. 1. Asymmetric point of bifurcation.

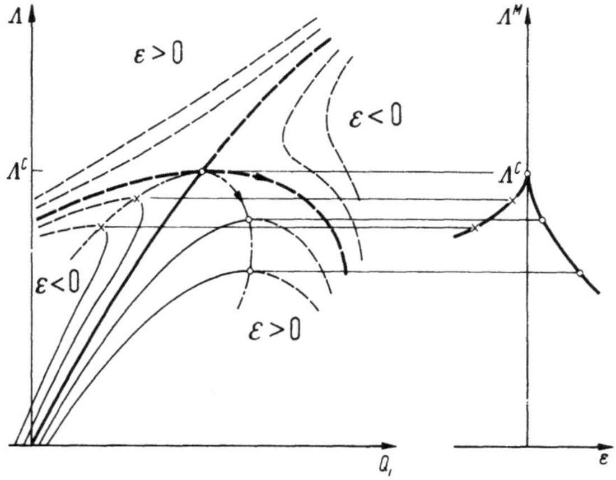

Fig. 2. Unstable symmetric point of bifurcation.

for the symmetric point of bifurcation this curve assumes a two thirds power-law cusp with an infinite slope at $\varepsilon = 0$; the subsequent asymmetry of this cusp is likewise produced by higher-order contaminations.

3. Perturbation patterns

Let us now look at the patterns developed by the perturbation processes shown in Fig. 3. Following precisely the work of Thompson [3], we consider a discrete conservative system described by the total potential energy V, which is a function of a set of n generalised coordinates Q_i, a loading parameter Λ, and an imperfection parameter ε. We suppose that the equilibrium equations present us with a single-valued fundamental solution and we define a sliding set of incremental

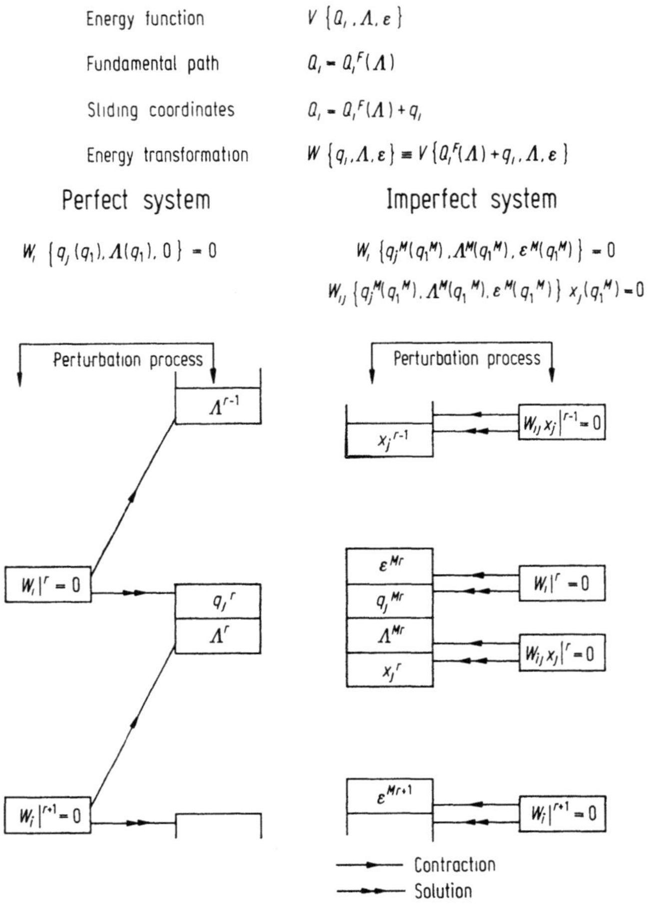

Fig. 3. Schematic representation of the perturbation processes.

coordinates q_i emerging from the fundamental path. We then define a transformed energy function W and, assuming the existence of a discrete critical point on the fundamental path, seek to express any post-buckling path of the perfect system emerging from this point in a parametric form using the first generalised coordinate as expansion parameter. A perturbation scheme for the perfect system is now developed by successive total differentiation of the equilibrium equations with respect to this parameter which produces a series of linear equations. These may be sequentially solved for the post-buckling path derivatives. It should be noted that a new notation has here been adopted in which a superscript indicates the number of times a particular expression has been differentiated. The contraction process (namely summation after multiplication by the post-buckling rates) plays a continuing role throughout the scheme and it may be noted that it performs precisely the same function as the "suppression of secular terms" in a corresponding continuum analysis [4].

We must now consider the perturbation scheme for imperfect systems. The original formulation of the problem is similar to the scheme for the perfect system, the "path" being written in parametric form with the first generalised coordinate as expansion parameter and the equations differentiated with respect to this parameter to develop a continuing process. However now it becomes necessary to consider two types of equations, firstly equilibrium equations and secondly critical state equations. The latter are associated with the snap-buckling of an imperfect system which is treated as a local eigenvalue problem with the eigenvector x_j appearing in the analysis [3]. The patterns developed by the process may be seen in Fig. 3. It is interesting to note that the "perfect" scheme produces an interaction between successive steps of the process which is apparent from the sloping contraction line. There

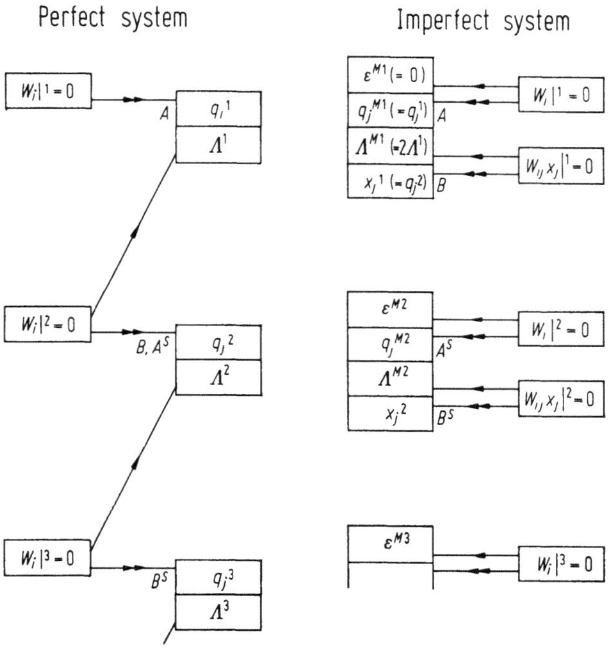

Fig. 4. Interplay at the initial stages of the schemes.

is however no corresponding phenomenon associated with the "imperfect" scheme.

Now let us consider the *initial* forms of the two schemes shown in Fig. 4. Although in the general form ($r > 3$ say) there is no interaction between the schemes, certain interrelationships may be found during the initial stages. A detailed study of the lower order equations reveals that, once the "perfect" scheme has been followed, some of the equations reappear in the "imperfect" scheme. For the asymmetric point of bifurcation this occurs twice only as indicated by the letters A and B on the solution flow lines (A^s and B^s refer to a similar interplay for the symmetric point of bifurcation). It therefore becomes possible to equate some of the path derivatives and these are shown by the bracketed equalities on the first set of imperfect derivatives. However as the scheme continues the equations become contaminated and one can no longer draw conclusions of this type.

4. Asymptotic equations of imperfection sensitivity

Taylor series expansions of the load and imperfection parameters allow them to be related to each other in the form of an asymptotic equation. For the asymmetric point of bifurcation this equation is

$$\Lambda^M = \Lambda^C + \alpha(\varepsilon^M)^{1/2} + \beta(\varepsilon^M) + \gamma(\varepsilon^M)^{1+1/2} + \cdots,$$

where

$$\alpha = \pm\left(\frac{2}{\varepsilon_{11}^{MC}}\right)^{1/2}\Lambda_1^{MC}, \quad \beta = \frac{\Lambda_{11}^{MC}}{\varepsilon_{11}^{MC}} - \frac{\Lambda_1^{MC}\varepsilon_{111}^{MC}}{3(\varepsilon_{11}^{MC})^2},$$

$$\gamma = \pm\left(\frac{2}{\varepsilon_{11}^{MC}}\right)^{1/2}\left\{\frac{\Lambda_{111}^{MC}}{3\varepsilon_{11}^{MC}} - \frac{\Lambda_{11}^{MC}\varepsilon_{111}^{MC}}{3(\varepsilon_{11}^{MC})^2} + \frac{\Lambda_1^{MC}}{36(\varepsilon_{11}^{MC})^3}\left[5(\varepsilon_{111}^{MC})^2 - 3\varepsilon_{11}^{MC}\varepsilon_{1111}^{MC}\right]\right\},$$

etc., in which the notation of THOMPSON [3] is followed. The sign of γ depends on the sign of α and each square root should be taken as positive. For the symmetric point of bifurcation this equation becomes

$$\Lambda^M = \Lambda^C + \alpha(\varepsilon^M)^{2/3} + \beta(\varepsilon^M) + \gamma(\varepsilon^M)^{4/3} + \cdots,$$

where

$$\alpha = \frac{\Lambda_{11}^{MC}}{2}\left(\frac{6}{\varepsilon_{111}^{MC}}\right)^{2/3}, \quad \beta = \frac{1}{2}\left\{\frac{2\Lambda_{111}^{MC}}{\varepsilon_{111}^{MC}} - \frac{\Lambda_{11}^{MC}\varepsilon_{1111}^{MC}}{(\varepsilon_{111}^{MC})^2}\right\},$$

$$\gamma = \left(\frac{6}{\varepsilon_{111}^{MC}}\right)^{1/3}\left\{\frac{\Lambda_{1111}^{MC}}{4\varepsilon_{111}^{MC}} - \frac{\Lambda_{111}^{MC}\varepsilon_{1111}^{MC}}{4(\varepsilon_{111}^{MC})^2} - \frac{\Lambda_{11}^{MC}\varepsilon_{11111}^{MC}}{10(\varepsilon_{111}^{MC})^2} + \frac{7\Lambda_{11}^{MC}(\varepsilon_{1111}^{MC})^2}{48(\varepsilon_{111}^{MC})^3}\right\},$$

etc.

5. Conclusions

The development of perturbation patterns has underlined the continuing role of the contraction mechanism of discrete analyses which we note corresponds to the suppression of secular terms in a continuum analysis. The *initial* interplay between "perfect" and "imperfect" schemes is thought to be significant because an *initial* estimate of imperfection sensitivity may be obtained without resort to a distinct imperfection analysis. The general forms of the asymptotic equations of imperfection sensitivity have been presented for the first time.

References

1. SEWELL, M. J.: J. Mech. Phys. Solids **13**, 247 (1965).
2. THOMPSON, J. M. T.: J. Mech. Phys. Solids **13**, 295 (1965).
3. THOMPSON, J. M. T.: J. Mech. Phys. Solids **17**, 1 (1969).
4. THOMPSON, J. M. T., HUNT, G. W.: Comparative perturbation studies of the elastica. Int. J. Mech. Sci. (to be published).

Non-conservative Effects Produced by Thrust of Jet Engine

By

A. Petre

Polytechnic Institute, Bucuresti, Roumania

It is well known that an aeroelastic phenomenon may occur if an interaction exists between aerodynamic and elastic forces, and in addition massic forces appear for an aeroelastic dynamic phenomenon.

Usually, from the two aerodynamic components, i.e. lift and drag, only the lift effect is taken into account, both components depending on the attack angle. A more accurate physical point of view is to consider two other components of the aerodynamic forces. One, a passive drag, independent in some limits of the attack angle, and another, an active dynamic force, depending on this angle, both attached to the lift surface. Therefore it is possible to refound the classic aeroelasticity by considering the aerodynamic forces as a follower system. Recently the effect of drag on bending-torsion flutter has been shown. It is interesting to point out that the shear force produced by drag could be backward or forward, depending on the position of the propeller engine. If the propelling engine is in the middle of the aeroplane, the shear forces are backward, and if the propelling engines are at the free end of the wing, the shear forces are forward. In the latter case, the engine being attached to the wing, follower forces are also produced.

Let us consider only the case of vertical take-off, case in which the aerodynamic forces of the wing vanish. In this case the wing becomes a simple cantilever beam, assumed to be loaded on the free end by a constant transversal force attached to it and orientated in the direction of the highest rigidity.

For solving the problem, consider the well-known differential equations of bending and torsion for such a beam, assumed for instance to have a constant cross-section

$$EI\frac{\partial^4 w}{\partial x^4}+m\frac{\partial^2 w}{\partial t^2}+P\frac{\partial^2}{\partial x^2}[(l-x)\varphi]=0,$$
$$GI_d\frac{\partial^2\varphi}{\partial x^2}-J\frac{\partial^2\varphi}{\partial t^2}-P(l-x)\frac{\partial^2 w}{\partial x^2}=0.$$

Assuming also that at the limit of the stability field the motion is periodic, we can set

$$w=w(x)\cdot e^{i\omega t},$$
$$\varphi=\varphi(x)\cdot e^{i\omega t}$$

and by changing the variable x and the unknown function w into x/l and w/l we obtain another dimensionless differential equations system

$$\frac{d^4w}{dx^4}-\frac{X}{B_1}w+\alpha\frac{d^2}{dx^2}[(1-x)\varphi]=0,$$
$$\frac{d^2\varphi}{dx^2}+X\varphi-\alpha B_2(1-x)\frac{d^2w}{dx^2}=0,$$

in which

$$B_2=\frac{EI}{GI_d};\quad B_1=B_2\left(\frac{i}{l}\right)^2;\quad X=\omega^2\frac{Jl^2}{GI_d}=\left(\frac{\omega}{\omega_\varrho}\right)^2;\quad \alpha=\frac{Pl^2}{EI},$$

X being the unknown dimensionless frequency and α the unknown critical end force, the beam being specified only by the two dimensionless parameters B_1 and B_2.

It is easy to observe that this equations do not contain end mass effects, and the fact that the end force is a follower force. To clarify the problem it is better to use integral equations. With the aid of Green functions this system becomes

$$w=\frac{X}{B_1}\left[\int_0^x\frac{\xi^2}{6}(3x-\xi)\,w\,d\xi+\int_x^1\frac{x^2}{6}(3\xi-x)\,w\,d\xi\right]$$
$$+\alpha\left[\int_0^x(x-\xi)(1-x)(\varphi-\varphi_n)\,d\xi\right]+\frac{X}{B_1}B_4\frac{x^2(3-x)}{6}w_n,$$

$$\varphi=X\left[\int_0^x\xi\varphi\,d\xi+\int_x^1x\varphi\,d\xi\right]$$
$$+\alpha X\frac{B_2}{B_1}\left\{\int_0^x\xi\left[(1-\xi)\int_\xi^1(\bar\xi-\xi)\,w\,d\bar\xi+B_4(1-\xi)^2w_n\right]d\xi\right.$$
$$\left.+\int_x^1x\left[(1-\xi)\int_\xi^1(\bar\xi-\xi)\,w\,d\bar\xi+B_4(1-\xi)^2w_n\right]d\xi\right\}$$
$$+\alpha^2B_2\left[\int_0^x\xi(1-\xi)^2(\varphi-\varphi_n)\,d\xi+\int_x^1x(1-\xi)^2(\varphi-\varphi_n)\,d\xi\right]$$
$$+XB_5x\varphi_n.$$

In this equations the fact that the end force is attached to the end cross-section is taken into account by the difference $\varphi - \varphi_n$, and in the last terms the end mass effect appears very clearly, in which

$$B_4 = \frac{M_1}{ml}\,;\ B_5 = \frac{J_1}{Jl} = B_4\left(\frac{i_1}{i}\right)^2,$$

M_1 being the end mass, and J_1 being the moment of inertia of the end mass.

For solving this equations it is necessary to be careful about the modes. It is better to use the extremal property of the eigenvalue in an iterating method. For numerical results, the integral form could be very easily transformed into a linear system, replacing the unknown function w and φ by two vectors with n components. The integral equations become

$$w = \frac{X}{B_1} w_{11} w + \alpha\, \varphi_{11} \varphi\,,$$
$$\varphi = \alpha X w_{21} w + X \varphi_{21} \varphi + \alpha^2 \varphi_{22} \varphi\,,$$

in which w and φ are column matrixes, and w_{11}, φ_{11}, w_{21}, φ_{21} and φ_{22} are square matrixes.

It is interesting to find the lower value of the critical force, no matter whether it corresponds to the first combinations of the modes. The way could be the same as the physical proof of the stability by an initial perturbation. We could take this initial deformation as normed vectors multiplied by an arbitrary amplitude as

$$w = \overline{w} w_n;\ \ \varphi = \overline{\varphi} \varphi_n.$$

From the linear system we could take the last equations

$$w_n = \frac{X}{B_1} \tilde{w}_{11} w_n + \alpha \tilde{\varphi}_{11} \varphi_n\,,$$
$$\varphi_n = \alpha X \tilde{w}_{21} w_n + X \tilde{\varphi}_{21} \varphi_n + \alpha^2 \tilde{\varphi}_{22} \varphi_n\,,$$

which yield the secular equation

$$X^2 \tilde{w}_{11} \tilde{\varphi}_{21} - X\left[\tilde{\varphi}_{21} + \tilde{w}_{11} + \alpha^2 (\tilde{\varphi}_{11} \tilde{w}_{21} - \tilde{w}_{11} \tilde{\varphi}_{22})\right] + 1 - \alpha^2 \tilde{\varphi}_{22} = 0\,.$$

For having stable solutions it is necessary to satisfy the condition

$$\left[\tilde{\varphi}_{21} + \tilde{w}_{11} + \alpha^2 (\tilde{\varphi}_{11} \tilde{w}_{21} - \tilde{w}_{11} \tilde{\varphi}_{22})\right]^2 - 4\, \tilde{w}_{11} \tilde{\varphi}_{21} (1 - \alpha^2 \tilde{\varphi}_{22}) = 0\,,$$

from which we find α, and also the two confounded solutions of X,

$$X = \frac{\tilde{\varphi}_{21} + \tilde{w}_{11} + \alpha^2 (\tilde{\varphi}_{11} \tilde{w}_{21} - \tilde{w}_{11} \tilde{\varphi}_{22})}{2\, \tilde{w}_{11} \tilde{\varphi}_{21}}\,.$$

With these results, the linear system formed by the two equations is compatible, and it is possible to find the ratio of the amplitudes

$$\frac{\varphi_n}{w_n} = \frac{1 - \dfrac{X}{B_1} \tilde{w}_{11}}{\alpha \tilde{\varphi}_{11}}\,.$$

Taking $w_n = 1$ and knowing now the whole right side of the complete linear system, we could find two new normed vectors $\bar{\varphi}$ and $\bar{w}$. Repeating this way, we may find with a predicted error the critical value α, the dimensionless frequency X, the modes in flexion and in torsion, and the ratio of the amplitudes of these modes.

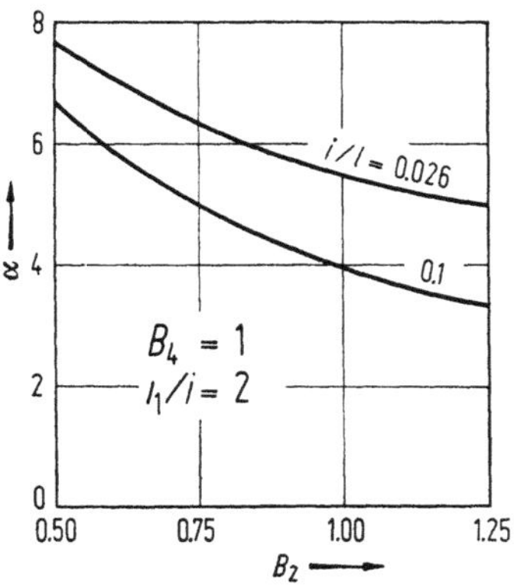

Fig. 1

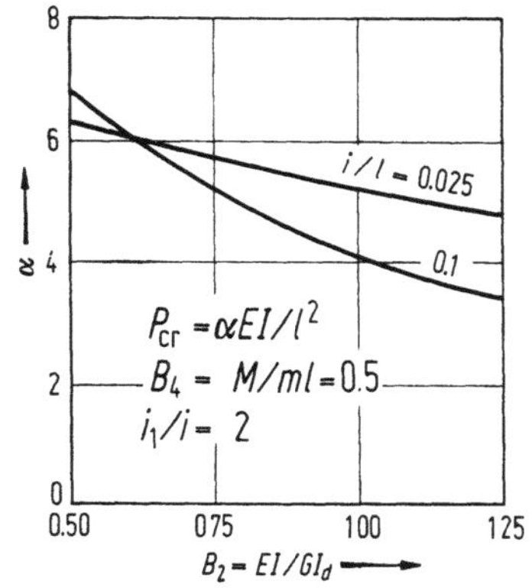

Fig. 2

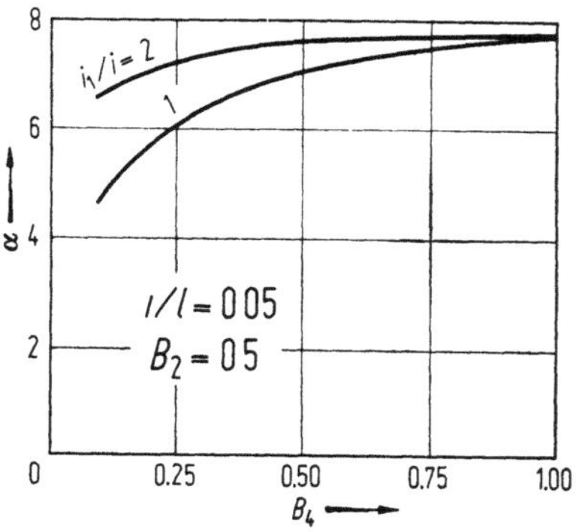

Fig. 3

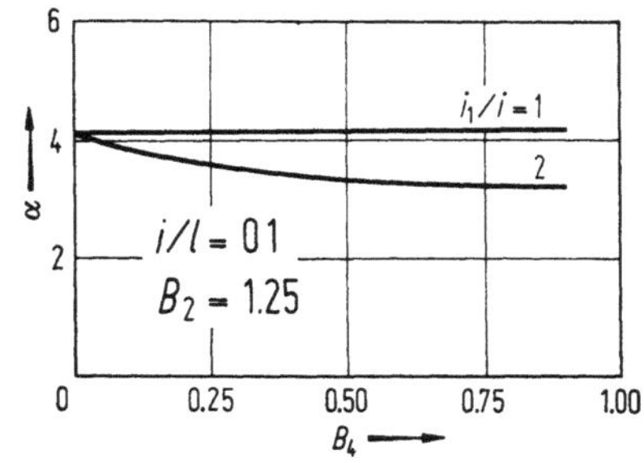

Fig. 4

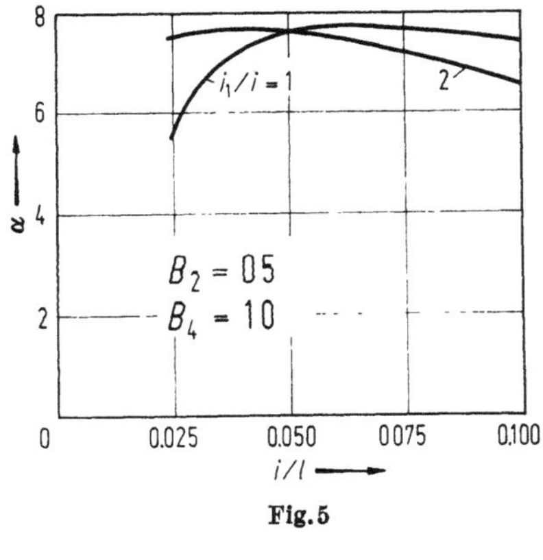

Fig. 5

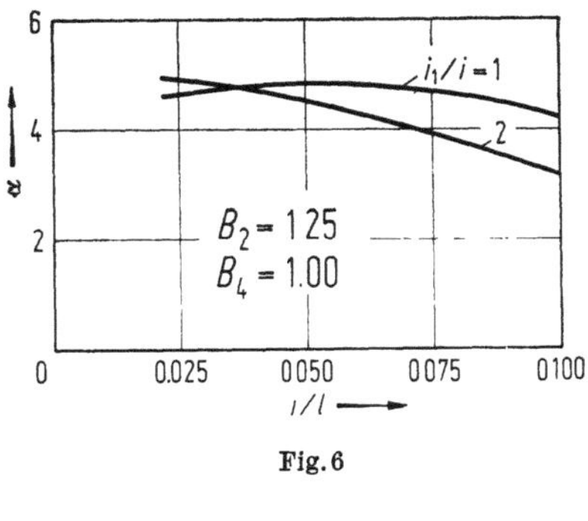

Fig. 6

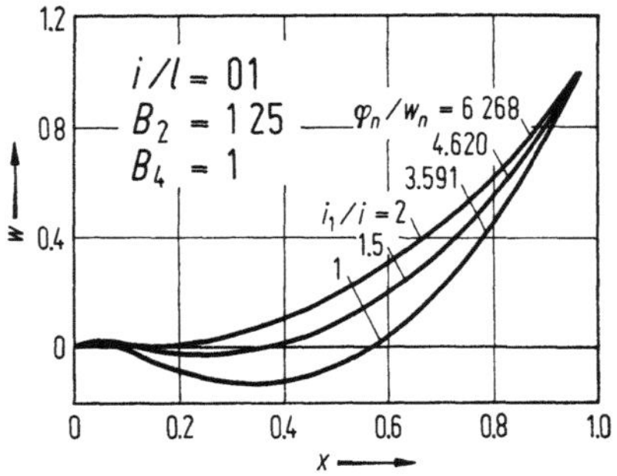

Fig. 7

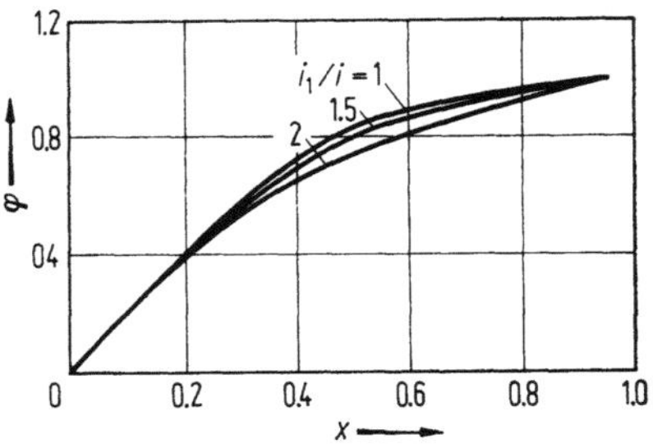

Fig. 8

Figs. 1–6 show some of the results obtained in this way, and Fig. 7 presents some modes in flexion and torsion. It is important to observe that the initial perturbation has no importance for the final results.

Stability of Viscoelastic Systems Subjected to Nonconservative Forces

By

V. V. Bolotin

U.S.S.R. Academy of Sciences, Moscow, U.S.S.R.

In the classical theory of elastic stability only potential forces (usually — the loads of the gravitational origin) were considered. With the development of technology the class of forces widened sufficiently. Many new problems were stated: stability of structures subjected to forces varying periodically in time (BELYAEV [1]), stability of structures in the presence of follower forces (NIKOLAI [2]) etc. Survey of works dealing with the nonconservative problems of elastic stability is given in papers by METTLER [3], EVAN-IWANOWSKI [4] and HERRMANN [5] and in books by BOLOTIN [6, 7] published firstly in 1956 and 1961.

Investigation of stability of potential system is usually based on the bifurcation method or on the consideration of the whole energy functional. Application of these methods to the nonconservative systems is, generally, incorrect. Beginning from BELYAEV's and NIKOLAI's papers, a dynamic method connected with the general theory of the stability of motion was used. This method consists in consideration of the set of motions neighbouring the undisturbed motion. When the stability of the equilibrium is considered, equations of disturbed motion are usually linearized. The obtained equations describe small vibrations of the system near the undisturbed equilibrium. So this method is also called small vibrations method.

Some features of the stability problems for elastic systems subjected to nonconservative forces are to be mentioned. The loss of stability of equilibrium can be both vibrational and unvibrational. In the first case critical loads depend not only on the rigidity distribution but also on the mass distribution. Usually the critical parameter decreases when the partial natural frequencies become closer. The other feature of the nonconservative problems is a strong influence of the damping on the value of critical parameters in the case of vibrational instability. This phenomenon was discovered firstly by Nikolai [2]. Dealing with the stability of an elastic bar with circular-cross section subjected to the follower torque, Nikolai showed that equilibrium is unstable for each magnitude of the torque. To obtain the finite critical torque, Nikolai took into account internal damping. As another example a problem of the parametric resonance of an elastic system can be mentioned. If damping is not considered, a sufficient part of the parameter space should be interpreted as an instability region. Consideration of damping permites us to determine a real region of dangerous parametric excitation.

Nevertheless, there is a number of papers dealing with nonconservative problems of elastic stability, where damping is not taken into account. Perhaps here an analogy with the classical problems of stability for potential systems was used. In potential systems the addition of small damping forces with complete dissipation always causes the stabilization of equilibrium. Similar statements in the presence of nonconservative forces, generally, are not valid. In the application to the follower forces this fact was shown by Ziegler [8]. Considering a double mathematical pendulum subjected on a free end to a follower force, Ziegler found that addition of a small damping can reduce the magnitude of the critical force in comparison with one found without taking into account the damping. This phenomenon is analogous to the well-known effect of destabilization due to the small viscosity in the earliest theory of hydrodynamic stability. Effect of small damping was investigated later by several authors. In paper [9] Bolotin considered the dependence of critical parameters on the ratio of partial damping coefficients. It was proved that reduction of critical parameters caused by addition of vanishing damping is absent only in a case of equal partial damping coefficients. An analogous effect on the combinational parametric vibration was indicated by Schmidt and Weidenhammer [10]. Leonov and Zory [11] considered the effect of damping on the stability of a cantilever bar subjected to follower and dead forces and having at its ends two equal masses located at some distance from each other. Herrmann and Jong [12] studied the effect of damping in Ziegler's model. Leipholz [13] considered a finite-degree-of-freedom system with a diagonal damping matrix. Some general results were obtained also by Herrmann and coauthors [14—17]. Recent papers and new results will be discussed later.

Let us consider a general statement of the stability problem for a linear viscoelastic body subjected to nonconservative forces. Usually, to get a judgement on stability of motion it is admissible to suppose the perturbations are sufficiently small and to treat the linearized equations of the disturbed motion. Here we show a procedure of obtaining the linearized equations for a case of equilibrium or slow (creep) motion of a viscoelastic body. We use the Volterra viscoelastic analogy and the nonlinear elasticity theory equations in the form proposed by NOVOZHILOV [18].

Let an undisturbed state of the viscoelastic body be characterized by displacement vector u_j, stress tensor σ_{jk}, bulk and surface force vectors X_j and p_j. Let us suppose that the body obtains small deviation from the undisturbed state and let us follow the time development of perturbations $\bar{u}_j$, $\bar{\sigma}_{jk}$, $\bar{X}_j$ and $\bar{p}_j$. Generally, the perturbations of bulk and surface forces depend on the displacements $\bar{u}_j$ and on the time t. Using smallness of the perturbations and linearizing we obtain equations

$$[\bar{\sigma}_{kl}(\delta_{jl} + u_{j,l}) + \sigma_{kl}\bar{u}_{j,l}]_{,k} + \bar{X}_j - \varrho u_{,tt} = 0 \tag{1}$$

and boundary conditions on the loaded surface:

$$[\bar{\sigma}_{kl}(\delta_{jl} + u_{j,l}) + \sigma_{kl}\bar{u}_{j,l}]\, n_k = \bar{p}_j. \tag{2}$$

Here $\bar{\sigma}_{jk} = \Lambda_{jklm}\bar{u}_{l,m}$ and Λ_{jklm} is the linear viscoelastic operator corresponding to the undisturbed state.

Eqs. (1) with the boundary conditions (2) are analogous to the "*équations aux variations*" in the classical Liapunov-Poincaré theory of stability. In classical theory theorems are proved concerning conditions under which the linearized equations yield the comprehensive solution of the stability problem in Liapunov's sense. About ten years ago mainly thanks to the works by KRASOVSKI, ZUBOV and RUMYANTSEV an analogue of the Liapunov theory in abstract spaces was developed. Thus the Liapunov methods were extended to continuous systems (see survey paper by RUMYANTSEV [19]). Some applications to the theory of elastic stability were presented by MOVCHAN [20] and KOITER [21]. It is very probable that well-known theorems on stability by first approximation are valid for continuous systems with the completely continuous Green operator and, consequently, with the discrete natural frequencies spectrum (when stability is defined using the potential energy metric). We accept later this statement as a postulate.

Let the undisturbed state be an equilibrium. Then Eqs. (1) and boundary conditions (2) are satisfied by substitution

$$\bar{u}_j = \varphi_j e^{st}, \tag{3}$$

where φ_j do not depend on time and s is a characteristic exponent. If the undisturbed state is a slow creep motion and the perturbations are rapidly varying time functions, so it is admissible to treat the time dependence of the disturbed components as a parametric one. Thus, we differ the "slow" time for the undisturbed motion and the "rapid" time for disturbances. The substitution (3) satisfies Eqs. (1) and (2) approximately and the better the slower the undisturbed motion is. The vector $\boldsymbol{\varphi} = (\varphi_1, \varphi_2, \varphi_3)$ is to be determined from a linear nonselfadjoint boundary problem. Written in the operator form, the equation of this problem is

$$[As^2 + C + \varepsilon D(s) + \beta B]\,\boldsymbol{\varphi} = 0. \tag{4}$$

Here A is a positive selfadjoint linear operator (gradient of kinetic energy of the system), $C + \varepsilon D(s)$ is a positive linear operator corresponding to viscoelastic operator (purely viscous part is separated with the factor ε), βB is a nonselfadjoint linear operator corresponding to nonconservative external forces given with the parameter β. In the general case operators C and B depend on the characteristic exponent s.

Eq. (4) corresponds to an eigenvalue problem with three parameters. They are: the characteristic exponent s, the load parameter β and the damping parameter ε. If $\beta = 0$, all the characteristic exponents are on the left half-plane of the complex variable and all motions (3) of the linearized system are decaying in time. Analogous to the classical stability theory this case will be qualified as an asymptotic stability of the undisturbed state.

When the parameter β varies continuously, the characteristic exponents vary continuously, too. At some value β_* at least one of the exponents will be imaginary. With further increasing of β the characteristic exponents with the positive real parts appear and consequently the instability occurs (Fig. 1a). The critical value β_* is determined from the condition that at $\beta > \beta_*$ among the characteristic exponents at

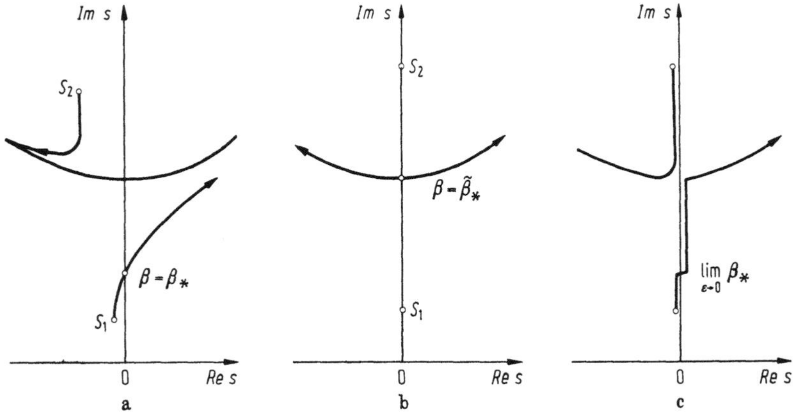

Fig. 1

least one exponent with the positive real part appears. If transition to the right half-plane through the value $s = 0$ takes place, so instability is a nonvibrational one. In other cases a vibrational instability takes place.

Now, supposing the operators C and B do not depend on s, let us consider a boundary problem

$$(As^2 + C + \beta B)\,\varphi = 0 \tag{5}$$

that follows from (4), if $\varepsilon = 0$. In many papers devoted to the nonconservative problems of elastic stability the problems of type (5) were considered. If $\beta = 0$, the boundary problem (5) is selfadjoint and all the eigenvalues s^2 are negative. Hence, all the characteristic exponents are purely imaginary. Let the parameter β increase monotonously. At the same value $\beta = \tilde{\beta}_*$ one or several couples of exponents become multiple. If β further increases, the exponents can become complex. Thus the characteristic exponents with a positive real part appear (Fig. 1b).

In former investigations on stability of elastic systems the presence of the characteristic exponents on the imaginary axis was qualified usually as stability, and the value $\tilde{\beta}_*$ as a critical one. In other words, stability was interpreted as a bounded character of motions of the linearized system. This is analogous to the questionable (critical) case in the Liapunov theory when "*équations aux variations*" do not answer the question about stability. Thus there is no sufficient reason for the application of the dynamic method in this version. Even an analogy with the stability theory of finite-degree-of-freedom systems does not hold.

Hence, analogous to the classical stability theory the case of purely imaginary characteristic exponents is to be qualified as a questionable one. When external forces are potential, an application of the direct Liapunov method in metric spaces yields a rigorous proof of stability at $\varepsilon = 0$. In this case an addition of arbitrarily small complete damping displaces all the characteristic exponents to the left half-plane. If the external forces are nonconservative, the addition of damping with complete dissipation eliminates the questionable case, too. But it was found that, generally, only a part of the segment $[0, \tilde{\beta}_*]$ belongs to the stability region. The vanishing of the dissipation does not imply necessarily $\beta_* \to \tilde{\beta}_*$. By a number of examples it was shown that a relation holds:

$$\lim_{\varepsilon \to 0} \beta_* < \tilde{\beta}_*. \tag{6}$$

The equality

$$\lim_{\varepsilon \to 0} \beta_* = \tilde{\beta}_* \tag{7}$$

takes place only if the partial damping coefficients are equal. The relations (6) and (7) are illustrated on Fig. 2 taken from paper by Bolotin [9].

Many previous authors interpreted the inequality (6) as a destabilization phenomenon caused by damping. It would be more correct to talk not about the destabilization but about the false conclusions based on the interpretation of the questionable case as the stability case. The

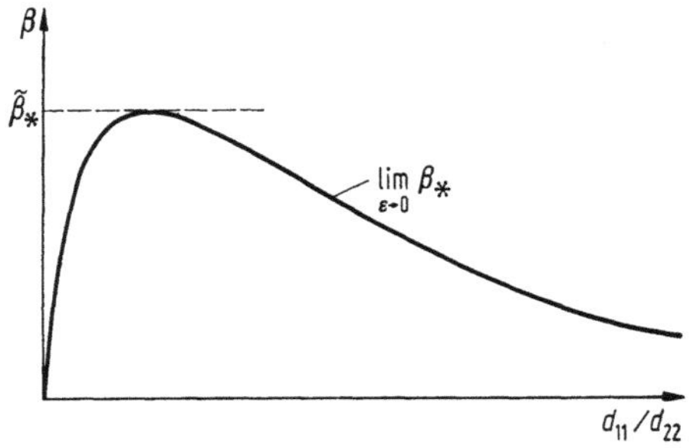

Fig. 2

"destabilization paradoxes" are consequences of noncritical application of the dynamic method. On the other side, the solutions obtained without the consideration of damping may have some physical and engineering sense for systems with sufficiently small damping. It is reasonable to introduce a terminology, by which these solutions obtain a proper place. Let us call *quasicritical* such a value $\tilde{\beta}_*$ that at $\beta = \tilde{\beta}_*$ all the characteristic exponents are purely imaginary and at $\beta > \tilde{\beta}_*$ at least one exponent comes to the right half-plane. The position of all the characteristic exponents on the imaginary axis will be called *quasistability*. We use the terms *critical* value and *stability* for the quantity β_* and for the asymptotic stability of the linearized system correspondingly. Formulae (6) and (7) describe the relation between the quasicritical value and the limit quantity of the critical value, if dissipation approaches to zero.

A discussion of the effect of small damping on the stability was presented by Herrmann, Nemat-Nasser and Prasad [14, 15]. Expanding the characteristic determinant in integer powers of damping parameter some conclusions were made about the behaviour of characteristic exponents of finite-degree-of-freedom system with small damping. An attempt was made to extend these results to continuous systems. But in the neighbourhood of the boundary of the quasistability region the integer powers expansions become invalid.

A rigorous proof of theorems on the relation between the critical and quasicritical parameters was published by Zhinzher [22]. Zhin-

ZHER considered the characteristic equation

$$F(s, \varepsilon, \beta) = 0 \tag{8}$$

connecting the characteristic exponent s, the damping parameter ε and the load parameter β. At any fixed β, Eq. (7) determines the inexplicit function $s(\varepsilon)$. To determine branches of the solutions of the Eq. (8) in the neighbourhood of the point $\beta = \tilde{\beta}_*$, $s = s_*$, $\varepsilon = 0$, an expansion in fractural powers of ε was applied. Here we consider a case when Eq. (8) has a double nonzero root at $\beta = \tilde{\beta}_*$, $\varepsilon = 0$. If a condition

$$\left.\frac{\partial F(s_*, \varepsilon, \tilde{\beta}_*)}{\partial \varepsilon}\right|_{\varepsilon=0} \neq 0 \tag{9}$$

is satisfied, the inequality (6) takes place, i.e. a phenomenon occurs called in earlier papers "destabilization due to small damping". The equality

$$\left.\frac{\partial F(s_*, \varepsilon, \tilde{\beta}_*)}{\partial \varepsilon}\right|_{\varepsilon=0} = 0 \tag{10}$$

is a necessary condition under which Eq. (7) holds.

The condition (10) is very special; it is satisfied by some restrictions on the viscoelastic properties of the system. Many earlier stated facts may be considered as special cases of the theory presented. For example, the condition (10) is satisfied for finite-degree-of-freedom systems with the scalar matrix of the damping forces. This condition is satisfied also in those panel flutter problems where disturbed pressure is introduced according to the linearized piston theory and structural damping is neglected. But generally the manyfold of systems for which the condition (10) is valid, is relatively poor. For viscoelastic continuous systems the fulfilment of the inequality (9) and, consequently, the inequality (6) is typical.

Development of the theory in application to continuous systems was given in paper by BOLOTIN and ZHINZHER [23]. In this paper both flutter and divergence instability are investigated, equations of the critical surface in a case of the infinitesimal damping are obtained and a method of numerical analysis adapted to the digital computers is proposed. Here we discuss a partial problem treated in the paper mentioned above. Let a cantilever bar made of the standard linear viscoelastic solid be subjected to the follower (tangential) force P and dead force Q (Fig. 3). The characteristic equation corresponding to the boundary eigenvalue problem has a form $F(s, \eta, \gamma, \alpha, \beta) = 0$. Here η is a damping parameter, γ is a ratio of the

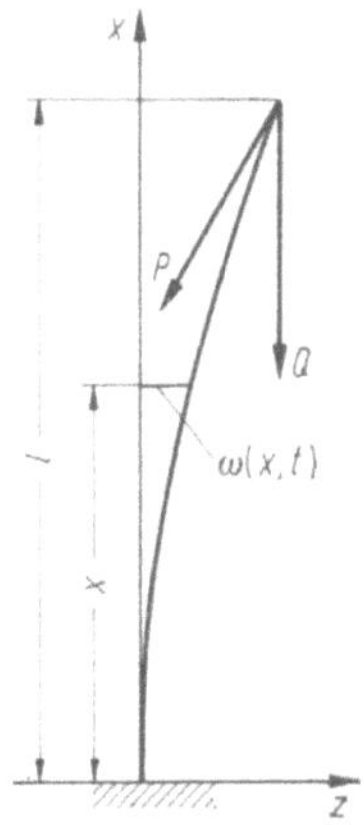

Fig. 3

relaxed modulus to the unrelaxed one, α and β are nondimensional parameters of the dead and the follower forces respectively.

Calculating the partial derivative incoming in the conditions (9) and (10), we find that this derivative is not equal to zero everywhere at $0 \leq \gamma < 1$ and $s_* \neq 0$. Thus, the boundary of the vibrational instability region in the presence of infinitesimal damping displaces everywhere to the smaller values of the parameter β. This statement is illustrated by results of numerical analysis of the characteristic equation. A typical diagram is presented on Fig. 4. Broken lines are the boundaries of the quasistability region, continuous lines the boundaries of the stability regions by different values of η. At $\eta \to 0$ a considerable part of the quasistability region is in fact the region of the vibrational instability.

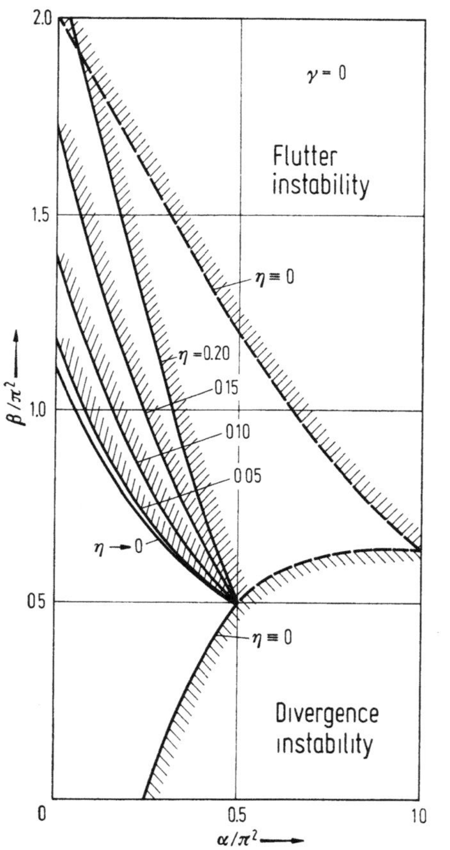

Fig. 4

Similar diagrams were obtained in a problem of stability of a rectangular plate made of viscoelastic solid in the gaz flow. Let the plate be compressed by thermal forces. An analogue of the parameter α used in the problem discussed above, is a parameter of thermal forces, an analogue of the parameter β is the velocity of the flow or the dynamic pressure. If the structural damping is sufficiently small, the real parts of the characteristic exponents in the mutual parts of the stability and quasistability regions are sufficiently small, too. It is possible that in these parts of the parameter space a "quiet" flutter occurs. For the other parts of the vibrational instability region a "violent" flutter is typical. The existence of the two types of flutter has been mentioned in some experimental works.

Stability of elastic systems in gaz flow is considered usually taking into account the aerodynamic damping which is enclosed naturally in expressions for the disturbed gaz pressure. Internal damping often is ignored. But in some cases the internal damping sufficiently effects on the critical velocities of the flow. In application to the infinitely long shells this effect was shown by Bolotin [24]. In another paper [25] it was shown that consideration of damping is necessary for correct solution of the nonlinear flutter problem using the Poincaré small parameter method. We later shall discuss some effects of length in nonconservative stability problems. Assuming the system as an unbounded one in the direction of the flow, essential simplifications of the numerical analysis are achieved. This way was used by Bolotin [26] and by Miles [27]. On the other side, some difficulties arise at the interpretation of the results.

One of the difficulties is connected with the application of the approximate aerodynamic formulae of "piston theory" or Ilyushin's "plane sections law" [28]. Formulae of this type, being applied widely to the calculation of the flutter velocities for plates and shells, have definite boundaries of application [29] which cannot be ignored. The Mach number must be sufficiently large in comparison with unity, the reduced frequency must be sufficiently small, and the characteristic length of the perturbations across the flow must be sufficiently small in comparison with those along the flow. In application to the perturbations of "travelling wave" type, two first conditions are equivalent to the requirement that flow velocity U is sufficiently large in comparison with the phase velocity of elastic waves $V(k)$ (here k is a wave number). In the problem of stability of infinitely long elastic cylindrical shell the critical velocity is

$$U_{*\infty} = \min_k V(k).$$

Thus, on the boundary of the stability region the gaz particles velocity relative to the elastic wave is equal to zero. This circumstance was used by Miles [30] to criticize some papers where the piston theory has been applied to infinitely long plates and shells.

As was shown in paper by Bolotin [24], the difficulty mentioned above vanishes or at least relaxes, if the internal loss of energy in the material of the shell is considered. For example, when a simplest mathematical model of the viscoelastic shell was considered, a formula was obtained:

$$U_{*\infty} = \min_k \{V(k)\,[1 + \varepsilon_0(k)/\varepsilon]\}. \tag{11}$$

As a rule, the material internal damping parameter ε_0 is large in comparison with the aerodynamic damping parameter ε. Thus a condition $U_{*\infty} \gg V(k)$ will be satisfied for a shell made of material with high damping capacity.

Introduction of the viscoelasticity in the structure model permits to solve another, more principal difficulty. As Movchan [31] has shown, the magnitude of the critical flow velocity for an infinitely long elastic plate differs sufficiently from the magnitude that would be obtained considering the bounded plate the length of which later tends to infinity. An analogous result was obtained for a circular cylindrical shell. For the shell with sufficiently great length l the critical speed U_* decreases as $l^{-5/3}$ or, maybe, as l^{-1} [32]. On the other side the length parameter is not involved in the formula for the critical velocity of the unbounded shell. Miles's remarks [30] on the boundaries of application of the "piston theory" do not solve this paradox.

The main question to be answered, is the following: at what conditions a structure of sufficiently large length can be treated as an unbounded one? In statics we often use unbounded or semi-unbounded models. It is reasonable to apply this approach to such problems where the influence of the boundary conditions at some parts of the boundary decay sufficiently rapid by moving off from the boundary. Then superposition of the effects due to boundary conditions in different sufficiently distant points of the boundary can be ignored. Saint-Venant's principle in the elasticity theory, Love's and Reissner's effects in the theory of shells etc. are examples of this approach. An analogous approach can be used in dynamics, too. A body of bounded size can be treated as an unbounded one only in the case, when boundary effects are negligible.

Now let us consider a wave propagation through the body. If the body is elastic and if its size is arbitrarily large but bounded, so each wave disturbance comes to the boundary and after reflection returns in some form. The influence of the boundary may be neglected only in the case when the material of the body is dissipative and the wave intensity decays rapidly enough moving off from the disturbance origin. Let ψ be a specific damping parameter and λ a characteristic scale of the disturbance (for example, the wave length). Wave disturbance in some point with the distance l from the origin can be assumed as sufficiently decaying, if a condition is satisfied

$$l \gg \frac{\lambda}{\psi^{1/n}}, \tag{12}$$

where n is a number of dimensions in wave propagation. The larger the energy losses are, the smaller is the characteristic size l starting from that a bounded but sufficiently large body can be simulated by an unbounded body. Similar considerations can be performed in a case of steady vibration. Then the condition (12) secures the sufficiently rapid decaying of boundary effects when we move into internal region.

As an illustrative example the critical velocity U_* for the mathematical model of shell was calculated. Dependence of the critical velocity U_* on the ratio $\varepsilon_0/\varepsilon$ and on the nondimensional length l/λ was studied. Here λ is a characteristic length parameter. The larger the ratio $\varepsilon_0/\varepsilon$ is, the smaller is the magnitude of the ratio l/λ by which the critical velocity U_* differs not too much from the value calculated by means of the formula (11).

In conclusion, it must be emphasized that a physically and mathematically correct treatment of stability of systems subjected to nonconservative forces is impossible without consideration of internal damping and other dissipation factors. Moreover because of the great sensitivity of the stability conditions to the law of internal damping, a proper attention to the damping properties of structures is necessary.

References

1. Беляев, Н. М.: Устойчивость призматических стержней под действием переменных продольных сил, Сб. "Инженерные сооружения и строительная механика", Ленинград: Изд. "Путь" 1924.
2. Николаи, Е. Л.: Об устойчивости прямолинейной формы равновесия сжатого и скрученного стержня. Изв. Ленингр. политехн. ин-та 1928, № 31; К вопросу об устойчивости скрученного стержня. Вестн. прикл. матем. мех. 1929, № 1. See also: Николаи, Е. Л.: Труды по механике, Москва: Гостехиздат 1955.
3. Mettler, E.: Dynamic buckling. Handbook of engineering mechanics, New York: McGraw-Hill 1962.
4. Evan-Iwanowski, R. M.: On the parametric response of structures. Appl. Mech. Rev. **18**, No. 9 (1965).
5. Herrmann, G.: Stability of equilibrium of elastic systems subjected to nonconservative forces. Appl. Mech. Rev. **20**, No. 2 (1967).
6. Болотин, В. В.: Динамическая устойчивость упругих систем, Москва: Гостехиздат 1956 (German translation, Berlin: Verlag der Wissenschaften 1961; English translation, San Francisco: Holden Day 1964).
7. Болотин, В. В.: Неконсервативные задачи теории упругой устойчивости, Москва: Физматгиз 1961 (English translation, Oxford: Pergamon Press 1963).
8. Ziegler, H.: Die Stabilitätskriterien der Elastomechanik. Ing.-Arch. **20**, No. 1 (1952).
9. Болотин, В. В.: О колебаниях и устойчивости стержней, находящихся под действием неконсервативых сил, Сб. "Колебания в турбомашинах", Москва: Изд. АН СССР 1959.
10. Schmidt, G., Weidenhammer, F.: Instabilität gedämpfter rheolinearer Schwingungen. Math. Nachr. **23**, No. 4—5 (1961).
11. Леонов, М. Я., Зорий, Л. М.: О влиянии трения на критическую силу сжатого стержня. Доклады АН СССР **145**, № 2 (1962).
12. Herrmann, G., Jong, I. C.: On the destabilizing effect of damping in nonconservative elastic systems. J. Appl. Mech. **32**, No. 3 (1965).
13. Leipholz, H.: Über den Einfluß der Dämpfung bei nichtkonservativen Stabilitätsproblemen elastischer Stäbe. Ing.-Arch. **33**, No. 5 (1964).
14. Nemat-Nasser, S., Prasad, S. N., Herrmann, G.: Destabilizing effect of velocity-dependent forces in nonconservative continuous systems. AIAA J. **4**, No. 7 (1966).

15. Nemat-Nasser, S., Herrmann, G.: Some general considerations concerning the destabilizing effect in nonconservative systems. ZAMP **17**, No. 2 (1966).
16. Herrmann, G., Jong, I. C.: On nonconservative stability problems of elastic systems with slight damping. J. Appl. Mech. **33**, No. 1 (1966).
17. Nemat-Nasser, S.: On the stability of the equilibrium of nonconservative continuous systems with slight damping. J. Appl. Mech. **34**, No. 3 (1967).
18. Новожилов, В. В.: Основы нелинейной теории упругости, Москва: Гостехиздат 1948.
19. Румянцев, В. В.: Метод функций Ляпунова в теории устойчивости, Сб. "Механика в СССР за 50 лет", т. 1, Москва: Изд. "Наука" 1969.
20. Мовчан, А. А.: О прямом методе Ляпунова в задачах устойчивости упругих систем. Прикладная математика и механика **23**, № 3 (1959).
21. Koiter, W. T.: The energy criterion of stability for continuous elastic bodies. Proc. Roy. Acad. Sci. Netherlands **68 B**, No. 4 (1965).
22. Жинжер, Н. И.: О дестабилизирующем влиянии трения на устойчивость неконсервативных упругих систем. Механика твердого тела 1968, № 3.
23. Bolotin, V. V., Zhinzher, N. I.: Effects of damping on stability of elastic systems subjected to nonconservative forces. Int. J. Solids and Structures **5**, 965 (1969).
24. Болотин, В. В.: Некоторые новые задачи динамики оболочек, Сб. "Расчеты на прочность", вып. 4, Москва: Машгиз 1959.
25. Болотин, В. В.: Нелинейный флаттер пластин и оболочек. Инженерный сборник **28** (1960).
26. Болотин, В. В.: Колебания и устойчивость упругой цилиндрической оболочки в потоке сжимаемого газа. Инженерный сборник **24** (1956).
27. Miles, J. W.: Supersonic flutter of a cylindrical shell. J. Aeronaut. Sci. **24**, No. 2 (1957); **25**, No. 5 (1958).
28. Ильюшин, А. А.: Закон плоских сечений в аэродинамике больших сверхзвуковых скоростей. Прикладная математика и механика **20**, № 6 (1956).
29. Болотин, В. В.: О применении "закона плоских сечений" для определения аэродинамических сил, действующих на колеблющиеся оболочки, Изв. АН СССР, ОТН, Механ. и машиностр. 1961, № 1.
30. Miles, J. W.: On supersonic flutter of long panels. J. Aeronaut. Sci. **27**, No. 6 (1960).
31. Мовчан, А. А.: Об устойчивости панели, движущейся в газе. Прикладная математика и механика **21**, № 2 (1957).
32. Григолюк, Э. И., Лампер, Р. Е., Шандаров, Л. Г.: Флаттер панелей и оболочек, Сб. "Итоги науки", Механика, 1963, Москва: Изд. ВИНИТИ 1965.

The Instability of Pipe Poiseuille Flow with Respect to Finite Amplitude Disturbances

By

M. Lessen, B. H.-P. Chen and **K. H. Lau**

University of Rochester, Rochester, N.Y., U.S.A.

Abstract. The stability with respect to finite amplitude disturbances of Hagen-Poiseuille flow is studied, and it is found that for a disturbance velocity amplitude of 5% (scaled to the unperturbed velocity in the pipe center) and a wave number of unity (scaled to the pipe radius), a wall mode of an azimuthally periodic disturbance of periodicity unity is neutrally stable at a Reynolds number of 2900. A linear wall mode is used to calculate the Reynolds stress and the modified steady flow thereby obtained is then examined for instability with respect to the same mode. Although it is found that a particular finite amplitude mode predisposes the flow towards instability with respect to the same mode, the same may be true for other modes and there may also exist important interactions between modes.

I. Introduction

At the end of the last century, Osborne Reynolds (1883) observed the phenomenon of transition from laminar to turbulent flow for Poiseuille flow in a pipe. Later, experimenters found that the critical Reynolds number varied from 2000 to 40000 depending on the roughness of the pipe and initial disturbance intensity.

The linearized stability theory with respect to axisymmetric disturbances of pipe Poiseuille flow has been studied by Sexl (1927), Pretsch (1941), Corcos and Sellars (1959), Pekeris (1948), and Davey and Drazin (1969). All concluded that the flow is stable subject to the assumptions made. Lessen, Fox, Bhat and Liu (1964) concluded

Acknowledgement. This work was supported in part by grants from the National Science Foundation and the Office of Naval Research.

that some three-dimensional disturbances are less stable than two-dimensional ones, which led to the theoretical investigation of LESSEN, SADLER and LIU (1968) where no instability was found. Later, BURRIDGE and DRAZIN (1969) qualitatively confirmed their result with asymptotic theory.

In the theory of infinitesimal disturbances, second and higher order disturbance terms are discarded and hence the effect of Reynolds stress on the mean motion is neglected; therefore, the parabolic velocity profile is used throughout as the basic flow. In the case of finite amplitude disturbances operating on the basic flow, the parabolic velocity profile is only an initial condition. As the disturbances decay they interact with the basic flow through Reynolds stresses and the initial velocity profile is distorted. It has been found that this slightly distorted basic flow profile has unstable eigenvalues even if the initial disturbances are fairly small.

II. Theoretical considerations

The Navier-Stokes equations and the equation of continuity for an incompressible, viscous fluid in cylindrical coordinates are used to describe the fluid motion. These equations are:

$$\frac{\partial U}{\partial t} + U\frac{\partial U}{\partial r} + \frac{V}{r}\frac{\partial U}{\partial \theta} + W\frac{\partial U}{\partial z} - \frac{V^2}{r} = -\frac{\partial P}{\partial r} + \frac{1}{R}\left[\Delta U - \frac{2}{r^2}\frac{\partial V}{\partial \theta} - \frac{U}{r^2}\right], \tag{1}$$

$$\frac{\partial V}{\partial t} + U\frac{\partial V}{\partial r} + \frac{V}{r}\frac{\partial V}{\partial \theta} + W\frac{\partial V}{\partial z} + \frac{UV}{r} = -\frac{1}{r}\frac{\partial P}{\partial \theta} + \frac{1}{R}\left[\Delta V + \frac{2}{r^2}\frac{\partial U}{\partial \theta} - \frac{V}{r^2}\right], \tag{2}$$

$$\frac{\partial W}{\partial t} + U\frac{\partial W}{\partial r} + \frac{V}{r}\frac{\partial W}{\partial \theta} + W\frac{\partial W}{\partial z} = -\frac{\partial P}{\partial z} + \frac{1}{R}\Delta W, \tag{3}$$

$$\frac{\partial U}{\partial r} + \frac{1}{r}\frac{\partial V}{\partial \theta} + \frac{\partial W}{\partial z} + \frac{U}{r} = 0,$$

$$\Delta = \frac{\partial^2}{\partial r^2} + \frac{1}{r}\frac{\partial}{\partial r} + \frac{1}{r^2}\frac{\partial^2}{\partial \theta^2} + \frac{\partial^2}{\partial z^2}, \tag{4}$$

where r, θ, z denote radial, azimuthal and axial coordinates and U, V, W their respective velocity components. It is further assumed that disturbances have exponential dependence in axial, azimuthal and time coordinates and they are:

$$\begin{aligned} U &= \qquad\qquad\qquad \varepsilon u(r) \exp[i(\alpha z + n\theta - ct)], \\ V &= \qquad\qquad\qquad \varepsilon v(r) \exp[i(\alpha z + n\theta - ct)], \\ W &= W(r, t) \qquad + \varepsilon w(r) \exp[i(\alpha z + n\theta - ct)], \\ P &= P(r, z, \theta, t) + \varepsilon p(r) \exp[i(\alpha z + n\theta - ct)], \end{aligned} \tag{5}$$

where α, n and c are the respective wave numbers.

The linearized disturbance equations are as follows:

$$u'' + \frac{1}{r} u' - \left(\frac{1+n^2}{r^2} + \alpha^2 - icR + i\alpha RW\right) u - \frac{2in}{r^2} v - Rp' = 0, \tag{6}$$

$$v'' + \frac{1}{r} v' - \left(\frac{1+n^2}{r^2} + \alpha^2 - icR + i\alpha RW\right) v + \frac{2in}{r^2} u - \frac{in}{r} Rp = 0, \tag{7}$$

$$w'' + \frac{1}{r} w' - \left(\frac{n^2}{r^2} + \alpha^2 - icR + i\alpha RW\right) w - RuW' - i\alpha Rp = 0, \tag{8}$$

$$u' + \frac{u}{r} + i\left(\alpha w + \frac{nv}{r}\right) = 0, \tag{9}$$

where prime denotes differentiation with respect to r.

The relevant boundary conditions on Eqs. (6)—(9) are

(a) at $r = 1$, for all values of n

$$u = v = w = p = 0,$$

(b) at $r = 0$, for $n \neq 1$

$$u = v = w = p = 0,$$

(c) at $r = 0$, for $n = 1$

$$w = p = 0, \quad u + iv = 0.$$

Eqs. (6)—(9) and the boundary conditions are explored and the eigenfunctions and eigenvalues obtained are used to obtain the Reynolds stresses associated with small finite amplitude disturbances. The Reynolds stresses are then used to calculate a distorted "mean" velocity distribution. It is assumed, of course, that the linear theory describes the initial finite amplitude disturbance to sufficient accuracy.

III. Distorted basic flow profile due to Reynolds stresses

The following analysis is similar to the mathematical treatment of turbulent flow, and the treatment is after the manner of MEKSYN and STUART (1951). Here, it is assumed that the disturbance velocity components have the form of (5). The average of a function is defined to be:

$$\bar{f} \equiv \frac{n\alpha}{4\pi^2} \int_0^{2\pi/n} d\theta \int_0^{2\pi/\alpha} f\, dz. \tag{10}$$

Substituting (5) into (1)—(4), and averaging over θ, z variables, one obtains:

$$\frac{\partial W}{\partial t} + \varepsilon^2 \left[\frac{d}{dr}(\overline{uw}) + \frac{\overline{uw}}{r}\right] = -\frac{\partial \bar{P}}{\partial z} + \frac{1}{R}\left[\frac{\partial^2 W}{\partial r^2} + \frac{1}{r}\frac{\partial W}{\partial r}\right], \tag{11}$$

$$\frac{1}{r}\frac{\partial \bar{P}}{\partial \theta} = -\varepsilon^2 \left[\frac{d}{dr}(\overline{uv}) + \frac{2\overline{uv}}{r}\right], \tag{12}$$

$$\frac{\partial \bar{P}}{\partial r} = -\varepsilon^2 \left[\frac{d}{dr}(\overline{uu}) + \frac{\overline{uu}}{r} - \frac{\overline{vv}}{r}\right]. \tag{13}$$

For small ε we shall assume that the steady pressure gradient remains unchanged. This can be justified physically, since pressure is applied externally (STUART (1957)). One can rewrite (11) as:

$$\left[\frac{\partial}{\partial t} - \frac{1}{R}\left(\frac{\partial^2}{\partial r^2} + \frac{1}{r}\frac{\partial}{\partial r}\right)\right] W = \frac{4}{R} - \varepsilon^2 e^{2\alpha c_i t}\left[\frac{1}{r}\frac{d}{dr}(r\overline{uw})\right] \tag{14}$$

with initial and boundary conditions:

$$\begin{aligned} W(r, 0) &= 1 - r^2, \\ W(1, t) &= 0, \\ W(0, t) &= \text{finite}. \end{aligned} \tag{15}$$

It is assumed that a theoretical pulse in Reynolds stress somewhat approximating reality may be fabricated from a decaying finite amplitude disturbance operating initially on a parabolic profile. The final velocity profile (distorted) is taken after an e-folding time in the Reynolds stress.

Eq. (14) can be solved by a zeroth order Hankel transformation and the complete solution is given by:

$$W(r, t) = 1 - r^2 + \sum_{m=1}^{\infty} \frac{J_0(\varrho_m r)}{J_1(\varrho_m)} \left\{ \frac{2k(\varrho_m)}{J_1(\varrho_m)\left[\frac{\varrho_m^2}{R} + 2\alpha c_i\right]} \left(e^{2\alpha c_i t} - e^{-\frac{\varrho_m^2}{R} t}\right) \right\} \tag{16}$$

where ϱ_m is the mth root of the zeroth order Bessel function and

$$k(\varrho_m) = -\varepsilon^2 \int_0^1 J_0(\varrho_m r) \frac{d}{dr}(r\overline{uw})\, dr.$$

The solution can be expanded into power series

$$W(r,t) = 1 - r^2$$
$$+ \sum_{m=1}^{\infty} \sum_{k=0}^{\infty} \frac{1}{J_1(\varrho_m)} \left\{ \frac{2k(\varrho_m)}{J_1(\varrho_m)\left[\frac{\varrho_m^2}{R} + 2\alpha c_i\right]} \left(e^{2\alpha c_i t} - e^{-\frac{\varrho_m^2}{R} t} \right) \right\} \frac{(-1)^k \varrho_m^{2k}}{2^{2k}(k!)^2} r^{2k} \quad (17)$$

for the Frobenius expansion start of the disturbance function integration.

Timewise instability is determined by choosing α, n real and c complex. If $c_i > 0$ for some α, n, R then the flow is said to be unstable; if $c_i = 0$ for some α, n, R then the flow is neutrally stable; if $c_i < 0$ for all α, n, R then the flow is said to be stable.

IV. Method of solution

Lessen, Sadler and Liu (1968) have solved several linear modes with azimuthal periodicity of one. These eigenfunctions are used as the initial disturbance for solving the distorted flow.

At low Reynolds numbers the solutions to the coupled linear differential equations are obtained by power series solution in the radius similar to the Frobenius method. For higher Reynolds numbers, the Frobenius solution is used as a start for a step by step (Taylor series) integration method. The readers are referred to the paper by Lessen, Sadler and Liu (1968) for detailed descriptions of these methods and the method of obtaining eigenvalues.

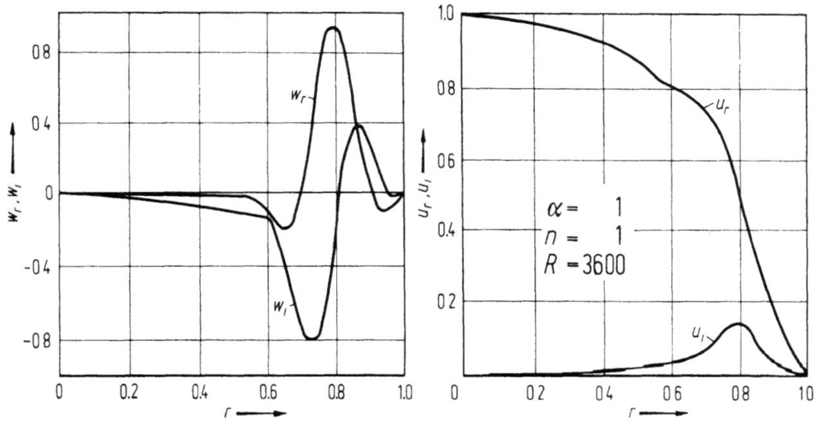

Fig. 1. Typical wall mode eigenfunctions.

The initial disturbance that has been chosen to calculate the distorted flow profile is a wall mode according to Lessen, Sadler and

Liu (1968). The typical eigenfunctions of this mode are shown in Fig. 1. Fig. 2 is a typical distorted flow profile; it should be noted that a point of inflection exists in all cases even with a fairly small disturbance intensity. The distorted flow is computed after an e-folding time in the decaying Reynolds stress due to the initial disturbance. The stability of the distorted velocity profile is then computed in the same manner as was used for the undistorted velocity profile. Thus, for a given wave number and intensity of disturbance, a Reynolds number is obtained such that the distorted profile is neutrally stable to the same mode of disturbance as the initial one.

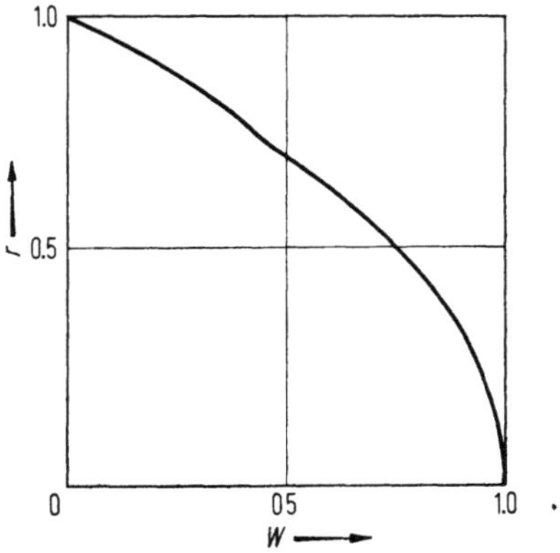

Fig. 2. Distorted velocity profile of $\varepsilon = 3.5\%$, $R = 6000$.

V. Results and discussion

Neutral stability in the sense of the previous section has been obtained for pipe Poiseuille flow. Calculations were carried out for the least stable (linear) wall mode for $n = 1$ and $\alpha = 1$. It was found that for disturbance intensities (ε), where the velocity and length scales were the velocity in the center of the pipe and the pipe radius respectively of 0, 3.5%, 4%, and 5%, the critical (neutrally stable) Reynolds numbers were ∞, 6000, 4200, and 2900 respectively (see Fig. 3).

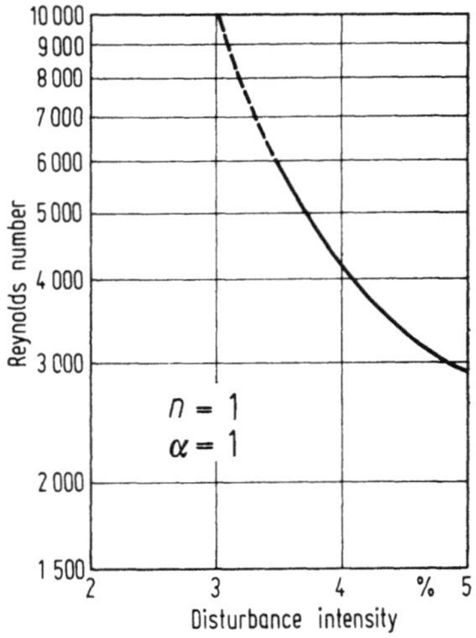

Fig. 3. Critical Reynolds number curve.

A number of approximations were made in this investigation. These include:

(a) the use of the linear eigenfunctions to obtain a finite Reynolds stress,

(b) the assumption that the shifting eigenvalues during the distortion process did not greatly affect the Reynolds stress,

(c) neglecting of still higher order effects.

It may be seen, from Eq. (16) that the characteristic time of decay of the initial Reynolds stress is very much smaller than the characteristic time of restoration of the parabolic profile from the distorted one by viscous forces. Thus, once a profile is distorted, it tends to remain so for a long time so that a linear instability can grow to large amplitude. If that instability is the same mode as the initial disturbance, the initial disturbance may be said to "lock in" as is the present case.

At the start of the present investigation, a "center" mode disturbance was first investigated; it was found that this mode merely flattened out the nose of the velocity profile but did not greatly affect its stability. It is possible that the lock-in phenomenon could also be found for this mode but at a much higher Reynolds number than that for the wall mode.

In looking back at the experimental data for the paper by Fox, Lessen and Bhat (1968), it seemed that immediately downstream of the perturbing wave, the disturbance amplitude was greatest at the center of the pipe, but further downstream (20 diameters) the disturbance amplitude was greatest near the wall. Also, preliminary observations on hydrogen bubble sheets generated by a diametral wire in a Reynolds experiment seem to indicate that the greatest disturbance amplitude is at the wall, near transition.

References

1. Burridge, D. M., Drazin, P. G.: Phys. Fluids **12**, 264 (1969).
2. Corcos, G. M., Sellars, J. F.: J. Fluid Mech. **5**, 97–113 (1959).
3. Davey, A., Drazin, P. G.: J. Fluid Mech. **36/2**, 209–218 (1969).
4. Fox, J. A., Lessen, M., Bhat, W. V.: Phys. Fluids **11**, 1–4 (1968).
5. Lessen, M., Fox, J. A., Bhat, W. V., Liu, T. Y.: Phys. Fluids **7**, 1384 (1964).
6. Lessen, M., Sadler, S. G., Liu, T. Y.: Phys. Fluids **11**, 1404 (1968).
7. Meksyn, D., Stuart, J. T.: Proc. Roy. Soc. London **A 209**, 517–529 (1951).
8. Pekeris, C. L.: Proc. Nat. Acad. Sci. U.S.A. **34**, 285–295 (1948).
9. Pretsch, J.: Z. angew. Math. Mech. **21**, 204–217 (1941).
10. Reynolds, O.: Scientific Papers, Vol. 2, London/New York: Cambridge University Press 1883, pp. 51–105.
11. Sexl, T.: Ann. Phys. **83**, 835–848 (1927).
12. Stuart, J. T.: J. Fluid Mech. **4**, 1–21 (1958).

The Structure of the Damping Disturbances in the Stability of Unbounded Laminar Flows

By

T. Tatsumi and **K. Gotoh**

University of Kyoto, Kyoto, Japan

Abstract. The structure of the disturbances of finite damping rate at very large Reynolds numbers is investigated. It is shown that the eigensolutions of the damping disturbances have the form of a concentrated wave-packet. As the consequence of this structure it is shown that the damping of the disturbance at infinite Reynolds number is actually caused by the viscous dissipation.

§ 1. Introduction

The nonlinear stability theory has to deal with the nonlinear interaction between different wave-number modes of the disturbances. Up to the present the nonlinear treatment of the stability has been mostly concerned with the near-critical problem, that is the situation slightly above or under the critical Reynolds number, in which the explicit treatment of the nonlinear coupling between different wave-number modes can be avoided (see STUART [1]). If, on the other hand, we deal with the nonlinear stability of unbounded laminar flows like jets, wakes, and free boundary layers, which usually have very low critical Reynolds numbers, we can no longer employ the near-critical approximation. The stability characteristics of unbounded flows at large Reynolds numbers do not appreciably change with the Reynolds number: those modes whose wave-numbers are less than a certain value are amplified whereas all modes of higher wave-numbers are damped. In actual situations the disturbance which first grows up spontaneously is the

Acknowledgement. One of the authors (T.T.) wishes to express his cordial thanks to Professor H. LEIPHOLZ of Karlsruhe University and Committee of the International Union of Theoretical and Applied Mechanics for the financial support for the travel expenses.

mode of the largest growth rate. When this mode is amplified to a finite amplitude there arises a train of higher harmonics due to the nonlinearity of the equation of motion, and most of harmonics belong to the damping mode. Thus we have to deal with the nonlinear interaction between the amplifying modes and the damping modes of the disturbance.

Unlike the amplifying modes, the functional behaviour of the damping modes has not been well clarified to date mostly due to the presence of the singular region in the latter. Since the wrong knowledge of the functional form of the disturbances sometimes leads to erroneous conclusions in the nonlinear treatment of stability it is necessary to investigate the structure of the damping modes in the singular region, and this is the subject of the present paper.

It will be shown as the result that at large Reynolds numbers the eigenfunction of the damping mode has quite different structure than that of the solutions of the inviscid equation, and in fact it takes the form of a concentrated wave-packet involving a large number of fine oscillations in it. As a consequence of this structure of the damping modes it will also be shown that the *inviscid damping*, or the damping of the disturbance at infinite Reynolds number, is actually caused by the viscous dissipation.

§ 2. Inner friction region

Let us denote the basic laminar flow by $(U(y))$, 0 and a Fourier component of the disturbance stream-function by $\varphi(y)\exp[i\alpha(x-ct)]$, where (x, y) represent the two-dimensional coordinates, t the time, $\alpha\ (>0)$ the wave-number, c a complex number $c = c_r + ic_i$, c_r and αc_i being the phase-velocity and the growth-rate of the disturbance respectively. The amplitude function $\varphi(y)$ is governed by the Orr-Sommerfeld equation

$$(U-c)(\varphi''-\alpha^2\varphi) - U''\varphi = \frac{1}{i\alpha R}(\varphi^{\mathrm{IV}} - 2\alpha^2\varphi'' + \alpha^4\varphi), \qquad (1)$$

where dashes denote derivatives with respect to y, R is the Reynolds number of the basic flow, and all quantities have been made nondimensional without changing notations.

It is well known that at extremely large values of αR the four particular solutions φ_i ($i = 1, 2, 3, 4$) of Eq. (1) may be classified into two groups (see LIN [2]). Two solutions which are usually called the "inviscid solutions", φ_1 and φ_2 say, are nearly identical with the solutions of the Rayleigh equation, or the inviscid limit of Eq. (1),

$$(U-c)(\varphi''-\alpha^2\varphi) - U''\varphi = 0, \qquad (2)$$

at almost all values of y except at the singular region which will be described later. The other two which are called the "viscous solutions", φ_3 and φ_4, on the other hand, are rapidly varying functions of y and vanish everywhere outside the thin layers adjacent to the solid boundaries. If we deal with unbounded basic flows, φ_3 and φ_4 do not appear in the boundary value problem and the eigensolution $\varphi(y)$ is expressed as a linear combination of φ_1 and φ_2 (see FOOTE and LIN [3]).

The singular region of φ_1 and φ_2 where they do not agree with the inviscid solutions is a sector in the complex y-plane with the vertex at the turning point of Eq. (1), $y = y_s$, at which $U(y_s) = c$. Since it can be shown that this sector covers a part of the real y-axis when $c_i < 0$ and excludes the whole real y-axis when $c_i > 0$, there always appears a singular region in the physical space for the damping mode ($c_i < 0$) and no such a region for the amplifying mode ($c_i > 0$). So long as we are interested in the calculation of eigenvalues of c, the presence of this singular region is immaterial because we can integrate Eq. (1) along a path avoiding the singular sector in the complex y-plane, and the eigenvalue is solely determined by the behaviour of φ_1 and φ_2 outside this singular region. If, on the other hand, we wish to know the eigenfunction $\varphi(y)$ of the damping mode along the real y-axis we have to find the structure of the solution at the singular region. This is the problem we shall deal with in the following.

§ 3. Solutions in the neighbourhood of the turning point

The turning point y_s of Eq. (1) is the singular point of Eq. (2) where the solution has a logarithmic singularity. Thus, although Eq. (2) can be used as an overall approximation of Eq. (1) for very large αR, we cannot neglect the highest derivative on the right-hand side of Eq. (1) in the neighbourhood of y_s. In the vicinity of y_s we may expand $U(y)$ into Taylor series: $U(y) - c = U'_s \cdot (y - y_s) + O((y - y_s)^2)$, $U''(y) = U''_s + O(y - y_s)$, where $U'_s = U'(y_s)$ and $U''_s = U''(y_s)$. Taking only first terms in the above expansion we may express the equation which governs $\varphi(y)$ in the vicinity of y_s as follows:

$$\varphi^{\mathrm{IV}} - i\alpha R\,[U'_s \cdot (y - y_s)\,\varphi'' - U''_s\varphi] = 0. \tag{3}$$

Then if we make the transformation $y - y_s = \lambda\eta$, where $|\lambda| = |U'_s/U''_s|$, $\arg\lambda = \frac{1}{3}\left(\frac{5\pi}{2} - \arg U'_s\right)$, $\Lambda^2 = \alpha R\,|U'_s\lambda^3|$, $|\sigma| = 1$, $\arg\sigma = \arg\lambda + \pi$, Eq. (3) is written as

$$\ddddot{\varphi}(\eta) + \Lambda^2\big(\eta\ddot{\varphi}(\eta) + \sigma\varphi(\eta)\big) = 0, \tag{4}$$

where dots denote derivatives with respect to η. If $\sigma = 1$, Eq. (4) is identical with the "special equation" dealt with by WASOW [4].

The solution of Eq. (4) may be expressed in the form of the complex Laplace transform:

$$\left.\begin{aligned} \varphi(n) &= \int_C F(\eta, \zeta)\, d\zeta, \\ F(\eta, \zeta) &= (1/\zeta^2) \exp\left[\eta\zeta - \sigma/\zeta + \zeta^3/(3\Lambda^2)\right], \end{aligned}\right\} \tag{5}$$

where C is one of the paths of integration shown in Fig. 1. Among seven particular solutions V, A_1, A_2, A_3, W_1, W_2, W_3, corresponding to different paths, only four are independent, and there exist three obvious relations between them:

$$A_1 + A_2 + A_3 = V, \quad W_3 - W_2 = A_1, \quad W_1 - W_3 = A_2. \tag{6}$$

Now we shall examine the asymptotic behaviour of the seven particular solutions for large values of αR.

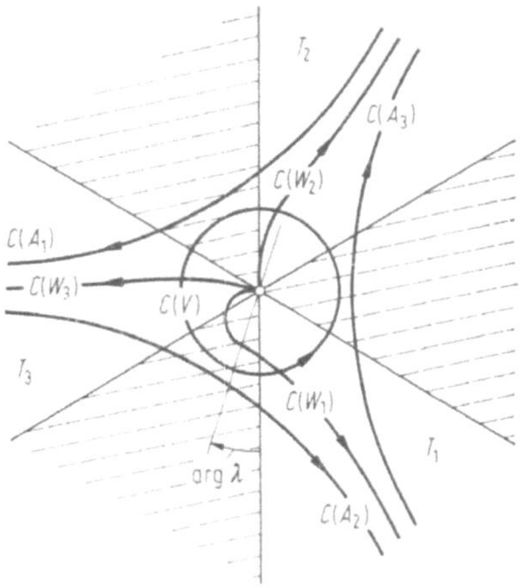

Fig. 1. Paths of integration in the ζ-plane.

3.1 $V(\eta)$. For large αR or Λ, $F(\eta, \zeta)$ may be expanded into inverse powers of Λ, and we have

$$\begin{aligned} V(\eta) &= \int_{C(V)} F(\eta, \zeta)\, d\zeta \\ &= 2\pi i \sqrt{\eta/\sigma}\, J_1(2\sqrt{\sigma\eta}) + O(1/\Lambda^2) \\ &= 2\pi i\eta + O(\eta^2) + O(1/\Lambda^2), \end{aligned} \tag{7}$$

where J_1 denotes the Bessel function of order one.

Thus it may be seen that $V(\eta)$ is a regular function of η in the neighbourhood of the singularity $\eta = 0$ $(y = y_s)$, and hence corresponds to the regular inviscid solution which is usually denoted as φ_1.

3.2 $A_k(\eta)$ $(k = 1, 2, 3)$. The asymptotic form of A_k for large Λ may be derived using the method of steepest descent as follows:

$$A_k(\eta) = \int_{C(A_k)} F(\eta, \zeta)\, d\zeta = (\sqrt{\pi}/\Lambda^{3/2})\,(-\eta)^{-5/4} \exp[(2/3)\,\Lambda(-\eta)^{3/2}]\,[1 + O(1/\Lambda)], \tag{8}$$

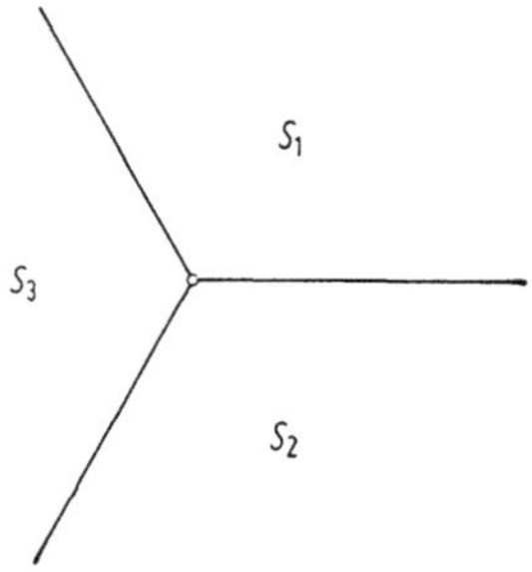

Fig. 2. Sectors S_k in the η-plane.

where a branch of $(-\eta)^{3/2}$ must be taken for each A_k such that $\mathrm{Re}[(-\eta)^{3/2}] < 0$ at $\eta \in S_k$, S_k being defined by $2(1-k)\,\pi/3 < \arg\eta < 2(2-k)\,\pi/3$ (Fig. 2). Thus A_k exponentially vanishes for large Λ at the sector S_k whereas it increases indefinitely at the remaining two sectors S_m $(m \neq k)$ where $\mathrm{Re}[(-\eta)^{3/2}] > 0$. This behaviour of A_k shows that A_k are identical with the viscous solutions φ_3 and φ_4 and hence irrelevant to the present problem.

3.3 $W_k(\eta)$. If $\eta \in S_m$ $(m \neq k)$, we can expand $F(\eta, \zeta)$ of W_k into inverse powers of Λ, and we have

$$\begin{aligned} W_k(\eta) &= \int_{C(W_k)} F(\eta, \zeta)\, d\zeta \\ &= \pi i \sqrt{\eta/\sigma}\, H_1^{(1)}(2\sqrt{\sigma\eta}) + O(1/\Lambda^2) \\ &= 1/\sigma - \eta \log\eta + O(\eta) + O(1/\Lambda^2), \end{aligned} \tag{9}$$

where $H_1^{(1)}$ denotes the Hankel function of the first kind and order one. This asymptotic behaviour of W_k for $\Lambda \to \infty$ shows that W_k are identical with the inviscid solution φ_2 which also has logarithmic singularity at $y = y_s$ in the limit of $\alpha R \to \infty$.

At $\eta \in S_k$, on the other hand, we cannot use the above expansion but can express W_k using the relation (6) as follows:

$$\begin{aligned} W_{1,2,3}(\eta) &= A_{2,3,1}(\eta) \\ &= (\sqrt{\pi}/\Lambda^{3/2})\,(-\eta)^{-5/4} \exp[(2/3)\,\Lambda(-\eta)^{3/2}]\,[1 + O(1/\Lambda)], \end{aligned} \tag{10}$$

where, according to the choice of the branch in (8), the proper branch of $(-\eta)^{3/2}$ for W_k is such that $\mathrm{Re}[(-\eta)^{3/2}] > 0$ in S_k. Thus each W_k shows singular behaviour at the sector S_k, which indeed corresponds to the singular domain of the inviscid solution φ_2.

§ 4. Structure of the eigensolution in the inner friction region

In some basic flows it is shown that all negative eigenvalues of c_i tend to zero in the limit of $\alpha R \to \infty$ (Grohne [5]). In these cases the inner friction region is reduced to an infinitely thin layer in the inviscid limit since then the turning point y_s is located on the real y-axis, and the behaviour of the eigensolution in the inner friction layer is well represented by the solutions derived in the previous section. When, on the other hand, there exist the damping modes with non-zero c_i even in the inviscid limit (see TATSUMI, GOTOH and AYUKAWA [6]), the real y-axis usually comes beyond the region of validity of these solutions, which were obtained under the approximation that $U(y)$ and $U''(y)$ are replaced by the first terms of the Taylor series. Thus, we need some other expression which is valid along the real y-axis.

For this purpose we employ the asymptotic solutions of Eq. (1) due to the WKB method,

$$\Phi(y) \simeq (U - c)^{-5/4} \exp\left[\int_{y_s} \{i\alpha R(U - c)\}^{1/2} dy\right], \tag{11}$$

where $\{i\alpha R(U - c)\}^{1/2}$ has two branches. In the neighbourhood of the turning point y_s the expression (11) becomes

$$\Phi(n) \simeq (-\lambda U_s')^{-5/4} (-\eta)^{-5/4} \exp[(2/3) \Lambda (-\eta)^{3/2}]. \tag{12}$$

It may be seen that Φ is identical with the expression (10) of W_k in S_k except for constant factor if the branch of $(-\eta)^{3/2}$ such that $\mathrm{Re}\,[(-\eta)^{3/2}] > 0$ is taken in S_k. Thus, if a certain W_k is chosen among W_1, W_2, W_3 for representing the inviscid solution φ_2 in the region $\eta \in S_m$ $(m \neq k)$, the asymptotic behaviour of φ_2, or W_k, in the singular region S_k must be expressed by Φ:

$$\varphi_2(y) \simeq \sqrt{\pi}(-\lambda U_s')^{5/4} \Lambda^{-3/2} (U - c)^{-5/4} \exp\left[\int_{y_s} \{i\alpha R(U - c)\}^{1/2} dy\right], \tag{13}$$

which is valid in $\eta \in S_k$ including the real y-axis.

Now we shall examine the behaviour of $\varphi_2(y)$ along the real y-axis. It may easily be seen that φ_2 expressed by (13) has maximum absolute value at the point $y = y_c$ where $U(y_c) = c_r$. Then if we normalize $\varphi_2(y)$ with respect to $\varphi_2(y_c)$ without changing notation, $\varphi_2(y)$ may be expressed as follows:

$$\varphi_2(y) \simeq \left[1 + i \frac{5U_c'}{4|c_i|}(y - y_c)\right] \times \exp\left[-\sqrt{\alpha R\,|c_i|}\left\{\frac{|U_c'|}{4|c_i|}(y - y_c)^2 + i \frac{U_c'}{|U_c'|}(y - y_c)\right\}\right], \tag{14}$$

where $U_c' = U'(y_c)$. For example, if we take free boundary layer $U(y) = \tanh y$ as the basic flow, then $y_c = 0$, $U_c' = 1$, and (14) is reduced to

$$\varphi_2(y) \simeq \left[1 + \frac{5i}{4|c_i|} y\right] \exp\left[-\sqrt{\alpha R\,|c_i|}\left\{\frac{y^2}{4|c_i|} + iy\right\}\right]. \tag{15}$$

It may be seen from (14) that $\varphi_2(y)$ has the form of a concentrated wave-packet whose amplitude decreases to both sides of y_c until it vanishes beyond the distance of order of $(\alpha R/|c_i|)^{-1/4}$ from y_c and composite waves have the wave-length of order of $(\alpha R\,|c_i|)^{-1/2}$ (see Fig. 3).

The eigenfunction $\varphi(y)$ of the boundary value problem is expressed as $\varphi(y) = C_1\varphi_1(y) + C_2\varphi_2(y)$ where C_1 and C_2 are numerical constants. Since, for every large αR, φ_1 and φ_2 become independent of R outside the inner friction region and also c does not change with R, the constants

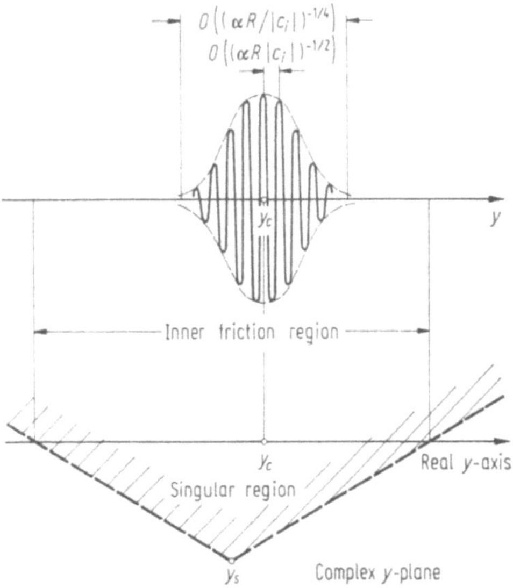

Fig. 3. Structure of the eigensolution in the inner friction region.

C_1 and C_2 must have finite values which are independent of R. Then it may be seen in view of the regularity of $\varphi_1(y)$ that the addition of the term $(C_1/C_2)\,\varphi_1$ to φ_2 does not change appreciably the asymptotic behaviour (14). Thus, (14) may be used as the asymptotic form of the eigenfunction $\varphi(y)$ for large αR. It should be noted that the width of the wave-packet represented by (14), which is of order of $(\alpha R)^{-1/4}$, is much narrower than that of the inner friction region, which is the length of the real y-axis cut off by the sector S_k and hence does not change with R. It is true that the eigenfunction $\varphi(y)$ which satisfies the boundary conditions $\varphi(\infty) = \varphi(-\infty) = 0$ has finite values outside the inner friction region and shows singular behaviour throughout this region, but when normalized with respect to its enormous maximum value at y_c all these structures are depressed and we have the eigenfunction $\varphi(y)$ concentrated within a thin layer around the point y_c.

§ 5. Energy balance in the damping disturbance

In the foregoing sections we have discussed the structure of the eigenfunction at very large αR and finite negative c_i. Although the existence of negative c_i in the inviscid limit is an undeniable mathematical result there has been lacking so far a simple physical explanation for the mechanism of this "inviscid" damping. A plausible explanation would be that the damping is caused by the feeding back of energy from the disturbance to the basic flow through the action of the Reynolds stress[1]. But we have had no means of checking this conjecture since we did not know the precise functional behaviour of the damping disturbance. Now that we have (14) as such a knowledge we can look for what is responsible to the damping by calculating the balance of energy of the disturbance.

We consider the two-dimensional disturbance (u, v) periodic in the x-direction. The change of energy of the disturbance averaged over one wave-length is governed by the following equation:

$$\frac{d}{dt}\left[\frac{1}{2}\int_{-\infty}^{\infty}(\overline{u^2}+\overline{v^2})\,dy\right] = -\int_{-\infty}^{\infty}\overline{uv}\,\frac{dU}{dy}\,dy - \frac{1}{R}\int_{-\infty}^{\infty}\left[\overline{\left(\frac{\partial u}{\partial x}\right)^2}+\overline{\left(\frac{\partial u}{\partial y}\right)^2}+\overline{\left(\frac{\partial v}{\partial x}\right)^2}+\overline{\left(\frac{\partial v}{\partial y}\right)^2}\right]dy, \tag{16}$$

where bars denote averaged values. The first term on the right-hand side represents the rate of conversion of energy from the basic flow to the disturbance through the action of the Reynolds stress, while the second term gives the dissipation of energy due to viscosity.

Substituting the expressions $u = \mathrm{Re}\,[\varphi'(y)\exp\{i\alpha(x-ct)\}]$ and $v = \mathrm{Re}\,[-i\alpha\varphi(y)\exp\{i\alpha(x-ct)\}]$ into (16) and retaining only dominant terms, we obtain

$$\frac{1}{2}\alpha c_i\int_{-\infty}^{\infty}|\varphi'|^2\,dy = \frac{\alpha}{2}\int_{-\infty}^{\infty}\mathrm{Im}\,(\varphi'\varphi^*)\,\frac{dU}{dy}\,dy - \frac{1}{2R}\int_{-\infty}^{\infty}|\varphi''|^2\,dy, \tag{17}$$

where asterisk denotes the complex conjugate. If we evaluate the integrals using the expression (14) for $\varphi(y)$, the left-hand side and the second term on the right-hand side are found to be exactly of the same order of magnitude, whereas the first term on the right-hand side is of order of $|U_c'|/(\alpha R\,|c_i|^3)^{1/2}$ compared to others. Thus we can conclude that the damping of the disturbance is caused by the viscous dissipation even in the limit of infinite Reynolds number. The first term on the right-hand side has negative sign showing that the Reynolds stress transfers energy from the disturbance to the basic flow, but its influence is of minor magnitude in comparison with the effect of viscosity.

References

1. Stuart, J. T.: J. Fluid Mech. **9**, 353 (1960).
2. Lin, C. C.: The Theory of Hydrodynamical Stability, Cambridge University Press 1955.
3. Foote, J. R., Lin, C. C.: Quart. Appl. Math. **8**, 265 (1956).
4. Wasow, W.: Ann. Math. **52**, 350 (1950).
5. Grohne, D.: Z. angew. Math. Mech. **34**, 344 (1954).
6. Tatsumi, T., Gotoh, K., Ayukawa, K.: J. Phys. Soc. Japan **19**, 1966 (1964).

[1] This possibility was suggested by Dr. J. T. Stuart in his private comment to the authors' previous paper [6].

Linear Dynamical Systems in Hilbert Space

By

L. J. F. Broer and **J. de Graaf**

Technological University of Eindhoven, Eindhoven, Netherlands

1. Definition of the solution

We consider the initial value problem:

$$\frac{du}{dt} - \mathscr{A} u = f(t), \tag{1}$$

$$u(0) = u_0, \tag{2}$$

where u_0 and $u(t)$, $f(t)$ for $0 \leq t \leq T$ are elements of a Hilbert space H. T is arbitrary but fixed. $\mathscr{A}$ is a linear operator. The differentiation with respect to t is meant in the generalised sense.

The meaning of (1) is obvious when $\mathscr{A}$ is defined everywhere in H, that is when $D(\mathscr{A}) = H$. The first problem is to define the solution when $D(\mathscr{A}) \subset H$. For this purpose we observe that $f(t)$, which represents a (not necessarily continuous) trajectory in H, also can be considered as an element in a Hilbert space H_T. The inner product in H_T is defined by:

$$(f, g)_T = \int_0^T (f, g)\, dt, \tag{3}$$

where (f, g) is the inner product in H. Note that $\mathscr{A}$ defines an linear operator $\mathscr{A}_T$ on a domain in H_T. A solution $u(t)$, hopefully, would be an element of H_T, uniquely determined by f in H_T and u_0 in H. We consider f, u_0 as an element of the Hilbert space $W = H_T \times H$. Therefore:

$$(f, u_0; g, v_0)_W = \int_0^T (f, g)\, dt + (u_0, v_0). \tag{4}$$

A solution of the initial value problem posed by (1) and (2) now is, per definition, provided by a mapping of $H_T(u)$ on $W(f, u_0)$, which satisfies the following requirements:

The mapping is one-to-one and closed.

Whenever $u(t)$ is in $D(\mathscr{A})$ Eq. (1) is satisfied.

$u(0) = u_0$.

2. Existence theorem

It is possible to prove that a solution as defined in the foregoing section exists when the operator $\mathscr{A}$ satisfies the following conditions:

(I) *$\mathscr{A}$ is a closed operator on H.*

(II) *$D(\mathscr{A})$ is dense in H.*

(III) *There exist real constants α_1, α_2 such that:*

$$\text{for all } u \text{ in } D(\mathscr{A}) \qquad \operatorname{Re}(u, \mathscr{A} u) \leq \alpha_1 (u, u),$$

$$\text{for all } v \text{ in } D(\mathscr{A}^*) \qquad \operatorname{Re}(v, \mathscr{A}^* v) \leq \alpha_2 (v, v),$$

where $\mathscr{A}^*$ is the adjoint of $\mathscr{A}$.

The proof will be published elsewhere. It is a generalisation of a proof, valid for a certain class of differential operators, given by DE GRAAF and PELETIER [1]. This work can in its turn, be considered as a generalisation of some results due to LADYSHENSKAYA [2].

The proof is direct. The idea is, roughly speaking, to define a mapping on W of the domain of the operator $d/dt - \mathscr{A}_T$ in H_T by means of (1) and (2). One then uses the conditions (I) to (III) to show that this domain is dense in H_T, that the range of the operator is dense in W and that the extension by closure of this mapping satisfies the requirements of Sec. 1.

3. Some properties of the solution

From the definition in Sec. 1 it follows immediately that the solution is unique whenever it exists in this sense.

The inverse mapping (W on H_T) is continuous, that is: the solution depends continuously on the initial and source data.

An immediate corollary of the existence proof is that $u(t)$ is a continuous trajectory in H.

More involved are the proofs of some persistence theorems. We mention here only two of them.

Persistence of initial conditions:

When $f \equiv 0$ and u_0 in $D(\mathscr{A}^n)$, n some natural number, then for

$0 \leq t \leq T$ $u(t)$ is in $D(\mathscr{A}^n)$ and the trajectory $u(t)$ in H is n times continuously differentiable.

Persistence of source conditions:

When $u_0 = 0$ and f in $D(\mathscr{A}_T^m)$ then $u(t)$ is in $D(\mathscr{A}^m)$.

As $u(t)$ depends linearly (and continuously) on the initial and source data we can write formally:

$$u(t) = R(t)\, u_0 + S_T(t)\, f, \tag{5}$$

where $R(t)$ is a time-dependent linear operator on H, S_T a linear operator in H_T. It then turns out that:

$$S_T(t)\, f = \int_0^t R(t-\tau)\, f(t)\, d\tau, \tag{6}$$

which means that Duhamel's principle is valid for our solution.

$R(t)$ is a bounded linear operator defined on H. It depends continuously on t. One can show that $R(t)$ is a semi-group. Of some special interest are problems in which $R(t)$ is a so-called holomorphic semi-group. This means that $u(t)$ is in any $D(\mathscr{A}^n)$ for all $t > 0$ and for all u_0 ($f \equiv 0$). Conditions on $\mathscr{A}$ which ensure this property have been found. These conditions and the proof will be given in the forthcoming paper [1] mentioned above. When H is an L^2 space, $\mathscr{A}$ a differential operator, these conditions will be sufficient for the solution to be continuously differentiable any number of times for all $t > 0$ provided u_0 is in L^2. The diffusion equation is a simple example of an equation having this property.

References

1. DE GRAAF, J., PELETIER, L.A.: Rend. Sem. Mat. Univ. Padova (to be published).
2. LADYSHENSKAYA, O. A.: Dokl. Akad. Nauk S.S.S.R. **102**, 207 (1955).

On the Stability of Constant Profile Waves

By

D. R. Bland

Cranfield Institute of Technology, Cranfield, Bedford, U.K.

Summary. Non-linear wave equations with convection and dissipation generally admit constant profile solutions. This paper considers various methods of proving stability of these solutions in the simplest case.

1. Introduction

A plane wave is one in which the dependent variables are functions of only one cartesian space variable, x say, and of the time t. If the dependent variables depend on x and t through $\xi = x - Vt$, where V is a constant, and if the variables tend to different fixed limits as $\xi \to \pm\infty$, then the plane wave is called a constant profile wave. The disturbance travels with velocity V without attenuation or distortion. If a particular non-linear wave equation for f_i admits a solution of the form $f_i(\xi)$, $\xi = x - Vt$, with f_i or its derivations tending to different limits as $\xi \to \pm\infty$, then this solution is called a constant profile solution. Clearly it represents a constant profile wave. We shall only consider waves travelling in the positive direction of x throughout this paper.

The simplest non-linear wave equation which admits such a solution is

$$\left(1 + \frac{2B}{E}\frac{\partial u}{\partial x}\right)\frac{\partial^2 u}{\partial x^2} + \frac{\nu}{E}\frac{\partial^3 u}{\partial t\,\partial x^2} = \frac{1}{c^2}\frac{\partial^2 u}{\partial t^2}, \tag{1}$$

with the constant profile solution

$$m = \frac{1}{2}m_1\left(1 - \tanh\frac{Bm_1\xi}{2\nu V}\right), \quad V = c\left(1 + \frac{B}{2E}m_1\right), \tag{2}$$

where

$$m = \frac{\partial u}{\partial x} \tag{3}$$

and

$$m \to 0 \text{ as } \xi \to +\infty \text{ and } m \to m_1 \text{ as } \xi \to -\infty. \tag{4}$$

This solution (2) is the so-called Taylor shock. It is well-known that this shock is stable. Any initial distribution at $t = 0$, $m(x, 0)$, satisfying conditions (4) gives (2) as the asymptotic solution for sufficiently large t. The standard proof, see e.g. LIGHTHILL [1] or BLAND [2], is now sketched.

If the dissipative term $\frac{\nu}{E} \frac{\partial^3 u}{\partial t\, \partial x^2}$ were not present in Eq. (1), the simple wave solution representing a wave travelling in the positive sense into an undisturbed medium would have the Riemann invariants, carried by the negative characteristics, all equal to zero. Hence if ν/E is small it is reasonable to suppose that s is second order in a typical measure of the relevant dependent variable; in this case m_1 suffices as such a measure. The Riemann invariants r, carried by the positive characteristics, are first order in m_1. Eq. (1) can be reduced to two equations in r and s. If ν/E is treated as first order in m_1 and if terms of third order and higher in m_1 are discarded, a reduced form of r satisfied the Burgers equation

$$\theta \frac{\partial \theta}{\partial \xi} + \frac{\partial \theta}{\partial \tau} = \frac{\beta}{2} \frac{\partial^2 \theta}{\partial \xi^2}, \qquad \beta \text{ const.} \tag{5}$$

The substitution

$$\theta = -\frac{\beta}{\psi} \frac{\partial \psi}{\partial \xi} \tag{6}$$

reduces (5) to the heat conduction equation. The solution of this last equation, subject to the relevant boundary conditions, can be written down. Asymptotic methods and back substitution enable the asymptotic solution for m to be found for large t. It is Eq. (2).

There are many other non-linear wave equations admitting constant profile solutions. These solutions represent a physical balance between the steeping effect of convection and the flattening effect of some dissipative mechanism. The method of treating stability outlined above breaks down in the vast majority of cases because substitutions do not exist in general which reduce non-linear partial differential equations to linear ones. It is the object of this paper to investigate methods of treating stability which might be expected to be more generally applicable. We shall treat Eq. (1) because it is the simplest example of a non-linear wave equation admitting a constant profile solution. In Sec. 2 an example in which Eq. (1) arises is given, followed by a quick method of deducing Eq. (5) and then the derivation of the perturbation equation (10). Sec. 3 and 4 give two methods of treating the perturbation equation, the first uses a Liapunov function and the second linearisation and eigenfunction expansion. It is hoped that one or other of these methods will be applicable to more complex wave equations.

2. Derivation of the perturbation equation

An example, where Eq. (1) arises, is for the displacement u in a dilatational wave in a non-linear Voigt type solid. Such a solid is mechanically equivalent to a non-linear spring and a linear dashpot in parallel with stress-strain relations $\sigma = E\varepsilon + B\varepsilon^2$ and $\sigma = \nu\dot{\varepsilon}$ respectively. u satisfies Eq. (1) with $c^2 = E/\varrho$, where ϱ is the density of the solid. Note that if $\nu = 0$, we have the non-linear elastic solid; if $\nu = B = 0$, the classical wave equation. We shall consider $\nu/E = O(m_1)$, or more precisely $m_1 = O(\nu/E)$ and work to second order in m_1. Hence we are using a second order theory. Strains are small but not infinitesimal. This is consistent with isentropic theory for non-linear elastodynamics. In the full theory the entropy change across a shock is $O(m_1^3)$ and neglect of entropy change is only justifiable correct to second order.

Eq. (1) arises in other branches of applied mathematics.

A quick method of reduction of Eq. (1) to Eq. (5) is possible. We are concerned with waves travelling in the positive sense so that the first order solution of (1) is $u = f(x - ct)$. If we change the independent variables from x and t to ξ and η, where

$$\xi = x - vt, \quad \eta = t \tag{7}$$

and $v = c(1 + O(m_1))$, then $\partial u/\partial\eta = O(m_1^2)$. If u and all its derivatives with respect to ξ are $O(m_1)$, it is consistent with this last equation to treat $\partial/\partial\eta$ as $O(m_1)$. Transform Eq. (1) to the variables ξ and η:

$$\begin{aligned} &\left(1 + \frac{2B}{E}\,\frac{\partial u}{\partial\xi}\right)\frac{\partial^2 u}{\partial\xi^2} + \frac{\nu}{E}\left(-\,v\,\frac{\partial^3 u}{\partial\xi^3} + \left[\frac{\partial^3 u}{\partial\xi^2\,\partial\eta}\right]\right) \\ &\quad = \frac{1}{c^2}\left(v^2\,\frac{\partial^2 u}{\partial\xi^2} - 2v\,\frac{\partial^2 u}{\partial\xi\,\partial\eta} + \left[\frac{\partial^2 u}{\partial\eta^2}\right]\right). \end{aligned} \tag{8}$$

The terms in brackets are third order and are neglected. Put

$$n = \frac{\partial u}{\partial\xi}, \quad 1 - \frac{v^2}{c^2} + \frac{2B}{E}\,m = \theta\,, \quad \frac{\nu v}{E} = \frac{2}{1}\,\beta, \quad \tau = \frac{c^2}{2v}\,\eta$$

and Eq. (5) is reproduced.

The constant profile solution of (8) is the same as that of (1):

$$m = m^*(\xi) = \frac{1}{2}\,m_1\left(1 - \tanh\frac{Bm_1\xi}{2\nu V}\right), \quad v = V = c\left(1 + \frac{B}{2E}\,m_1\right), \tag{2}$$

where we have chosen $v = V$, which is permissible since $V = c(1 + O(m_1))$. To investigate stability, we substitute

$$m(\xi, \eta) = m^*(\xi) + m_1\mu(\xi, \eta) \tag{9}$$

into (8). Now we wish to determine whether every profile for which $m(\xi, \eta) \to 0$ as $\xi \to +\infty$ and $m(\xi, \eta) \to m_1$ as $\xi \to -\infty$ adopts the Taylor profile; i.e. does $\mu(\xi, \eta) \to 0$ as $\eta \to \infty$ for all ξ given that $\mu(\xi, \eta) \to 0$ as $\xi \to \pm\infty$ for all η?

The algebra can be tidied. Put $\lambda = \frac{B m_1}{\nu c}$, $\omega = \frac{\nu c^2}{2E}$ and $x(\xi) = m^*(\xi)/m_1$ in (8), (2) and (9) and change the independent variable ξ to x $(0 \leq x \leq 1)$ to give

$$(x - x^2)\frac{\partial^2 \mu}{\partial x^2} + 2\mu \frac{\partial \mu}{\partial x} + 2\mu = \frac{\gamma}{x - x^2}\frac{\partial \mu}{\partial \eta}, \quad \gamma = \frac{1}{\omega \lambda^2}. \tag{10}$$

The question now is: if $\mu(x, \eta) = 0$ for $x = 0, 1$, does $\mu(x, \eta) \to 0$ as $\eta \to \infty$? In other words, does the perturbation of the constant profile solution die out? (10) is the equation for the perturbation $\mu(x, \eta)$.

3. Liapunov function method

Consider

$$n(\eta) = \int_{-\infty}^{\infty} \mu^2 \, d\xi = \int_0^1 \frac{\mu^2}{x - x^2} \, dx. \tag{11}$$

We assume $n(0)$ exists, we wish to show that $n(\eta) \to 0$ as $\eta \to \infty$. Multiply Eq. (10) by μ and integrate with respect to x from 0 to 1. After one integration by parts, using $\mu = 0$ at $x = 0, 1$,

$$\frac{\gamma}{2}\frac{dn(\eta)}{d\eta} = \frac{\gamma}{2}\frac{\partial}{\partial \eta}\int_0^1 \frac{\mu^2}{x - x^2}\, dx = -\int_0^1 \left\{(x - x^2)\left(\frac{\partial \mu}{\partial x}\right)^2 - \mu^2\right\} dx. \tag{12}$$

Note that the non-linear term has gone out. This is a consequence of the particular choice of explicit x-dependence of the Liapunov function $n(\eta)$ defined by Eq. (11). Now $n(\eta) > 0$ unless $\mu \equiv 0$, the author assumes it is sufficient to show that $dn/d\eta < 0$, i.e.

$$I = \int_0^1 \left\{(x - x^2)\left(\frac{\partial \mu}{\partial x}\right)^2 - \mu^2\right\} dx > 0, \tag{13}$$

for all $\mu(x)$ satisfying (14) except $\mu(x) \equiv 0$. Since the left-hand side of (10) can be written

$$\frac{\partial}{\partial x}\left\{(x - x^2)\frac{\partial \mu}{\partial x} - (1 - 2x)\,\mu + \mu^2\right\},$$

integration of both sides of (10) with respect to x gives $\frac{\partial}{\partial \eta}\int_0^1 \frac{\mu}{x - x^2}\, dx = 0$ or

$$\int_0^1 \frac{\mu}{x - x^2}\, dx = \text{const.} = 0, \tag{14}$$

if the origin of ξ is suitably chosen. (14) is a profile centring condition and is a restriction on $\mu(x, \eta)$ considered as a function of x at any time η.

We shall first treat the integral in (13) by expressing it as the sum of a perfect differential and a perfect square. It gives

$$(x-x^2)\left(\frac{\partial\mu}{\partial x}\right)^2-\mu^2=-\frac{\partial}{\partial x}(p\mu^2)+\left(\sqrt{x-x^2}\,\frac{\partial\mu}{\partial x}+\frac{p}{\sqrt{x-x^2}}\,\mu\right)^2, \quad (15)$$

where

$$p=-(x-x^2)\frac{\theta_1'(x)+c\theta_2'(x)}{\theta_1(x)+c\theta_2(x)}$$

and θ_1, θ_2 are two independent solutions of $(x-x^2)\dfrac{d^2\theta}{dx^2}+(1-2x)\dfrac{d\theta}{dx}+\theta=0$. The substitution $x=\frac{1}{2}(\zeta-1)$ enables θ_1 and θ_2 to be expressed as Legendre functions of order $\frac{1}{2}(\sqrt{5}-1)$. Each of these functions has at least one zero in the range $0\le x\le 1$. If $C=0$ and $\theta_1(x)$ is the Legendre function of the first kind, p has only one singularity. But

$$\int_0^1\frac{\partial}{\partial x}(p\mu^2)\,dx \neq (p\mu^2)\big|_0^1=0$$

and the method breaks down. However we can add a term $\chi\dfrac{\mu}{x-x^2}$ to the right-hand side of (15) and choose χ to eliminate the one singularity. Unfortunately this introduces singularities at $x=0$ and 1 and so the method of completing the square fails.

We can treat (13) differently: substitute $x=\frac{1}{2}(1-\zeta)$, so that $-1\le\zeta\le 1$, expand $\mu(\zeta)$ in Legendre polynomials, i.e. $\mu(\zeta)=\sum_{r=0}^{\infty}A_rP_r(\zeta)$, then the integral in (13) is given by

$$2I=\sum_{r=0}^{\infty}\frac{r^2+r-1}{2r+1}A_r^2c-A_0^2+\frac{1}{3}A_1^2+A_2^2+\frac{11}{7}A_3^2+\cdots.$$

The boundary conditions give $\sum_{r\,\text{odd}}A_r=0$ and $\sum_{r\,\text{even}}A_r=0$. The condition (14) gives

$$\sum_{r\,\text{even}}B_rA_r=0 \quad\text{where}\quad B_r=\frac{2}{3}\sum_{\substack{t=2\\ t\,\text{even}}}^{r}\frac{2t-1}{t(t-1)}.$$

The sum of the odd terms is positive definite. The sum of the even terms can be written, on elimination of A_0 and A_2 as

$$I_e=\sum_{\substack{r,s=4\\ r,s\,\text{even}}}^{\infty}(C_r\delta_{rs}+B_r+B_s-1)\frac{\alpha_r\alpha_s}{rs}, \quad\text{where}\quad C_r=\frac{r^2+r-1}{2r+1}$$

and $\alpha_r=rA_r$.

$\alpha_r = O(1)$ provided $\mu(\zeta)$ is of bounded variation in -1, 1. The Euclidean norm of the coefficients is bounded. The matrix for I_e has been put on a digital computer and its algebraically least eigenvalue calculated. For a 60×60 matrix, it was found to be positive. It is reasonable to support that I_e and hence (13), subject to (14), is positive definite.

4. Linear stability, eigenfunction expansion

Eq. (10) linearises to

$$(x - x^2)^2 \frac{\partial^2 \mu}{\partial x^2} + 2(x - x^2)\,\mu - \gamma \frac{\partial \mu}{\partial \eta} = 0. \tag{16}$$

Put $\mu = e^{-p\eta} f_p(x)$:

$$(x - x^2)^2 \frac{d^2 f_p}{dx^2} + (2(x - x^2) + \varphi)\, f_p = 0,\ \varphi = \gamma p,\ f_p = 0 \text{ at } x = 0, 1.$$

The eigenfunction expansion of the solution of (16) is

$$\mu(x, \eta) = \left(\int_0^1 \frac{h(t)}{t(1-t)}\,dt\right) x(1-x) - \frac{1}{2\pi}\int_{1/4}^{\infty} \{g_1(\varphi)\, x^\theta (1-x)^{1-\theta}(\theta - x)$$
$$+ g_2(\varphi)\, x^{1-\theta}(1-x)^\theta\big(\theta - (1-x)\big)\} e^{-\frac{\varphi}{\gamma}\eta}\, d\varphi,$$

where

$$g_1(\varphi) = \frac{1}{\varphi\sqrt{4\varphi - 1}} \int_0^1 h(t)\, t^{-1-\theta}(1-t)^{\theta-2}\big(\theta - (1-t)\big)\, dt,$$

$$g_2(\varphi) = \frac{1}{\varphi\sqrt{4\varphi - 1}} \int_0^1 h(t)\, t^{\theta-2}(1-t)^{-1-\theta}(\theta - t)\, dt,$$

$h(x) = \mu(x, 0)$, the given initial values,

and $\theta = \frac{1}{2} - \frac{i}{2}\sqrt{4\varphi - 1}$. The first term in the expansion represents a shift in the centre of the profile; as in (14) we can choose the origin of ξ so that $\int_0^1 \frac{h(t)}{t(1-t)}\,dt = 0$. The least value of φ in the spectrum is $\varphi = 1/4$. Hence $\mu \to 0$ as $\eta \to \infty$ for all x, the constant profile solution is stable. The time taken for the disturbance to reach $1/e$ of its initial value is $\leq 4\gamma$. This value agrees with that found by the particular method described in the introduction.

References

1. Lighthill, M. J.: Surveys in Mechanics, Cambridge University Press 1956, pp. 250–351.
2. Bland, D. R.: J. Inst. Math. Applics. **1**, 56 (1965).

A Theory of Elasto-plastic Buckling of Structures

By

M. Como

University of Naples, Naples, Italy

The formulation of a theory of elasto-plastic buckling of structures is a research of a theoretical and applied interest. Such a problem requires the evaluation of the plastic response of the strain-hardening materials under a sudden change of loading and here the real difficulties hide. Next we sketch a brief outline of the slip theory and of some of its developments. In fact, to our opinion, for the aim of this research the slip theory seems to be sufficiently precise in its predictions.

1. Plastic behaviour of the strain-hardening materials

1.1 Some basic concepts of the slip theory. The Bauschinger effect and the evaluation of the subsequent yield loci

Let us consider a small region that surrounds in a body a point P and let P_{x_i} $(i = 1, 2, 3)$ be a reference system of axes and σ_{ij} the corresponding stress components (Fig. 1). Let $Pabc$ be another system of axes and let $\tau_{ab} = \tau_{ba}$ be the shear stress acting between a, b. Plastic strains of the region around P are due to the superimposition of all the effects of the active local slidings between all the couples a, b of planes where the shear stress τ_{ab} overcomes the limit value τ_L [1]. Because of the Bauschinger effect there are two different values τ_L^+

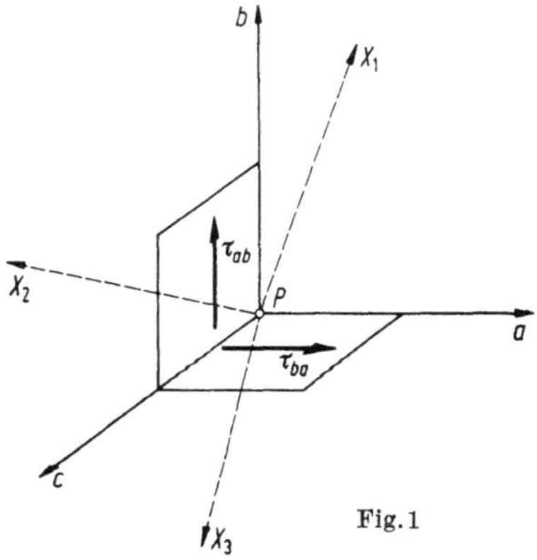

Fig. 1

Acknowledgement. This work has been sponsored by C.N.R. (Consiglio Nazionale delle Ricerche, Italy).

and τ_L^- of the limit shear stress [2]. Strain increments are given by

$$\dot{\varepsilon}_{ij} = \int\limits_{\Omega(a,b)} c_{ab} \frac{1}{G_{t_{ab}}} f_{ij}^{ab} \dot{\tau}_{ab} \, d\Omega(a, b), \tag{1}$$

where c_{ab} is a function, sensitive of the local Bauschinger effect, that defines the coming up of the local active shear increment $\dot{\gamma}_{ab}^{ab}$ between a, b; $G_{t_{ab}}$ is the local tangent shear modulus; $\Omega(a, b)$ the width of the whole couple of the orthogonal straight lines a, b through P and f_i^a the function

$$f_{ij}^{ab} = \frac{1}{2}(n_i^a n_j^b + n_j^a n_i^b) \tag{2}$$

with n_i^a, n_i^b direction cosines of a, b with respect to the P_{x_i} ($i = 1, 2, 3$) system.

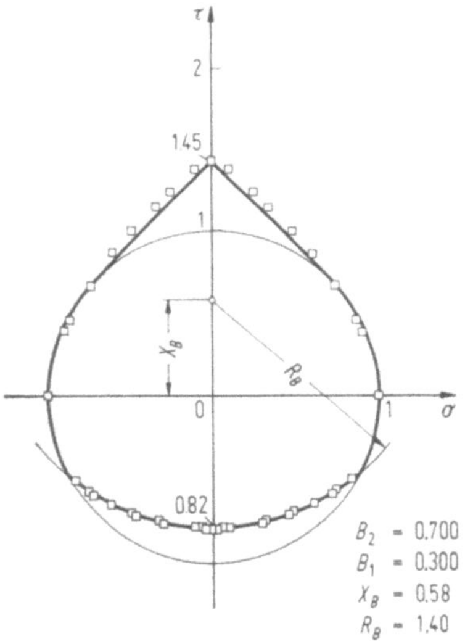

Fig. 2

From (1) several properties of the plastic strains, relative to the assumed model, follow. Besides it is worthwhile to emphasize that singular points can be present on the yield surface and that the stability of the material is insured if $G_{t_{ab}} \geq 0$. Systematic studies along this trend have been developed [3—9]; Fig. 2 gives a significant picture of the comparison among the subsequent theoretical yield surfaces and the experimental ones in $\tau - \sigma$ plane [5, 10]. Besides, the influence of the offset on the experimental evaluation of the subsequent yield loci has been thoroughly examined [5].

1.2 The proportional loading. Evaluation of the plastic strains by means of the continuous shear stress-active shear strain law

Let us consider now the case of the proportional loading and therefore let the principal stresses be given by

$$\sigma_1 = \lambda\bar{\sigma}_1, \quad \sigma_2 = \lambda\bar{\sigma}_2, \quad \sigma_3 = \lambda\bar{\sigma}_3, \tag{3}$$

if λ is the loading multiplyer. In order to evaluate plastic strains it is convenient to assimilate the discontinuous shear stress-active shear strain law by means of a continuous one [7] (Fig. 3). Let us write therefore

$$\gamma_{ab}^{ab} = \frac{a}{2n-1} \tau_{ab}^{2n-1}, \tag{4}$$

where a and n are constants of the material. Strain components then can be easily obtained because now $c_{ab} = 1$ in Eq. (1) for the whole region $\Omega(a, b)$. We have therefore [7]

$$\varepsilon_i = \frac{a}{2n-1} J_2^{n-1} F_i(\mu)\, s_i, \tag{5}$$

where s_i are the principal deviatoric stresses, J_2 their second invariant and $F_i(\mu)$ some functions of the Lode parameter of stress $\mu = (2\sigma_2 - \sigma_3 - \sigma_1)/(\sigma_1 - \sigma_3)$ (if $\sigma_1 > \sigma_2 > \sigma_3$). $F_i(\mu)$ functions have been evaluated for different values of the index n.

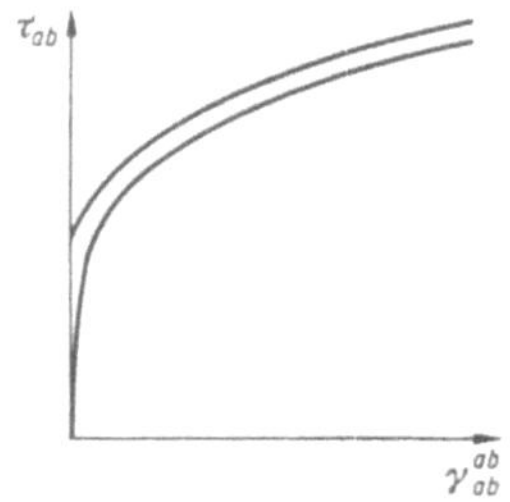

Fig. 3

We can notice that the change of the $F_i(\mu)$ functions with μ in the interval $0 \leq \mu \leq 1$ increases with n, i.e. with softer materials. An isotropic theory (PRANDTL-REUSS, NADAI etc.) gives $F_1 = F_2 = F_3 =$ const. while Eq. (5) gives such a result only if $n = 1, 2$ [7]. The stress vector is not parallel, therefore, to the strain vector.

If we introduce the Lode parameter of strain $\nu = (2\varepsilon_2 - \varepsilon_3 - \varepsilon_1)/(\varepsilon_1 - \varepsilon_3)$ (if $\varepsilon_1 > \varepsilon_2 > \varepsilon_3$) there is $\mu \neq \nu$ generally. Fig. 4 gives a picture of the function $\nu = f(\mu)$ that follows from (5); we can notice a good agreement with experiment [11, 12].

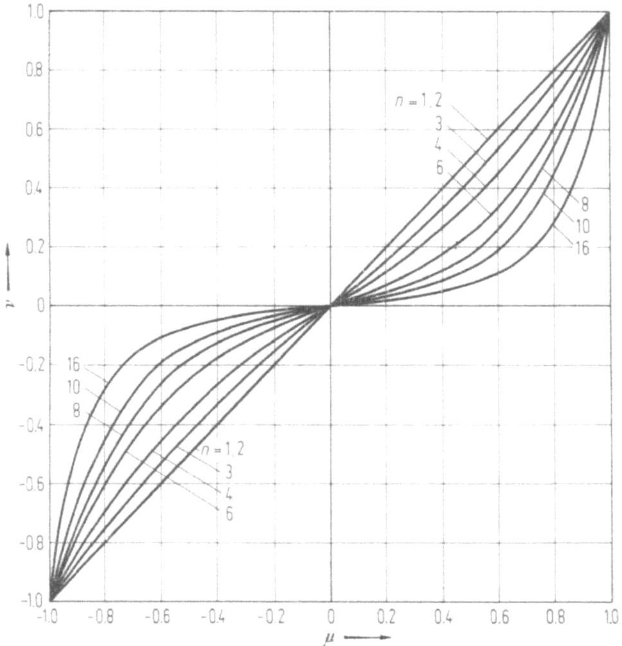

Fig. 4

From Eq. (5) we obtain the stress-strain intensities law corresponding to (4):

$$e_i = A(\mu)\,\sigma_i^{2n-1}, \tag{6}$$

where

$$\sigma_i = \frac{\sqrt{2}}{2}\left[(\sigma_1 - \sigma_2)^2 + (\sigma_2 - \sigma_3)^2 + (\sigma_3 - \sigma_1)^2\right]^{1/2}, \tag{7}$$

$$e_i = \sqrt{\frac{2}{3}}\,(\varepsilon_1^2 + \varepsilon_2^2 + \varepsilon_3^2)^{1/2}, \tag{8}$$

and $A(\mu)$ is a function evaluated in [7].

After all, that is the influence of the invariant J_3 that appears. The elimination of such an effect gives Nadai equations.

1.3 Total loading

After the proportional loading let us apply to the small region around P the stress increments $\dot{\sigma}_{ij}$. If all the couples a, b of planes that were previously slided, suffer active slidings again, we have the "total loading" condition. Let us indicate by means of $\hat{\dot{\sigma}}_{ij}$ such an increment and let $\hat{\dot{s}}_{ij}$ be the corresponding deviatoric stress rates. Let $\overline{\Omega}$ be the width of the bundle of all the couples a, b of planes that suffer active shears; because it is

$$\dot{\tau}_{ab} = \hat{\dot{s}}_{ij} f_{ij}^{ab}, \tag{9}$$

it will be

$$\hat{\dot{\varepsilon}}_{ij} = \hat{\dot{s}}_{pq} \mathscr{N}_{ijpq}, \tag{10}$$

where

$$\mathscr{N}_{ijpq} = \int_{\overline{\Omega}} \frac{1}{G_{t_{ab}}} f_{ij}^{ab} f_{pq}^{ab}\, d\Omega(a, b). \tag{11}$$

Likewise we also have

$$\hat{\dot{s}}_{ij} = \hat{\dot{\varepsilon}}_{pq} \mathscr{M}_{ijpq} \tag{12}$$

with the coefficients $\mathscr{M}_{ijpq}$ given by reversal of (10); it will be

$$\mathscr{N}_{ijpq} = \mathscr{N}_{pqij},\quad \mathscr{M}_{ijpq} = \mathscr{M}_{pqij}. \tag{13}$$

Let us now apply increments to the $\hat{\dot{s}}_{ij}$ in total loading. Then, because

$$\frac{\partial \hat{\dot{\varepsilon}}_{ij}}{\partial \hat{\dot{s}}_{pq}} = \frac{\partial \hat{\dot{\varepsilon}}_{pq}}{\partial \hat{\dot{s}}_{ij}},\qquad \frac{\partial \hat{\dot{s}}_{ij}}{\partial \hat{\dot{\varepsilon}}_{pq}} = \frac{\partial \hat{\dot{s}}_{pq}}{\partial \hat{\dot{\varepsilon}}_{ij}}, \tag{14}$$

the quantities

$$d\hat{\dot{\Phi}} = \hat{\dot{s}}_{ij}\, d\hat{\dot{\varepsilon}}_{ij},\quad d\hat{\dot{\Phi}}^* = \hat{\dot{\varepsilon}}_{ij}\, d\hat{\dot{s}}_{ij} \tag{15}$$

are total differentials and we can integrate

$$\hat{\Phi} = \frac{1}{2} \mathcal{M}_{ijpq} \hat{\varepsilon}_{pq} \hat{\varepsilon}_{ij}, \quad \hat{\Phi}^* = \frac{1}{2} \mathcal{N}_{ijpq} \hat{s}_{pq} \hat{s}_{ij}. \tag{16}$$

These results can be generalized to include total loading finite stress increments [9]. The response of the plastic strain-hardening material in total loading is that of a particular non-linear elastic body.

1.4 The total loading strain rates evaluated by means of the continuous model. Significance of the differentiated Nadai equations

In order to evaluate the strain increments $\hat{\varepsilon}_{ij}$, by using the continuous model (4), we can disregard the effect of the active slidings in the region $\Omega - \bar{\Omega}$ as we did in the case of the proportional loading. The influence of the region $\Omega - \bar{\Omega}$ on the integrals (11) is very small because there the local tangent shear modulus G_{tab} is very large in comparison with the values of G_{tab} in the region $\bar{\Omega}$ of the continuous curve $\tau_{ab} - \gamma_{ab}^{ab}$ approaching that discontinuous one.

We then have in explicit form

$$\hat{\varepsilon}_{11} = \mathcal{N}_{1111} \hat{s}_{11} + \mathcal{N}_{1122} \hat{s}_{22} + \mathcal{N}_{1133} \hat{s}_{33}, \quad \hat{\varepsilon}_{12} = \mathcal{N}_{1212} \hat{s}_{12} \tag{17}$$
$$\cdots\cdots\cdots\cdots \qquad \cdots\cdots$$

and by reversal

$$\hat{s}_{11} = \mathcal{M}_{1111} \hat{\varepsilon}_{11} + \mathcal{M}_{1122} \hat{\varepsilon}_{22} + \mathcal{M}_{1133} \hat{\varepsilon}_{33}, \quad \hat{s}_{12} = \mathcal{M}_{1212} \hat{\varepsilon}_{12} \tag{18}$$
$$\cdots\cdots\cdots\cdots \qquad \cdots\cdots$$

where $\mathcal{N}_{ijpq}(\mu)$ and $\mathcal{M}_{ijpq}(\mu)$ have been evaluated in [13] for different values of the index n. If we disregard the influence of J_3 the increments $\hat{\varepsilon}_{ij}$ (or $\hat{s}_{ij}$) can be evaluated by differentiation of the Nadai equations. We obtain, instead of (17) and (18),

$$\hat{\varepsilon}_{11} = N_{1111} \hat{s}_{11} + N_{1122} \hat{s}_{22} + N_{1133} \hat{s}_{33}, \quad \hat{\varepsilon}_{12} = N_{1212} \hat{s}_{12} \tag{19}$$
$$\cdots\cdots\cdots\cdots \qquad \cdots\cdots$$

and

$$\hat{s}_{11} = M_{1111} \hat{\varepsilon}_{11} + M_{1122} \hat{\varepsilon}_{22} + M_{1133} \hat{\varepsilon}_{33}, \quad \hat{s}_{12} = M_{1212} \hat{\varepsilon}_{12}. \tag{20}$$
$$\cdots\cdots\cdots\cdots \qquad \cdots\cdots$$

A systematic comparison between coefficients $\mathcal{N}_{ijpq}$, $\mathcal{M}_{ijpq}$ and N_{ijpq}, M_{ijpq} in the interval $0 \leq \mu \leq 1$ has been developed in [13]. Eqs. (17) and (19) can be easily generalized to include the elastic strain rates.

2. A variational condition of the plastic buckling of structures

By means of the previous results we can formulate a theory of the plastic buckling of structures that improves a previous one [14] and attempts to explain such a controversial matter [15—20].

Let us consider therefore a body S subjected to a system of forces λF increasing with λ. If the material of S exhibits strain-hardening we can admit, with good approximation, that the internal stresses increase with λ, too [22].

Let us suppose now that, while loads *are increasing*, the primary equilibrium of S bifurcates: a new system of strains appears because of the buckling of S. We will assume now that the passage of S from the primary position C_0 of S, immediately before buckling, to the buckled position C, occurs without unloading, i.e. in condition of *total loading*; the plastic flow is therefore sustained by the continuously increasing loads. Such a hypothesis is an extension of Shanley's formulation on the buckling of columns and the corresponding buckling load is the lowest load under which the equilibrium of S bifurcates.

Under such a hypothesis a potential function of stress then exists and we can use a static approach.

Let $\dot{u}_i$ $(i = 1, 2, 3)$ be the components of the velocity of the points of S during its motion from C_0 to C; they can be decomposed with

$$\dot{u}_i = \dot{u}_i^c + \dot{u}_i^b, \tag{21}$$

where the $\dot{u}_i^c$ give the velocity of the primary motion of S, independent of the buckling of S and due to the continuously increasing loads, $\dot{u}_i^b$ the velocity of the merely deviating motion while S buckles. If we impose the equilibrium of S in both positions C_0 and C, because of the orthogonality between systems $\dot{u}_i^c$ and $\dot{u}_i^b$ of velocities [14] we can obtain the following variational condition:

$$\delta(\dot{\mathscr{L}}_2^* + \dot{U}_2 + \dot{W}_p) = 0, \tag{22}$$

where only the $\dot{u}_i^b$ components appear.

In Eq. (22) functionals $\dot{\mathscr{L}}_2^*$ and $\dot{U}_2$ are independent of the plastic state. More precisely $\dot{\mathscr{L}}_2^*$ is the second order work

$$\dot{\mathscr{L}}_2^* = \int_V (\sigma_{ij})_{C_0} \dot{\varepsilon}_{ij}^{(2)} \, dV, \tag{23}$$

where $(\sigma_{ij})_{C_0}$ are the stress components at C_0 and

$$\dot{\varepsilon}_{ij}^{(2)} = \frac{1}{2}(\dot{\varepsilon}_{ip}\dot{\omega}_{pj} + \dot{\varepsilon}_{jp}\dot{\omega}_{pi}) + \frac{1}{2}\dot{\omega}_{ip}\dot{\omega}_{jp}, \tag{24}$$

with

$$\dot{\varepsilon}_{ij} = \frac{1}{2}\left(\frac{\partial \dot{u}_i^b}{\partial x_j} + \frac{\partial \dot{u}_j^b}{\partial x_i}\right), \quad \dot{\omega}_{ij} = \frac{1}{2}\left(\frac{\partial \dot{u}_i^b}{\partial x_j} - \frac{\partial \dot{u}_j^b}{\partial x_i}\right), \tag{25}$$

are the second order components of the pure strain velocity in $\dot{u}_i^b$ components. $\dot{U}_2$ is the second order rate of the potential energy of loads during the deviating motion $C_0 - C$.

The functional $\dot{W}_p$ is, instead, sensitive of plastic state of S and is given by:

$$\dot{W}_p = \frac{1}{2}\int_V \mathscr{M}_{ijpq}\dot{\varepsilon}_{pq}\dot{\varepsilon}_{ij} \, dV, \tag{26}$$

where $\dot{\varepsilon}_{ij}$ are the first order strain rates (25). $\dot{W}_p$ is therefore the rate of the internal work done by the stress rates, due to the deviating deformation, for the corresponding strain rates. The coefficients $\mathcal{M}_{ijpq}$ correspond to Eqs. (18) provided that also the effects of the elastic strain rates are included there.

On the other hand with a more approximate approach we can evaluate the internal work $\dot{W}_p$ by using the Nadai differentiated equations (20); we will have, therefore:

$$\dot{W}_p = \frac{1}{2}\int_V \left\{\frac{2}{3}\dot{\varepsilon}_{ij}\dot{\varepsilon}_{ij} + \left(K - \frac{2}{9}\frac{\sigma_i}{e_i}\right)_{C_0}(\dot{\varepsilon}_{hh})^2 + \left[\frac{4}{9e_i^2}\left(\frac{d\sigma_i}{de_i} - \frac{\sigma_i}{e_i}\right)\right]_{C_0}\left[(\varepsilon_{ij})_{C_0}\dot{\varepsilon}_{ij} - \frac{1}{3}(\varepsilon_{hh})_{C_0}\dot{\varepsilon}_{hh}\right]^2\right\} dV, \tag{27}$$

where σ_i, e_i are the intensities of stress and strain (7) and (8) and K is the bulk modulus.

Eqs. (26) and (27) give the same results in various problems of plastic buckling of thin beams. In more complex problems of plastic buckling of plates and shells their corresponding results seem to be very close.

References

1. Batdorf, S. B., Budiansky, B.: N.A.C.A., T.N. 1871 (1949).
2. Como, M., D'Agostino, S.: Meccanica **4**, 146—158 (1969).
3. Como, M., D'Agostino, S.: Fondazione Politecnica Mezzogiorno Italia. Quad. 38 (1969).
4. Como, M., Grimaldi, A.: Meccanica **4**, 286—297 (1969).
5. Como, M., D'Agostino, S., Grimaldi, A.: Fondazione Politecnica Mezzogiorno Italia. Quad. 41 (1969).
6. Como, M.: Rend. Cl. Sci. Fis. e Mat. Acc. Sci. Lett. Arti di Napoli 72—86 (1969).
7. Como, M., Grimaldi, A.: La Ricerca, Ist. Edit. del Mezzog. 1, 2 (1969).
8. Como, M., Grimaldi, A.: Meccanica **5**, 117—125 (1970).
9. Como, M., Grimaldi, A.: La Ricerca, Ist. Edit. del Mezzog. 1, 2 (1970).
10. Naghdi, P. M., Essemburg, F., Koff, W.: J. Appl. Mech. **25**, 201—209 (1958).
11. Lode, W.: Z. Physik **36**, 913—939 (1926).
12. Taylor, G. I., Quinney, H.: Phil. Trans. Roy. Soc. London **A 230**, 323—362 (1931).
13. Como, M., Grimaldi, A.: La Ricerca, Ist. Edit. del Mezzog. 3 (1970).
14. Como, M.: Costruzioni Metalliche 2, 84—93; 4, 252—265; 5, 329—340; 6, 439—447 (1966); 2, 113—119 (1967).
15. Bijlaard, P. P.: J. Aeronaut. Sci. **26**, 405—408 (1959).
16. Handelman, G. H., Prager, W.: N.A.C.A., T.N. 1530 (1947).
17. Pearson, C. E.: J. Aeronaut. Sci. **17**, 417 (1950).
18. Cicala, P.: J. Aeronaut. Sci. **18** (1951).
19. Stowell, E. Z.: N.A.C.A., T.N. 1556 (1948).
20. Lee, L. H. N.: J. Aeronaut. Sci. **29**, 87—95 (1962).
21. Cicala, P.: Rend. Acc. Lincei, Serie VIII, Vol. IX, Fasc. 1—2 (1950).
22. Iliuschin, A.: Plasticité, Paris: Eyrolles 1956.
23. Biot, M.: Incremental deformations, New York: J. Wiley 1965.
24. Como, M.: Teoria della stabilitá dell' equilibrio elastico, Napoli: Liguori 1967.

Buckling and Postbuckling Behavior of Initially Imperfect Orthotropic Cylindrical Shells under Axial Compression and Internal Pressure

By

N. S. Khot, V. B. Venkayya and **L. Berke**

Air Force Flight Dynamics Laboratory, Wright-Patterson Air Force Base, Ohio, U.S.A.

Introduction

In recent years orthotropic cylindrical shells such as stiffened or fiber reinforced cylinders have been used extensively in aerospace structures. These structures are often subjected to axial compression and internal pressure. The thin cylindrical shells are known to be imperfection sensitive, and their buckling loads are substantially reduced by the presence of small initial deviations from the circular shape of the shell. The behavior of imperfect cylindrical shells has been studied by various investigators [1—5]. Both isotropic and anisotropic shells have been considered in these structures. In most of these investigations Karman-Donnell strain-displacement equations were used.

In the present investigation Sanders [6] nonlinear strain-displacement relations are used. They are modified to include initial geometric imperfections in the radial direction. It is generally agreed that Sanders strain-displacement relations are more accurate than the Karman-Donnell relations. The principal advantage of using Karman-Donnell relations is that it admits an Airy stress function to satisfy the inplane equilibrium equations and as a result the number of unknowns in the energy function can be reduced.

The solution to the problem is obtained by the application of the principle of stationary potential energy. In the present study, orthotropic cylindrical shells subjected to axial compression and internal pressure are considered. The results at buckling and at a point beyond buckling are presented for various radius to thickness ratios of the cylinder. The results obtained by using Sanders strain-displacement relations are compared with those obtained by Karman-Donnell relations.

Analytical formulation

R, L and h are the mean radius, length and the thickness of the cylindrical shell respectively. U, V and W are the axial, tangential and radial displacement components in the three coordinate directions x, y, z respectively. $\overline{W}$ is the initial deviation of the shell in the radial direction. The radial displacement due to the applied load is $(W - \overline{W})$.

The midsurface strain-displacement and the change of curvature-displacement relations as given by SANDERS [6] and modified to include initial imperfection can be written as

$$\begin{aligned} \varepsilon_x &= U_{,x} + \frac{1}{2}(W_{,x})^2 - \frac{1}{2}(\overline{W}_{,x})^2, \\ \varepsilon_y &= V_{,y} + \frac{1}{2}(W_{,y} + \underline{V/R})^2 - \frac{1}{2}(\overline{W}_{,y})^2 - W/R + \overline{W}/R, \qquad (1) \\ \varepsilon_{xy} &= U_{,y} + V_{,x} + W_{,x}(W_{,y} + \underline{V/R}) - \overline{W}_{,x}\overline{W}_{,y}, \end{aligned}$$

$$\begin{aligned} k_x &= W_{,xx} - \overline{W}_{,xx}, \quad k_y = W_{,yy} + \underline{V_{,y}/R} - \overline{W}_{,yy}, \\ k_{xy} &= 2(W_{,xy} - \overline{W}_{,xy}) + \underline{V_{,x}/R}. \end{aligned} \qquad (2)$$

The underlined terms in Eqs. (1) and (2) drop out in the Karman-Donnell equations.

The constitutive relations for an orthotropic cylindrical shell are given by

$$[\boldsymbol{N}] = [\boldsymbol{A}]\,[\boldsymbol{\varepsilon}], \quad [\boldsymbol{\varepsilon}] = [\boldsymbol{a}]\,[\boldsymbol{N}], \quad [\boldsymbol{M}] = [\boldsymbol{D}]\,[\boldsymbol{k}], \qquad (3)$$

where $[\boldsymbol{N}]$ and $[\boldsymbol{M}]$ are the stress and moment resultants. These relations are valid for homogeneous orthotropic cylinders as well as stiffened cylindrical shells when the effect of stiffeners is assumed to be smeared.

The extensional and bending strain energy in the orthotropic cylindrical shell are given by

$$\pi_1 = \frac{1}{2}\int_0^L \int_0^{2\pi R} [\boldsymbol{\varepsilon}]^T [\boldsymbol{A}]\,[\boldsymbol{\varepsilon}]\,dx\,dy, \quad \pi_2 = \frac{1}{2}\int_0^L \int_0^{2\pi R} [\boldsymbol{k}]^T [\boldsymbol{D}]\,[\boldsymbol{k}]\,dx\,dy \qquad (4)$$

respectively.

The potential energy of the axial compressive load when applied symmetrically at the ends of the shell, is given by

$$\pi_3 = -\int_0^{2\pi R} [N_x]_{x=L}\,dy \int_0^L \frac{\partial U}{\partial x}\,dx \qquad (5)$$

and the potential energy of the internal pressure $\bar{p}$ is given by

$$\pi_4 = -\int_0^L \int_0^{2\pi R} \{-\bar{p}\} W \, dx \, dy. \tag{6}$$

The total potential energy of the system is

$$\pi = \pi_1 + \pi_2 + \pi_3 + \pi_4. \tag{7}$$

Method of solution

The following trigonometric functions were selected to represent the components of the displacements U, V, W and the initial imperfection $\overline{W}$:

$$\begin{aligned}
n\left(\frac{U}{h}\right) &= U_1 \sin Mx \cos Ny + U_2 \sin 2Mx \cos 2Ny + U_3 \sin 2Mx \\
&\quad + U_4 \sin 3Mx \cos Ny + U_5 \sin 4Mx - \bar{\varepsilon}\left(\frac{n}{h}\right) x, \\
n\left(\frac{V}{h}\right) &= V_1 \cos Mx \sin Ny + V_2 \cos 2Mx \sin 2Ny + V_3 \sin 2Ny \\
&\quad + V_4 \cos 3Mx \sin Ny, \\
\left(\frac{W}{h}\right) &= W_1 \cos Mx \cos Ny + W_2 \cos 2Mx + W_0, \\
\left(\frac{\overline{W}}{h}\right) &= \overline{W}_1 \cos Mx \cos Ny + \overline{W}_2 \cos 2Mx, \\
M &= \frac{m\pi}{L}, \quad N = \frac{n}{R},
\end{aligned} \tag{8}$$

where $U_1 \cdots U_5$; $V_1 \cdots V_4$; $\overline{W}_1$, $\overline{W}_2$ are the unknown coefficients, W_0 is the inward radial displacement; $\overline{W}_1$ and $\overline{W}_2$ are the known or specified amplitudes of the initial geometric imperfection. $2m$ and n are the number of waves in the longitudinal and circumferential directions. The expressions for U and V in Eq. (8) are the same as those obtained from Karman-Donnell strain-displacement relations by utilizing the Airy stress function, except that the coefficients are left here as unknowns.

The relation between average end shortening $\bar{\varepsilon}$ and the applied load $\bar{\sigma}$ can be expressed as

$$\begin{aligned}
\frac{\bar{\varepsilon} R}{h} - \frac{A_{12}}{A_{11}A_{22} - A_{12}^2}\left(\frac{\bar{p}R^2}{h}\right) &= \frac{1}{8}\left[(W_1 + \overline{W}_1)(W_1 - \overline{W}_1)\right. \\
&\quad \left. + 8(W_2 + \overline{W}_2)(W_2 - \overline{W}_2)\right] u^2 \eta + \sigma,
\end{aligned} \tag{9}$$

where

$$\sigma = \left(\frac{\bar{\sigma} R}{h}\right) \frac{A_{22}}{A_{11}A_{22} - A_{12}^2}, \quad u = \frac{m}{n}, \quad \eta = n^2 \frac{h}{R}.$$

The left side of Eq. (9) is the effective end shortening parameter and is denoted as "EPS". The total potential energy of the system can be obtained by utilizing Eqs. (1)—(9). The complete expression for the energy can be found in Ref. [7]. The only unknown quantities in the expression for the total potential energy are the coefficients of the assumed displacement functions. By applying the principle of stationary potential energy the coefficients of the displacement functions can be determined. Two approaches are considered for finding the extremum. In the first approach the total potential is minimized with respect to the unknown coefficients and the resulting set of nonlinear algebraic equations are solved by the Newton-Raphson iterative method. In the second approach the search for the minimum energy is carried out by the direct function minimization procedure of FLETCHER and POWELL [8].

Numerical results

The numerical results are obtained for a cylindrical shell of radius 6.0 in. and length 15 in. The thickness of the shell is varied to get the desired radius to thickness ratios. The elastic stiffness constants of the shell are calculated with the assumption that the material has the following properties: $E_x = 7.27 \times 10^6$ psi, $E_y = 2.01 \times 10^6$ psi, $\nu_{12} = 0.272$, and $G = 1.06 \times 10^6$ psi. The shell is subjected to axial compression and internal pressure $\bar{p}$ (lbs./in.^{2}) which is specified in terms of the parameter $p = \frac{\bar{p} R^2}{2} \sqrt{\frac{a_{22}}{d_{22}}}$. For evaluating the classical buckling load σ_{cl}, the number of longitudinal half waves m, and the circumferential waves n are taken to be integer numbers. The initial imperfection is assumed to have the same wave form as the classical buckling mode. For isotropic shell it is noticed that the imperfections with the wave numbers equal to those of the classical buckling mode are most critical [3—5]. The initial imperfection parameter $\overline{W}^*$ is defined as the sum of the initial imperfection amplitudes $\overline{W}_1$ and $\overline{W}_2$ in Eq. (8). The amplitude $\overline{W}_1$ is taken to be equal to $4\overline{W}_2$.

Table 1 shows the comparison between the results obtained by utilizing Sanders and Karman-Donnell strain-displacement relations. The results are given for a point beyond buckling and for the point of buckling. Comparison of these results reveals that the values and σ_{cl} obtained by Sanders equations are lower than those obtained by Karman-Donnell equations. However, the ratios $\sigma/\sigma_{\text{cl}}$ given by Sanders equations are greater than those of Karman-Donnell equations.

Fig. 1 shows the load-shortening curves for an orthotropic shell with imperfection parameter 0.02 and pressure parameters 0.0, 0.05 and 0.2. These curves are given for Sanders strain-displacement relations. The difference between Karman-Donnell and Sanders equations

Table 1. *Buckling and postbuckling loads. All the results are for* $\overline{W}^* = 0.02$

	p	EPS	Solution by Sanders equations σ_{cl}	σ	$\frac{\sigma}{\sigma_{cl}}$	$\frac{R}{h}$	Solution by Karman-Donnell equations σ_{cl}	σ	$\frac{\sigma}{\sigma_{cl}}$	m, n
Point beyond buckling	0.0	0.4	0.23744	0.15117	0.63666	20	0.24529	0.15419	0.62861	2, 4
			0.24327	0.14220	0.58454	50	0.24645	0.14332	0.58157	3, 6
			0.24294	0.16141	0.66440	100	0.24429	0.16178	0.66228	5, 9
			0.24421	0.16093	0.65900	1000	0.24434	0.16096	0.65876	16, 28
	0.08	0.39649	0.25752	0.22068	0.85693	20	0.26291	0.22144	0.84226	3, 4
			0.26093	0.23499	0.90059	50	0.26276	0.23513	0.89484	5, 6
			0.26221	0.23193	0.88450	100	0.26322	0.23213	0.88186	7, 9
			0.26218	0.21878	0.83445	1000	0.26229	0.21878	0.83413	21, 27
At buckling	0.0	variable	0.23744	0.20562	0.86599	20	0.24529	0.21061	0.85861	2, 4
			0.24327	0.19621	0.80654	50	0.24645	0.19624	0.79630	3, 6
			0.24294	0.20889	0.85984	100	0.24429	0.20975	0.85861	5, 9
			0.24421	0.20997	0.85981	1000	0.24434	0.21010	0.85849	16, 28
	0.08	variable	0.25752	0.23176	0.89994	20	0.26291	0.23541	0.89539	3, 4
			0.26093	0.23867	0.91467	50	0.26276	0.23987	0.91290	5, 6
			0.26221	0.23786	0.90713	100	0.26322	0.23856	0.90632	7, 9
			0.26218	0.23514	0.89685	1000	0.26229	0.23521	0.89676	21, 27

is found to be too small to indicate it on the curves. As expected, reducing the number of terms in the expressions for the displacements U and V, Eq. (8), indicated stiffer postbuckling behavior.

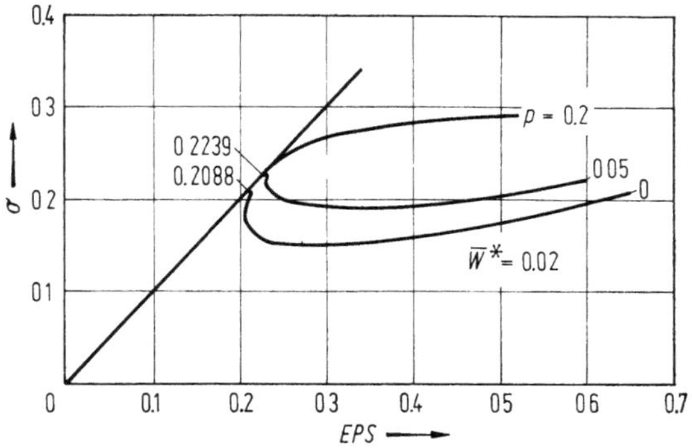

Fig. 1. Load-shortening curve for orthotropic shell

p	σ_{cl}	σ_{max}	σ_{max}/σ_{cl}
0.0	0.2429	0.2088	0.8598
0.05	0.2556	0.2239	0.8758
0.2	0.2777	—	—

The variations in the ratio of the buckling load of an imperfect shell to the classical buckling load, with the change in the internal pressure are presented in Figs. 2 and 3 for R/h values of 100 and 1000 respectively. The curves in Fig. 3 show a gradual change compared to those in Fig. 2. This appears to be due to the change in the values of the number of longitudinal half waves m. For $R/h = 1000$, the value of m changes for each value of p, while for $R/h = 100$ it remains the same for more

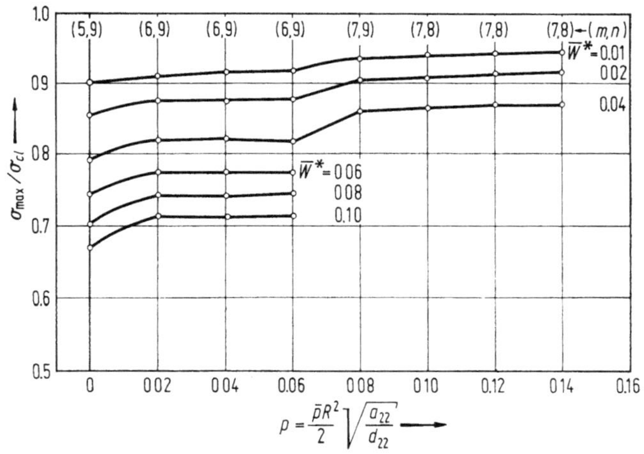

Fig. 2. Variation of σ_{max}/σ_{cl} with internal pressure, $R/h = 100$.

than one value of p. The sudden change in the curves is observed in Fig. 2 whenever the value of m changes.

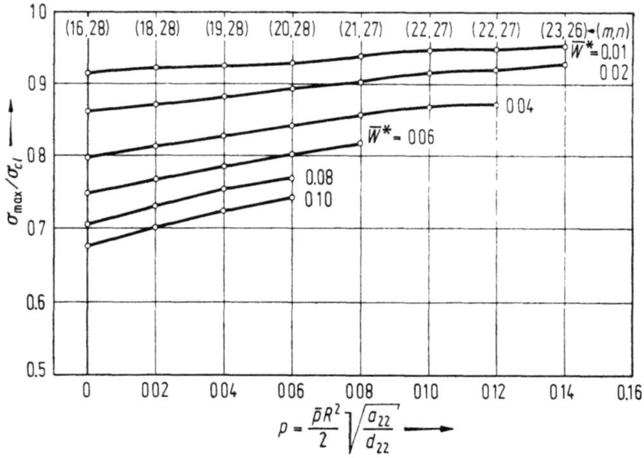

Fig. 3. Variation of σ_{max}/σ_{cl} with internal pressure, $R/h = 1000$.

The curves in Figs. 2 and 3 are shown only for those cases in which snap-buckling is observed. As the values of imperfection parameters increase, the range of snap-buckling seems to reduce. The plausible explanation for this behavior is the inadequacy of the number of terms used in the assumed displacement functions.

As mentioned earlier, the results were obtained by two approaches and both approaches give the same buckling and postbuckling loads except for the following differences. In the Newton-Raphson iterative methods, convergence is faster and the required computational time is small. However, the iteration must start with an approximate solution quite close to the actual solution otherwise it may not converge at all. In the Fletcher-Powell minimization scheme, much larger computational times are required, but the starting solution need not be as close to the actual result.

References

1. Donnell, L. H., Wan, C.: J. Appl. Mech. **17/18**, 73 (1950).
2. Koiter, W. T.: Nonlinear Problems, ed. R. E. Langer, University of Wisconsin Press 1963, p. 257.
3. Madsen, W. A., Hoff, N. J.: SUDAER No. 227, Stanford University 1965.
4. Hutchinson, J.: AIAA J. **3**, 1461 (1965).
5. Khot, N. S.: Paper presented at AIAA 7th Aero. Sci. Meeting, New York 1969, No. 69—93.
6. Sanders, J. L.: Quart. Appl. Math. **21**, 21 (1963).
7. Khot, N. S., Venkayya, V. B., Berke, L.: Orthotropic Cylindrical Shells with Initial Imperfections. AFFDL-TR (in preparation).
8. Fletcher, R., Powell, M. J. D.: Computer J. **6**, 163 (1963).

Instabilität der Ruhelage für ein System mit zwei Freiheitsgraden

Von

A. Huaux

Centre Interuniversitaire d'Etude des Phénomènes non-linéaires
und Ecole Technique Supérieure de l'Etat d'Anderlecht, Brüssel, Belgien

Zusammenfassung. Colombo [2, 3] und Banfi [1] haben verallgemeinerte Systeme von van der Pol studiert, d.h. endliche Systeme von zwei Differentialgleichungen der zweiten Ordnung. Es wird hier gezeigt, daß die Ruhelage dieses Systems instabil ist; die Bedingungen sind die von Colombo und Banfi, jedoch benutzen wir die direkte Methode von Liapunov, die mit Hilfe von zwei Verallgemeinerungen einer Liapunovschen Funktion zuerst für eine Differentialgleichung der Astronautik gefunden wurde (Huaux [8]).

§ 1. Einführung

Colombo und Banfi haben die Schwingungen des Systems

$$\left.\begin{aligned} \ddot{x}_1 + a_1(x_1)\,\dot{x}_1 + c_{11}x_1 + c_{12}x_2 = 0, \\ \ddot{x}_2 + a_2(x_2)\,\dot{x}_2 + c_{21}x_1 + c_{22}x_2 = 0 \end{aligned}\right\} \tag{1.1}$$

studiert; sie nehmen an, daß

1. hinreichende Bedingungen die Existenz, die Eindeutigkeit und die stetige Abhängigkeit von den Anfangsgrößen sichern, $\forall(x_1, x_2)$, $x_1 \in (\alpha_{11}, \alpha_{12})$, $x_2 \in (\alpha_{21}, \alpha_{22})$, $\alpha_{11} < 0$, $\alpha_{12} > 0$, $\alpha_{21} < 0$, $\alpha_{22} > 0$;

2. $a_1(x_1) < 0$, $x_1 \in (\beta_{11}, \beta_{12})$, $\alpha_{11} \leq \beta_{11} < 0 < \beta_{12} \leq \alpha_{12}$, $a_2(x_2) < 0$, $x_2 \in (\beta_{21}, \beta_{22})$, $\alpha_{21} \leq \beta_{21} < 0 < \beta_{22} \leq \alpha_{22}$;

3. c_{11}, c_{12}, c_{21} und c_{22} positive Konstanten sind, die außerdem die Bedingungen (2.3) erfüllen; es ist besser,

$$c_{12} \neq c_{21} \tag{1.2}$$

anzunehmen, um ein allgemeineres Ergebnis zu erhalten.

§ 2. Satz

Wir beweisen nun den folgenden Satz:

Die triviale Lösung oder Ruhelage des Systems (1.1) *ist instabil, wenn die folgenden hinreichenden Bedingungen erfüllt sind*:

$$1.\ \textit{die obigen Bedingungen } 1., 2. \textit{ und } 3.; \tag{2.1}$$

$$2.\ c_{11} > 0,\ c_{12} > 0,\ c_{21} > 0,\ c_{22} > 0; \tag{2.2}$$

$$3.\ c_{11} \cdot c_{22} - c_{12} \cdot c_{21} > 0. \tag{2.3}$$

Zuerst ersetzt man (1.1) durch das folgende äquivalente System von vier Differentialgleichungen der ersten Ordnung:

$$\left.\begin{aligned} \dot{x}_1 &= y_1, \\ \dot{y}_1 &= -c_{11}x_1 - c_{12}x_2 - a_1(x_1)y_1, \\ \dot{x}_2 &= y_2, \\ \dot{y}_2 &= -c_{21}x_1 - c_{22}x_2 - a_2(x_2)\,y_2. \end{aligned}\right\} \tag{2.4}$$

Es sei die Funktion gegeben:

$$\begin{aligned} v = {} & \frac{\lambda}{2}\,(c_{21}y_1^2 + c_{12}y_2^2 + c_{11}c_{21}x_1^2 + c_{22}c_{12}x_2^2 + 2c_{12}c_{21}x_1x_2) \\ & + \mu\left(c_{21}x_1y_1 + c_{12}x_2y_2 + c_{21}\int_0^{x_1} a_1(s)\,s\,ds + c_{12}\int_0^{x_2} a_2(s)\,s\,ds\right). \end{aligned} \tag{2.5}$$

Man berechnet

$$\dot{v} = \frac{dv}{dt} \tag{2.6}$$

mit Hilfe der Differentialgleichungen (2.4); man findet

$$\begin{aligned} \dot{v} = {} & -c_{21}\left(\lambda a_1(x_1) - \mu\right) y_1^2 - c_{12}\left(\lambda a_2(x_2) - \mu\right) y_2^2 \\ & -\mu\,(c_{11}c_{21}x_1^2 + c_{22}c_{12}x_2^2 + 2c_{12}c_{21}x_1x_2); \end{aligned} \tag{2.7}$$

λ and μ sind beliebige Parameter. Man wählt

$$\lambda > 0,\ \mu < 0 \tag{2.8}$$

und λ groß genug, um

$$\left.\begin{aligned} \lambda a_1(x_1) - \mu < 0 \\ \lambda a_2(x_2) - \mu < 0 \end{aligned}\right\} \forall\,(x_1, x_2),\ x_1 \in (\beta_{11}, \beta_{12}),\ x_2 \in (\beta_{12}, \beta_{22}) \tag{2.9}$$

zu haben; die Bedingungen gewährleisten, daß $v_{(2.5)}$ indefinit oder positiv definit ist, und daß $\dot{v}_{(2.7)}$ immer positiv definit ist; nach einem Liapunovschen Satz (LIAPUNOV [19], S. 262) ist die triviale Lösung von (1.1) (oder (2.4)) instabil.

§ 3. Konstruktion der Liapunovschen Funktion

Die direkte Methode von LIAPUNOV gibt Auskunft über die Stabilität oder Instabilität einer konstanten Lösung oder Ruhelage und auch einer periodischen Lösung, untersucht mit Hilfe einer Störungsgleichung oder einer Variationsgleichung (z. B. HAHN [6]). Es müssen Liapunovsche Funktionen konstruiert werden, und bequeme Verfahren dazu sind selten. Eine allgemeine Mitteilung gaben DEARMAN und LE MAY [4] mit Ergebnissen bis 1964, d. h. eine Beschreibung von einigen Konstruktionsverfahren; weitere Verfahren stammen von KU [14, 15], KU und PURI [16], KU, MEKEL und SU [17], INFANTE und CLARK [12], WALKER und CLARK [21], PONZO [20].

Das folgende Verfahren ist eine Verallgemeinerung der Ponzoschen Methode.

Es sei ein System von n autonomen Differentialgleichungen der ersten Ordnung gegeben:

$$\left.\begin{aligned} \dot{x}_1 &= f_1(x_1, x_2, \ldots, x_n), \\ \dot{x}_2 &= f_2(x_1, x_2, \ldots, x_n), \\ &\ldots\ldots\ldots\ldots\ldots\ldots \\ \dot{x}_n &= f_n(x_1, x_2, \ldots, x_n). \end{aligned}\right\} \tag{3.1}$$

Es seien n Funktionen $g_j(x_1, x_2, \ldots, x_n)$. Man berechnet die n^2 Ausdrücke

$$M_{ij} = \int_{t_0}^{t} \big(\dot{x}_i - f_i(x_1, x_2, \ldots, x_n)\big) \cdot g_j(x_1, x_2, \ldots, x_n)\, dt. \tag{3.2}$$

Man beachtet

$$A = \sum_{i,j=1}^{n} a_{ij} M_{ij} = 0; \tag{3.3}$$

die a_{ij} sind n^2 Konstanten. Man kann

$$A = A_1(x_1, x_2, \ldots, x_n) + A_2(x_1, x_2, \ldots, x_n) \tag{3.4}$$

schreiben. In A_1 ist die Integration beendet, jedoch nicht in A_2; man hat

$$A_2 = \int_{t_0}^{t} B(x_1, x_2, \ldots, x_n)\, dt \tag{3.5}$$

mit $B \in C$, $\forall t \geq t_0 > 0$ und für die Werte $x_1, x_2, \ldots, x_n$, die in (3.1) definiert werden. Man kann

$$v = A_1(x_1, x_2, \ldots, x_n) \tag{3.6}$$

und

$$\dot{v} = -B(x_1, x_2, \ldots, x_n) \tag{3.7}$$

setzen; es genügt, die Konstanten a_{ij} und die Funktionen $g_j(x_1, x_2, \ldots, x_n)$ günstig zu wählen, um ein positiv definites oder negativ definites oder verschwindendes $\dot{v}$ zu haben; die Berechnung von v ist dann eine einfache Folge der Berechnung von $\dot{v}$. In der Berechnung von $v_{(3.6)}$ hat man Integrationskonstanten; man setze diese Konstanten gleich Null. Natürlich sind die Berechnungen mit $n \geq 3$ langwierig.

Beispiele dieser Methode haben wir in anderen Mitteilungen gegeben (Huaux [8, 9]); es seien

$$1.\ \ddot{x} + a(x)\,\dot{x} + \varphi(x) = 0\ \big(\varphi(0) = 0\big), \tag{3.8}$$

was gleichbedeutend ist mit

$$\left.\begin{aligned} \dot{x} &= y, \\ \dot{y} &= -\varphi(x) - a(x)\,y \end{aligned}\right\} \tag{3.9}$$

oder mit

$$\left.\begin{aligned} \dot{x} &= y - \int_0^x a(s)\,ds = y - A(x), \\ \dot{y} &= -\varphi(x); \end{aligned}\right\} \tag{3.10}$$

$$2.\ \dddot{x} + a(\dot{x})\,\ddot{x} + b\dot{x} + cx = 0, \tag{3.11}$$

was gleichbedeutend ist mit

$$\left.\begin{aligned} \dot{x} &= y, \\ \dot{y} &= z, \\ \dot{z} &= -cx - by - a(y)\,z; \end{aligned}\right\} \tag{3.12}$$

vgl. auch La Salle [18];

$$3.\ \dddot{x} + a\ddot{x} + b(x)\,\dot{x} + cx = 0, \tag{3.13}$$

was gleichbedeutend ist mit

$$\left.\begin{aligned} \dot{x} &= y, \\ \dot{y} &= z, \\ \dot{z} &= -cx - b(x)\,y - az; \end{aligned}\right\} \tag{3.14}$$

vgl. auch Ezeilo [5].

Für (3.9) hat man (Huaux [8])

$$v = \frac{1}{2}\big(y + A(x)\big)^2 + \frac{y^2}{2} + 2\phi(x) \tag{3.15}$$

mit

$$\phi(x) = \int_0^x \varphi(s)\,ds, \quad \alpha_1 < s < \alpha_2, \tag{3.16}$$

und

$$\dot{v} = -A(x)\,\varphi(x) - a(x)\,y^2. \tag{3.17}$$

Aus (3.15) und (3.17) folgt, daß die triviale Lösung von (3.9) (und (3.8)) die folgenden Eigenschaften hat:

a) *Sie ist asymptotisch stabil, wenn*

$$\left.\begin{aligned} x\cdot\varphi(x) > 0,\ x \neq 0 \\ a(x) > 0 \end{aligned}\right\} x \in (\beta_1, \beta_2),\ \alpha_1 \leq \beta_1 < 0 < \beta_2 \leq \alpha_2$$

b) *sie ist instabil, wenn* (3.18)

$$\left.\begin{aligned} x\cdot\varphi(x) > 0,\ x \neq 0 \\ a(x) < 0 \end{aligned}\right\} x \in (\beta_1, \beta_2),\ \alpha_1 \leq \beta_1 < 0 < \beta_2 \leq \alpha_2.$$

Die Liapunovsche Funktion $v_{(2.5)}$ ist nur eine Verallgemeinerung von $v_{(3.15)}$. Eine Modifikation ist unentbehrlich, denn wegen der Bedingung (1.2) ist keine unmittelbare Verallgemeinerung möglich.

Um eine kurze Berechnung zu haben, beachten wir nur das System

$$\left.\begin{aligned} \ddot{x}_1 &= c_{11}x_1 + c_{12}x_2, \\ \ddot{x}_2 &= c_{21}x_1 + c_{22}x_2, \end{aligned}\right\} \tag{3.19}$$

das in das folgende System transformiert werden kann:

$$\left.\begin{aligned} \dot{x}_1 - y_1 &= 0, \\ \dot{y}_1 + c_{11}x_1 + c_{12}x_2 &= 0, \\ \dot{x}_2 - y_2 &= 0, \\ \dot{y}_2 + c_{21}x_1 + c_{22}x_2 &= 0. \end{aligned}\right\} \tag{3.20}$$

Wir beachten die Funktion

$$\begin{aligned} A = \int_{t_0}^{t} \Big(& a_{11}(\dot{x}_1 - y_1)\,x_1 + a_{22}(\dot{y}_1 + c_{11}x_1 + c_{12}x_2)\,y_1 \\ & + a_{33}(\dot{x}_2 - y_2)\,x_2 + a_{44}(\dot{y}_2 + c_{21}x_1 + c_{22}x_2)y_2 \\ & + a_{13}(\dot{x}_1 - y_1)\,x_2 + a_{13}(\dot{x}_2 - y_2)\,x_1\Big)\,dt = 0. \end{aligned} \tag{3.21}$$

Es wird

$$\begin{aligned} & \frac{1}{2}(a_{11}x_1^2 + a_{22}y_1^2 + a_{33}x_2^2 + a_{44}y_2^2) + a_{13}x_1x_2 \\ & = \int_{t_0}^{t} \Big(a_{11}x_1y_1 - a_{22}(c_{11}x_1 + c_{12}x_2)\,y_1 + a_{33}x_2y_2 \\ & \quad - a_{44}(c_{21}x_1 + c_{22}x_2)\,y_2 + a_{13}(x_2y_1 + x_1y_2)\Big)\,dt. \end{aligned} \tag{3.22}$$

Man setzt

$$v \equiv a_{11}x_1^2 + a_{22}y_1^2 + a_{33}x_2^2 + a_{44}y_2^2 + 2a_{13}x_1x_2, \tag{3.23}$$

und

$$\begin{aligned} \dot{v} \equiv 2\int_{t_0}^{t} \Big(& a_{11}x_1y_1 - a_{22}(c_{11}x_1 + c_{12}x_2)\,y_1 + a_{33}x_2y_2 \\ & - a_{44}(c_{21}x_1 + c_{22}x_2)\,y_2 + a_{13}(x_2y_1 + x_1y_2)\Big)\,dt. \end{aligned} \tag{3.24}$$

Man verlangt noch

$$\dot{v} \equiv 0, \tag{3.25}$$

und so hat man

$$\left.\begin{aligned} a_{11} - a_{22}c_{11} &= 0, \\ -a_{44}c_{21} + a_{13} &= 0, \\ -a_{22}c_{12} + a_{13} &= 0, \\ a_{33} - a_{44}c_{22} &= 0. \end{aligned}\right\} \tag{3.26}$$

Man findet, in der folgenden Ordnung,

$$\left.\begin{aligned} a_{22} &= c_{21}, \\ a_{44} &= c_{12}, \\ a_{13} &= c_{12} \cdot c_{21}, \\ a_{11} &= c_{11} \cdot c_{21}, \\ a_{33} &= c_{22} \cdot c_{12}. \end{aligned}\right\} \tag{3.27}$$

Die Beziehungen (3.23) und (3.27) ergeben

$$v = c_{11}c_{21}x_1^2 + c_{21}y_1^2 + c_{22}c_{12}x_2^2 + c_{12}y_2^2 + 2c_{21}c_{12}x_1x_2. \tag{3.28}$$

Wenn man von (3.20) nach (2.5) gehen will, ist es nicht schwierig, von (3.28) nach (2.5) zu gehen.

Bemerkung 3.1. Man kann eine andere Verallgemeinerung erhalten; beachten wir das System

$$\left.\begin{aligned} \ddot{x}_1 &= c_{11}(x_1)\, x_1 + c_{12}x_2, \\ \ddot{x}_2 &= c_{21}x_1 + c_{22}(x_2)\, x_2, \end{aligned}\right\} \tag{3.29}$$

das gleichbedeutend ist mit dem folgenden System

$$\left.\begin{aligned} \dot{x}_1 - y_1 &= 0, \\ \dot{y}_1 + c_{11}(x_1)\, x_1 + c_{12}x_2 &= 0, \\ \dot{x}_2 - y_2 &= 0, \\ \dot{y}_2 + c_{21}x_1 + c_{22}(x_2)\, x_2 &= 0. \end{aligned}\right\} \tag{3.30}$$

Für (3.30) erhält man aus (3.28)

$$v = 2c_{21}\int_0^{x_1} c_{11}(s)\, s\, ds + 2c_{12}\int_0^{x_2} c_{22}(s)\, s\, ds + 2c_{12}c_{21}x_1x_2 + c_{21}y_1^2 + c_{12}y_2^2 \tag{3.31}$$

und ferner

$$\dot{v} \equiv 0. \tag{3.32}$$

Die triviale Lösung von (3.29) (oder (3.30)) ist schwach stabil, wenn die Voraussetzungen des Satzes von § 2 erfüllt sind.

Bemerkung 3.2. Man kann dieselbe Theorie für ein nichtautonomes Differentialgleichungssystem aufstellen, d.h. für ein System der Form

$$\dot{x}_i - f_i(x_1, x_2, \ldots, x_n, t) = 0 \quad (i = 1, 2, \ldots, n). \tag{3.33}$$

Wegen der Funktionen $g_j(x_1, x_2, \ldots, x_n)$ beachtet man die Funktionen $g_j(x_1, x_2, \ldots, x_n, t)$ und den Ausdruck

$$A = \sum_{i,j=1}^{n} a_{ij} \int_{t_0}^{t} \left(\dot{x}_i - f_i(x_1, x_2, \ldots, x_n, t)\right) \cdot g_j(x_1, x_2, \ldots, x_n, t)\, dt. \tag{3.34}$$

Die Folge der Beweisführung ist dieselbe wie nach (3.3).

Literatur

1. Banfi, C.: Sulle oscillazioni di un sistema non lineare in due gradi di libertà. Atti della Accademia delle Scienze di Torino **98** (1963/64), Sitzung vom 5.2.1964, 19 S.
2. Colombo, G.: Sulle oscillazioni non lineari con due gradi di libertà. Rend. Sem. Mat. Univ. Padova **19**, 413–441 (1950).
3. Colombo, G.: Moti di regime di un sistema non lineare autonome in due gradi di libertà con debole accoppiamento capacitivo. Rend. Sem. Mat. Univ. Padova **24**, 400–420 (1955).
4. Dearman, C. C., Jr., Le May, A. R.: A survey of methods for generating Liapunov functions. In: Mathematische Methoden der Himmelsmechanik und Astronautik, Hrsg. E. Stiefel, Mannheim: Bibliographisches Institut 1964/65, S. 213–255.
5. Ezeilo, J. O. C.: A stability result for the solution of certain third order differential equations. J. London Math. Soc. **37**, 405–409 (1962).
6. Hahn, W.: Stability of Motion, Übers. A. P. Baartz (Die Grundlehren der mathematischen Wissenschaften in Einzeldarstellungen, Band 138), Berlin/Heidelberg/New York: Springer 1967, 446 S.
7. Huaux, A.: On the construction of Liapunov functions. IEEE Trans. Automatic Control **AC-12**, 465 (1967).
8. Huaux, A.: Sur une équation de la dynamique spatiale. Mém. Soc. Sci. Phys. Nat. Bordeaux, vol. spécial Congrès de l'Association Française pour l'Avancement des Sciences, 1968, S. 19–22.
9. Huaux, A.: Sur la stabilité des solutions constantes de l'équation autonome de Liénard. Exemples et généralisations. Bulletin de l'Institut Polytechnique de Jassy **10 (19)**, 171–180 (1964).
10. Huaux, A.: Stabilität der Ruhelage der Differentialgleichung $\dddot{x} + a\ddot{x} + b(x)\,\dot{x} + cx = 0$. ZAMM, Sonderheft 1965, T 117–T 118.
11. Huaux, A.: Sur la construction des fonctions de Liapounoff. 94e Congrès National des Sociétés Savantes, Pau, 1969.
12. Infante, E. F., Clark, L. G.: A method for the determination of the domain of stability of second-order nonlinear autonomous systems. J. Appl. Mech., Paper 63-WA-57, **31**, 315–320 (1964).

13. INGWERSON, D. R.: A modified Lyapunov method for nonlinear stability analysis. IRE Trans. Automatic Control **AC-6**, 199—210 (1961).
14. KU, Y. H.: Formulation of Liapunov functions of nonlinear systems for stability studies. Proc. First Annual Allerton Conference on Circuit and System Theory, Nov. 15, 16 and 17, 1963, S. 309—313.
15. KU, Y. H.: Lyapunov function of a fourth-order system. IEEE Trans. Automatic Control **AC-9**, 276—278 (1963).
16. KU, Y. H., PURI, N. N.: On Liapunov functions of high order nonlinear systems. J. Franklin Inst. **276**, 349—364 (1963).
17. KU, Y. H., MEKEL, R., SU, C. C.: Stability and design of nonlinear control systems via Liapunov's criterion. IEEE International Convention, New York, N.Y., March 23—26, 1964, Paper 212.
18. LASALLE, I., LEFSCHETZ, S.: Stability by Liapunov's Direct Method, New York/London: Academic Press 1961, pp. 70, 71.
19. LIAPOUNOFF, A.: Problème général de la stabilité du mouvement; Übers. E. DAVAUX. Ann. Fac. Sci. Toulouse, 2e série, **9**, 203—474 (1907).
20. PONZO, P. J.: On the stability of certain nonlinear differential equations. IEEE Trans. Automatic Control **AC-10**, 470—472 (1965).
21. WALKER, J. A., CLARK, L. G.: An integral method of Liapunov function generation for nonlinear autonomous systems. J. Appl. Mech., Trans. ASME, Series E, **32**, 569—575 (1965).

Substantiation of the Theory of Stability of Cylindrical Shells on the Basis of the Gauss Principle

By

N. A. Kilchevsky and **S. N. Nikulinskaya**

Academy of Sciences of the Ukrainian S.S.R., Kiev, U.S.S.R.

1.

The movement of the elements of a shell during the instability, i.e. during the "snap", is subject to the general principles of mechanics, in particular, to the principle of the least constraint (the Gauss principle).

According to this principle, the movements of the elements of the shell should maximally weaken the influence of the outer factors that lead to the loss of stability. To prove it, let us analyse the expression for the constraint for a discrete system:

$$Z = \frac{1}{2}\sum_{i=1}^{n} m_i \left| \boldsymbol{w}_i - \frac{\boldsymbol{F}_i}{m_i} \right|^2 = \frac{1}{2}\sum_{i=1}^{n} \frac{\boldsymbol{R}_i^2}{m_i}. \tag{1.1}$$

Here, m_i — masses of the points of the system, $\boldsymbol{w}_i$ — their acceleration, $\boldsymbol{F}_i$ — the resultant of the active forces, applied to i-th point of the system, $\boldsymbol{R}_i$ — the resultant of the reaction of ideal couplings.

As can be seen, the constraint diminishes as the values of $\boldsymbol{R}_i$ are diminishing ($\boldsymbol{R}_i$ — reactions of couplings, that are the outward manifestation of the constraint).

Let us generalize the expression (1.1) for a continuum. To that effect, it is sufficient to analyse the reactions of the couplings imposed on an element of the medium.

In order to obtain the equations of couplings and then to use the method of Lagrange multipliers, we confine our attention to the properties of the continuity of the field of displacements and on the conditions of simultaneity of the Saint-Venant strains.

We divide the body in two parts with an arbitrary surface S. All the characteristics of the continuum, relative to one part, we denote by a superscript (1), and to the other part by a superscript (2). Then, the conditions of continuity of the field of displacements lead to the following equalities:

$$u_i^{(1)}(x^1, x^2, x^3, t) = u_i^{(2)}(x^1, x^2, x^3, t) \quad (i = 1, 2, 3). \tag{1.2}$$

Here, x^k — the Lagrange coordinates of the point of the medium, t — time, U_i — the components of the displacement vector.

From (1.2) follows:

$$\frac{\partial u_i^{(1)}}{\partial x^k} = \frac{\partial u_i^{(2)}}{\partial x^k} \quad (i, k = 1, 2, 3). \tag{1.3}$$

In order to use the method of Lagrange multipliers, which in this case will be functions of coordinates and time, we express them in a finite-difference form, vary them and multiply by the Lagrange multipliers λ_j ($j = 1, 2, \ldots, 9$). Then we add these equations to the general equation of dynamics (expressed also in a finite-difference form) and make once again the limiting transition. Taking the coefficients at the variations $\delta u_i^{(k)}$ as equal to zero, we find:

$$\begin{aligned} &x_1^{(1)} - \varrho \frac{\partial^2 u_1^{(1)}}{\partial t^2} - \frac{\partial \lambda_1}{\partial x^1} - \frac{\partial \lambda_4}{\partial x^2} - \frac{\partial \lambda_7}{\partial x^3} = 0, \\ &\cdots\cdots\cdots\cdots\cdots\cdots \\ &x_1^{(2)} - \varrho \frac{\partial^2 u_1^{(2)}}{\partial t^2} + \frac{\partial \lambda_1}{\partial x^1} + \frac{\partial \lambda_4}{\partial x^2} + \frac{\partial \lambda_7}{\partial x^3} = 0. \end{aligned} \tag{1.4}$$

By comparing (1.4) with the equations of the movement of an element of continuum, we come to the conclusion that the field of the Lagrange multipliers coincides with the Cauchy stress field. We obtained a generalization of the result found by A. SOMMERFELD [1].

From the properties of couplings (1.3) follows that the Lagrange multipliers, i.e. *the stresses, are reactions of the couplings, acting on an element of continuum.*

The second group of equations of couplings consists of the equations of the simultaneity of the Saint-Venant strains. Once again we transform them into the finite-difference form, vary and multiply by the Lagrange multipliers. Then we add the equations obtained to the general equation of dynamics, expressed in the finite-difference form, with the field of variables coinciding with the components of the strain tensor. Taking the coefficients at the $\delta\varepsilon_{ik}$ as equal to zero, we obtain the equations of movement with Lagrange multipliers.

From a comparison of these equations with the conclusions that follow from the equations of the movement of an element of continuum, we conclude that in that case also the Lagrange multipliers have the dimensions of stresses, but form a field that does not coincide with the Cauchy strain field [2]. As this field is generated on the basis of the condition of simultaneity of strains, it is natural to assume that it is connected also with various sources of the inner stresses — with dislocations, inclusions, etc. [3]. The analysis of this stress field is out of the scope of this communication.

2.

Let us examine for the case of circular cylindrical shells the constraint Z, expressed by the reactions of couplings (1.2). The constraint is defined by the reactions of ideal couplings. The stresses may be assumed as belonging to the reactions of ideal couplings only in that case when the work done by them on the possible displacements of a body is not negative.

The sign of the value of the virtual work is opposite to the sign of change of the potential energy of the deformed body. Upon the loss of stability, the change of the potential energy is equal to zero or negative. Consequently, the work done by reactions of the couplings in this case is equal to zero or positive. That demonstrates that upon the loss of stability of the shell the stresses may be taken as a reaction of ideal couplings.

Note the two possible forms of constraint for the analysed particular case:

$$Z_{(1)} = 2\pi a \int_{-h}^{+h} T_1^2 \, dx^3 + O(h^3), \tag{2.1}$$

$$Z_{(2)} = \int_{x_0}^{x_0+\Delta l} T_2^2 \, dx^1 . \tag{2.2}$$

Here, T_1, T_2 — longitudinal and annular forces, $O(h^3)$ — members of relative order h^3, x_0 — the coordinate of an arbitrary point of the generatrix, Δl — the length of buckle, $2h$ — the thickness of the shell, a — radius.

The constraint $Z_{(1)}$ has a minimum value at the least value of the longitudinal force T_1. It follows that the condition of minimum $Z_{(1)}$ is at the same time a condition of the appearance of the critical state of the shell. The value of $Z_{(1)\min}$ depends on $T_1 = T_{cr}$. In its turn, T_{cr} strongly depends on the boundary conditions. In [4] the authors proved that the least critical value of the load at the axisymmetrical form of the loss of stability is obtained when the following conditions are satisfied:

$$u_3'' = 0,\ Bu_3''' - T_1 u_3' = 0 \quad (x^1 = 0,\ x^1 = l), \tag{2.3}$$

where l — the length of the shell, B — the cylindrical stiffness.

These are conditions on so-called free gliding butt-ends of the shell, maximally free from couplings. At the loss of stability the elements of the shell which have a relatively great compliance in the radial direction, are displaced so that the critical value of load that leads to a snap, assumes the least possible value.

During that process the buckles of a certain length are formed, and the boundary conditions are approaching the conditions (2.3) on the boundaries of the region filled with buckles.

For example, with $2h: a = 0.005$ the least value of the constraint will be with the relation $\Delta l: a = 0.3$ (a — radius of the middle surface of the shell). The value of $T_{cr} = 0.43\, T_{cr}^0$, where T_{cr}^0 corresponds to the hinged support of the borders.

The analysis of the constraint $Z_{(2)}$ shows that the convex side of the buckles should be directed to the axis of the tube.

Taken as a whole, the foregoing leads to the conclusion that the principle of the least constraint and the forms of constraint $Z_{(1)}$ and $Z_{(2)}$ confirm the hypothesis of A. V. Pogorelov on the form of the loss of stability as a kind of approximation to the isometric transformation of the middle surface of the shell.

References

1. Зоммерфельд, А.: Механика деформируемых сред, Москва: ИЛ 1954.
2. Кильчевский, Н. А., Никулинская, С. Н.: Принцип Гаусса в механике сплошной среды, Прикладная механика, т. V, в. 2, Киев: издательство "Наукова думка" 1969.
3. Kröner, E.: Kontinuumstheorie der Versetzungen und Eigenspannungen, Berlin/Göttingen/Heidelberg: Springer 1958.
4. Кильчевский, Н. А., Никулинская, С. Н.: Об осесимметричной форме потери устойчивости круговой цилиндрической оболочкой, Прикладная механика, т. I, в. 2, Киев: издательство "Наукова думка" 1965.

On Structural Instability Due to Strainsoftening

By

G. Maier

Politecnico di Milano, Milan, Italy

Strainsoftening (unstable) flexural behavior (Fig. 1a) has been observed in triangulated steel frames, reinforced concrete and composite beams and joints in precast concrete constructions; its strong implications in the overall structural response have been emphasized, and a systematic study often advocated but so far only begun (cf.

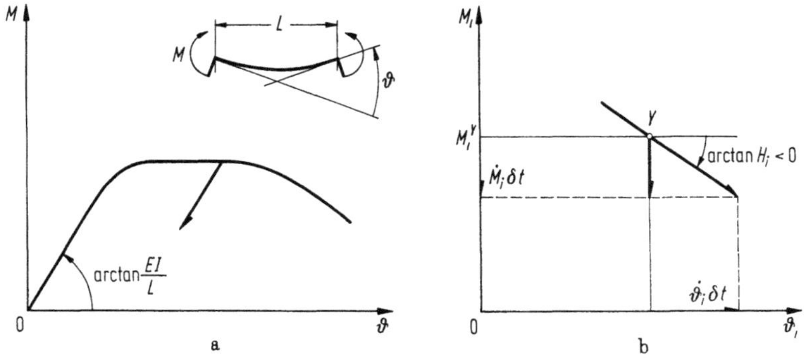

Fig. 1

[1—7]). The aim of this presentation is to supply a basis for an elastoplastic beam and frame theory allowing for unstable bending characteristics. Nonflexural deformations, interactions and geometrical second order effects are neglected. The notion of "falling-moment plastic hinge" is assumed, i.e. plastic deformations are supposed to consist of relative rotations ϑ in some sections. Some previous results [6, 8, 9] are specialized or extended; the new ones admit generalizations not pursued here. Important related topics, such as dynamics after the onset of instability and combined presence of physical (softening) and geometrical unstabilizing effects, will be discussed elsewhere.

Basic rate relations

Consider a beam or frame in a known static situation S under (conservative) concentrated loads. Let n sections be plastic at S: $M_i = M_i^Y$ ($i = 1, \ldots, n$). The flow-rule of the i-th plastic hinge can be expressed in the form (Fig. 1b)[1]

$$\dot{\varphi}_i = (\text{sign } M_i^Y)\, \dot{M}_i - H_i\dot{\lambda}_i,\ \dot{\varphi}_i \leq 0,\ \dot{\lambda}_i \geq 0, \tag{1 a}$$

$$\dot{\varphi}_i\dot{\lambda}_i = 0,\ \dot{\vartheta}_i = (\text{sign } M_i^Y)\, \dot{\lambda}_i \tag{1 b}$$

and the moment rate as the sum of the elastic responses $\dot{M}_i^E$, $\dot{M}_i^p$ to load and plastic rotation rates:

$$\dot{M}_i = \dot{M}_i^E + \dot{M}_i^p = \dot{M}_i^E + \sum_1^n{}_j z_{ij}\dot{\vartheta}_j. \tag{2}$$

Substitute (2) in (1) and condense for $i = 1, \ldots, n$ into matrices and vectors: thus the plastic response to any set of straining effect rates is governed by the following relations in the unknown vectors $\dot{\boldsymbol{\lambda}}$, $\dot{\boldsymbol{\varphi}}$:

$$\dot{\boldsymbol{\varphi}} = \tilde{\boldsymbol{D}}\dot{\boldsymbol{M}}^E - \boldsymbol{A}\dot{\boldsymbol{\lambda}},\ \dot{\boldsymbol{\varphi}} \leq \boldsymbol{0},\ \dot{\boldsymbol{\lambda}} \geq \boldsymbol{0},\ \tilde{\dot{\boldsymbol{\varphi}}}\dot{\boldsymbol{\lambda}} = 0, \tag{3a—d}$$

$$\dot{\boldsymbol{\vartheta}} = \boldsymbol{D}\dot{\boldsymbol{\lambda}}, \tag{4}$$

where

$$\boldsymbol{A} \equiv \boldsymbol{H} - \tilde{\boldsymbol{D}}\boldsymbol{Z}\boldsymbol{D} = \tilde{\boldsymbol{A}}. \tag{5}$$

Stability conditions

We adopt the following terminology and (static) stability criterion (cf. [10—12]). Let $S \to S_1$ denote an infinitesimal configuration change leading from S to another equilibrium situation S_1; let δW be the work done by the additional external forces required by $S \to S_1$: S is *unstable* if $\delta W < 0$ for some $S \to S_1$; *stable* otherwise, *strictly stable* if $\delta W > 0$ for all $S \to S_1$ (a structure stable but not strictly is in "neutral" equilibrium). It is easily seen that, under the present hypotheses:

$$\delta W = \frac{1}{2}\left(\int_s \frac{\dot{M}^2}{EI}\, dx + \tilde{\dot{\boldsymbol{M}}}\dot{\boldsymbol{\vartheta}}\right)\delta t^2, \tag{6}$$

where the integral extends to the whole structure. We give below some statements useful for solving problems of the kind in hand and for a mechanical insight into them.

Theorem I. *The structure is stable in S if and only if $\boldsymbol{A}$ is copositive, i.e.*

$$\tilde{\dot{\boldsymbol{\lambda}}}\boldsymbol{A}\dot{\boldsymbol{\lambda}} \geq 0 \quad \textit{for all} \quad \dot{\boldsymbol{\lambda}} \geq \boldsymbol{0}. \tag{7}$$

---

[1] See at the end (notation) key for symbols not explained in the text.

Proof of sufficiency. Using (2), (3a), (4), (3d) and the virtual work principle we may write (6) in the forms

$$\delta W = \frac{1}{2}\left[\int_s \frac{\dot{M}^{E2}}{EI}\,dx + \tilde{\dot{\boldsymbol{\vartheta}}}(\dot{\boldsymbol{M}} - \dot{\boldsymbol{M}}^p)\right]\delta t^2 = \frac{1}{2}\left(\int_s \frac{\dot{M}^{E2}}{EI}\,dx + \tilde{\dot{\boldsymbol{\lambda}}}\boldsymbol{A}\dot{\boldsymbol{\lambda}}\right)\delta t^2. \tag{8}$$

As the integral in (8) is nonnegative, the conclusion follows.

Proof of necessity. Suppose $\tilde{\dot{\boldsymbol{\lambda}}}\boldsymbol{A}\dot{\boldsymbol{\lambda}} < 0$ for a $\dot{\boldsymbol{\lambda}}' \geq \boldsymbol{0}$. Vector $\dot{\boldsymbol{\lambda}}'$ defines $\dot{\vartheta}_i'$ and $\dot{M}_i'$ through (1), $\dot{M}_i^{p\prime}$ through (2) ($i = 1, \ldots, n$). Consider the particular disturbance pattern $S \to S_1'$ that is compatible with the plastic rotations $\dot{\boldsymbol{\vartheta}}'\,\delta t$ and equilibrated by pairs of couples $(\dot{M}_i' - \dot{M}_i^{p\prime})\,\delta t$ applied as shown in Fig. 2 at each plastic hinge where $\dot{\lambda}_i' > 0$, in the absence of any other load increments. For this $S \to S_1'$ the integral in (8) vanishes; hence $\delta W' < 0$, which means instability of S (q.e.d.).

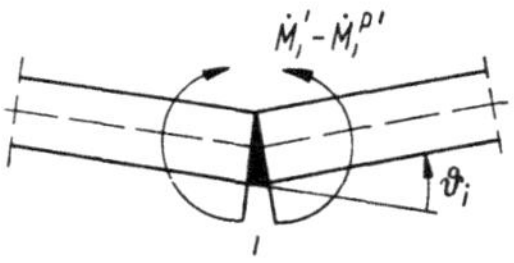

Fig. 2

Determinantal criteria for copositiveness are available [13].

Analogous conditions for instability and strictstability and stronger (only sufficient) stability conditions follow clearly from Theorem I. In particular:

Theorem I′. *The structure is stable in S if $\boldsymbol{A}$ is positive semidefinite, i.e.*

$$\tilde{\dot{\boldsymbol{\lambda}}}\boldsymbol{A}\dot{\boldsymbol{\lambda}} \geq 0 \text{ for all } \dot{\boldsymbol{\lambda}}. \tag{9}$$

Theorem II. *The structure is strictly stable in S if and only if $\boldsymbol{A}$ is strictly copositive* ($\tilde{\dot{\boldsymbol{\lambda}}}\boldsymbol{A}\dot{\boldsymbol{\lambda}} > 0$ *for* $\boldsymbol{0} \neq \dot{\boldsymbol{\lambda}} \geq \boldsymbol{0}$).

Theorem II′. *It is so if $\boldsymbol{A}$ is positive definite.*

By Theorems I′, II′, the nonnegativeness (or positiveness) of any lower bound for the least eigenvalue α_m of $\boldsymbol{A}$ represents a stability test for S, which is often of practical interest. From a "comparison theorem" for eigenvalue spectra (of hermitian operators), it follows that *α_m cannot increase when further plastic hinges form* in the loading process, and in fact it generally decreases. Therefore α_m can be regarded, in a sense, as a "deterioration index" of the structure.

Beams with softening flexural properties were tested imposing displacements by "stiff loading devices"; abrupt destructive configuration change with energy release was observed ("explosive failure" or "intrinsic instability") [2, 3]. We note that, if, instead of loads, straining actions like constraint movements or other dislocations lead to S, all the above considerations hold unaltered. Therefore we may affirm that:

Theorem I*. *Intrinsic instability cannot occur if and only if* (7) *holds with* $\boldsymbol{A}$ *containing, according to* (5), *the influence matrix* $\boldsymbol{Z}^*$ *calculated for the system: structure plus loading device.*

Incremental solutions and minimum properties

Theorem III. *Any plastic response* $\dot{\boldsymbol{\vartheta}}^\circ = \boldsymbol{D}\dot{\boldsymbol{\lambda}}^\circ$ *to given rates of external actions, is characterized by*

$$\text{minimum } Q(\dot{\boldsymbol{\lambda}}) \equiv \tilde{\dot{\boldsymbol{\lambda}}}\boldsymbol{A}\dot{\boldsymbol{\lambda}} - \tilde{\dot{\boldsymbol{\lambda}}}\boldsymbol{D}\dot{\boldsymbol{M}}^E \tag{10a}$$

$$\text{subject to } \tilde{\boldsymbol{D}}\dot{\boldsymbol{M}}^E - \boldsymbol{A}\dot{\boldsymbol{\lambda}} \leq \boldsymbol{0},\ \dot{\boldsymbol{\lambda}} \geq \boldsymbol{0} \tag{10b}$$

provided that this minimum Q_m *be zero.*

Proof. Note that $Q \geq 0$ for any $\dot{\boldsymbol{\lambda}}$ which satisfies (10b), and, by virtue of (3), $Q = -\tilde{\dot{\boldsymbol{\varphi}}}\dot{\boldsymbol{\lambda}}$. If $Q_m = 0$ the corresponding vector $\dot{\boldsymbol{\lambda}}$ fulfils all (3): hence it defines $\dot{\boldsymbol{\vartheta}}^\circ$. Conversely, any solution of (3) makes Q zero and minimum in the vector set defined by (10b) (q.e.d.).

Theorem IV. *If the condition of Theorem I′ is satisfied, any solution* $\dot{\boldsymbol{\lambda}}^\circ$

$$\text{maximizes } Q_{\mathrm{I}}(\dot{\boldsymbol{\lambda}}) \equiv -\frac{1}{2}\tilde{\dot{\boldsymbol{\lambda}}}\boldsymbol{A}\dot{\boldsymbol{\lambda}} + \tilde{\dot{\boldsymbol{\lambda}}}\tilde{\boldsymbol{D}}\dot{\boldsymbol{M}}^E \text{ subject to } \dot{\boldsymbol{\lambda}} \geq \boldsymbol{0}, \tag{11}$$

$$\text{minimizes } Q_{\mathrm{II}}(\dot{\boldsymbol{\lambda}}) \equiv \frac{1}{2}\tilde{\dot{\boldsymbol{\lambda}}}\boldsymbol{A}\dot{\boldsymbol{\lambda}} \text{ subject to } \tilde{\boldsymbol{D}}\dot{\boldsymbol{M}}^E - \boldsymbol{A}\dot{\boldsymbol{\lambda}} \leq \boldsymbol{0}. \tag{12}$$

Proof (see details in [8, 9]). If (9) holds, $-Q_{\mathrm{I}}$ is convex. The Kuhn-Tucker local conditions [14, 15] of the minimization problem (11) are both necessary and sufficient, and turn out to identify with (3); hence problems (11) and (3) are equivalent. By "dualizing" problem (11) in programming theory sense [15], we find (12), which, hence, is solved by any solution of (11) (q.e.d.).

Theorem IV implies that, when (9) holds (it certainly does in the absence of softening [8]), problem (3) can efficiently be solved, or found to be unsolvable, in form (11) by (convex) quadratic programming algorithms [15]; or in form (3) by LEMKE's method [16]. Strainsoftening may invalidate (9) and, hence, Theorem IV, as the Kuhn-Tucker optimality conditions are then only necessary. When condition (9) is violated or not easy to check, the incremental problem can be solved in form (10) as a possibly nonconvex quadratic program by RITTER's method [17].

Theorem V. *There is a unique response to any load rates if and only if A is positive definite.*

This statement flows from the uniqueness theorem proved in [18] for mathematical problems of form (3). The sufficiency is already known [8]. A remarkable consequence of Theorem V, through Theorem II′ is

Theorem VI. *If there is a unique response to any load rates, the structure is strictly stable in S.*

Bounds to instantaneous compliances

It can be of practical interest to estimate the stiffness k the structure exhibits at S in a point P and a direction d: $k = \dot{F}/\dot{u}$, $\dot{F} = 1$ being a unit force rate applied in P along d, $\dot{u}$ the consequent corresponding deflection rate there. Bracketing the compliance $c = k^{-1}$ may save solving completely the incremental problem. In the elastic range, let $c^E = \dot{u}/\dot{F}$ be the compliance and $\dot{M}_u^E(x)$ the bending moments due to $\dot{F} = 1$.

Theorem VII. *If A is positive semidefinite, if $\dot{\boldsymbol{\lambda}}^*$ denotes an arbitrary nonnegative vector and $\dot{\boldsymbol{\lambda}}'$ any vector satisfying the inequality in* (12) *for $\dot{\boldsymbol{M}}^E = \dot{\boldsymbol{M}}_u^E$, the following inequalities hold:*

$$2\tilde{\dot{\boldsymbol{\lambda}}}^*\tilde{D}\dot{M}_u^E - \tilde{\dot{\boldsymbol{\lambda}}}^*A\dot{\boldsymbol{\lambda}}^* \le c - c^E \le \tilde{\dot{\boldsymbol{\lambda}}}'A\dot{\boldsymbol{\lambda}}'. \tag{13}$$

Proof. Using (8), if $\dot{\boldsymbol{\lambda}}_u^0$ defines the actual solution for $\dot{F} = 1$, we may write:

$$c\dot{F}^2 \equiv \dot{F}\dot{u} = \int_s \frac{\dot{M}_u^{E2}}{EI}\,dx + \tilde{\dot{\boldsymbol{\lambda}}}_u^0 A\dot{\boldsymbol{\lambda}}_u^0 = c^E\dot{F}^2 + \tilde{\dot{\boldsymbol{\lambda}}}_u^0 A\dot{\boldsymbol{\lambda}}_u^0. \tag{14}$$

Duality of the programming problems (11), (12) ensures that

$$Q_{\mathrm{I}}(\dot{\boldsymbol{\lambda}}_u^0) = Q_{\mathrm{II}}(\dot{\boldsymbol{\lambda}}_u^0) = \frac{1}{2}\,\tilde{\dot{\boldsymbol{\lambda}}}_u^0 A\dot{\boldsymbol{\lambda}}_u^0. \tag{15}$$

Combining (14), (15), and Theorem IV, we note that the maximum property (11) supplies the lower bound in (13), the minimum property (12) the upper bound (q.e.d.).

As an obvious corollary of Theorem VII, *any vector $\dot{\boldsymbol{\lambda}}''$ satisfying both inequalities* (10b) *for $\dot{\boldsymbol{M}}^E = \dot{\boldsymbol{M}}_u^E$, alone provides both bounds* (13) *for $c - c^E$.*

The "phase I" procedure of DANTZIG's simplex method [14] supplies a $\dot{\boldsymbol{\lambda}}''$ or shows that none exists. In this regard we may state in general:

Theorem VIII. *If inequalities* (10b) *are inconsistent for some $\dot{\boldsymbol{M}}^E$, the structure cannot be strictly stable at S.*

Proof. It has been proved in mathematics [16] that an inequality set structured like (10b) is not solvable if and only if there exists a $\dot{\boldsymbol{\lambda}} \geq \mathbf{0}$ satisfying $\boldsymbol{A}\dot{\boldsymbol{\lambda}} \leq \mathbf{0}$ and $\tilde{\dot{\boldsymbol{\lambda}}}\tilde{\boldsymbol{D}}\dot{\boldsymbol{M}}^E > 0$. The necessity part of this result, combined with Theorem II, yields Theorem VIII (q.e.d.).

On a deformation theory

By proportional loading plastic hinge rotations are likely to never cease once they are formed. This suggests to assume (cautiously) reversible $M \div \vartheta$ laws and solve the analysis problem in total quantities, avoiding step-by-step procedures ("deformation theory").

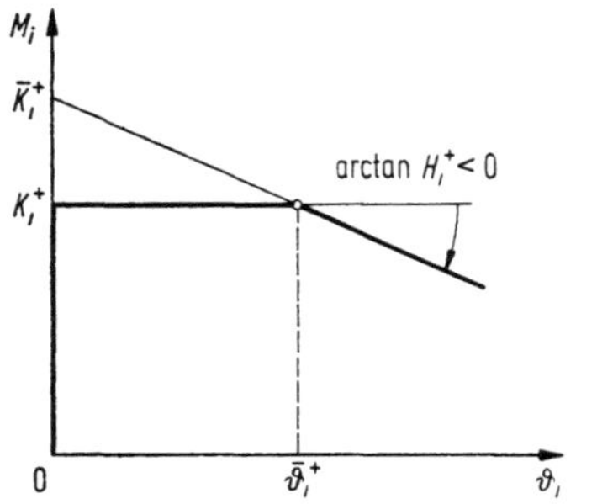

Fig. 3

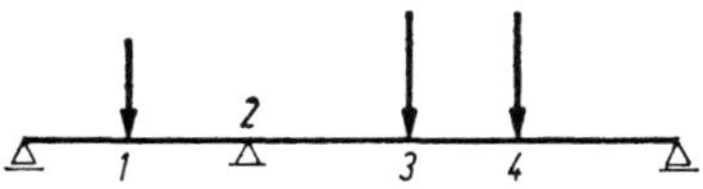

Fig. 4

It is easy to realize that the laws depicted in Fig. 3 can be represented by the relations

$$\begin{bmatrix} \varphi_i^+ \\ \bar{\varphi}_i^+ \end{bmatrix} = \begin{bmatrix} 1 \\ 1 \end{bmatrix} M_i - \begin{bmatrix} K_i^+ \\ \bar{K}_i^+ \end{bmatrix} - \begin{bmatrix} 0 & 0 \\ H_i^+ & H_i^+ \end{bmatrix} \begin{bmatrix} \lambda_i^+ \\ \bar{\lambda}_i^+ \end{bmatrix}, \tag{16a}$$

$$\begin{bmatrix} \varphi_i^+ \\ \bar{\varphi}_i^+ \end{bmatrix} \leq \mathbf{0}; \quad \begin{bmatrix} \lambda_i^+ \\ \bar{\lambda}_i^+ \end{bmatrix} \geq \mathbf{0}; \quad [\varphi_i^+ \bar{\varphi}_i^+] \begin{bmatrix} \lambda_i^+ \\ \bar{\lambda}_i^+ \end{bmatrix} = 0, \tag{16b}$$

$$\vartheta_i = \lambda_i^+ + \bar{\lambda}_i^+ . \tag{16c}$$

If superscript $+$ is replaced by $-$ and 1 by -1, (16) expresses a similar law in the negative quadrant. Let N be the number of the "critical" sections, where plastic hinges are expected to form (see e.g. Fig. 4). By suitable condensing for $i = 1, \ldots, N$, in matrix notation, relations like (16) and (2) written for total (undotted) variables, the finite problem can be given a formulation in vectors $\boldsymbol{\varphi}$, $\boldsymbol{\lambda}$ fully analogous to (3), except that matrix $\boldsymbol{A}$ *is no longer symmetric*. Due to this lack of symmetry, solutions, if any, enjoy only a minimum property of the kind of Theorem IV and can be obtained by RITTER's algorithm [17].

Notation

Bold face letters denote matrices and column-vectors, a tilde transpose, a dot rate. The main symbols are listed below in the order in which they appear.

i	plastic hinge index
M	bending moment
M_i^Y	current yield moment at the i-th plastic hinge
ϑ	plastic hinge rotation
$\dot{\varphi}$	slack variable interpreted as plastic potential rate
	sign $M_i^Y = 1$ if $M_i^Y > 0$, $= -1$ if $M_i^Y < 0$
H	strainhardening (≥ 0) or softening (< 0) coefficient (always finite)
λ	plastic multiplier or activation coefficient
M^E, M^p	elastic moment response to external actions and plastic rotations, respectively
z_{ij}	influence coefficient for moment in section i due to rotational dislocation in section j

$\tilde{\dot{\boldsymbol{\varphi}}} \equiv [\dot{\varphi}_1 \cdots \dot{\varphi}_n]$; vectors $\dot{\boldsymbol{\lambda}}$, $\dot{\boldsymbol{\vartheta}}$, $\dot{\boldsymbol{M}}$ analogously defined

$\boldsymbol{D} \equiv \operatorname{diag}[\text{sign } M_i^Y]$; $\boldsymbol{H} \equiv \operatorname{diag}[H_i]$; $\boldsymbol{Z} \equiv [z_{ij}]$

δW	disturbance work performed by an external agency
EI	elastic flexural stiffness
u	displacement
F	load
c	current local compliance
K	positive constant

References

1. Francis, A. J., Stevens, L. K.: I.E. Austr., Civ. Eng. Trans. **3**, 1 (1961).
2. Rosenblueth, E., Diaz de Cossio, R.: ASCE-ACI Symp., Flex. Mech. Reinf. Concrete, Miami 1965, p. 439.
3. Barnard, P. R., Johnson, R. P.: Proc. Inst. Civ. Engrs. **32**, 577 (1966).
4. Wood, R. H.: Design Phil. Precast Concr. Struct., AIPC Symp., London, Cem. Concr. Ass. 1967, p. 31.
5. Cranston, W. B.: Cem. Concr. Ass., Tech. Rep., 1965, p. 22.
6. Maier, G.: Ist. Lomb., Rend. Sci. **A 102**, 648 (1968).
7. Wood, R. H.: In: Engin. Plasticity, Cambridge Univ. Press 1968, p. 665.
8. Maier, G.: Meccanica **3**, 121 (1968).
9. Maier, G.: J. Méc. **8**, 5 (1969).
10. Drucker, D. C.: In: Struct. Mech., Oxford: Pergamon 1960, p. 407.
11. Horne, M. R.: In: Progr. Sol. Mech., vol. II, Amsterdam: North-Holland 1961, p. 277.
12. Mandel, J.: In: IUTAM Symp., Rheol. Soil Mech., Berlin/Heidelberg/New York: Springer 1966, p. 58.
13. Hall, M., Newman, M.: Proc. Camb. Phil. Soc. **59**, 329 (1963).
14. Dantzig, G. B.: Linear Programming and Extensions, Princeton Univ. Press 1963.
15. Collatz, L., Wetterling, W.: Optimierungsaufgaben, Berlin/Heidelberg/New York: Springer 1967.
16. Lemke, C. K.: Manag. Sci. **11**, 681 (1965).
17. Ritter, K.: Z. Wahrsch. verw. Geb. **4**, 340 (1966).
18. Ingleton, A. W.: Proc. London Math. Soc. **16**, 519 (1966).

Asymptotic Stability of Travelling Waves

By

L. A. Peletier

University of Sussex, Falmer Brighton, Sussex, U.K.

1. Introduction

In this note we shall consider solutions of equations of the form

$$u_t + \varphi(u)\, u_x = u_{xx} \quad (\varphi'(u) > 0) \tag{1}$$

in the half space $t \geq 0$. In particular we shall be interested in travelling wave solutions, i.e. solutions of the form $u(x, t) = f(x - ct)$, where c is a constant: the speed of propagation. With the rising interest in non-linear waves, such waves have had considerable attention [1, 2].

Here we shall be interested in the convergence of solutions of the Cauchy problem of (1) to travelling wave solutions. Il'in and Oleinik [3] demonstrated some time ago that, subject to some conditions on the initial data, solutions indeed converge to travelling waves. Their method was based upon maximum principles. In the present note we shall prove a similar result, but use a different method, based on an energy inequality. The main interest will therefore be in the method. It is best demonstrated on the Burgers equation,

$$u_t + uu_x = u_{xx}. \tag{2}$$

2. Travelling waves of (2)

Substitution of $u(x, t) = f(x - ct)$ into (2) yields

$$(-c + f)\, f' = f'',$$

where the prime denotes differentiation. If we require $f, f' \to 0$ as $x \to \infty$ and put $f''(0) = 0$ to dispose of an arbitrary phase factor, we find

$$f(x) = c\left(1 - \tanh \frac{1}{2} cx\right).$$

Thus, for each c we have a uniquely determined travelling wave. Henceforth we shall keep c fixed and positive.

3. Preliminaries

I. We shall consider the Cauchy problem of (2) in the strip $H_T = (-\infty, \infty) \times (0, T)$; T is an arbitrary finite number. Let at $t = 0$,

$$u(x, 0) = u_0(x) \tag{3}$$

and let $u_0(x)$ be a smooth function which tends to 0 as $x \to \infty$ and to $2c$ as $x \to \infty$. Then there exists a unique smooth solution of the Cauchy problem (2), (3) in H_T [4].

II. Let u and v be smooth solutions of the Cauchy problem (2), (3) in H_T, and let $u(x, 0) \leq v(x, 0)$. Then $u(x, t) \leq v(x, t)$ in H_T. This follows at once from a maximum principle due to KRZYZANSKI [5]. Thus, when at $t = 0$, u lies between two travelling wave profiles:

$$f(x + \gamma_1) \leq u_0(x) \leq f(x + \gamma_2) \quad (\gamma_2 \leq \gamma_1),$$

then, for $0 \leq t < T$

$$f(x + \gamma_1 - ct) \leq u(x, t) \leq f(x + \gamma_2 - ct).$$

In particular, for $0 \leq t < T$, $u \to 0$ as $x \to \infty$ and $u \to 2c$ as $x \to -\infty$.

III. Let u and v be smooth solutions of the Cauchy problem (2), (3). Then

$$\int_{-\infty}^{\infty} (u - v)\, dx = \text{const}.$$

To see this one need only integrate the equation for $z = u - v$:

$$z_t + \frac{\partial}{\partial x} \frac{1}{2} (u^2 - v^2) = z_{xx}.$$

It is now expedient to transform to a coordinate frame which moves with the wave. Set $\xi = x - ct$, $\tau = t$. Then (2) becomes

$$u_\tau + (u - c)\, u_\xi = u_{\xi\xi}. \tag{4}$$

Although u stands for a different function here, no confusion will ensue when we do not change the notation.

IV. Suppose $u(\xi, \tau)$ is a solution of (4) in H_∞ and let

(i) $$\int_{-\infty}^{\infty} \{u(\xi, 0) - f(\xi)\}\, d\xi = 2\alpha,$$

(ii) $$\lim_{\tau \to \infty} \int_{-\infty}^{\infty} \cosh^2\left(\frac{1}{2} c\xi\right) \{u(\xi, \tau) - f(\xi - \gamma)\}^2\, d\xi = 0.$$

Then $\gamma = \alpha$.

Proof. Set $u(\xi, \tau) - f(\xi - \gamma) = cv(\xi, \tau)$. Then

$$\int_{-\infty}^{\infty} \{u(\xi,\tau) - f(\xi)\}\, d\xi = c \int_{-\infty}^{\infty} v\, d\xi + \int_{-\infty}^{\infty} \{f(\xi - \gamma) - f(\xi)\}\, d\xi. \tag{5}$$

Because

$$\int_{-\infty}^{\infty} v\, d\xi \leq \left\{ \int_{-\infty}^{\infty} \cosh^{-2}\left(\frac{1}{2} c\xi\right) d\xi \cdot \int_{-\infty}^{\infty} \cosh^2\left(\frac{1}{2} c\xi\right) v^2\, d\xi \right\}^{1/2}$$

and

$$\int_{-\infty}^{\infty} \{f(\xi - \gamma) - f(\xi)\}\, d\xi = 2\gamma,$$

(5) yields in the limit as $\tau \to \infty$: $2\alpha = 2\gamma$.

4. The main result

Let $u(\xi, \tau)$ be a smooth solution of the Cauchy problem (2), (3) in H_τ. Then if at $\tau = 0$

$$f(\xi + \gamma_1) \leq u(\xi, 0) \leq f(\xi + \gamma_2)$$

and $0 \leq \gamma_1 - \gamma_2 < 2.4/c$, there exist positive constants M and β, such that for $0 \leq t < T$,

$$\int_{-\infty}^{\infty} \cosh^2\left(\frac{1}{2} c\xi\right) \{u(\xi, \tau) - f(\xi + \gamma)\}^2\, d\xi \leq M e^{-\beta\tau} \tag{6}$$

for some $\gamma \in [\gamma_2, \gamma_1]$.

Proof. It is clear from II that if $u(\xi, \tau) \to f(\xi + \gamma)$ for some γ, then $\gamma \in [\gamma_2, \gamma_1]$. There will be no loss in generality if we take $\gamma = 0$. Then, according to III and IV:

$$\int_{-\infty}^{\infty} \{u(\xi, \tau) - f(\xi)\}\, d\xi = 0. \tag{7}$$

Set $u(\xi, \tau) = f(\xi) + cv(\xi, \tau)$ and subsequently $v(\xi, \tau) = \cosh^{-1}\left(\frac{1}{2} c\xi\right) w(\xi, \tau)$. Put $\frac{1}{2} c\xi = \eta$. Then the equation becomes

$$L(w) + N(w) = \frac{4}{c^2} w_\tau, \tag{8}$$

where

$$L(w) = w_{\eta\eta} + \left(-1 + \frac{2}{\cosh^2 \eta}\right) w,$$

$$N(w) = -\frac{2}{\cosh \eta} w (w_\eta - \tanh \eta w).$$

When we multiply (8) by w and integrate with respect to x, we obtain

$$\frac{2}{c^2}\frac{d}{d\tau}\int_{-\infty}^{\infty} w^2\,d\eta = \int_{-\infty}^{\infty} wL(w)\,d\eta + \int_{-\infty}^{\infty} wN(w)\,d\eta. \tag{9}$$

We claim that

$$\int_{-\infty}^{\infty} wL(w)\,d\eta \le -\int_{-\infty}^{\infty} w^2\,d\eta, \tag{10}$$

i.e.

$$\int_{-\infty}^{\infty}\left(w_\eta^2 - \frac{2}{\cosh^2\eta}w^2\right)d\eta \ge 0,$$

subject to the constraint (7)

$$\int_{-\infty}^{\infty} \cosh^{-1}\eta w\,d\eta = 0. \tag{11}$$

Accepting this inequality for the moment, (9) yields

$$\begin{aligned}\frac{d}{d\tau}\int_{-\infty}^{\infty} w^2\,d\eta &\le -\frac{1}{2}c^2\int_{-\infty}^{\infty}\{w^2 - wN(w)\}\,d\eta\\ &= -\frac{1}{2}c^2\int_{-\infty}^{\infty}\left(1-\frac{4}{3}\tanh\eta v\right)w^2\,d\eta \qquad (12)\\ &\le -\frac{1}{2}c^2\,\varepsilon\int_{-\infty}^{\infty} w^2\,d\eta,\end{aligned}$$

if there exists an $\varepsilon > 0$ such that $1-\frac{4}{3}\tanh v \ge \varepsilon$ for $0 \le \tau < T$. By II this will be so if

$$f(\xi+\gamma_1) \le u(\xi, 0) \le f(\xi+\gamma_2)$$

and the two travelling waves are not too far apart, i.e. $\gamma_1 - \gamma_2 < 2.4/c$. It then follows from (12) that

$$||w(\tau)|| \le ||w(0)||\,e^{-c^2\varepsilon\tau/4},$$

where $||w(\tau)||^2 = \int_{-\infty}^{\infty}\{w(\eta,\tau)\}^2\,d\eta$.

Thus it remains to prove inequality (10). We expand w in terms of the eigenfunctions of L. For the details of this expansion we can refer to TITCHMARSH [6] where it is discussed at length. The spectrum of L consists of the point $\{0\}$ and the interval $(-\infty, -1]$, and the expansion can be written as

$$\begin{aligned}w(\eta,\tau) = c_0(\tau)\,\theta_0(\eta) &+ \frac{1}{\pi}\int_{-\infty}^{-1}\theta(\eta,\lambda)\,c(\tau,\lambda)\,\xi'(\lambda)\,d\lambda\\ &+ \frac{1}{\pi}\int_{-\infty}^{-1}\varphi(\eta,\lambda)\,d(\tau,\lambda)\,\zeta'(\lambda)\,d\lambda\end{aligned}$$

in the notation of TITCHMARSH. $\theta(\eta, \lambda)$ and $\varphi(\eta, \lambda)$ are the even and the odd eigenfunction of L associated with the eigenvalue λ. $\theta_0 = \cos^{-1}\eta$ is the even eigenvalue associated with $\lambda = 0$; the odd one is not square-integrable and must therefore be discarded. ξ' and ζ' are positive functions of λ.

Because of the constraint, (11), w and θ_0 are orthogonal. Therefore $c_0 = 0$ and

$$\int_{-\infty}^{\infty} wL(w)\, d\eta = \frac{1}{\pi} \int_{-\infty}^{-1} \lambda c^2 \xi'\, d\lambda + \frac{1}{\pi} \int_{-\infty}^{-1} \lambda\, d^2 \zeta'\, d\lambda$$

$$\leq -\frac{1}{\pi} \int_{-\infty}^{-1} c^2 \xi'\, d\lambda - \frac{1}{\pi} \int_{-\infty}^{-1} d^2 \zeta'\, d\lambda$$

$$= -\int_{-\infty}^{\infty} w^2\, d\eta,$$

which proves the inequality.

References

1. WHITHAM, G. B.: Proc. Roy. Soc. **A 283,** 238 (1965).
2. ZABUSKY, N.: Nonlinear Partial Differential Equations, ed. W. F. AMES, New York: Academic Press 1967, p. 223.
3. IL'IN, A. M., OLEINIK, O. A.: Mat. Sbornik **51,** 191 (1960).
4. OLEINIK, O. A., KRUZHKOV, S. N.: Russ. Math. Surveys **16,** 105 (1961).
5. KRZYZANSKI, M.: Bull. Acad. Polon. Sci., Ser. Math. Ast. Phys. **7,** 131 (1959).
6. TITCHMARSH, E. C.: Eigenfunction Expansions, vol. I, 2nd ed. Oxford: Oxford University Press 1962.

721/26/70

MIX
Papier aus verantwortungsvollen Quellen
Paper from responsible sources
FSC® C105338

If you have any concerns about our products,
you can contact us on
ProductSafety@springernature.com

In case Publisher is established outside the EU,
the EU authorized representative is:
Springer Nature Customer Service Center GmbH
Europaplatz 3, 69115 Heidelberg, Germany

Printed by Libri Plureos GmbH
in Hamburg, Germany